AF783990

A Monograph of *Codonopsis* and Allied Genera (Campanulaceae)

A Monograph of *Codonopsis* and Allied Genera (Campanulaceae)

De-Yuan HONG
State Key Laboratory of Systematic and Evolutionary Botany, Institute of Botany, Chinese Academy of Sciences, Beijing

In association with

Qiang WANG and Kai-Yu PAN

Science Press
Beijing

ELSEVIER

AMSTERDAM • BOSTON • HEIDELBERG • LONDON • NEW YORK • OXFORD
PARIS • SAN DIEGO • SAN FRANCISCO • SINGAPORE • SYDNEY • TOKYO
Academic Press is an imprint of Elsevier

Academic Press is an imprint of Elsevier
125 London Wall, London EC2Y 5AS, UK
525 B Street, Suite 1800, San Diego, CA 92101-4495, USA
225 Wyman Street, Waltham, MA 02451, USA
The Boulevard, Langford Lane, Kidlington, Oxford OX5 1GB, UK

Copyright © 2015 China Science Publishing & Media Ltd. Published by Elsevier Inc. All rights reserved.

No part of this publication may be reproduced or transmitted in any form or by any means, electronic or mechanical, including photocopying, recording, or any information storage and retrieval system, without permission in writing from the publisher. Details on how to seek permission, further information about the Publisher's permissions policies and our arrangements with organizations such as the Copyright Clearance Center and the Copyright Licensing Agency, can be found at our website: www.elsevier.com/permissions.

This book and the individual contributions contained in it are protected under copyright by the Publisher (other than as may be noted herein).

Notices
Knowledge and best practice in this field are constantly changing. As new research and experience broaden our understanding, changes in research methods, professional practices, or medical treatment may become necessary.

Practitioners and researchers must always rely on their own experience and knowledge in evaluating and using any information, methods, compounds, or experiments described herein. In using such information or methods they should be mindful of their own safety and the safety of others, including parties for whom they have a professional responsibility.

To the fullest extent of the law, neither the Publisher nor the authors, contributors, or editors, assume any liability for any injury and/or damage to persons or property as a matter of products liability, negligence or otherwise, or from any use or operation of any methods, products, instructions, or ideas contained in the material herein.

ISBN: 978-0-12-801933-7

British Library Cataloguing-in-Publication Data
A catalogue record for this book is available from the British Library

Library of Congress Cataloging-in-Publication Data
A catalog record for this book is available from the Library of Congress

For information on all Academic Press publications
visit our website at http://store.elsevier.com/

ELSEVIER Book Aid International Working together to grow libraries in developing countries
www.elsevier.com • www.bookaid.org

Publisher: Janice Audet
Acquisition Editor: Simon Tian
Editorial Project Manager: Simon Tian
Production Project Manager: Julia Haynes
Cover Designer: Victoria Pearson

Typeset by TNQ Books and Journals
www.tnq.co.in

Dedication

Dedicated to my parents:
Brave and diligent father, Guan-Hua HONG (1894–1959);
Intelligent and diligent mother, Shi-Ru HU (1901–1971)

Contents

Synopsis

The author has worked on the Campanulaceae s. s. since 1970's. Recently after he had partially finished his work on *Paeonia* worldwide he has been concentrating his energy on studies of *Codonopsis* and its allies, which is a group very controversial among authors in taxonomy and systematics. He and his assistants made expeditions to the Himalaya (S and SE Tibet and N Pakistan), to the Hengduan Mountains (NW Yunnan and SW and W Sichuan), and to E Qinghai, where most members of this group are distributed. During the field work they conducted population observations, population sampling, and collection of materials for chromosome counting and DNA analyses in addition to collection of vouchers and herbarium specimens. The author examined and identified all specimens available in 27 herbaria. Based on the material collected in the field and studied in herbaria they observed gross morphology, seed morphology, pollen grains, and chromosomes, and made intensive DNA sequence analyses. As a result, they have established three new genera, described seven new species and two new subspecies, clarified the circumscription of *Codonopsis* and its relationships with allied genera, and created a new three-tribal classification system of the Campanulaceae s. s. The monograph assembles the results achieved in the author's work, and discusses evolution of the group, particularly the evolutionary trends of characters such as ovary position, fruit morphology, and basic number of chromosomes. The major part of the monograph is the taxonomic revision of *Codonopsis* and allied genera: *Codonopsis* (46 spp.), *Cyclocodon* (3 spp.), *Echinocodon* (1 sp.), *Himalacodon* (1 sp.), *Pankycodon* (1 sp.), and *Pseudocodon* (8 spp.). It clarifies the relationships between genera and between species, provides an useful key to genera and to species for each genus with more than one species, and gives a concise but diagnostic description for each species. Every species is provided with a fine line illustration, and most species with photos taken in the fields. Specimen citation and distribution map are presented for each species. Forty-three collections are designated as types (lectotypes, isolectotypes, or neotypes). There is a list of all the herbarium specimens examined and their species designations.

Preface

Nearly all species of *Codonopsis* are medicinal plants. For example, *Codonopsis pilosula* (Franch.) Nannf., "Dang Shen" in Chinese, is one of the most famous Chinese traditional medicinal plants, and is widely cultivated in China. The raw extract of its roots promotes digestion, strengthens the immune system, dilates peripheral blood vessels and inhibits adrenal cortex activity. *Codonopsis* and its allies are also important in horticulture. According to Grey-Wilson (1990), 32 species of *Codonopsis* in the broad sense (64% of the entire genus) are cultivated throughout the world as ornamentals for their beautiful flowers, even though the great majority of species are foetid.

The taxonomy of *Codonopsis*, however, has been extremely controversial on several issues.

First, there is no stable classification system of the Campanulaceae s. s. and thus how many subdivisions should be recognized, three or four, is still in dispute. Closely related to this issue is the systematic position of *Codonopsis* and allied genera, which until 1997 had been placed with *Wahlenbergia*, but the former has colpate pollen, whereas the latter possess porate pollen. Takhtajan's (1997) classification system of the Campanulaceae s. s. recognized four subfamilies: Campanuloideae, Cyananthoideae, Canarinoideae and Ostrowskioideae. He separated *Codonopsis* and its allies from *Wahlenbergia*, which is a good step towards a natural classification of the family, but at the same time he separated *Codonopsis* from some of its allies, e. g. *Canarina* and *Cyclocodon*. Therefore, the systematic position of *Codonopsis* and allied genera, and relationships among these are still unclear.

Second, the circumscription of the genus *Codonopsis* has been controversial. Ten generic names have been involved in the taxonomic history of the complex: *Codonopsis*, *Campanula* L. *Campanumoea* Bl., *Campanopsis* Kuntze, *Cyclocodon* Griff. ex Hook. f. & Thomson, *Echinocodon* D. Y. Hong, *Glossocomia* D. Don, *Leptocodon Lem.*, *Numaeacampsa* Gagnep., and *Wahlenbergia* Schrad. ex Roth. The recent literature focuses on two questions. One is whether *Campanumoea*, *Cyclocodon* and *Leptocodon* should be kept separate from *Codonopsis* as genera or merged with *Codonopsis* (Moeliono, 1960; Lammers and Klein in Hong et al., 2011, in notes), while the other is whether *Codonopsis*, even in the narrow sense (Hong, 1983; Hong et al., 2011), is homogeneous or heterogeneous (Eddie et al., 2003; Haberle et al., 2009).

Third, the taxonomy of a number of species complexes in *Codonopsis* and its allies has been questionable or controversial. For example, in the *Codonopsis pilosula* (Franch.) Nannf. complex, eight species have been described, but their identities and relationships are unclear. The taxonomy of *Codonopsis convolvulacea* Kurz in subg. *Pseusocodonopsis* is even more problematic and has been very controversial among authors. The number of species recognized in the complex has changed from one, three, four, six to nine (Shen & Hong, 1983: Hong in Hong et al., 2011; Hong & Ma, 1992; Grey-Wilson, 1995; Lammers & Klein, 2010). Which treatment is most rational? Grey-Wilson (1995) expressed the difficulty: "no treatment of *Pseudocodonopsis* is going to be entirely satisfactory and any treatment will have its critics". Therefore, taxonomic revision of *Codonopsis* and its allies is a great challenge.

Fourth, some of the taxonomic controversies of *Codonopsis* and its allies mentioned above come from the evaluation of characters. For example, Kolakovsky (1987 & 1994) established a new subfamily, Canarinoideae with two genera, *Canarina* and *Campanumoea*, just because these two genera share baccate fruit. A justifiable answer to the question whether *Campanumoea* should be merged with *Codonopsis* or kept separate from it as an independent genus depends on how to evaluate the systematic significance of the baccate fruit. The position of ovary, dehiscence of fruit, and pollen morphology, apertures in particular, have been used frequently in the taxonomy of the Campanulaceae s. s.

In addition, the distribution patterns of the platycodonoid group (= the tribe Cyanantheae), where *Codonopsis* and its allies are placed, are very interesting and a number of questions arise from this.

The great economic importance of *Codonopsis* and allied genera, a number of challenging questions mentioned above, and the interesting distribution patterns attracted my attention. In the present study I took an integrative approach to taxonomy, making extensive field observation and collection in the Himalaya and its adjacent Hengduan Mts., covering N Pakistan, S and SE Tibet, NW Yunnan, W and SW Sichuan, E Qinghai, and a number of other sites, and critically examining thousands of specimens in 27 herbaria in Austria, France, Germany, India, Japan, Nepal, Pakistan, Sweden, Switzerland, the United States, and the United Kingdom, in addition to China. With great help from my assistants and colleagues I observed the gross morphology, pollen morphology, seed-coat sculpture, and chromosomes, and sequenced four chloroplast genes and one nuclear gene, and subsequently constructed phylogenetic trees based on the molecular data. In doing so, my intention was to address the questions mentioned above. As a result, we proposed a three-tribal classification system of the Campanulaceae s. s. and established six subtribes for one of three tribes, the Cyanantheae, so the systematic position of *Codonopsis* and its allies has been clarified. Based on molecular phylogenetic analyses, combined with data from gross morphology, pollen morphology and cytology, we have proposed three new genera and clarified the circumscription of *Codonopsis* and its relationships with allied genera. Combined molecular data, population observations and sampling, critical examination of a large amount of specimens and statistical analysis have resulted in improved taxonomy of the *Pseudocodon* (*Codonopsis*) *convolvulaceus* complex and other groups. We have traced evolutionary trends of the characters which were often used for studies on phylogeny and taxonomy. Finally we also inferred the origin and the formation of distribution pattern of the tribe Cyanantheae.

Acknowledgements

I could not have completed this complex work without assistance from numerous persons. First, I should send my sincere gratitude to Dr. D. Geltman in the Komarov Institute of Botany, Russian Academy of Sciences, for providing us with DNA material of *Ostrowskia magnifica*, and to Prof. Dr. J. Murata in the Botanical Garden, Tokyo University, for adding localities of three species in Japan on the distribution maps. My special thanks are due to Dr. D. Boufford of Harvard University and Dr. M. Gilbert at the Royal Botanic Gardens, Kew, for significantly improving the English. Mr. Bing LI provided us with DNA material of *Echinocodon draco* (Pamp.) D. Y. Hong (*Echinocodon lobophyllus* D. Y. Hong). Mr. Xin-Tang MA helped me as head of my expeditions to Tibet, and took many beautiful photos. Miss Ai-Li LI prepared all the illustrations, which are of a high standard from both scientific and aesthetic point of view. Miss Xiao-Lin GONG processed the manuscript. Sincere thanks are due to Mr. Min LI and Miss Yan-Li XUE who assisted in preparation of colour plates. I am grateful to Professor Hang SUN, Dr. Huan-Chong WANG, Mr. Yang YANG and Mr. Ji-Pei YUE for their help in field trips to NW Yunnan in 2010, to professor Xing-Jin HE and Mr. Xiang-Guang MA for their help in the field work in SW Sichuan in 2011, and professor Shang-Wu LIU and Mr. Qing-Bo GAO for their help in the expedition to Datong, Qinghai, in 2012. I should send my gratitude to Professors Hang SUN and Dun-Yan TAN, Drs. Fu-Sheng YANG, Shu-Ren ZHANG, Xiao-Hua JIN, and Mr. Lian-Zhong FU, for generously providing me with photos. My sincere thanks are due to curators of the herbaria A, B, BM, CAL, CAS, CDBI, E, G, GXMI, HAST, HNWP, IFP, ISL, K, KATH, KUH, KUN, LD, LE, P, PE, PMNH, SM, SZ, TI, W, WU, for their permission to examine their specimens, and particularly to curators of the herbaria A, K, B, P, KUN, CDBI, SM and SZ for sending me many important specimens on loan. I should also thank many anonymous persons who have contributed to this work. The present work was financially supported by the National Natural Science Foundation of China (grant no. 31170175), and the National Publication Foundation of China.

Chapter 1

Introduction: The Taxonomic History and Questions to be Addressed

Readers need to pay attention to three issues when reading this chapter. One is the circumscription of *Codonopsis*, which has been greatly changed and is still in dispute; second is the systematic position of *Codonopsis* and its allies in the Campanulaceae s. s., which involves the classification system of the family as a whole and the phylogeny of the tribe Cyananteae in particular, and the third is the great controversy over taxonomic treatments of certain species complexes, particularly those species related to *Codonopsis convolvulacea* Kurz (= *Pseudocodon convolvulaceus*).

DISCOVERIES IN *CODONOPSIS* AND ALLIED GENERA

Wallich was the first botanist to report on *Codonopsis* when he established it as a new in 1824, and described three species, *C. viridis*, *C. purpurea* and *C. thalictrifolia*, all from the Himalaya.

Roxburgh (1824) described *Campanula lancifolia* from what is now Bangladesh as new. The species was transferred to *Cyclocodon* by Kurz (1872), transferred to *Campanumoea* by Clarke (1881), and treated as a synonym in *Codonopsis* by Moeliono (1960). With the restoration of *Cyclocodon* Griff. ex Hook. f. & Thomson, it is now placed back in *Cyclocodon* (Hong & Pan, 1998).

In 1825, Don described a new genus and species from Nepal, *Glossocomia tenera*, which was, however, the same taxon as Wallich's *Codonopsis thalictrifolia*.

Another new genus, *Campanumoea*, with two new species, *C. javanica* Bl. and *C. celebica* Bl. was described by Blume (1826) from Indonesia. The status of *Campanumoea*, its circumscription, and its relationship with *Codonopsis* in particular, has been a controversial issue. *Campanumoea javanica* Bl. as a species has never been questioned, whereas the fate of *C. celebica* is quite different. It was transferred to *Codonopsis* by Miquel (1835), and then into *Campanula* by Dietrich (1839), but placed back in *Codonopsis* as a subspecies of *Codonopsis lancifolia* by Moeliono (1960). It was transferred by the present author to *Cyclocodon* (Hong in Hong & Pan, 1998). This action was accepted by Lammers (1998) and Morris and Lammers (1999), but they treated it again as a subspecies, *Cylcocodon lancifolia* subsp. *celebicus* (Blume) K. E. Morris & Lammers.

De Candolle (1830) published two new species from India, *Codonopsis parviflora* Wall. ex A. DC. and *Codonopsis truncata* Wall. ex A. DC., based on plants discovered by Wallich. The former was transferred to *Cyclocodon* by Hooker and Thomson (1857), then placed in *Campanumoea* by Bentham and Hooker (1876), but finally returned to *Cyclocodon* by Hong (Hong & Pan, 1998). *Codonopsis truncata* was transferred to *Cyclocodon* by Hooker and Thomson (1857), then placed in *Campanumoea* by Diels (1901), but was treated as conspecific with *Campanumoea celebica* by Clarke (1881), with *Campanumoea lancifolia* by Tsoong (1935) and Hong (1983), and with *Codonopsis lancifolia* by Moeliono (1960). Finally, it was treated as a synonym of *Cyclocodon lancifolius* (Roxb.) Kurz by Hong (Hong & Pan, 1998; Hong et al., 2011). De Candolle (1839) described another species, *Campanumoea roylei*, which was reduced to the synonymy of *Codonopsis ovata* by Hooker and Thomson (1857).

Bentham (1834) described two species as new from the western Himalaya and Karakoram, *Codonopsis rotundifolia* and *C. ovata*. The former has never been questioned as a distinct species, while the latter has been recognized by all authors except Nasir (1984), who treated it as conspecific with *C. clematidea*.

Campanumoea lanceolata Sieb. and Zucc. (1835) was transferred to *Codonopsis* by Trautvetter in 1879, while Lindley's (1839) *Codonopsis lurida* is a synonym of *C. rotundifolia*. Schrenk (1841) discovered a new species of *Codonopsis* in Middle Asia, but he placed it in *Wahlenbergia*, as *W. clematidea*. *Codonopsis albiflora* was described by Griffith (1854), but was reduced by Clarke (1881) to the synonymy of *Campanumoea celebica*. In 1856, Hasskarl described *Campanumoea cordata* from Sumatra, Indonesia, but it was found to be another name for *Campanumoea javanica*.

Hooker and Thomson (1855) described two new species, *Codonopsis gracilis* and *C. inflata*, but one year later Lemaire (1856) separated the former as a new genus, *Leptocodon* Lem., which was subsequently accepted by Hooker and Thomson (1857).

The year 1857 is important in the history of studies on *Codonopsis*. Hooker and Thomson published an article containing four distinct new species, *C. affinis*, *C. benthamii*,

A Monograph of *Codonopsis* and Allied Genera (Campanulaceae). http://dx.doi.org/10.1016/B978-0-12-801933-7.00001-5
Copyright © 2015 China Science Publishing & Media Ltd. Published by Elsevier Inc. All rights reserved.

C. foetens and *C. subsimplex*, all from the Himalaya. In addition, the article includes a description of Griffith's (1854) unpublished genus, *Cyclocodon* Griff. ex Hook. f. & Thomson, which contained two species with baccate fruit. This genus has been variously treated since then. Clarke (1881) treated it as a section of *Campanumoea*, also with baccate fruit. Moeliono (1960) merged all the species of *Campanumoea* with *Codonopsis*, but Hong (in Hong & Pan, 1998) separated *Cyclocodon* from *Campanumoea* and restored it to generic status. Also in 1857, Ruprecht described two new species; *Glossocomia ussuriensis* (Ruprecht & Maximowicz, 1857), from "Manshuria", the Far East, which was transferred to *Codonopsis* by Hemsley (1889), and *Glossocomia hortensis* Rupr. (Ruprecht 1857), which was treated as a synonym of *Codonopsis lanceolata* by Hemsley (1889). Again in 1857, another new species, *Codonopsis leucocarpa,* was described by Miquel (1857), but it was treated as a synonym of *Campanumoea celebica* by Clarke (1881).

In 1863 Morren published *Campanumoea japonica* Sieb. ex Morr., another synonym of *Codonopsis lanceolata.*

Kurz (1873) described *Codonopsis convolvulacea* from southern Yunnan, which is distinct from other members of *Codonopsis* in having the corolla cleft nearly to the base, and thus rotate, and in having globose or ellipsoid root tubers. Komarov (1908) later established a new subgenus, *Codonopsis* subg. *Pseudocodonopsis* Kom., based on this species.

Unaware of the name *Campanumoea japonica* Sieb. ex Morr. (1863), Maximowicz described *Campanumoea japonica* Maxim. in 1867, which is, of course, an invalid name. This plant was then treated as a variety of *Campanumoea javanica* by Makino (1908), *Campanumoea javanica* var. *japonica* Makino. It was transferred to *Codonopsis* as a subspecies, *Codonopsis javanica* subsp. *japonica* by Lammers (1992).

The year of 1881 is also important, for it was then Maximowicz described *Codonopsis viridiflora* from southeastern Gansu, and Clarke described two new species from the Himalaya, *Wahlenbergia dicentrifolia* and *Codonopsis griffithii.* Although the latter species is a superfluous name of *Codonopsis viridis* Wall., the first species is very distinct. It was transferred to *Codonopsis* by Smith (1913a), and treated by Hong (1980) as the single member of *Codonopsis* subg. *obconicicapsula* D. Y. Hong based on its obconical capsule and compressed seeds.

Three years late Franchet found a medicinal plant cultivated in the vicinity of Beijing which he described as a new species of *Campanumoea, C. pilosula* (Franchet, 1884). It was transferred to *Codonopsis* by Nannfeldt (1930) as *Codonopsis pilosula* (Franch.) Nannf. This is one of China's most commonly used medicinal plants and is widely cultivated in China.

Oliver (1891) described two new species from western Hubei of China, *Codonopsis henryi* and *C. tangshen.* The former has never been questioned as a distinct species. The specific epithet '*tangshen*' is derived from the name of a famous Chinese medicinal plant, "Dang Shen," the same name given to *Codonopsis pilosula.* I tend to consider them to be two forms of the same species, *Codonopsis pilosula.*

Two very important articles were published in 1908 after a silence of nearly 20 years. In a review of the entire genus Komarov (1908) recognized a total of 23 species, including five new ones, *C. cardiophylla* Diels ex Kom., *C. cordifolia, C. subscaposa, C. tubulosa,* and *C. vinciflora.* Subgenera were proposed for *Codonopsis* for the first time: *Eucodonopsis* (= subg. *Codonopsis*) and *Pseudocodonopsis* Kom. Within subg. *Codonopsis* Komarov grouped all the species with twining stems into series *Volubiles* with three subseries, while the others with erect stems were placed in series *Erectae.* Later that year Chipp (1908) also provided a taxonomic revision of *Codonopsis,* which recognized 22 species, including four new, *C. deltoidea, C. micrantha, C. mollis,* and *C. pilosa.* The first two are still maintained as distinct species whilst *C. mollis* is now treated as a subspecies of *C. thalictrifolia* and *C. pilosa* is a synonym of *C. tubulosa.* Interesting is that he described a new variety, *Codonopsis ovata* var. *nervosa* Chipp based on a collection from Kangding (Tachielu), Sichuan, China, far from the distribution area of *C. ovata* (NW Pan-Himalaya and S Middle Asia, see Map IV 1–8). That is actually *Codonopsis foetens* subsp. *nervosa* (Chipp) D. Y. Hong. He did not accept Komarov's subgeneric divisions, nor make his own subdivision.

Pampanini (1910) described a new species, *Codonopsis draco* Pamp. from Hubei of China, which was ignored until 2014. This plant was rediscovered in Hubei over 70 years later and described as a new genus, *Echinocodon lobophylla* D. Y. Hong (1984). The new combination, *Echinocodon draco* (Pamp.) D. Y. Hong, has been made (Hong, 2014b).

Hayata (1911) described *Codonopsis kawakamii* from Taiwan, the only species of the genus in the narrow sense in Taiwan.

Diels identified five further species, one of which was published by Komarov (1908), *Codonopsis cardiophylla.* The remaining four species were published by Diels (1912): *Codonopsis macrocalyx, C. forrestii, C. bulleyana* and *C. meleagris. Codonopsis macrocalyx* was reduced to a synonym of *C. benthamii* by Hong (2010); *C. forrestii* has been treated variously by previous taxonomists. Ballard (1939) and Shen (1983) treated it as a variety of *Codonopsis convolvulacea,* Hong and Ma (1992) and Hong (Hong in Hong et al., 2011) treated it as a subspecies of *C. convolvulacea,* whereas Grey-Wilson (1995) and Lammers and Klein (2010) restored it to specific status. There has been no dispute among taxonomists on the other two species.

Smith (1913a) transferred *Wahlenbergia dicentrifolia* to *Codonopsis,* but it is very isolated in that genus. In the same year he described two new species, both from Yunnan, *Codonopsis subglobosa* and *C. efilamentosa* (Smith, 1913b). The latter is a member of *Codonopsis* subg. *Pseudocodonopsis* Kom.

Interestingly, Léveillé described eight species from Guizhou and Yunnan under *Codonopsis* and other generic names between 1904 and 1916 (Léveillé, 1904, 1914, 1915, 1916). In a review of the material in Léveillé's herbarium, Chamberlain (1977) synonymised all but one of these, *C. graminifolia* H. Lév. (1915). The status of *C. graminifolia* has been controversial. It was treated as a synonym of *C. limprichtii* Lingelsh. & Borza (1914) by a number of authors, e.g. Lammers & Klein (2010), but we (Hong & Ma, 1994; Hong in Hong et al., 2011) found it to be distinct in *Codonopsis* subg. *Pseudocodonopsis* Kom., covering all forms with linear leaves (see section 4 in this chapter for more).

Lingelsheim and Borza (1914) described *Codonopsis limprichtii* as new from Dali, Yunnan Province. It was treated as a variety of *C. convolvulacea* by Anthony (1926), but Lammers and Klein (2010) argued for its status as a good species.

Nakai (1915) described *Codonopsis minima* from the Korean Peninsula, which was reduced to a variety of *C. ussuriensis* by Lee (1986). According to my observations of the holotype (TI) and isotypes (E & K), it is a simple synonym of *C. ussuriensis*.

Handel-Mazzetti (1924) published four varieties of *Codonopsis* based on his own collections from SW Sichuan and NW Yunnan, *C. foetens* var. *major*, *C. macrocalyx* var. *coerulescens*, *C. limprichtii* var. *hirsuta*, and *C. limprichtii* var. *pinifolia*. The third has been raised to specific rank, *C. hirsuta* (Hand.-Mazz.) D. Y. Hong & L. M. Ma (1992), while the last one is a synonym of *C. graminifolia*.

Anthony (1926) provided a key to the world species of *Codonopsis*, treating 35 species but not including four species not seen by him. In the article he described two new species, *C. farreri* and *C. chimiliensis* from the border between China and Myanmar. The two species are both distinct, and no later author has challenged their status as species.

Nannfeldt (1930, 1931, 1936, 1940, 1950) contributed much to the study of *Codonopsis*. In his first publication Nannfeldt (1930) described *C. modesta* from N Sichuan as new and recognized *C. nervosa* (Chipp) Nannf. (= *C. ovata* var. *nervosa* Chipp) from NW Sichuan. He found the second species to be very common in W Sichuan and N Yunnan. More importantly, he transferred Franchet's *Campanumoea pilosula* to *Codonopsis*, creating the new combination, *Codonopsis pilosula* (Franch.) Nannf. In his second publication, Nannfeldt (1931) described two new species, *C. alpina* and *C. macrantha*, both from NW Yunnan. The first is distinct, whereas the second is merely a form of *C. nervosa* with larger flowers. Nannfeldt (1931. fig.1) emphasized with drawings the value of the shape of both calyx lobes and corolla for delimitation of some species. In his third publication Nannfeldt (1936) recorded 17 species of *Codonopsis* from China without describing any new species . In the fourth publication Nannfeldt (1940) described three species, *C. canescens* from W Sichuan, and *C. glaberrima* and *C. volubilis*, both from Shanxi, but unfortunately only *C. canescens* is distinct, the other two are synonyms of *C. pilosula* (Franch.) Nannf.

P. C. Tsoong, my supervisor when I was a graduate student from 1962 to 1966, was the first Chinese author to produce an overall taxonomic revision of the Campanulaceae in China (Tsoong, 1935). For *Codonopsis* he recognized 20 species, including two new ones, *C. argentea* from Guizhou (Fanjing Shan) and *C. cordifolioidea* from NW Yunnan (Fugong). Both have proven to be distinct species with very narrow distributions.

Gagnepain (1948) described a scandent herb from Laos (*A. F. G. Kerr 21186*, P) as a new species in a new genus, *Numaeacampa kerrii* Gagnep., and stated that it was closely related to *Campanumoea*. Takhtajan (1997) and Mabberley (2008) treated *Numaeacampa* as a synonym of *Codonopsis*. Lammers (2007b), however, considered the identity of this taxon to be a mystery; he stated, based on Gagnepain's description, that the plant is unlike anything else in the Campanulaceae. Therefore, he writes: "*Numaeacampa* Gagnep. = ?" and "*Numaeacampa kerrii* Gagnep. = ? ". Thanks to the images of this specimen from P, we determined it as *Lycianthes lysimachioides* (Wall.) Bitter, a species of the Solanaceae (Wang & Hong, unpublished). The ovary of the genus *Lycianthes* looks to be inferior, but is actually superior, surrounded by a cupular calyx.

There were few studies on the taxonomy of *Codonopsis* from 1940 to 1965. Wu (1965) broke the silence when he published his report on the floristic survey of Yunnan (Campanulaceae), recording five species of *Codonopsis*, of which two were new, *C. chlorocodon* from Dêqên County and *C. gombalana* from Gongshan County. They were both recognized at specific rank by Shen & Hong (1983) and by Hong et al. (2011), but later I found that the second one is actually a synonym of *C. farreri* J. Anthony.

Ludlow, who spent many years on plant collection in the Himalaya, particularly in SE Tibet and Bhutan, described *Codonopsis bhutanica* as new based on eight collections from the border area between Bhutan and Tibet (Ludlow, 1972). My *C. xizangensis* D. Y. Hong (1980) from Cona, S Tibet, proved to be a synonym of this species.

Shen and his colleagues (Shen et al., 1975) revised the taxonomy of *Codonopsis* in Sichuan, recognizing 24 species, of which one, *C. levicalyx* from N Sichuan, was considered to be new. *Codonopsis levicalyx* was recognized by Shen & Hong (1983) and Hong et al. (2011), but after critically examining the holotype and a large number of specimens it is here treated as a synonym of *C. henryi*.

Hara (1978) described *Codonopsis nepalensis* from Nepal, but after critically examining a large number of specimens of *C. rotundifolia* ranging from N Pakistan to Nepal, I found Hara's species to be within the wide variation range of *C. rotundifolia*.

In 1980, I described three new species from Tibet, *Codonopsis longifolia*, *C. xizangensis*, and *Leptocodon hirsutus* (Hong, 1980). The former two proved to be synonyms, since I was then unaware of Ludlow's (1972) and Hara's (1978) publications, while the last was transferred to *Codonopsis* by Lammers (2001).

Huang (1984) described *Codonopsis pianmaensis* and *C. farreri* var. *grandiflora* from NW Yunnan, but the latter has been reduced (Hong in Hong et al., 2011), while the former is an extreme form of the variable *C. benthamii* Hook. f. & Thomson.

Studies on the taxonomy of *Codonopsis* appears to have entered another active period since 1990. First, Grey-Wilson (1990) published an article "A survey of *Codonopsis* in cultivation", where he used a broadly defined *Codonopsis*, including *Leptocodon* and *Campanumoea*. He recognized and described 33 species, of which two from central Nepal, *C. bragaensis* and *C. nepalensis* Grey-Wilson (1990, non Hara, 1978), were new. The first species was said to be very distinct, while the second was renamed as *C. grey-wilsonii* by Shaw [1996, = *Pseudocodon grey-wilsonii* (J. M. H. Shaw) D. Y. Hong].

Wang and Xu (1993) described two species as new, *C. microtubulosa* from N Sichuan and *C. retroserrata* from S Sichuan. They both were reduced to synonymy in 'Flora of China' (Hong et al., 2011). However, due to examination of more materials, field observation, and molecular analysis, they were found to be distinct and thus are restored in the present work.

Lammers and Klein (2010) revised *Codonopsis* subg. *Pseudocodonopsis* Kom., adding one more species of uncertain provenance to the genus: *Codonopsis macrophylla*. It appears to me a form of the very variable *Codonopsis convolvulacea* subsp. *forrestii* [= *Pseudocodon convolvulaceus* (Kurz) D. Y. Hong & H. Sun subsp. *forrestii* (Diels) D. Y. Hong].

Dash and Mao (2011) published *Codonopsis vadsea* from the Yarlung Zangbo-Brahmaputra region in the eastern Himalaya, which, based on their figure and description, I treat here as a subspecies of *C. tubulosa*.

The most active period in the study of *Codonopsis* and allied genera was from 1900 to 1949, when 33 new species, nearly half of the total, were described. Although only 15 new species were described in 60 years from 1950 to 2011, several of them are distinct. Among them, *Codonopsis chlorocodon* C. Y. Wu (1965), *C. nepalensis* Grey-Wilson (1950) (=*C. grey-wilsonii* J. M. H. Shaw) and *C. bhutanica* Ludlow deserve particular mention.

Recently a number of new species have been described; Hong (2014a) published six new species from the Pan-Himalaya and one new subspecies from W Sichuan. They are *C. bomiensis*, *C. campanulata*, *C. elliptica*, *C. hemisphaerica*, *C. lixianica*, *C. reflexa*, and *C. cardiophylla* subsp. *megaphylla*. At the same year, Wang and Hong (2014) described a distinct species as new from Gaoligong Shan, NW Yunnan, *Codonopsis gongshanica*.

From Wallich (1824) to the present, a total of 101 species have been described under *Codonopsis* or first under other generic names and later transferred to *Codonopsis*. In addition, numerous subspecific and particularly varietal taxa have been published. Ten generic names have been used to accommodate these 101 specific names besides *Cyananthus* and *Melothria*, which were wrongly used by Léveillé for two of his new species. It is clear from the above that frequent changes have been made in the taxonomic history of *Codonopsis* and its allies and many generic names have been involved in the taxonomy of the group. These facts reflect the intricate relationships among *Codonopsis* and its allies and imply that to clarify the circumscription of *Codonopsis* is a challenging but very important task.

CIRCUMSCRIPTION OF *CODONOPSIS*

Since the establishment of the genus *Codonopsis* by Wallich in 1824, its circumscription has been questioned and changed greatly, as mentioned at the end of the previous section. Ten generic names have been involved in the circumscription of *Codonopsis*.

Don (1825) used a new generic name, *Glossocomia*, to accommodate his new plant, *Glossocomia terera* from Nepal. This was treated as a synonym of Wallich's *Codonopsis thalictrifolia* by Hooker and Thomson (1857). *Glossocomia ussuriensis* Rupr. & Maxim. (1857) was transferred to *Codonopsis* by Hemsley (1889), while *Glossocomia hortensis* Rupr. (1857) was treated also by Hemsley as a synonym of *Codonopsis lanceolata*. Since then no one used the generic name *Glossocomia* for any new species.

De Candolle (1839) described *Wahlenbergia roylei*, Schrenk (1841) described *W. clematidea*, and Clarke (1881) used the name *Wahlenbergia* with a question marker for his new species, *W.* (?) *dicentrifolia*. All three of these species were transferred to *Codonopsis*, the former two by Clarke (1881) and the last one by Smith (1913a). Since then, no species of *Codonopsis* or its allies have been described under *Wahlenbergia*.

Sprengel (1825) transferred Wallich's three species of *Codonopsis*, *C. viridis*, *C. purpurea* and *C. thalictrifolia*, to *Campanula* L., but his action has never been followed.

Campanumoea was established by Blume (1826) for species with baccate fruit. Siebold and Zuccarini's (1835) *Campanumoea lanceolata* from Japan lacking this was transferred to *Codonopsis* by Trautvetter (1879). *Campanumoea pilosula* Franch. (1884) was transferred to *Codonopsis* by Nannfeldt (1930). *Campanumoea japonica* Sieb. ex Morr. (1863) was treated as a synonym of *Codonopsis lanceolata* by Komarov (1908). All these three taxa have a loculicidal capsule. The relationship between *Campanumoea* sensu Blume and *Codonopsis* has been in dispute. Clarke (1881) expanded the circumscription of *Campanumoea* by including not only *Codonopsis inflata* Hook. f., but also *Cyclocodon* Griff. ex Hook. f. & Thomson (1857). He then divided the newly defined genus into two sections, sect. *Campanumoea* and sect. *Cyclocodon* (Griff. ex Hook. f. & Thomson) C. B. Clarke (1881). The two sections share baccate fruit. Clarke's (1881) concept of *Campanumoea* was followed by Tsoong (1935) and Hong (1983). Moeliono (1960) merged *Campanumoea* sensu Clarke into *Codonopsis*. This action was followed by Lammers (1992a & b), who transferred *Campanumoea javanica* subsp. *japonica* (Makino) D. Y. Hong into *Codonopsis* and made a new combination accordingly, stating that a single character, berry or capsule, is insufficient to justify generic status. In another direction, however, Hong & Pan (1998), based on LM and SEM observations of pollen and seed coats, revealed that the five species of *Campanumoea* sensu Clarke (1881) form two distinct groups that correspond with Clarke's (1881) two sections. The pollen in sect. *Cyclocodon* is 3-colporate and the seed coat has irregularly shaped lumina with rope-like secondary ornamentation on the radial walls, in addition to the plants having erect stems. In sect. *Campanumoea* the pollen is 5–8-colpate, the seed coat has regular and polygonal lumina with bead-like secondary ornamentation on the radial walls, and the habit is twining. Based on these observations, Hong & Pan (1998) restored *Cyclocodon* to generic status. Furthermore, *Cyclocodon* was considered to be more closely related to *Platycodon* than to *Campanumoea*. That is, *Codonopsis* as circumscribed by Moeliono (1960) and his followers is polyphyletic. The action of Hong & Pan (1998) were followed by Lammers (1998 & 1999). Recent molecular data (Borsch et al., 2009; Wang et al., 2013) clearly indicate that *Cyclocodon* is more closely related to *Platycodon* than to *Campanumoea* and *Codonopsis*. The status of *Campanumoea* in the narrow sense, whether it should be maintained or merged with *Codonopsis*, is still controversial (see *Fl. China* 19: 505. 2011).

Leptocodon (Hook. f. & Thomson) Lem. (Lemair, 1856) was established based on *Codonopsis gracilis* Hook. f. (Hooker, 1855). Its generic status was recognized by Hooker and Thomson (1857), Clarke (1881), Tsoong (1935), and Hong (1983). Grey-Wilson (1990), however, merged *Leptocodon* with *Codonopsis*, a move followed by Lammers (1999 & 2001). The pollen of *Leptocodon* is similar to that in core *Codonopsis* (Erdtman, 1952; Avetisjan, 1986; Morris & Lammers, 1997; Hong & Pan, 2012).

Pampanini's (1910) *Codonopsis draco* has been transferred to *Echinocodon*, and the new combination made (Hong, 2014b).

In *Codonopsis* s. s., Komarov (1908) separated two species with rotate corolla as subgenus *Pseudocodonopsis*; Hong (1980) treated *Codonopsis dicentrifolia* (C. B. Clarke) W. W. Smith as the only representative of subgenus *Obconicicapsula* D. Y. Hong, based on its obconical ovary and fruit and compressed seeds. As a result of rather extensive observations on pollen morphology of *Codonopsis* and its allies Morris and Lammers (1997) recognized four types of pollen morphology: Type I for pollen of *Codonopsis* subg. *Codonopsis*, *Leptocodon* and *Campanumoea* sect. *Campanumoea*, Type II for pollen of *Campanumoea* sect. *Cyclocodon* (= *Cyclocodon*), Type III for pollen of *Codonopsis* subg. *Pseudocodonopsis* (= *Pseudocodon*), and Type IV for pollen of *Codonopsis* subg. *Obconicicapsula* (= *Himalacodon*). Hong and Pan (2012) made even more extensive observation on the pollen morphology of this group. Our results confirmed the conclusions of Morris and Lammers' (1997) and showed that the pollen of *Campanumoea* s. str. differs from that of subg. *Codonopsis* in having shorter colpi (colpus length/polar axis length = 0.54–0.57 vs. ≥ 0.7) and a granular colpus membrane (vs. spinulose). Thus Hong and Pan (2012) defined this type of pollen as the *Codonopsis* Type (Type I) *Campanumoea* subtype. Their observations also revealed that *Codonopsis* subg. *Codonopsis* is heterogeneous in pollen morphology, i.e. the pollen of *C. purpurea* and *C. chimiliensis* is basically similar to Morris and Lammers' (1997) Type IV, but with a smooth colpus membrane and they thus assigned it to the *Obconicicapsula* Type (= Type IV) *Purpurea* subtype.

According to Shen & Hong (1983) and Hong (in Hong et al., 2011), *Codonopsis* comprises three subgenera, subgs. *Codonopsis*, *Obconicicapsula* D. Y. Hong and *Pseudocodonopsis* Kom. In the ITS trees presented by Eddie et al. (2003) and Haberle et al. (2009), *Codonopsis*, even in Hong's (in Hong et al., 2011) narrow sense, is not monophyletic; *Codonopsis dicentrifolia*, the only species of subg. *Obconicicapsula*, forms a clade with *Cyananthus lobatus* Wall. ex Benth. Therefore, *Codonopsis* is polyphyletic from palynological and molecular data.

According to the above review of the history, further multidisciplinary studies should be conducted on *Codonopsis* and its allies to address the following questions.

1. Should *Campanumoea* and *Leptocodon* be included in or kept separate from *Codonopsis*?
2. Is *Codonopsis* in the narrow sense monophyletic or polyphyletic?
3. What are the phylogenetic relationships within *Codonopsis* and its allies?
4. How should *Codonopsis* be circumscribed?

SYSTEMATIC POSITION OF *CODONOPSIS* AND ALLIED GENERA

There has been a common view that *Codonopsis* and allied genera belong to the Campanulaceae. The circumscription of the Campanulaceae, however, has involved different views. De Candolle (1839) treated it in a narrow sense, separating it from the Cyphiaceae (including *Lobelia*), but including *Pentaphragma* with a question marker. Bentham and Hooker (1876) were the first to circumscribe the family broadly, including Lobelieae and Cyphieae. From then on botanists have been divided into two schools. Endlicher (1841), Baillon (1880), Wettstein (1924), Hutchinson (1973), Thulin (1975), Kovanda (1978), Dahlgren (1983), Takhtajan (1987, 1997), Lammers (1992), Cosner et al. (1994), and Gustafsson and Bremer (1995) hold the narrow view, whereas Schönland (1889), Bessey (1915), Wagenitz (1964), Cronquist (1981), Thorne (1992), Lammers (1998, 2007), and Brummitt (2007) hold the broad view.

According to Gustafsson and Bremer (1995) the Campanulaceae clade includes five families: Campanulaceae, Nemacladaceae, Cyphiaceae, Cyphocarpaceae and Lobeliaceae. Lammers (1998, 2007), however, treated these five families as five subfamilies in the Campanulaceae s. l.

Regardless of whether one accepts the family broadly or narrowly, the Campanulaceae (Gustafsson & Bremer, 1995) or the Campanuloideae (Lammers, 1998, 2007) is a monophyletic group. Therefore, it is a matter of taste to take either view. As we prefer to recognize the family in the strict sense, *Codonopsis* and its allies are members of the Campanulaceae s. s. *Codonopsis* and its allies have been placed in the Campanulaceae since de Candolle (1830, 1839).

Phylogenetic relationships of the genera within the Campanulaceae s. s., however, are far from well established. As stated by Kovanda (1978), "The family Campanulaceae……is rather natural and homogeneous but its subdivision presents serious problems……". De Candolle (1839) divided the family into three tribes, Wahlenbergieae Endl., Campanuleae G. Don and Mercieae A. DC., and placed *Campanumoea*, *Codonopsis*, *Canarina* and *Platycodon* together with *Wahlenbergia* in Wahlenbergieae. His *Codonopsis* is actually *Cyclocodon*, while he treated true *Codonopsis*, including the first described three species of *Codonopsis*, i.e. *C. viridis, C. thalictrifolia* and *C. purpurea*, as members of *Wahlenbergia*. Two issues have been involved in the systematic position of *Codonopsis* and allied genera. First, since de Candolle (1839), *Codonopsis* and its allies have been placed together with *Wahlenbergia* primarily because they share a loculicidal capsule. Second, the circumscription of *Codonopsis* and its phylogenetic relationships with its allies have been unclear. They were distantly separated and placed in different groups. Schönland (1889) divided the tribe Campanuleae (= Campanulaceae s. s.) into three subtribes, with *Codonopsis*, *Campanumoea*, *Leptocodon* and *Cyananthus* along with *Wahlenbergia* in the Wahlenberginae, *Platycodon*, *Microcodon* and *Musschia* in the Platycodinae, and *Canaria* and *Ostrowskia* together with *Campanula* and *Adenophora* in the Campanulinae. Some authors went even further. For example, Kolakovsky (1987 & 1994) partitioned *Codonopsis* and its allies into two groups and placed *Codonopsis*, *Leptocodon* and *Platycodon*, all with capsular fruit, together with *Wahlenbergia* in the Wahlenbergioideae, while he placed *Canarina* and *Campanumoea* with baccate fruit in the separate subfamily, Canarinoideae. Although Takhtajan (1997) was the first botanist who separated *Codonopsis*, *Leptocodon*, *Campanumoea*, *Cyananthus*, *Platycodon* from *Wahlenbergia*, he also separated *Canarina* and *Ostrowskia* as two independent subfamilies, Canarinoideae and Ostrowskioideae. Recent molecular work (Eddie et al., 2003; Haberle et al., 2009) have provided insights into the phylogenetic relationships within the Campanulaceae s. s., but unfortunately sampling in those studies was not adequate enough to clarify the systematic position of *Codonopsis* and its relationships with its allies, which remains our present task. Chapter III provides a detailed discussion of this issue.

TAXONOMICAL REVISIONS OF *CODONOPSIS* AND ALLIED GENERA

The first significant taxonomic revision of *Codonopsis* and its allies was made by Hooker and Thomson (1857). Although their work was regional, involving only elements within British India, it covered the majority of the species described up to that time, 10 in *Codonopsis*, 2 in *Campanumoea*, 2 in *Cyclocodon* and 1 in *Leptocodon*. They properly transferred the species described under *Glossocomia*, *Wahlenbergia*, and *Campanula* to *Codonopsis*, which basically established the concept of *Codonopsis*. They reduced *Glossocomia tenera* D. Don, *Wahlenbergia roylei* A. DC. and *Codonopsis lurida* Lindl. to synonymy under *Codonopsis thalictrifolia* Wall., *C. ovata* Benth. and *C. rotundifolia* Benth. respectively, but their reduction of *Wahlenbergia clematidea* Schrenk to synonymy under *Codonopsis ovata* Benth. is unjustifiable. Clarke's (1881) work followed Hooker and Thomson's (1857) treatment with minor revisions; he transferred *Codonopsis inflata* Hook. f. & Thomson into *Campanumoea* and reduced *Cyclocodon* to *Campanumoea* sect. *Cyclocodon*.

Komarov (1908) made the most significant taxonomic revision up to that time, recording 23 species, of which five were new. Komarov's work is excellent in a number of aspects. First, he clearly defined the species, of which the great majority (19/23) have stood until now; of the five new species he described only *C. cordifolia* has become a synonym (of *Campanumoea javanica* Bl.). Second, he provided the first subgeneric division, dividing the genus into two subgenera, *Eucodonopsis* (= subg. *Codonopsis*) and *Pseudocodonopsis*, and then subdividing the former into two series, *Volubiles* with twining stems and *Erectae* with erect stems. He further divided series *Volubiles* into three subseries according to phyllotaxy and corolla shape. Third, the first key to species of *Codonopsis* appeared in this revision. Fourth, he listed 29 synonyms that were previously used under *Campanula*, *Campanumoea*, *Codonopsis*, *Glossocomia* and *Wahlenbergia*.

In the same year but following Komarov, Chipp (1908) published another worldwide taxonomic revision of

Codonopsis, recording 22 species. The differences between Chipp's taxonomy of 22 species and Komarov's of 23 species are: (1) Chipp added four new species; (2) he recognized three of Komarov's new species, while reducing the other two, *C. cordifolia* and *C. vinciflora*. Among Chipp's four new species, the specific status of *C. deltoidea* and *C. micrantha* has never been questioned, *C. mollis* is now treated as a subspecies of *C. thalictrifolia*, while *C. pilosa* is a synonym of *C. tubulosa*. *Codonopsis vinciflora* and *C. clematidea* were placed by Chipp in synonymy under *C. convolvulacea* and *C. ovata* respectively, but these two species are actually distinct. Chipp correctly reduced three additional species, *C. cordifolia* Kom. (= *Campanumoea javanica*), *C. griffithii* (= *C. viridis*) and *C. silvestris* (= *C. pilosula*), but did not provide subgeneric divisions.

Eighteen years late Anthony (1926) presented a key covering 39 species of *Codonopsis* worldwide, the highest number of species recorded for the genus until then. His major contribution is the addition of 15 species published after 1908, including two of his own new species, *C. chimiliensis* and *C. farreri*. Another remarkable feature of his work is a comprehensive list of botanical names (39) and synonyms (42). Anthony's two new species have never been questioned.

Nannfeldt's (1931) "some notes on the genus *Codonopsis* Wall." is virtually a revision of some species groups. For example, he further clarified the differences between *C. ovata* Benth. and *C. clematidea* (Schrenk) C. B. Clarke after Chipp (1908), indicating that the former has a funnel-shaped corolla and is confined to SW Kashmir and Chamba, while the latter has a campanulate corolla and is widely distributed, even extending to the Tianshan (Thian Shan). Unfortunately he failed to note the differences in two characters of the calyx. He also properly circumscribed *Codonopsis cardiophylla* Diels ex Kom., excluding several specimens cited under that name by Anthony (1926) and treating them as components of *Codonopsis alpina* Nannf. (new) and *C. macrantha* Nannf. (= *C. foetens* subsp. *nervosa*) respectively.

Although Chamberlain's (1977) paper is not a taxonomic revision in the strict sense, it is rather important for the taxonomy of *Codonopsis*. He reviewed eight species described by Léveillé (1914–1916), four in *Codonopsis* and four in three other genera, based on examination of the specimens in the Léveillé herbarium. He treated *Cyananthus mairei* H. Lév. (1915) as a synonym of *Codonopsis bulleyana* Forrest ex Diels, *Codonopsis graminifolia* H. Lév. (1915) as a synonym of *C. convolvulacea*, *C. mairei* H. Lév. (1915) as a synonym of *C. forrestii* Diels, *C. bodinieri* H. Lév. as a synonym of *C. lanceolata* (Sieb. & Zucc.) Trautv., *C. accrescenticalyx* H. Lév. (1915) as a synonym of *C. tubulosa* Kom., *Campanumoea violifolia* H. Lév. (1915) and *Melothria violifolia* H. Lév. (1916) both as synonyms of *C. micrantha* Chipp, and *Campanumoea labordei* H. Lév. as a synonym of *Campanumoea javanica* Bl.

Shen and Hong's (1983) work contained a number of taxonomic changes in *Codonopsis*. It was the first treatment of *Codonopsis* comprising three subgenera, subgs. *Codonopsis*, *Obconicicapsula* and *Pseudocodonopsis*. Second, the taxonomy of the *Codonopsis convolvulacea* Kurz complex was greatly changed to include seven varieties, vars. *convolvulacea*, *efilamentosa* (= *C. efilamentosa* W. W. Smith), *forrestii* (= *C. forrestii* Diels), *hirsuta* (= *C. limprichtii* var. *hirsuta* Hand.-Mazz.), *limprichtii* (= *C. limprichtii* Lingelsh. & Borza), *pinifolia* (= *C. limprichtii* var. *pinifolia* Hand.-Mazz.) and *vinciflora* (= *C. vinciflora* Kom.). In addition, *Codonopsis handeliana* Nannf., *C. modesta* Nannf., and *C. volubilis* Nannf. were reduced to varietal rank within *C. pilosula*; *C. mollis* Chipp and *C. macrantha* Nannf. were reduced to varieties of *C. thalictrifolia* Wall. and *C. nervosa* Nannf. respectively.

Hong and Ma (1992) revised *Codonopsis* in Sichuan, recognizing 22 species. Their 22 species are somewhat different from the 24 species of Shen et al. (1975). They critically revised the taxonomy of the *C. convolvulacea* group, recognizing three species and two subspecies: *C. graminifolia* H. Lév., *C. hirsuta* (Hand.-Mazz.) D. Y. Hong & L. M. Ma, *C. convolvulacea* subsp. *forrestii* (Diels) D. Y. Hong & L. M. Ma, and *C. convolvulacea* subsp. *vinciflora* (Kom.) D. Y. Hong.

Grey-Wilson's (1990) worldwide survey of *Codonopsis* in cultivation recognized *Codonopsis* in the broadest sense, merging both *Campanumoea* and *Leptocodon* into it. He grouped the 32 species he recognized into three informal groups and described a new species, *Codonopsis bragaensis*. Five years later, Grey-Wilson (1995) published a taxonomic revision of the *Codonopsis convolvulacea* Kurz complex (= subg. *Pseudocodonopsis* Kom., excluding *C. rosulata*), in which he recognized six species, including one new species, *C. nepalensis* Grey-Wilson (1995, non Hara, 1978) (= *C. grey-wilsonii* J. M. H. Shaw, 1996).

Hong (2010) published several notes on the taxonomy of *Codonopsis*, where he reduced six specific, one subspecific and eight varietal names to synonymy: *Codonopsis macrocalyx* Diels and its varieties *parviloba* J. Anthony and *coerulescens* Hand.-Mazz. to *C. benthamii* Hook. f. & Thomson, *C. xizangensis* D. Y. Hong (1980) to *C. bhutanica* Ludlow, *C. longifolia* D. Y. Hong to *C. rotundifolia* Benth., *C. bicolor* Nannf. to *C. viridiflora* Maxim., *C. microtubulosa* Z. T. Wang & G. J. Xu to *C. pilosula* (Franch.) Nannf., *C. retroserrata* Z. T. Wang & G. J. Xu to *C. convolvulacea* subsp. *forrestii* (Diels) D. Y. Hong & L. M. Ma, *C. nervosa* (Chipp) Nannf. subsp. *macrantha* (Nannf.) D. Y. Hong & L. M. Ma to *C. foetens* Hook. f. & Thomson subsp. *nervosa* (Chipp) D. Y. Hong, *C. levicalyx* L. T. Shen var. *hirsuticalyx* L. T. Shen (1983) and *C. rotundifolia* Benth. var. *angustifolia* Nannf. (1950) to *C. levicalyx* L. T. Shen, *C. pilosula* (Franch.) Nannf. var. *volubilis* (Nannf.) L. T. Shen, var. *glaberrima* (Nannf.) P. C. Tsoong ex L. T.

Shen (1975), and var. *modesta* (Nannf.) L. T. Shen (1983) to *C. pilosula* (Franch.) Nannf. He also reduced three species to subspecific rank; *Codonopsis grey-wilsonii* J. M. H. Shaw to a subspecies of *C. convolvulacea* Kurz, *C. tangshen* Oliv. (1891) to a subspecies of *C. pilosula* (Franch.) Nannf., and *C. nervosa* (Chipp) Nannf. (1930) to a subspecies of *C. foetens* Hook. f. & Thomson. Since then Hong found that *C. levicalyx* L. T. Shen is a synonym of *C. henryi* Oliv., while *C. rotundifolia* var. *angustifolia* Nannf. represents a new species, *C. bomiensis* D. Y. Hong

Lammers and Klein (2010) further revised *Codonopsis* subg. *Pseudocodonopsis* Kom. after Grey-Wilson (1995). For the *C. convolvulacea* Kurz complex (excluding *C. rosulata*) they recognized nine species, three more than Grey-Wilson (1995), adding *C. mairei* H. Lév., *C. limprichtii* Lingelsh. & Borza, and *C. macrophylla* Lammers & L. L. Klein. Thus, extremely controversial views exist for the taxonomy of the *Codonopsis convolvulacea* Kurz complex; one species with seven varieties (Shen in Shen & Hong, 1983), three species with two subspecies〔not including *C. grey-wilsonii* (Hong & Ma, 1992)〕, four species with two subspecies (Hong & Ma, 1994), three species with four subspecies (Hong in Hong et al., 2011), six species (Grey-Wilson, 1995), and nine species (Lammers & Klein, 2010). As Grey-Wilson admitted (1995), "no treatment of *Pseudocodonopsis* is going to be entirely satisfactory and any treatment will have its critics." However, in my opinion, there should be one rational classification, but it must be based on extensive field observations, critical examination of a large number of specimens, and statistical analysis.

The latest taxonomic account of the *Codonopsis* complex by Hong, Lammers and Klein (2011), treated 40 species, the largest number of species ever recognized in *Codonopsis*, plus two species of *Campanumoea*, three species of *Cyclocodon* and two species of *Leptocodon*. Two issues were exposed, however, to be remarkably controversial between Hong and the two co-authors, Lammers and Klein. One involved the circumscription of *Codonopsis*, whether it should include *Campanumoea* and *Leptocodon*, or the two should be recognized separately. The other concern was with the taxonomy of *Codonopsis* subg. *Pseudocodonopsis* Kom., as mentioned above.

One can see from facts presented above that since Komarov (1908) and Chipp (1908) there has been no comprehensive taxonomic revision of *Codonopsis*; Grey-Wilson's (1990) survey was confined to plants in cultivation, while Hong et al.'s (2011) work was confined to the members in China. Considering the controversial issues in the taxonomy of *Codonopsis* and its allies and the lack of a comprehensive worldwide taxonomic revision, a rational taxonomic revision of this fascinating group is badly needed.

Chapter 2

Biology

GROSS MORPHOLOGY

Morphology is a term covering a wide range of sense; in botany it may cover anatomy, pollen morphology, chromosome morphology, and so forth. Here in this section we focus on gross morphology, which may be observed by naked eyes or with lens.

General Morphology

Plants of all members of the Campannulaceae s.s. are lactiferous. Flowers are perfect and actinomorphic, mostly 5-merous but 4-merous and 6–9-merous calyx and corolla also occur. Calyx is usually connate but become free due to division to the base in some cases, for example, in *Cyclocodon parviflorus* (*Codonopsis parviflora*) and *Codonopsis javanica*. It is usually epigynous but perigynous or even hypogynous cases also occur, e.g. in *Codonopsis pilosula*. Corolla is always connate, but in *Codonopsis* subg. *Pseudocodonopsis* Kom. (= *Pseudocodon* D. Y. Hong & H. Sun) it is rotate and looks apopetalous; it is always epigynous. Stamens usually 5, but 4 or 6–9 may be found. Ovary mostly 3- or 5-, less frequently 4-, 6–9-locular, inferior, but superior in *Cyananthus* or inferior to corolla but from superior to half-superior to calyx in *Codonopsis pilosula* subsp. *tangshen*, *Codonopsis javanica* (Bl.) Hook. f. (*Campanumoea javanica* Bl.), and *Cyclocodon*. Placentation axile. Plants of *Codonopsis* and its allies, like other members of the Campanulaceae s. s., are always herbaceous, and perennial. Fruit a capsule, 3- or 5-loculicidal, dehiscent by apical valves (above the insertion position of corolla). However, in *Cyclocodon* and *Campanumoea* (both included in *Codonopsis* in sense of Moeliono, 1960), fruit a berry.

In the family, plants are usually outbreeding and protandrous. When corolla is still closed, and stigma is still closed and situated below anthers, anthers shed their pollen grains. When the style elongates, nearly all pollen grains disperse on dense hairs at the upper part of style. Only after the anthers have sheded all pollen grains, the corolla opens. At this stage stigma is still closed, so that pollen grains are taken by insects to other flowers. Only one or two days late, stigma segments open, recurve, and become receptive. The stigma segments continue to recurve until their receptive surfaces touch the hairs on the style, which makes them able to collect remaining pollen grains on the hairs (Figure II 1:11h). Therefore, the outbreeding mechanism is rather perfect, but self-pollination may take place if cross pollination fails.

Roots

The root is diverse in *Codonopsis* and allied genera. Four major types may be readily recognized, according to our observations. The most common one is the Carrot-shaped Type, whose roots are carrot-shaped, non-branched or branched only at the lower part, gray, and with numerous horizontal stripes. This type is represented by all species of *Codonopsis* subg. *Codonopsis* except for *Codonopsis purpurea* Wall. (= *Pankycodon purpureus*) (Figure II 1:1a–c). *Codonopsis ussuriensis* has roots which differ from the others of this type in having ellipsoid shape (Figure II 1:1d). The root in the Smooth Type, represented by *Codonopsis purpurea* Wall., is attenuate, white, branched, without horizontal stripes (Figure II 1:1e). The Globose Type, roots of which are globose or ovoid, rarely ellipsoid, pale yellow-brown, non-branched, without horizontal stripes, is represented by *Codonopsis* subg. *Pseudocodonopsis* Kom. (Figure II 1:1f). The Branched Type, represented by *Cyclocodon axillaris* (*Campanumoea axillaris*), has roots attenuate, branched, more or less lignified, without horizontal stripes (Figure II 1:1g). The features of roots in *Codonopsis* and its allies are correlated to some extent with the other features of external morphology. For example, the Globose Type is highly correlated with rotate corolla, greatly dilated and lanate filament base, and long-spinulose pollen exine, which supports the sepapation of *Codonopsis* subg. *Pseudocodonopsis* Kom. as genus (*Pseudocodon* D. Y. Hong & H. Sun) from the rest of *Codonopsis* (see relevant sections in this chapter).

Stems

Usually several to numerous stems arise from one caudex in *Codonopsis* subg *Codonopsis* (= *Codonopsis*) (Figure II 1:2d & e). A typical example is found in *Codonopsis canescens* Nannf. In *Codonopsis* subg. *Pseudocodonopsis* Kom. (= *Pseudocodon* D. Y. Hong & H. Sun), however, usually only a single stem comes out from one caudex (Figure II 1:2 a–c). Therefore, the number of stems from one caudex is characteristic of a particular group in *Codonopsis* and its allies.

A Monograph of *Codonopsis* and Allied Genera (Campanulaceae). http://dx.doi.org/10.1016/B978-0-12-801933-7.00002-7
Copyright © 2015 China Science Publishing & Media Ltd. Published by Elsevier Inc. All rights reserved.

FIGURE II 1:1 Various forms of roots in *Codonopsis* and allied genera. **a**, *Codonopsis bomiensis* D. Y. Hong, from Tibet: Mainling, by Xin-Tang MA & De-Yuan HONG; **b**, *Codonopsis lanceolata* (Sieb. & Zucc.) Trautv., from Beijing: Miyun, by De-Yuan HONG; **c**, *Codonopsis pilosula* (Franch.) Nannf., subsp. *tangshen* (Oliv.) D. Y. Hong, from Hubei: Shennongjia, by Zhuo ZHOU; **d**, *Codonopsis ussuriensis* (Rupr. & Maxim.) Hemsl., from Jilin: Yanbian, by Yao ZHOU; **e**, *Pankycodon purpureus* (Wall.) D. Y. Hong & X. T. Ma (*Codonopsis purpurea* Wall.), from Tibet: Nyalam, by Xin-Tang MA & De-Yuan HONG; **f**, *Pseudocodon convolvulaceus* (Kurz) D. Y. Hong & H. Sun subsp. *forrestii* (Diels) D. Y. Hong [*Codonopsis convolvulacea* Kurz subsp. *forrestii* (Diels) D. Y. Hong & L. M. Ma], from Yunnan: NE Kunming, by De-Yuan HONG; **g**, *Cyclocodon axillaris* (Oliv.) W. J. de Wilde & Duyfjes, from Sichuan: Emei, by De-Yuan HONG.

The stem in *Codonopsis* and its allies is usually branched but it varies from simple to multi-branched and from branched only at the lower part to branched at various parts. The simple stem can be found in *Codonopsis elliptica* D. Y. Hong and *C. lixianica* D. Y. Hong (Figure IV 1:2A), while all other members of the group are found to have branches (Figure II 1:2 f–i). Guite a large number of species in *Codonopsis* subg. *Codonopsis* have a special feature of stem branching. Many slender branches come from the lower part of main stem, and they have leaves usually nearly opposite, and thus such a branch appears a pinnate leaf (Figure II 1:2i). They are typical of *Codonopsis foetens*, *C. bulleyana*, *C. thalictrifolia*, *C. viridiflora*, *C. tsinlingensis*, *C. chlorocodon*, and are also found in *C. bhutanica* and so forth.

FIGURE II 1:2 **a–e**: The number of stems from one caudex, **a–c**, a single stem; **d** & **e**, three and four stems. **a**, *Codonopsis javanica* subsp. japonica (Makino) Lammers, from Sichuan: Emei, by De-Yuan HONG; **b**, *Codonopsis benthamii* Hook. f. & Thomson, from Tibet: Yadong, by Xin-Tang MA; **c**, *Pseudocodon rosulatus* (W. W. Smith) D. Y. Hong (*Codonopsis rosulata* W. W. Smith), from Sichuan: Yanbian, by De-Yuan HONG. **f–i**: Branching of stems, **f**, *Codonopsis viridiflora* Maxim., from Qinghai: Datong, by De-Yuan HONG; **g**, *Codonopsis clematidea* (Schrenk) C. B. Clarke, from Xinjiang: Ürümqi, by Dun-Yan TAN; **h**, *Codonopsis viridiflora* Maxim., from Qinghai: Datong, by De-Yuan HONG; **i**, *Codonopsis chlorocodon* C. Y. Wu, from Sichuan: Garzê, by You-Sheng CHEN.

a
b
c
d
e
f
g
h
i

The stem in *Codonopsis* and its allies is usually herbaceous, but slightly lignified at the lower part in *Cyclocodon*. The habit of stems is characteristic of some members of the group. They are nearly always twining, very rarely scandent or ascending in *Codonopsis* subg. *Pseudocodonopsis* Kom., often pendent from rocks in *Codonopsis dicentrifolia*, the only species of *Codonopsis* subg. *Obconicicapsula* D. Y. Hong, and *Codonopsis purpurea* Wall. It varies greatly in *Codonopsis* subg. *Codonopsis*, from twining, scandent, ascending, to erect. Komarov (1908) used the habit of stems for dividing *Codonopsis* subg. *Eucodonopsis* (= subg. *Codonopsis*) (including all speices out of subg. *Pseudocodonopsis* Kom.) into two series, Series I. *Volubiles* Kom. with twining stems and Series II. *Erectae* Kom. with erect stems. Hong (1983) raised them to sectional rank, sect. *Codonopsis* (= series *Volubiles* Kom.) and sect. *Erectae* (Kom.) D. Y. Hong. Although this division is functional in most cases, caution ought to be taken in identification, because the variation mentioned above may be found within species. This phenomenon occurs particularly in *Codonopsis viridiflora* Kom., which possesses ascending, scandent and even twining stems (Figure II 1:2h). In *Codonopsis benthamii* Hook. f. & Thomson stems are usually ascending or scandent but sometimes twining, whereas in *C. tubulosa* Kom. they are usually twining but rarely ascending. Therefore, to split *Codonopsis* subg. *Codonopsis* into two groups based only on the habit of stems is questionable.

Leaves

Leaves are always simple in *Codonopsis* and its allies. The shape and size are often used for distinguishing species. For example, *Codonopsis deltoidea* Chipp is easily recognized for its large and deltoid leaves with a rather long petiole, while *Codonopsis canescens* Nannf. has consistently the smallest leaves in this group. However, the utility of leaf characters for delimitating species sometimes causes controversial issue. For example, when *Codonopsis forrestii* Diels (1912) was described as new, the author stated that it differed from *C. convolvulacea* Kurz "by the broader leaves ("1.5–3 cm lata") with a long petiole ("0.5–1 cm long")......"; *Codonopsis efilamentosa* W. W. Smith was recognized as species by later authors (Grey-Wilson, 1990, 1995; Lammers & Klein, 2010) because of "leaves broad with a heart-shaped or truncated base" (Grey-Wilson, 1995) or "wide leaves (15–33 mm vs. 2–14 mm) on longer petioles (5–22 mm vs. 1–3 mm in *Codonopsis convolvulacea*)" (Lammers & Klein, 2010). However, all these forms of leaves have been observed by us in a population in NE Kunming, Yunnan Province (Figure II 1:3).

Inflorescences

Flowers in *Codonopsis* and its allies are usually solitary and terminal on main stems or also on branches, less frequently in a dichasium, or form a cyme of 2–4 (Figure II 1: 4). The inflorescence is rarely used for delimitation of species in the group. Grey-Wilson (1995) used this character, however, for distinguishing *Codonopsis efilamentosa* W. W. Smith from the other species because its flowers "are borne in short lateral inflorescences, quite unlike the other species within the subgenus" or its "flowers mostly in axillary clusters (Lammers & Klein, 2010). We sampled a population in NW Kunming, Yunnan, and found that solitary flowers, two flowers in a cluster (a cyme of two flowers) and three or even four flowers in a cluster (a cyme of three or four) coexisted in the population. Therefore, the variation in inflorescence may be a polymorphism in *Codonopsis* and its allies.

Numbers of Floral Structures

This character is variable in the *Codonopsis* complex. Flowers in *Cyclocodon* Griff. ex Hook. f. & Thomson (*Codonopsis* in sense of Moeliono, 1960) are nearly isomerous. In *Cyclocodon axillaris* and *C. lancifolius*, calyx is (4–) 5 (–7)-merous, corolla 5–6- merous, stamens 5–6-merous, and ovary (4–) 5–6-locular, but in *C. parviflorus* all the floral parts are tetramerous. In all three subgenera of *Codonopsis* and *Leptocodon* (= *Codonopsis*) flowers are always heteromerous; the calyx, corolla and androecium are pentamerous, whereas the pistil is usually trimerous, with 3-fid stigma and 3-locular ovary. Two species of *Campanumoea* (= *Codonopsis*) are different in this respect. *Campanumoea inflata* is similar to *Codonopsis* with 5-merous calyx, corolla and androecium, and 3-merous pistil, but *C. javanica* is different; all parts of its flowers are usually pentamerous. The number of floral structures is found varying even among flowers on one individual. In *Campanumoea javanica* and *Codonopsis purpurea* we found that 4-merous and 5-merous flowers coexisted on the same individual (Figure II 1:5).

Variation in Position of Ovary

This is an interesting and unique question rarely found in angiosperms. The Campanulaceae has been described as having ovary inferior. However, the genus *Cyananthus*, which is distributed in the Himalaya and its adjacent regions, has superior ovary. Hong and Ma (1991) once considered the genus to be the most primitive in the family based on its superior ovary and basic chromosome number $x = 7$. However, in the molecular trees of the Campanulaceae s. s. constructed by Eddie et al. (2003) using ITS sequences of nuclear ribosomal DNA, and by Haberle et al. (2009) based on DNA sequences of three

FIGURE II 1:3 Various forms of leaves within a population in NE of Kunming, Yunnan [D. Y. Hong et al. *H10012* (PE)] in *Pseudocodon convolvulaceus* (Kurz) D. Y. Hong & H. Sun subsp. *forrestii* (Diels) D. Y. Hong taken from seven individuals.

chloroplast genes, *Cyananthus* is not at the basal position in the platycodonoid group (= tribe Cyanantheae). Our phylogenetic analysis of the group using ITS sequences of nuclear ribosomal DNA and sequences of four chloroplast genes shows that *Cyananthus* and the *Codonopsis* group (colpate pollen alliance) form a major clade, with *Platycodon*, *Canarina*, and *Cyclocodon* (colporate pollen alliance) at the basal position (Wang et al., 2013). Therefore, it is justifiable to infer that the superior ovary in *Cyananthus* is derived. In *Codonopsis* and its allies the position of ovary to the calyx varies greatly. The ovary is usually inferior (actually half-inferior) to both calyx and corolla, that is to say, calyx and corolla are inserted at the same position on the hypanthium, which envelops the lower half of ovary. However, in many cases the calyx and corolla are inserted at different positions of the hypanthium. Such a situation is found in all three species of *Cyclocodon* (*Codonopsis* in sense of Moeliono, 1960). The corolla is always epigynous, that is to say, it is inserted at the top of hypanthium, but the calyx is not. The calyx tube is adnate to the hypanthium only up to the middle or the lower part, and thus there is a distance between the base of corolla and the base of calyx lobes in *Cyclocodon axillaris* and *C. lancifolius* (Figure II 1:6 g–i). An extreme and peculiar case is found in *C. parviflorus*, where the calyx is inserted on the pedicel, and thus there is a distance (anthophore) between the calyx and the other floral parts. Therefore, in the genus the calyx is half-epigynous or even hypogynous

FIGURE II 1:4 Types of inflorescences in *Codonopsis* and allied genera. a & b, solitary flower terminal only on main stem, **a**, *Codonopsis gongshanica* Q. Wang & D. Y. Hong, from Yunnan: Gongshan, by Qiang WANG; **b**, *Codonopsis bulleyana* Forrest ex Diels, from Yunnan: Shangri-la, by Hang SUN; c & d, solitary flowers terminal on both main stem and branches; **c**, *Himalacodon dicentrifolius* (C. B. Clarke) D. Y. Hong & Q. Wang, from Tibet: Nyalam, by De-Yuan HONG; **d**, *Codonopsis clematidea* (Schrenk) C. B. Clarke, from Xinjiang: Ürümqi, by Dun-Yan TAN; e & f, Dichasium/cyme, **e**, a dichasium in *Cyclocodon axillaris* (Oliv.) W. J. de Wilde & Duyfjes, from Chongqing: Three Gorges, by Xin-Tang MA; **f**, a cyme of 2 flowers in *Pseudocodon vinciflorus* (Kom.) D. Y. Hong, from Tibet: Gongbo'gyamda, by De-Yuan HONG.

FIGURE II 1:5 Variation in number of floral structures. **a** & **b**, *Codonopsis javanica* (Bl.) Hook. f. subsp. *japonica* (Makino) Lammers, from Sichuan: Emei, by De-Yuan HONG; a, a 4-merous flower; b, a 5-merous flower. **c** & **d**, *Pankycodon purpureus* (Wall.) D. Y. Hong & X. T. Ma (*Codonopsis purpurea* Wall.), from Tibet: Nyalam, by Xin-Tang MA & De-Yuan HONG; c, a 4-merous flower; d, a 5-merous flower.

(Figure II 1:6c–f). *Campanumoea* (= *Codonopsis*) *javanica* and *C. inflata* were treated as the only two members of *Campanumoea* (Hong in Hong et al., 2011), but they are different in respect of ovary position. In *C. inflata*, the calyx and corolla are both epigynous but in the other the corolla is epigynous, while the calyx is hypogynous (Figure II 1:6a & b, and Figure IV 1: 45B).

In the *Codonopsis pilosula* / *C. tangshen* complex, the corolla is epigynous, but the position of calyx varies from typically hypogynous, and thus calyx looks like five free sepals, gradually to half-epigynous (Figure II 1:6f and Figure IV 1: 23B d-g).

Corolla: Shape

The corolla is rather polytypical in *Codonopsis* and its allies. In *Cyclocodon* Griff. ex Hook. f. & Thomson (*Codonopsis* in sense of Moeliono, 1960) it is nearly rotate, i.e. limb is

a
b
c
d
e
f
g
h
i

spread (Figure II 1:7k & l). In *Codonopsis* subg. *Codonopsis* it is usually campanulate (bell-like), but tubular, tubular-campanulate, and funnel-shaped forms can be found (Figure II 1:7a–c; f–i). The shape of corolla may be used for delimitation of species. For example, *Codonopsis bulleyana* and *C. foetens* are very similar to each other, but the corolla of the latter is typically bell-like, while that of the former is tubular in the lower half, but suddenly expanded and spread from the middle, becoming funnel-shaped. These two forms are rather stable, so that specimens with flowers can be easily identified (Figure II 1:7d).

The corolla of *Codonopsis* subg. *Pseudocodonopsis* Kom. (= *Pseudocodon* D. Y. Hong & H. Sun) is distinct. It is neither campanulate nor tubular, but instead it is cleft to near the base, with corolla lobes horizontally spread, and thus the corolla of all members in this group is rotate (Figure II 1:7j).

Corolla: Colour

The corolla colour is extremely diverse in *Codonopsis* and allied genera. It varies from white, yellow, green, yellow-green, rose, red, purple-red, purple, and blue (Figure II 1:8). Fascinating with corolla colour in the group is that most species have a colourful corolla; species with single-coloured corolla are much less frequently found. In a number of species the corolla is of pale colour but with much darker veins. For example, in *Codonopsis subscaposa* (Figure IV 1:6B) the corolla is purple-red but with dark purple veins; in *C. farreri* (Figure II 1:8e) it is yellow but with purple veins. More interesting is that a corolla has two or more colours. In *Codonopsis bomiensis* D. Y. Hong, *Pseudocodon grey-wilsonii* (J. M. H. Shaw) D. Y. Hong, and *P. graminifolius* (H. Lév.) D. Y. Hong, the corolla is purely purple or white but with a dark red ring (Figure II 1:8). The most colourful corolla occurs in *Codonopsis clematidea* (Schrenk) C. B. Clarke and *Codonopsis subglobosa* W. W. Smith. Their corolla has several distinct colours which are distributed at different parts (Figure II 1:8). The corolla colour of *Codonopsis clematidea* and some others appears a very fascinating mosaic.

According to our extensive field observation, colour of corolla is species-specific in the group under study. One exceptional case is found in *Pseudocodon grey-wilsonii*, which has both purple (Tibet: Nyalam) and white corolla (Tibet: Gyirong) (Figure IV 4: 3B). It is worthwhile to conduct extensive observations on types and variation of the corolla in the group for studying its relationships with systematics, pollination biology, and evolution.

Androecium

Variation of stamens in the *Codonopsis* complex is mainly in shape of filaments and in indumentum. Filaments in *Codonopsis* subg. *Obconicicapsula* D. Y. Hong [with only one species, *Codonopsis dicentrifolia* (C. B. Clarke) W. W. Smith] (= *Himalacodon* D. Y. Hong & Q. Wang) are slightly dilated at the base and glabrous. In *Codonopsis* subg. *Codonopsis* filaments are all slightly dilated at the base, but vary from glabrous to more or less villous-ciliate at the dilated part. Filaments in *Codonopsis* subg. *Pseudocodonopsis* Kom. (= *Pseudocodon* D. Y. Hong & H. Sun) are very specialized and characteristic; they are short, and extremely dilated at the base, where they become semi-orbicular and lanate at the periphere on adaxial surface (Figure II 1: 9). This kind of filaments is found in all the species in this group.

Disc/Glands

The *Codonopsis* complex is heterogeneous in this respect. Disc/glands are absent in *Cyclocodon* Griff. ex Hook. f. & Thomson (*Codonopsis* in sense of Moeliono 1960). *Codonopsis* in Hong's narrow sense (Hong 1983; Hong in Hong et al., 2011) is also heterogeneous in the presence or absence of disc/glands. They are found only in subg. *Codonopsis* but not in subgs. *Pseudocodonopsis* Kom. and *Obconicicapsula* D. Y. Hong. Disc is also present in *Campanumoea* (= *Codonopsis* sect. *Campanumoea*); glands occur only in *Leptocodon* (= *Codonopsis* sect. *Leptocodon*).

The disc is polytypical in the group under discussion. It is ring-like or pentagonal in *Codonopsis* subg. *Codonopsis* (Figure II 1:10). In *Leptocodon*, we could not find a typical disc, but instead we found five glands which are alternate to stamens (Figure II 1:10f).

Gynoecium

We have mentioned the position of ovary to calyx and corolla, the number of ovary locules, and stigma segments before, and thus we here just show the shape of stigma segments. They are clavate in *Cyclocodon* (*Codonopsis*

FIGURE II 1:6 Variation in position of ovary in *Codonopsis* and allied genera. **a** & **b**, *Codonopsis inflata* Hook. f. & Thomson, from Tibet: Mêdog, by De-Yuan HONG, showing ovary inferior to both calyx and corolla; **c** & **d**, *Codonopsis bomiensis* D. Y. Hong, from Tibet: Mainling, by Xin-Tang MA & De-Yuan HONG, showing ovary nearly inferior to corolla but half-inferior to calyx; **e**, *Codonopsis thalictrifolia* Wall. subsp. *mollis* (Chipp) D. Y. Hong, from Tibet: Nyinchi, by Xin-Tang MA & De-Yuan HONG, showing ovary half-inferior to both calyx and corolla; **f**, *Codonopsis pilosula* (Franch.) Nannf. subsp. *tangshen* (Oliv.) D. Y. Hong (*Codonopsis tangshen* Oliv.), from Chongqing: Nanchuan, by Qiang WANG, showing ovary superior to calyx but inferior to corolla; **g**, *Cyclocodon axillaris* (Oliv.) W. J. de Wilde & Duyfjes, from Chongqing: Three Gorges, by Xin-Tang MA, showing ovary nearly superior to calyx but inferior to corolla; **h** & **i**, *Cyclocodon lancifolius* (Roxb.) Kurz, from Tibet: Mêdog, by Xin-Tang MA & De-Yuan HONG, showing ovary half-superior to calyx but typically inferior to corolla.

a
b
c
d
e
f
g
h
i
j
k
l

in sense of Moeliono 1960), but broad-elliptic in the rest members of *Codonopsis* and its allies (Figure II 1:11 a–c & g) or nearly triangular in *Himalacodon* and *Pankycodon* (Figure II 1:11 e & f). Stigma segments are entire, but with an exception in *Codonopsis subglobosa* W. W. Smith, where each of the three segments is shallowly 2- lobed (Figure II 1:11d).

Fruit

In *Codonopsis* and its allies, the three subgenera of *Codonopsis* and *Leptocodon* all have a 3-loculicidal capsule, whereas all the members in *Cyclocodon* and *Campanumoea* are of baccate fruit (Figure II 1:12).

The features of fruit have been emphasized by many authors. For example, Schönland (1889) divided the tribe Campanuleae (= Campanulaceae s. s.) into three subtribes, Campanulinae, Wahlenberginae and Platycodinae. He pulled *Platycodon* in East Asia together with *Microcodon* in South Africa and *Musschia* in Madeira (E Atlantic) into the Platycodinae, but separated *Codonopsis*, *Cyananthus* and *Leptocodon* from *Platycodon*, and placed them together with *Wahlenbergia*, *Roella*, *Prismatocarpus* into the Wahlenberginae. His action is understandable because the only source of data available for him is morphological. As I will mention, Erdtman's (1952) milestone work produced rich pollen data, which are very heuristic for the taxonomy of the Campanulceae s. s. in particular. However, Fedorov (1957) did not use his data, and still pulled *Codonopsis* and *Platycodon* into his Wahlenbergieae. Klakovsky (1987, 1994) went further. He even pulled *Campanumoea* and *Canarina* together into his new subfamily, Canarinoideae Kolak., obviously just because they share baccate fruit. Before molecular data become available, palynological data should be considered first as Thulin (1975) indicated.

We will evaluate the significomce of fruit features in taxonomy of *Codonopsis* and its allies based on a rather solid molecular tree in the next chapter.

Seeds

Seed biology involves gross morphology and microscopical structure, so that we deal with it in a separate section.

SEED MORPHOLOGY

Recent phylogenetic analyses based on molecular data (Eddie et al., 2003; Haberle et al., 2009; Wang et al., 2013) show that the Campanulaceae s. s. comprises three monophyletic groups: the campanuloids, wahlenbergioids and platycodonoids. Hong and Wang (2015) conducted further phylogenetic analyses of the family based on molecular data and survey on pollen morphology and fruit features. As a result, Hong (in Hong & Wang, 2015) established a three-tribal classification system of the Campanulaceae s. s., the Campanuleae, Wahlenbergieae, and Cyanantheae. The Cyanantheae is a natural group which has colpate or colporate pollen with elongate apertures (Hong & Pan, 2012) and a loculicidal capsule. The tribe, however, is polytypical in pollen morphology with nine types (Hong & Pan, 2012) and chromosome numbers with 2n = 10, 12, 14, 16, 18 and 34. Therefore, the significance of seed morphology in systematics of the group under study attracted our attention.

Belyayev (1986) investigated the anatomy and surface of seed-coat of five species in four genera of the Campanulaceae s. s.: *Berenice*, *Cephalostigma* (= *Wahlenbergia*), *Roella* and *Sergia*. This work shows that the seed-coat of all the genera examined has only one layer of cells. Shetler and Morin (1986) conducted extensive observations on the seed-coat surface of 34 North-American species (35 in total) in four genera of the Campanulaceae s. s. using scanning electron microscopy (SEM). In addition, sections of the seeds of 11 species in these four genera were observed also with SEM. They described and compared characteristics of seeds and their surface cells. Their conclusion is that the seed-coat morphology is relatively uniform, but there are recognizable generic patterns and a number of distinctive individual species. Haridasan and Mukherjee (1987) observed the seed surface of 27 Indian species of the Campanulaceae s. s. (including *Sphenoclea zeylanica*), with 13 species in the Cyanantheae (9 in *Codonopsis*, 2 in *Cyclocodon*, and 2 in *Cyananthus*). They found variation of seed-coats in shape of the areoles and secondary ornamentations on radial walls. Their conclusion is, however, "SEM ornamentations being species-specific can not be used as generic marker". Shrestha and Kravtsova (1992) made an extensive investigation on the seed-coat surface and section of 17 species in *Cyananthus*, a genus confined to the Himalaya and the adjacent Hengduan Mts. According to descriptions

FIGURE II 1:7 Corolla shape of *Codonopsis* and allied genera. **a**, *Codonopsis gracilis* Hook. f. & Thomson, from Tibet: Dinggyê, by Lei XIE; **b**, *Codonopsis benthamii* Hook. f. & Thomson, from Yunnan: Gongshan, by Qiang WANG; **c**, *Codonopsis thalictrifolia* Wall. subsp. *thalictrifolia*, from Tibet: Nyalam, by Fu-Sheng YANG; **d**, *Codonopsis bulleyana* Forrest ex Diels, from Sichuan: Muli by You-Sheng CHEN; **e**, *Codonopsis inflata* Hook. f. & Thomson, from Tibet: Mêdog, by De-Yuan HONG; **f**, *Codonopsis subscaposa* Kom., from Sichuan: Xiaojin, by You-Sheng CHEN; **g**, *Codonopsis canescens* Nannf., from Tibet: Jomda, by Fu-Sheng YANG; **h**, *Himalacodon dicentrifolius* (C. B. Clarke) D. Y. Hong & Q. Wang, from Tibet: Nyalam, by Xin-Tang MA & De-Yuan HONG; **i**, *Pankycodon purpureus* (Wall.) D. Y. Hong & X. T. Ma, from Tibet: Nyalam, by Xin-Tang MA & De-Yuan HONG; **j**, *Pseudocodon grey-wilsonii* (J. M. H. Shaw) D. Y. Hong, from Tibet: Nyalam, by Xin-Tang MA & De-Yuan HONG; **k**, *Cyclocodon axillaris* (Oliv.) W. J. de Wilde & Duyfjes, from Sichuan: Emei, by De-Yuan HONG; **l**, *Cyclocodon lancifolius* (Roxb.) Kurz, from Tibet: Mêdog, by De-Yuan HONG.

a
b
c
d
e
f
g
h
i
j
k
l

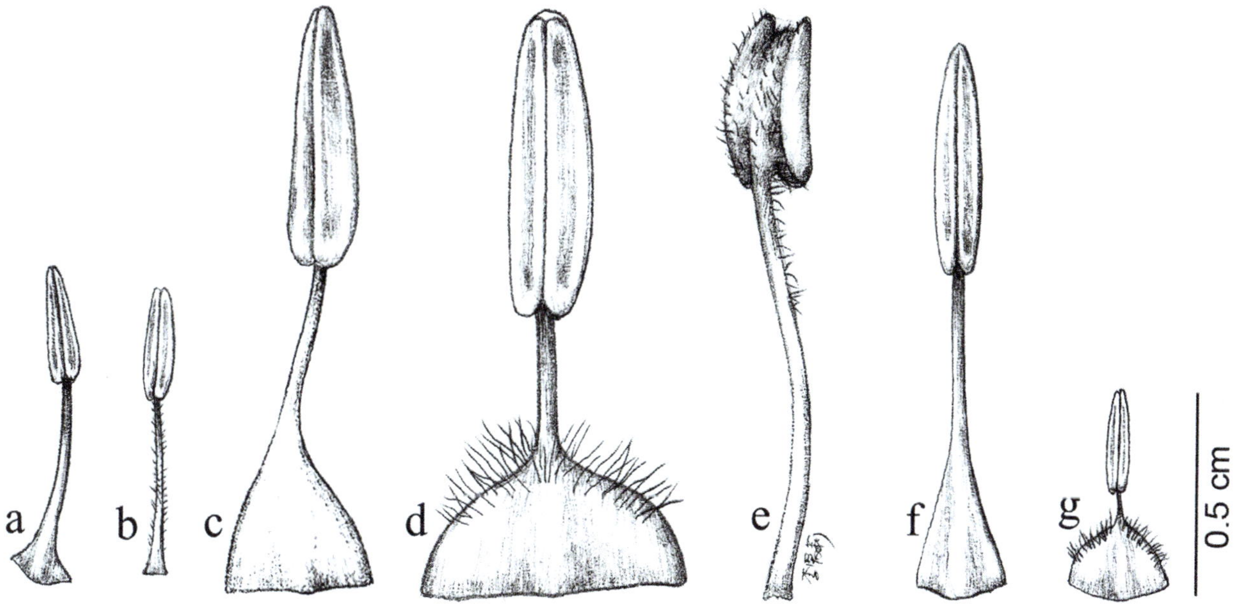

FIGURE II 1:9 Stamens with a special reference to filaments in *Codonopsis* and allied genera. **a**, *Codonopsis deltoidea* Chipp; **b**, *Codonopsis cordifolioidea* P. C. Tsoong; **c**, *Codonopsis campanulata* D. Y. Hong; **d**, *Codonopsis reflexa* D. Y. Hong; **e**, *Codonopsis tubulosa* Kom. subg. *tubulosa*; **f**, *Himalacodon dicentrifolius* (C. B. Clarke) D. Y. Hong & Q. Wang; **g**, *Pseudocodon convolvulaceus* (Kurz) D. Y. Hong & H. Sun subsp. *convolvulaceus*.

and photos they provided, the seed-coat is quite uniform; slight variation concerns mainly the exotesta, and cells on the surface view, which are from rectangular to striate. They inferred the trend of seed-coat evolution in *Cyananthus*. To address the controversial views on taxonomic treatment of *Cyclododon* Griff. ex Hook. f. & Thomson, Hong and Pan (1998) conducted examination of pollen and seed-coat morphology of four species, two in *Cyclocodon* and the other two in *Campanumoea* (also in *Codonopsis* in the sense of Moeliono, 1960). We found that the former two have 3-colporate pollen correlated with spiral ornamentations on the radial walls of the seed-coat, whereas the latter two have 5–8-colpate pollen correlated with bead-like ornamentations on the radial walls. As a result they restored *Cyclocodon* to generic level. This action was soon accepted by Lammers (1998, 1999).

The seed develops from an ovule within the ovary. Seed morphology should therefore be relatively slightly influenced by changes in environments. Shetler and Morin's (1986) work and ours (Hong & Pan, 1998) indicate that the seed-coat may serve as generic or other taxonomic markers. Unfortunately, however, from what we have mentioned above it is clear that there have been no extensive and comprehensive investigations of seed morphology in the Campanulaceae s. s.

We conducted a comprehensive survey on the seed morphology including the surface sculpture and section using SEM and based on 38 species representing all 10 genera of the tribe Cyanantheae on a world scale. For *Codonopsis* and allied genera, we sampled 28 species: *Codonopsis* (19), *Cyclocodon* (3), *Himalacodon* (= *Codonopsis* subg. *obconicicapsula* D. Y. Hong) (1), *Pankycodon* (= *Codonopsis* p. p.) (1), and *Pseudocodon* (= *Codonopsis* subg. *Pseudocodonopsis* Kom.) (4) (Table II 2: 1).

The seed of *Codonopsis* and its allies is uniform in having only one layer of palisade cells in the seed-coat. They are, however, very diverse in shape, presence or absence of wing, shape of the lumina, the secondary ornamentation of the radial walls, the transection, and so forth. Four types and two undetermined types could be recognized:

Type 1. Seeds oblong or ovoid, wingless. Transection oblong. Lumina nearly isodiametric, polygonal, with diameter larger than width of radial walls. Radial walls with beaded ornamentations. Tangential walls thin, easily broken. In transection radial walls elevated,

FIGURE II 1:8 Colour of corolla in *Codonopsis* and allied genera. **a**, *Codonopsis hongii* Lammers, from Tibet: Cona, by Xin-Tang MA & De-Yuan HONG; **b**, *Codonopsis bomiensis* D. Y. Hong, from Tibet: Mainling, by Xin-Tang MA & De-Yuan HONG; **c & d**, *Codonopsis subglobosa* W. W. Smith; c, from Sichuan: Muli, by De-Yuan HONG; d, from Sichuan: Daocheng, by Fu-Sheng YANG; **e**, *Codonopsis farreri* J. Anthony, from Yunnan: Gongshan, by Qiang WANG; **f**, *Codonopsis tubulosa* Kom., from Sichuan: Muli, by Zhuo ZHOU; **g**, *Codonopsis subglobosa* W. W. Smith, from by Bo XU; **h**, *Codonopsis clematidea* (Schrenk) C. B. Clarke, from Xinjiang, by Dun-Yan TAN; **i**, *Codonopsis lanceolata* (Sieb. & Zucc.) Trautv., from Beijing: Miyun, by Lian-Zhong FU; **j**, *Himalacodon dicentrifolius* (C. B. Clarke) D. Y. Hong & Q. Wang, from Tibet: Nyalam, by Xin-Tang MA & De-Yuan HONG; **k**, *Pseudocodon grey-wilsonii* (J. M. H. Shaw) D. Y. Hong, from Tibet: Gyirong, by Xin-Tang MA; **l**, *Pseudocodon vinciflorus* (Kom.) D. Y. Hong, from Tibet: Nyinchi, by Hong-Ya GU.

FIGURE II 1:10 Disc/glands of *Codonopsis*. a–e, diverse forms of disc: **a**, *Codonopsis inflata* Hook. f. & Thomson, from Tibet: Mêdog, by De-Yuan HONG; **b**, *Codonopsis lanceolata* (Sieb. & Zucc.) Trautv., from Beijing: Miyun, by Ai-Li LI; **c**, *Codonopsis clematidea* (Schrenk) C. B. Clarke, from Xinjiang, by Dun-Yan TAN; **d**, *Codonopsis foetens* Hook. f. & Thomson subsp. *foetens*, from Tibet: Nyinchi, by De-Yuan HONG; **e**, *Codonopsis subglobosa* W. W. Smith, from Sichuan: Muli, by De-Yuan HONG; **f**, *Codonopsis* hongii Lammers, from Tibet: Cona, by Xin-Tang MA & De-Yuan HONG, showing 5 glands alternate with stamens.

tangential walls depressed, cell cavity much wider than radial walls.

Observed samples: *Codonopsis cordifolioidea* P. C. Tsoong, *C. deltoidea* Chipp, *Codonopsis* (*Campanumoea*) *inflata* Hook. f. & Thomson, and *Codonopsis* (*Campanumoea*) *javanica* subsp. *japonica* (Makino) Lammers (Figures II 2: 1, a–p).

Type 2. Seeds rounded, wingless. Transection elliptic. Lumina irregularly polygonal or nearly orbicular, more or less isodiametric, with diameter nearly equal to width of radial walls. Radial walls with spiral ornamentations. In transection radial walls elevated, tangential walls depressed, cell cavity much wider than radial walls.

FIGURE II 1:11 The stigma of *Codonopsis* and allied genera. **a**, *Codonopsis javanica* subsp. *japonica* (Makino) Lammers, from Sichuan: Emei, by De-Yuan HONG; **b**, *Codonopsis foetens* Hook. f. & Thomson subsp. *foetens*, from Tibet: Nyalam, by De-Yuan HONG; **c–f**, *Codonopsis subglobosa* W. W. Smith, from Sichuan: Muli, by De-Yuan HONG; **g**, *Himalacodon dicentrifolius* (C. B. Clarke) D. Y. Hong & Q. Wang, from Tibet: Nyalam, by Xin-Tang MA & De-Yuan HONG; **h**, *Pankycodon purpureus* (Wall.) D. Y. Hong & X. T. Ma, from Tibet: Nyalam, by Xin-Tang MA & De-Yuan HONG; **i**, *Pseudocodon vinciflorus* (Kom.) D. Y. Hong subsp. *vinciflorus*, from Tibet: Nyinchi, by De-Yuan HONG; **j**, *Pseudocodon graminifolius* H. Lév. from Sichuan: Muli, by De-Yuan HONG; **k**, *Cyclocodon axillaris* (Oliv.) W. J. de Wilde & Duyfjes, from Sichuan: Emei, by De-Yuan HONG, showing 8 clavate stigma-segments which have recurved downwards to touch anthers: **l**, *Cyclocodon lanciflorus* (Roxb.) Kurz, from Tibet: Mêdog, by Xin-Tang MA & De-Yuan HONG.

FIGURE II 1:12 Fruit of *Codonopsis*. **a**, a 3-locilicidal capsule of *Codonopsis viridiflora* Maxim., from Qinghai: Datong, by De-Yuan HONG; **b**, two berries of *Codonopsis javanica* (Bl.) Hook. f. subsp. *japonica* (Makino) Lammers, from Guangxi: Xing'an, by Xin-Tang MA.

TABLE II 2:1 Materials for Observations on Seed Morphology of *Codonopsis* and Allied Genera

Taxon	Origin	Voucher
***Codonopsis* Wall.**		
C. affinis Hook. f. & Thomson	China, Tibet, Nyalam	*Xizang Chinese Mater. Medic. & Herbs Exped. 1579* (PE)
C. benthamii Hook. f. & Thomson	China, Yunnan, Gongshan	*T. T. Yü 23257* (PE)
C. bulleyana Forrest ex Diels	China, Sichuan, Muli	*D. Y. Hong et al. H11028* (PE)
C. canescens Nannf.	China, Sichuan, Dawu	*D. Y. Hong et al. H12070* (PE)
C. chlorocodon C. Y. Wu	China, Sichuan, Dawu	*D. Y. Hong et al. H12071* (PE)
C. cordifolioidea P. C. Tsoong	China, Yunnan, Gongshan	*Drungjiang Invest. Team 1446* (CAS)
C. deltoidea Chipp	China, Sichuan, Hongya	*Q. Wang H12086* (PE)
C. foetens Hook. f. & Thomson	China, Tibet, Cona	*D. Y. Hong et al. H11007* (PE)
C. hongii Lammers	China, Yunnan, Gongshan	*K. M. Feng 8673* (PE)
C. inflata Hook. f. & Thomson	China, Tibet, Mêdog	*Qinghai-Xizang Exped. 74-4988A* (PE)
C. javanica subsp. *japonica* (Makino) Lammers	China, Guangdong, Lianxian	*P. C. Tam 59973* (PE)
C. lanceolata (Sieb. & Zucc.) Trautv.	China, Beijing, Miyun	*D. Y. Hong et al. H11045* (PE)
C. pilosula subsp. *handeliana* (Nannf.) D. Y. Hong & L. M. Ma	China, Yunnan, Zhongdian	*D. Y. Hong et al. H10017* (PE)
C. subglobosa W. W. Smith	China, Sichuan, Muli	*D. Y. Hong et al. H11029* (PE)
C. subscaposa Kom.	China, Sichuan, Muli	*J. F. Rock 24596* (PE)
C. subsimplex Hook. f. & Thomson	China, Tibet, Yadong	*Qinghai-Xizang Exped. 74-2509* (PE)
C. thalictrifolia Wall.	China, Tibet, Gongbo'gyamda	*D. Y. Hong et al. H10008* (PE)
C. ussuriensis (Rupr. & Maxim.) Hemsl.	China, Jilin, Tonghua	*J. Sato 8983* (PE)
C. viridis Wall.	India, Uttaranchal, Kuthi	*Strachey & Winterboton No. 3* (PE)

TABLE II 2:1 Materials for Observations on Seed Morphology of Codonopsis and Allied Genera—cont'd

Taxon	Origin	Voucher
***Cyclocodon* Griff. ex Hook. f. & Thomson**		
C. lancifolius (Roxb.) Kurz	China, Yunnan, Gongshan	*s. coll. 24514* (PE)
C. axillaries (Oliv.) D. Y. Hong	China, Fujian, Wanjin	*G. D. Ye 2287* (PE)
C. parviflorus (Wall. ex A. DC.) Hook. f. & Thomson	China, Yunnan, Luchun	*D. D. Tao 586* (PE)
Himalacodon dicentrifolius (C. B. Clarke) D. Y. Hong & Q. Wang	China, Tibet, Nyalam	*C. C. Ni 1906* (PE)
Pankycodon purpureus *(Wall.)* D. Y. Hong & X. T. Ma	China, Tibet, Nyalan	*D. Y. Hong et al. H 11003* (PE)
***Pseudocodon* D. Y. Hong & H. Sun**		
P. convolvulaceus subsp. *forrestii* (Diels) D. Y. Hong	China, Sichuan, Luding	*W. K. Hu & Z. He 11584* (PE)
P. graminifolius (H. Lév.) D. Y. Hong	China, Yunnan, Dali	*Chen & Grey-Wilson 85-37* (PE)
P. hirsutus (Hand.-Mazz.) D. Y. Hong	China, Sichuan, Yanyuan	*D. Y. Hong et al. H 11023* (PE)
P. vinciflorus (Kom.) D. Y. Hong subsp. *vinciflorus*	China, Tibet, Maizhokunggar	*D. Y. Hong et al. H10001* (PE)

Observed samples: all three species of *Cyclocodon* (*Codonopsis* in the sense of Moeliono, 1960) (Figures II 2: 1, q–t; Figures II 2: 2, a–h).

Type 3. Seeds ovoid or oblong, wingless. Transection elliptic or oblong. Lumina from rectangular to striate, with length/width much over 5. Radial walls smooth, without secondary ornamentations. In transection radial walls elevated, tangential walls depressed, cell cavity wider than, or nearly equal to, radial walls.

Observed samples: *Codonopsis affinis* Hook. f. & Thomson, *C. benthamii* Hook. f. & Thomson, *C. canescens* Nannf., *C. chlorocodon* C. Y. Wu, *C. foetens* subsp. *foetens*, *C. bulleyana* Forrest ex Diels, *C. pilosula* subsp. *handeliana* (Nannf.) D. Y. Hong & L. M. Ma, *C. subglobosa* W. W. Smith, *C. subscaposa* Kom., *C. subsimplex* Hook. f. & Thomson, *C. thalictrifolia* Wall. (Figures II 2: 2, i–v; Figures II 2: 3, a–l), *C. hongii* Lammers (*Leptocodon hirsutus* D. Y. Hong) (Figures II 2: 4, g–j), *Pseudocodon convolvulaceus* subsp. *forrestii* (Diels) D. Y. Hong, *P. graminifolius* (H. Lév.) D. Y. Hong, *P. hirsutus* (Hand.-Mazz.) D. Y. Hong, and *P. vinciflorus* (Kom.) D. Y. Hong. (Figures II 2: 3, m–t; Figures II 2: 4, a–f).

Type 4. Seeds ovoid-ellipsoid, winged, with wing at the distal end of hilum. Transection of seed body elliptic. Lumina striate, with length/width over 5. Radial walls wider than lumina, smooth, without secondary ornamentations. In transection radial walls elevated, tangential walls depressed, radial walls as wide as cell cavity.

Observed samples: *Codonopsis lanceolata* (Sieb. & Zucc.) Trautv. and *C. ussuriensis* (Rupr. & Maxim.) Hemsl. (Figures II 2: 4, k–p).

In addition, two incompletely known types were observed. They are both confined to alpine mountains of the Himalaya. We have not obtained mature seeds though we made two expeditions to S Tibet. Therefore, we conducted observations on immature seeds.

1. Seeds narrowly ovoid. Lumina irregularly striate with length/width ratio over 5; radial walls with secondary ridge-like thickening.

Observed sample: *Himalacodon dicentrifolius* (C. B. Clarke) D. Y. Hong & Q. Wang [*Codonopsis dicentrifolia* (C. B. Clarke) W. W. Smith] (Figures II 2: 4, q & r).

2. Seeds lunate. Lumina striate with length/width ratio over 5. Radial walls with secondary ridge-like ornamentations.

Observed sample: *Pankycodon purpureus* (Wall.) D. Y. Hong & X. T. Ma (*Codonopsis purpureus* Wall.) (Figures II 2: 4, s & t).

The seeds in these two samples are characterized by the striate lumina and ridge-like secondary ornamentations on the radial walls. These two samples are different in seed morphology; the lumina are irregular in the former, but regular in the latter; the seed shape in *Pankycodon purpureus* is lunate, while in the other it is straight and narrowly ovoid. These two samples may represent one or two types of seed morphology.

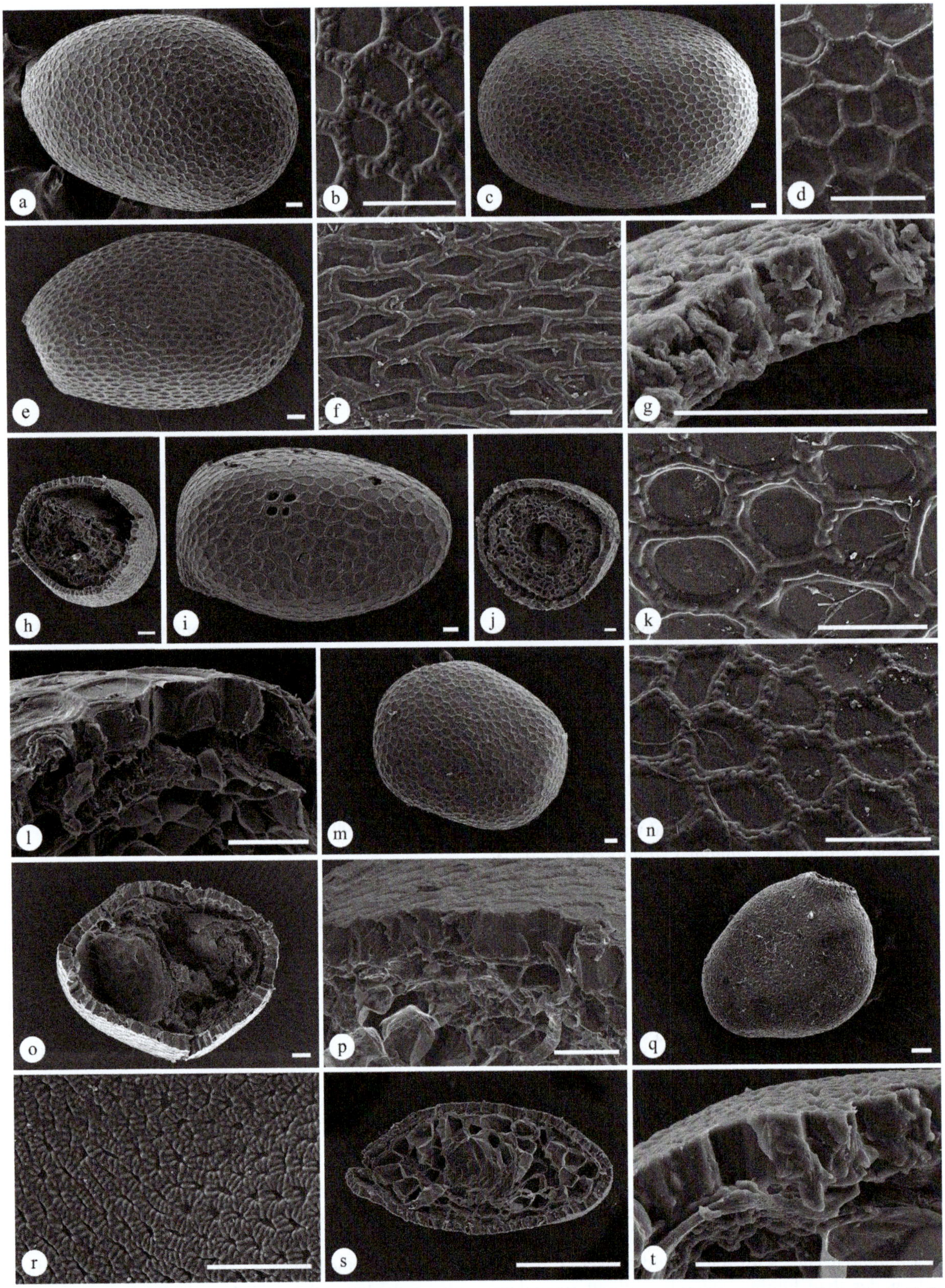

FIGURE II 2:1 a–t. SEM Photographs of seed-coat morphology of *Codonopsis* and *Cyclocodon*. **a** & **b**, *Codonopsis cordifolioidea* P. C. Tsoong; **c** & **d**, *Codonopsis deltoidea* Chipp; **e–h**, *Codonopsis viridis* Wall.; **i–l**, *Codonopsis javanica* (Bl.) Hook. f. subsp. *japonica* (Makino) Lammers; **m–p**, *Codonopsis inflata* Hook. f. & Thomson; **q–t**, *Cyclocodon parviflorus* (Wall. ex A. DC.) Hook. f. & Thomson. Scale bars = 50 μm.

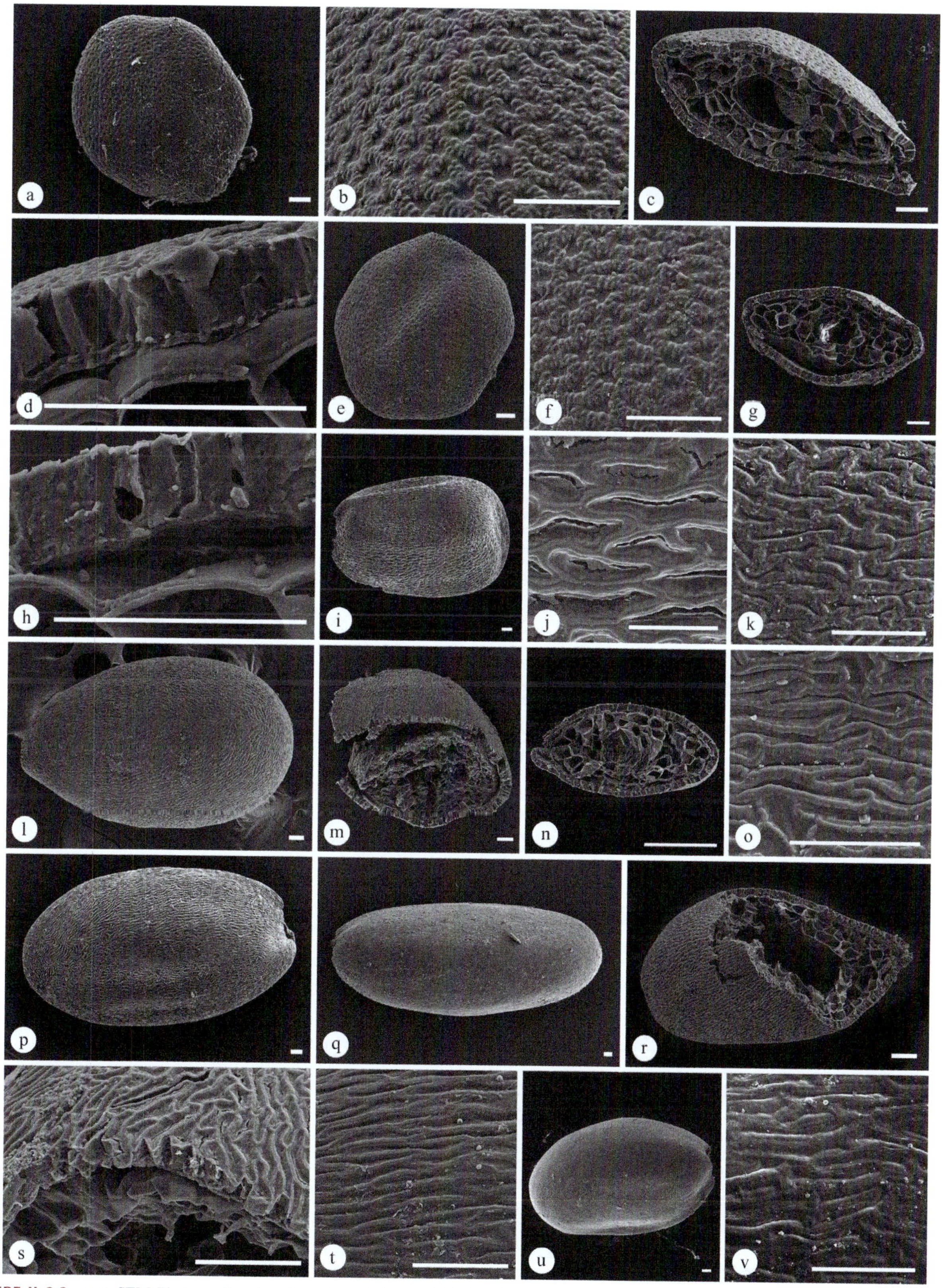

FIGURE II 2:2 a–v. SEM Photographs of seed-coat morphology of *Cyclocodon* and *Codonopsis*. **a–d**, *Cyclododon lancifolius* (Roxb.) Kurz; **e–h**, *Cyclocodon axillaris* (Oliv.) W. J. de Wilde & Duyfjes; **i** & **j**, *Codonopsis affinis* Hook. f. & Thomson; **k–n**, *Codonopsis benthamii* Hook. f. & Thonson; **o** & **p**, *Codonopsis subsimplex* Hook. f. & Thomson; **q–t**, *Codonopsis foetens* Hook. f. & Thomson subsp. *foetens*; **u** & **v**, *Codonopsis bulleyana* Forrest ex Diels. Scale bars = 50 μm.

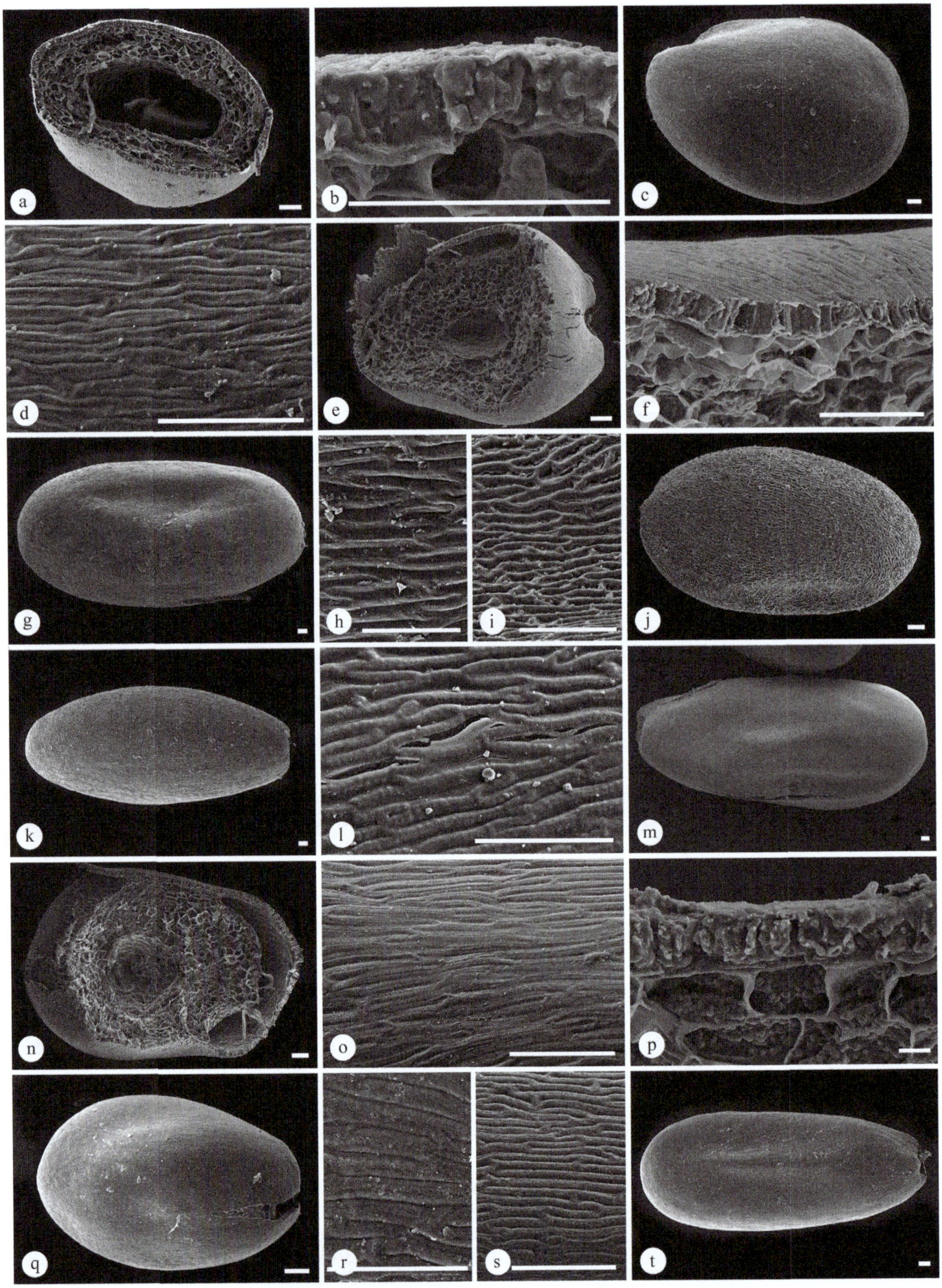

FIGURE II 2:3 a–t. SEM Photographs of seed-coat morphology of *Codonopsis* and *Pseudocodon* (*Codonopsis* subg. *Pseudocodonopsis* Kom.). **a** & **b**, *Codonopsis subscaposa* Kom.; **c** & **d**, *Codonopsis thalictrifolia* Wall. subsp. *thalictrifolia*; **e** & **f**, *Codonopsis canescens* Nannf.; **g** & **h**, *Codonopsis chlorocodon* C. Y. Wu; **i** & **j**, *Codonopsis subglobosa* W. W. Smith; **k** & **l**, *Codonopsis pilosula* (Franch.) Nannf. subsp. *handeliana* (Nannf.) D. Y. Hong; **m–p**, *Pseudocodon convolvulaceus* (Kurz) D. Y. Hong & H. Sun subsp. *forrestii* (Diels) D. Y. Hong; **q** & **r**, *Pseucododon hirsutus* (Hand. -Mazz.) D. Y. Hong; **s** & **t**, *Pseudocodon vinciflorus* (Kom.) D. Y. Hong. Scale bars = 50 μm.

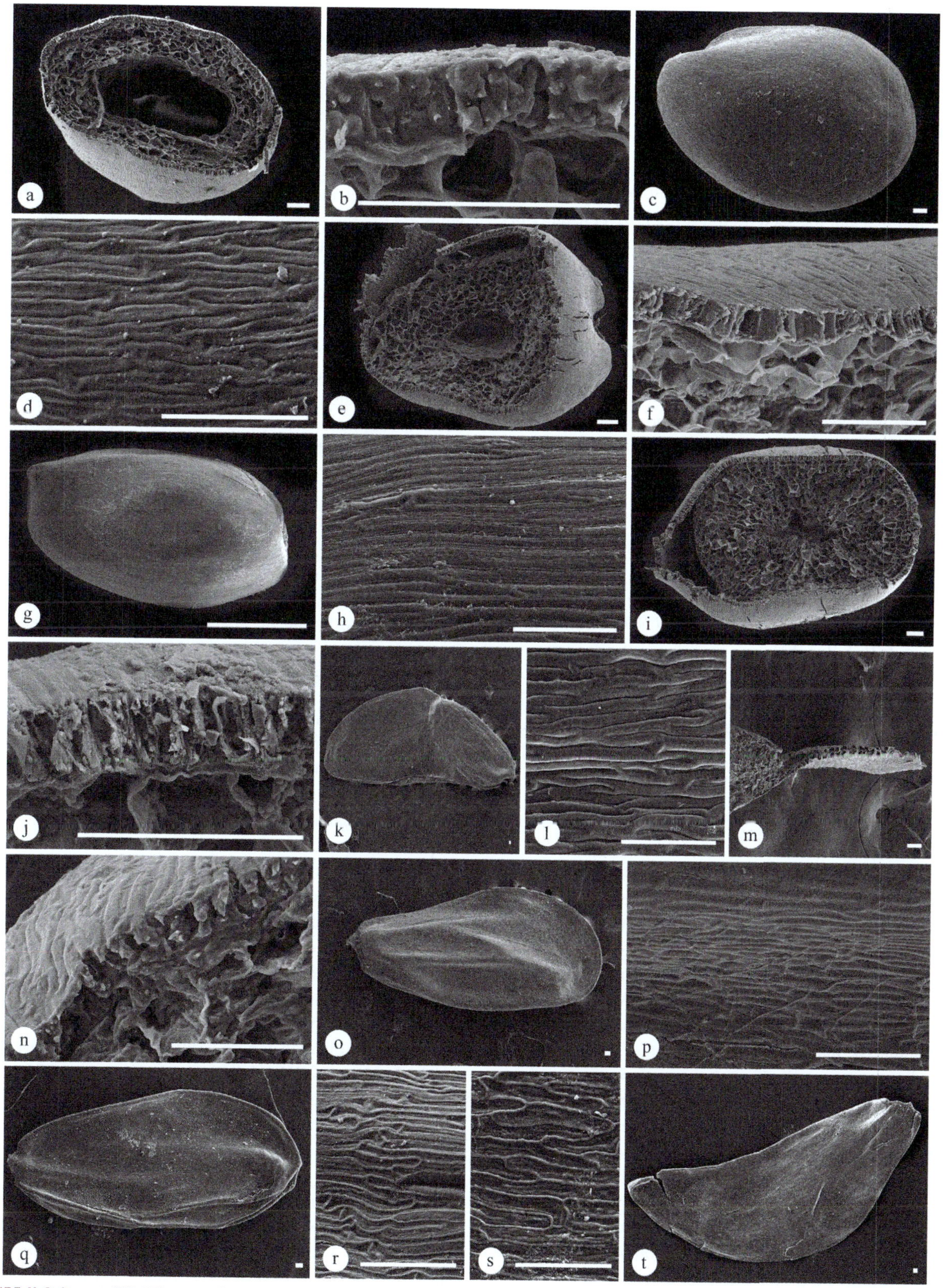

FIGURE II 2:4 a–t. SEM Photographs of seed-coat morphology of *Pseudocodon*, *Codonopsis*, *Himalacodon* (*Codonopsis* subg. *Pseudocodonopsis* Kom.), and *Pankycodon* (*Codonopsis purpurea* Wall.). **a** & **b**, *Pseudocodon vinciflorus* (Kom.) D. Y. Hong, continued from Figure II 2:3; **c–f**, *Pseudocodon graminifolius* (H. Lév.) D. Y. Hong; **g–j**, *Codonopsis hongii* Lammers (*Leptocodon hirsutus* D. Y. Hong); **k–n**, *Codonopsis lanceolata* (Sieb. & Zucc.) Trautv.; **o** & **p**, *Codonopsis ussuriensis* (Rupr. & Maxim.) Hemsl.; **q** & **r**, *Himalacodon dicentrifolius* (C. B. Clarke) D. Y. Hong & Q. Wang; **s** & **t**, *Pankycodon purpureus* (Wall.) D. Y. Hong & X. T. Ma. Scale bars = 50 μm.

From our observations on seed morphology on *Codonopsis* and allied genera we may indicate that seed morphology of the group under study has two characteristic features.

Diversity

We recognize in *Codonopsis* and allied genera four types of seed morphology plus two incompletely known types, which implies that the seed morphology is diverse and polytypic. The diversity is expressed in the shape and appendices of seeds, the shape of the testa cells in both surface view and section view, and various secondary ornamentations. Two species of *Codonopsis* are found to have a wing at the distal end of the hilum, very wide in *C. lanceolata* but narrower in *C. ussuriensis*. Lumina in the surface view vary greatly in the shape from nearly polygonal and isodiametric in *Codonopsis* sect. *Campanumoea* and several other species, and in all the species of *Cyclocodon*, to regularly or irregularly long-striate with length/width over 10 in several species of *Codonopsis*, like *C. hongii*.

Great variation occurs in the radial walls of testa cells, with various types of ornamentation, ridge-like, e. g. in *Codonopsis foetens* Hook. f. & Thomson (Figures II 2: 2, q–t), bead-like, e. g. in *Codonopsis inflata* (Figures II 2: 1, m–p) and spiral, e. g. three species of *Cyclocodon* (Figures II 2: 1, q–t; II 2: 2, a–h).

The great diversity in seed morphology of *Codonopsis* and allied genera, we think, must be significant and merits further studies on seed dispersal and adaptation. For example, the wide wing in *Codonopsis lanceolata* should have adaptive significance, but its function is still a question.

Significance in Systematics

As mentioned in the introduction of this section, authors have controversial views in this issue. Shetler and Morin (1986) stated that there are recognizable generic patterns and a number of distinctive individual species based on their observations on 34 species in four genera of the Campanulaceae s. s. in North America, whereas Haridasan and Mukherjee (1987) considered based on their observations on Indian groups of the Campanulaceae that SEM ornamentations of seed-coat can not be used as generic marker though they are species-specific.

Among the four types of seed morphology we recognized, Type 2 is characteristic of *Cyclocodon*. The systematic position of *Cyclocodon* has been problematic since 1857, when the genus was established. It was treated by Clarke (1881) as a section in *Campanumoea*. Moeliono (1960) merged *Campanumoea* (including *Cyclocodon*) with *Codonopsis*. Hong and Pan (1998) restored *Cyclocodon* as genus based on observations on pollen grains and seed-coat of two species in *Campanumoea* sect. *Campanumoea* and two species in *Campanumoea* sect. *Cyclocodon*. Recent molecular works (Eddie et al. 2003; Haberle et al. 2009; Wang et al. 2013) indicated that the closest relative of *Cyclocodon* is *Platycodon*, not *Codonopsis*. Our observations in the present work show that the three species of *Cyclocodon* are consistent in seed morphology and deserve an independent seed type (Figures. II 2: 21–32). The data from seed morphology is highly correlated with pollen morphology (Hong & Pan, 2012) and molecular data (Eddie et al., 2003; Haberle et al., 2009; Wang et al., 2013), and thus seed morphology further supports the restoration of *Cyclocodon*.

Our investigation shows that *Codonopsis* is polytypic in seed morphology. For example, *Codonopsis lanceolata* and *C. ussuriensis* both possess a winged seed with the wing at the distal end of hilum. Correlated with this is the phyllotaxy of the two species. Their leaves are alternate, but nearly always there is a distinctive verticil-like group of four or three leaves aggregated at the tip of each branch. A new section for these two species based on the two correlated characters should be justifiable.

It is clear from our extensive observations, that the seed morphology in *Codonopsis* and allied genera of the Cyanantheae is species-specific, and more importantly it may serve as a generic marker in some cases, or sectional markers in the others. Therefore, the seed morphology is significant for studies on biodiversity and adaptation, and even more significant for systematic pursuits.

POLLEN MORPHOLOGY

As we have reviewed in the second section of the next chapter, the Campanulaceae s.s. is a well established monophyletic group, but its subdivision was unclear before this century. Eddie et al. (2003) and Haberle et al. (2009) conducted phylogenetic analyses of the family based on ITS sequences of nuclear ribosomal DNA and based on DNA sequences of three chloroplast genes respectively. Their works have demonstrated that the family comprises three major monophyletic clades, campanuloids, platycodonoids and wahlenbergioids, which may be considered as three primary subdivisions of the Campanuluceae s.s. However, their works also show that phylogenetic relationships within each primary subdivision was unclear, and no attempt has been made to arrange genera within these three subdivisions since their works. We observed pollen grains of 25 accessions representing 24 species of *Codonopsis* and its allies, and took pollen morphology of the other genera in the platycodonoids observed by previous authors into considerations of phylogenetic relationships within the group (Hong & Pan, 2012). The aim of this section is to review the diversity of pollen morphology and the relationships between pollen types, and to discuss further the phylogenetic relationships and systematic position of *Codonopsis* and its allies in the platycodonoid group from palynological data.

Erdtman's (1952) pioneer work on pollen morphology of the Campanulaceae reveals that Schönland's (1889)

Campanuleae (= Campanulaceae s.s.) has two major pollen types:

1. Apertures poroid, and pollen porate: *Adenophora*, *Campanula*, *Roella*, *Wahlenbergia*, and many other genera.
2. Apertures elongate, and pollen colpate or colporate:
   a. pollen colpate: *Codonopsis*, *Cyananthus*, *Leptocodon*;
   b. pollen 3-colporate: *Campanumoea* (= *Cyclocodon*) *parviflora*, *Canarina*;
   c. pollen 5–6-colporate: *Platycodon*.

The above data show that pollen grains are polytypic in the Campanulaceae s. s., but they are basically uniform within genus. Erdtman was wise when he stated "for several reasons, including pollen morphological, the subtribes of Campanulaceae - Campanuloideae do not seem to be homogeneous. Thus eastern and southeastern Asiatic species belonging to *Campanumoea*, *Codonopsis*, *Cyananthus*, *Leptocodon* and *Platycodon* are characterized by pollen grains with elongate apertures. In Schönland's classification, however, *Platycodon* is placed close to the African genera *Micrododon* and *Musschia* which are provided with porate pollen grains." Erdtman's work and his statement should be heuristic, but, unfortunately, plant taxonomists did not listen to him. Fedorov (1957) still pulled the *Codonopsis* complex together with *Wahlenbergia* in his new system of the Campanulaceae s.s.

Dunbar (1975) observed pollen grains of 61 species of 18 genera in the Campanulaceae - Campanuloideae (= Campanulaceae s.s.). His results are nearly the same as Erdtman's (1952), that is, pollen grains are polytypical in the family but rather uniform within genus. He recognizes two major pollen types:

A. Porate pollen (with poroid apertures): *Adenophora*, *Asyneuma*, *Edraianthus*, *Githopsis*, *Jasione*, *Prismatocarpus*, *Phyteuma*, *Roella*, *Symphyandra*, *Triodanis*, *Wahlenbergia*. This type of pollen is relatively uniform.
B. Colpate/colporate pollen (with elongate apertures): This type of pollen is polytypical.
   7–8-colpate with spinulose exine: *Codonopsis*
   8–9-colpate with verrucose exine: *Cyananthus*
   6–7-colpate with rounded or elongated verrucae on exine: *Ostrowskia*
   3-colporate pollen: *Canarina*, *Campanumoea* (= *Cyclocodon*) *lancifolia*
   5–6-colporate pollen: *Platycodon*

Avetisjan (1986) continued her work on pollen morphology of the Campanulaceae s.s. with the same result as Erdtman's (1952) and Dunbar's (1975), but she classified pollen in the family into three groups and seven types:

Palynogroup I.
- Cyananthus Type: *Cyananthus*, *Codonopsis*, *Leptocodon*.
- Ostrowskia Type: *Ostrowskia*

Palynogroup II.
- Subgroup A
  - Platycodon Type: *Platycodon*
  - Canarina Type: *Canarina*
- Subgroup B
  - Campanumoea (= Cyclocodon) Type: *Campanumoea* (*Cyclocodon*) *parviflora*

Palynogroup III
- Subgroup A
  - Campanula Type: *Adenophora, Astrocodon, Asyneuma, Azorina, Berenice, Brachycodonia, Campanula, Cephalostigma, Cryptocodon, Cylindrocarpa, Edraianthus, Gadellia, Heterochaenia, Heterocodon, Jasione, Jasionella, Legousia, Lightfootia, Merciera, Michauxia, Microcodon, Musschia, Popoviocodonia, Roella, Sergia, Specularia, Symphyandra, Trachelium, Tracheliopsis, Triodanis, Wahlenbergia.*
  - Rhigiophyllum subtype: *Rhigiophyllum*
  - Githopsis subtype: *Githopsis*
- Subgroup B
  - Campanulastrum Type: *Campanulastrum*

From her classification of pollen morphology of this family it is clear that the pollen with poriform apertures is relatively uniform, and found in the great majority of genera, whereas the pollen with elongate apertures, colpate or colporate, is polytypical and found in a small number of genera.

Avetisjan (1986) also inferred the evolutionary trends of pollen morphology in the family: from colpate to colporate, and then to zonoporate, and finally to pantoporate.

Recent pollen investigation by Morris and Lammers (1997) should be particularly mentioned. Their research focused on *Codonopsis* and the allied genera (the *Codonopsis* complex) with relatively adequate sampling, the two species of *Campanumoea* sect. *Campanumoea*, the three species of *Campanumoea* sect. *Cyclocodon*, 17 species of *Codonopsis* subg. *Codonopsis*, the only species of *Codonopsis* subg. *Obconicicapsula*, four species of *Codonopsis* subg. *Pseudocodonopsis*, and the two species of *Leptocodon*. They recognize four types in *Codonopsis* and allied genera, two more than Murthy's (1982) observation.

Type I. 4–10-colpate, with long colpi, short exine spinules, and thin nexine:
   Codonopsis subg. *Codonopsis*, *Leptocodon*, and *Campanumoea* sect. *Campanumoea*

Type II. 3-colporate, with long colpi, long exine spinules and thick nexine.
   Campanumoea sect. *Cyclocodon*

Type III. 4–6-colpate, with short colpi, long spinules, and thin nexine.

Codonopsis subg. *Pseudocodonopsis* Kom. (= *Pseudocodon* D. Y. Hong & H. Sun)

Type IV. 5–7-colpate, with medium-long colpi, small spinules, and thin nexine.

Codonopsis subg. *Obconicicapsula* D. Y. Hong (= *Himalacodon* D. Y. Hong & Q. Wang)

For their excellent investigation five points should be indicated: 1) *Codonopsis* even in Hong's (1983; Hong in Hong et al., 2011) strict sense is heterogeneous in pollen morphology, *Codonopsis* subg. *Codonopsis* belonging to Type I, subg. *Pseudocodonopsis* Kom. belonging to Type III, while subg. *Obconicicapsula* D. Y. Hong belonging to Type IV; the three subgenera each has a distinct pollen type; 2) *Campanumoea* s. l. (Clarke, 1881) is also heterogeneous in this respect, sect. *Campanumoea* belonging to Type I, while sect. *Cyclocodon* having Type II; 3) *Leptocodon* and *Campanumoea* sect. *Campanumoea* possess the same type of pollen as that of *Codonopsis* subg. *Codonopsis*; 4) no diversity was detected in *Codonopsis* subg. *Codonopsis* in their work; 5) the pollen data they provided imply that the current taxonomy of the group needs to be significantly revised.

Hong and Pan (1998) restored *Cyclocodon* Griff. ex Hook. f. & Thomson (*Campanumoea* sect. *Cyclocodon*) to generic status based on pollen and seed-coat morphology, combined with external morphology.

Due to two considerations, the significance of pollen morphology for taxonomic revision of *Codonopsis* and allied genera, and the pollen diversity in the platycodonoid group as revealed by the previous observations on pollen grains (Erdtman, 1952; Avetisjan 1969, 1986; Dunbar, 1975; Morris & Lammers, 1997; Hong & Pan, 1998), we conducted intensive observation on pollen grains of the group and then used all available palynological data for phylogenetic analyses of the platycodonoid group (Hong & Pan, 2012). We sampled the two species of *Campanumoea*, 23 species of *Codonopsis* subg. *Codonopsis*, four species and four subspecies of *Codonopsis* subg. *Pseudocodonopsis*, all three species of *Cyclocodon*, and the two species of *Leptocodon*.

Our work confirmed Morris and Lammers' (1997) observation and their classification of pollen types in *Codonopsis* and allied genera, but we found that *Codonopsis* subg. *Codonopsis* is also heterogeneous in pollen morphology. *Codonopsis purpurea* Wall. and *C. chimiliensis* J. Anthony, which were not sampled by Morris and Lammers (1997), have 6-colpate pollen with short colpi and small exine spinules, similar to that of subg. *Obconicicapsula*, and thus we treated it as *Obconicicapsula* Type *Purpurea* subtype.

As a result of analyses within the platycodonoid group using all available data of pollen morphology, we proposed a classification of pollen morphology in the group: nine types and four subtypes.

Key to pollen types and subtypes of the platycodonoid group (alignment of genera is indicated in parentheses):

1a. Pollen colporate.
  2a. Pollen 5–6-colporate........ Platycodon Type (*Platycodon*)
  2b. Pollen 3-colporate.
    3a. Sexine verrucose........ Canarina Type (*Canarina*)
    3b. Sexine spinulose........ Cyclocodon Type (*Cyclocodon*)
1b. Pollen colpate.
  4a. Apertures elongate (colpus length/polar axis length ≥ 0.54).
    5a. Sexine verrucose, verrucae rounded or elongated, ca. 2 μm high........ Ostrowskia Type (*Ostrowskia*)
    5b. Sexine spinulose or verrucose, spinules or verrucae not rounded, ≤ 1.5 μm high.
      6a. Sexine verrucose, colpus membrane verrucose........ Cyananthus Type (*Cyananthus*)
      6b. Sexine spinulose, colpus membrane spinulose or granulose........ Codonopsis Type
        7a. Apertures long (colpus length/polar axis length ≥ 0.7); colpus membrane spinulose........ Codonopsis subtype (*Codonopsis* s. s. + *Leptocodon*)
        7b. Apertures medium-long (colpus length/polar axis length = 0.54–0.57); colpus membrane granulose........ Campanumoea subtype 〔*Campanumoea* s. s. (= *Codonopsis*)〕
  4b. Apertures short (colpus length/polar axis length ≤ 0.4).
    8a. Spinules divided at base........ Echinocodon Type (*Echinocodon*)
    8b. Spinules entire at base.
      9a. Sexine spinules ≥ 2μm high...... Pseudocodon Type [*Pseudocodon* (*Codonopsis* subg. *Pseudocodonopsis*]
      9b. Sexine spinules ≤ 1μm high........ Himalacodon Type
        10a. Colpus membrane granulose........ Himalacodon subtype [*Himalacodon* (*Codonopsis* subg. *Obconicicapsula*]
        10b. Colpus membrane smooth........ Pankycodon subtype [*Panlycodon purpureus* (*Codonopsis purpurea*) & *C. chimiliensis*]

Based on all observations on pollen morphology and molecular data (Eddie et al. 2003; Haberle et al. 2009; Wang et al., 2013) in the platycodonoid group (= tribe Cyanantheae) we may state:

1. A particular pollen type is generally characteristic of genus.
2. *Codonopsis* even in Hong's (1983, Hong in Hong et al., 2011) narrow sense is heterogeneous, subg. *Pseudocodonopsis* Kom. and subg. *Obconicicapsula* D. Y. Hong each possessing a particular pollen type; whilst subg. *Codonopsis* is also heterogeneous, with two pollen types, the Codonopsis Type and the Obconicicapsula Type Purpurea Subtype.
3. Pollen morphology in the platycodonoid group is unusually polytypical.
4. The pollen types in the platycodonoid group are highly correlated with external morphology. For example, the Cyananthus Type is correlated with a superior ovary, and the Codonopsis Type is correlated with the presence of a disc or glands.
5. The pollen types in the Campanulaceae s. s. are closely correlated with molecular data presented by Eddie et al. (2003), Haberler et al. (2009) and ourselves (Wang et al. 2013).
6. The closest relative of *Codonopsis* and its allies (excluding *Cyclocodon*) may well be *Cyananthus*; they share multi-colpate pollen.
7. The results of observations on pollen grains in the platycodonoid group imply that pollen morphology should be seriously considered in the phylogenetic analysis of the group.

At the end I show some SEM photographs of pollen grains of representative species in *Codonopsis* and its allies for the following purposes (Figures II 3: 1–4):

1. to demonstrate that pollen morphology in this group is polytypical, generally correlated with molecular data and gross morphology, and is very significant in systematics;
2. to explain based on palynology: A) It is justified to separate *Cyclocodon* from *Codonopsis* as an independent genus; B) Pollen morphology supports the establishment of three new genera, *Himalacodon*, *Pankycodon* and *Pseudocodon*; and C) There is no strong evidence against the merging of *Campanumoea* and *Leptocodon* with *Codonopsis*.

CHROMOSOMES

It is redundant to mention the significance of chromosomal information in studies on taxonomy and evolution of plants. We here provide chromosomal data of *Codonopsis* and allied genera, with a general survey of chromosome numbers in the Campanulaceae s. s., and then briefly analyse the relationship between chromosomal data and systematics.

General Survey of Chromosomes in *Codonopsis* and Allied Genera

Codonopsis subg. *Pseudocodonopsis* Kom. (= *Pseudocodon* D. Y. Hong & H. Sun) and *Obconicicapsula* D. Y. Hong (= *Himalacodon* D. Y. Hong & Q. Wang) had no cytological data before our work. We made expeditions to S and SE Tibet, and to SW Sichuan (Yanyuan and Muli), harvesting seeds or fixing flower buds. Our observations on chromosomes gave following results. Two species of subg. *Pseudocodonopsis*, i. e. *Codonopsis convolvulacea* subsp. *forrestii* and *C. hirsuta* both have chromosome number 2n = 14 (Wang et al., 2014a) (Figure II 4:1c & d). *Codonopsis dicentrifolia* in subg. *Obconicicapsula* possesses chromosome number also 2n = 14. Thus these two subgenera have the same chromosome number, 2n = 14 (Wang et al., 2014a) (Figure II 4:1a). Seven species in subg. *Codonopsis* have been reported to have chromosome number 2n = 16. Interesting is that according to our observation, *Codonopsis purpurea* Wall. [= *Pankycodon purpureus* (Wall.) D. Y. Hong & X. T. Ma] in *Codonopsis* subg. *Codonopsis* also has chromosome number 2n = 14 (Wang et al., 2014a) (Figure II 4:1b). Now *Codonopsis* and its allies have had relatively adequate chromosome information, which may be used for consideration on classification and evolution (Table II 4:1).

Relationships between Chromosomal Data and Taxonomy

Codonopsis subg. *Pseudocodonopsis* Kom. (= *Pseudocodon*) and subg. *Obconicicapsula* D. Y. Hong (= *Himalacodon*) have the same chromosome number, 2n = 14, and their pollen grains are short-colpate, with copus length/polar axis length ≤ 0.4, whereas the seven species in *Codonopsis* subg. *Codonopsis*, whose chromosome numbers had been reported before, are all of 2n = 16 (Table II 4:1), and their pollen grains are long/medium-colpate, with colpus length/polar axis length ≥ 0.54 (Hong & Pan, 2012). The difference in chromosome number between these two subgenera on one side and *Codonopsis* subg. *Codonopsis* on the other is closely correlated with the difference in pollen morphology. They are distinctly different from each other also in external morphology. Furthermore, in our molecular trees they each forms a monophyletic clade with strong support (Wang et al., 2013). Therefore, chromosomal data support separation of these two subgenera from *Codonopsis*.

Our observation on pollen grains of *Codonopsis* and its allies has revealed that the genus *Codonopsis* s.s. is heterogeneous in pollen morphology as mentioned above, and that even *Codonopsis* subg. *Codonopsis* is also heterogeneous in this respect. Pollen grains of *Codonopsis purpurea* Wall. are short-colpate, similar to the Obconicicapsula Type with short colpi, but distinctly different from the Codonopsis Type with long colpi, and thus assigned to the Obconicicapsula Type Purpurea subtype. Coincidently, chromosome

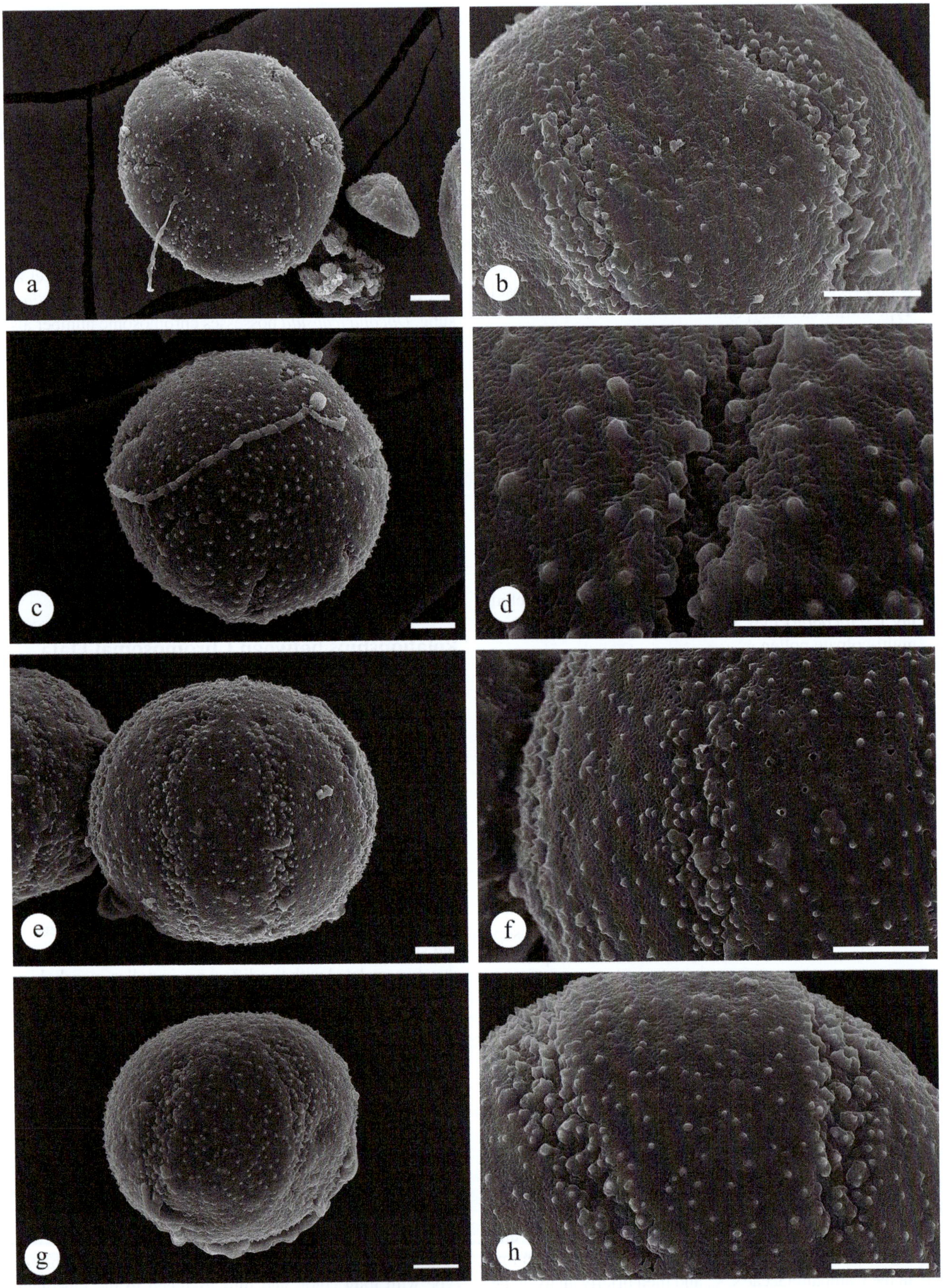

FIGURE II 3:1 SEM photographs of pollen grains of *Codonopsis*. **a** & **b**, *Codonopsis benthanii* Hook. f. & Thomson; **c** & **d**, *Codonopsis viridis* Wall.; **e** & **f**, *Codonopsis subscaposa* Kom.; **g** & **h**, *Codonopsis thalictrifolia* Wall.

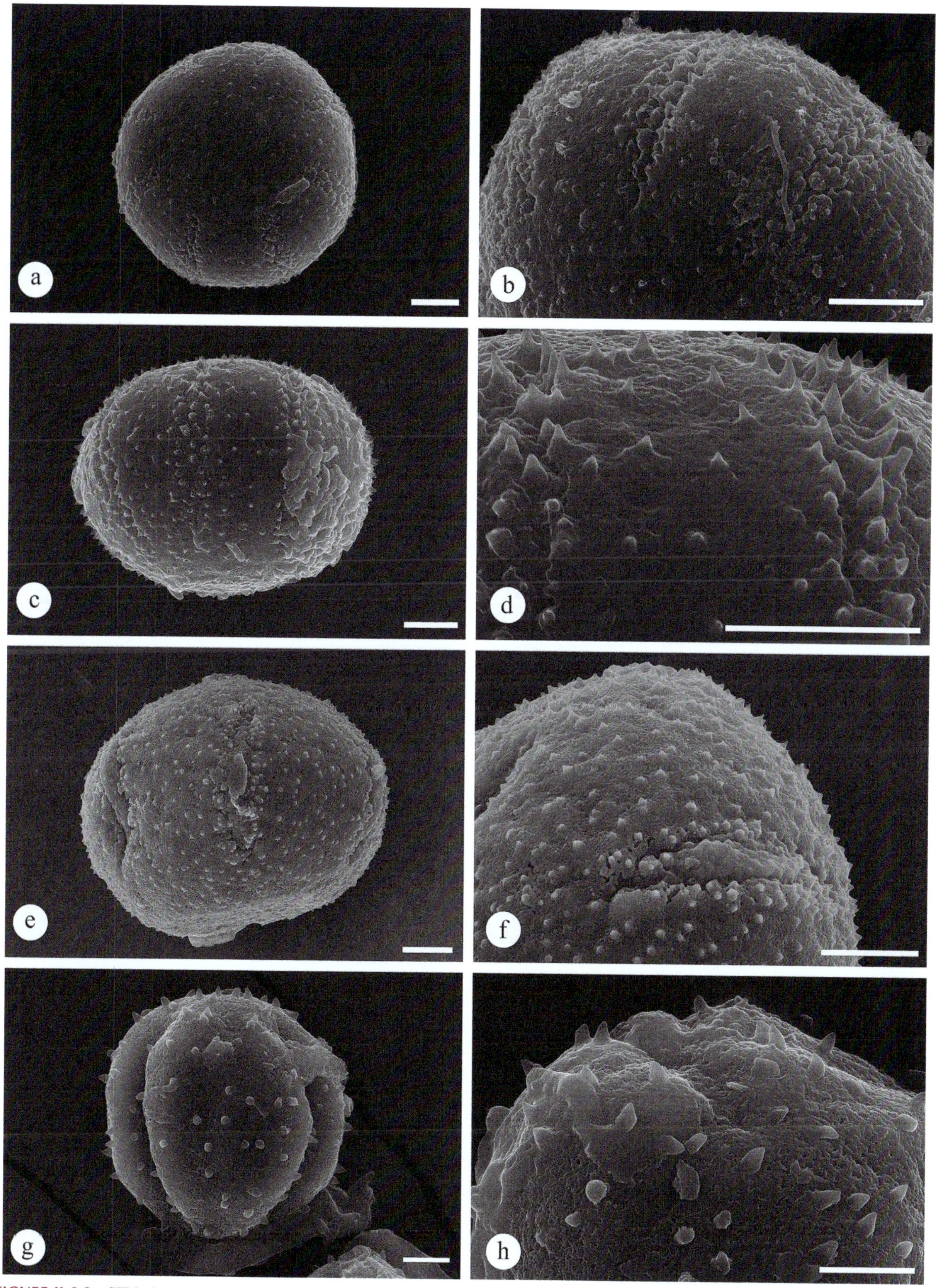

FIGURE II 3:2 SEM photographs of pollen grains of *Codonopsis* and *Cyclocodon*. **a & b**, *Codonopsis gracilis* Hook. f. & Thomson [*Leptocodon gracilis* (Hook. f. & Thomson) Lem.]; **c & d**, *Codonopsis hongii* Lammers (*Leptocodon hirsutus* D. Y. Hong); **e & f**, *Codonopsis javanica* (Bl.) Hook. f. (*Campanumoea javanica* Bl.); **g & h**, *Cyclocodon parviflorus* (Wall. ex A. DC.) Hook. f. & Thomson (*Codonopsis parviflora* Wall. ex A. DC.).

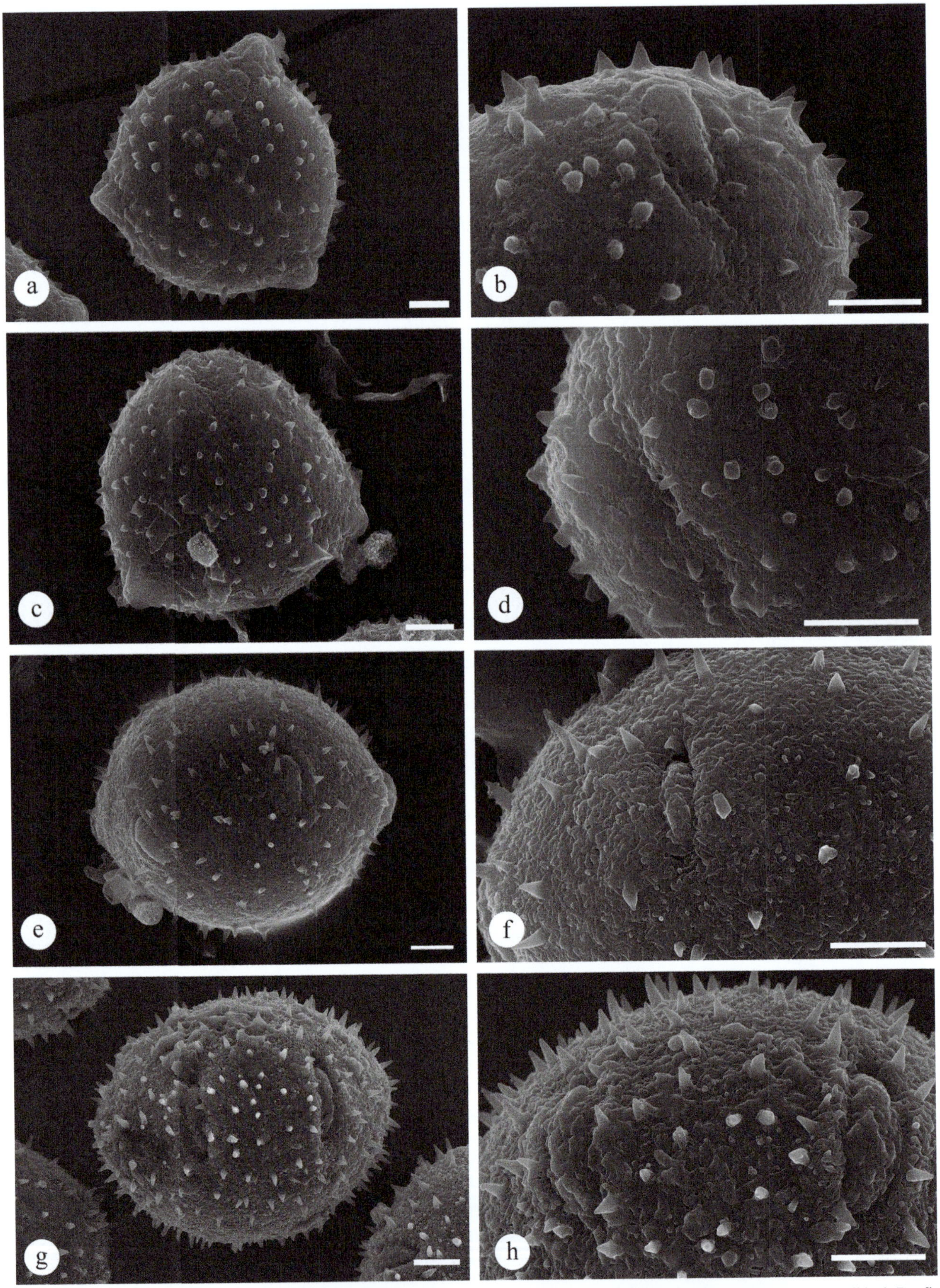

FIGURE II 3:3 SEM photographs of pollen grains of *Cyclocodon* and *Pseudocodon*. **a** & **b**, *Cyclocodon axillaris* (Oliv.) W. J. de Wilde & Duyfjes (*Campanumoea axillaris* Oliv.); **c** & **d**, *Cyclocodon lanciflorus* (Roxb.) Kurz [*Codonopsis lanciflora* (Roxb.) Moeliono] ; **e** & **f**, *Pseudocodon convolvulaceus* (Kurz) D. Y. Hong & H. Sun (*Codonopsis convolvulacea* Kurz) ; **g** & **h**, *Pseudocodon vinciflorus* (Kom.) D. Y. Hong (*Codonopsis vinciflora* Kom.).

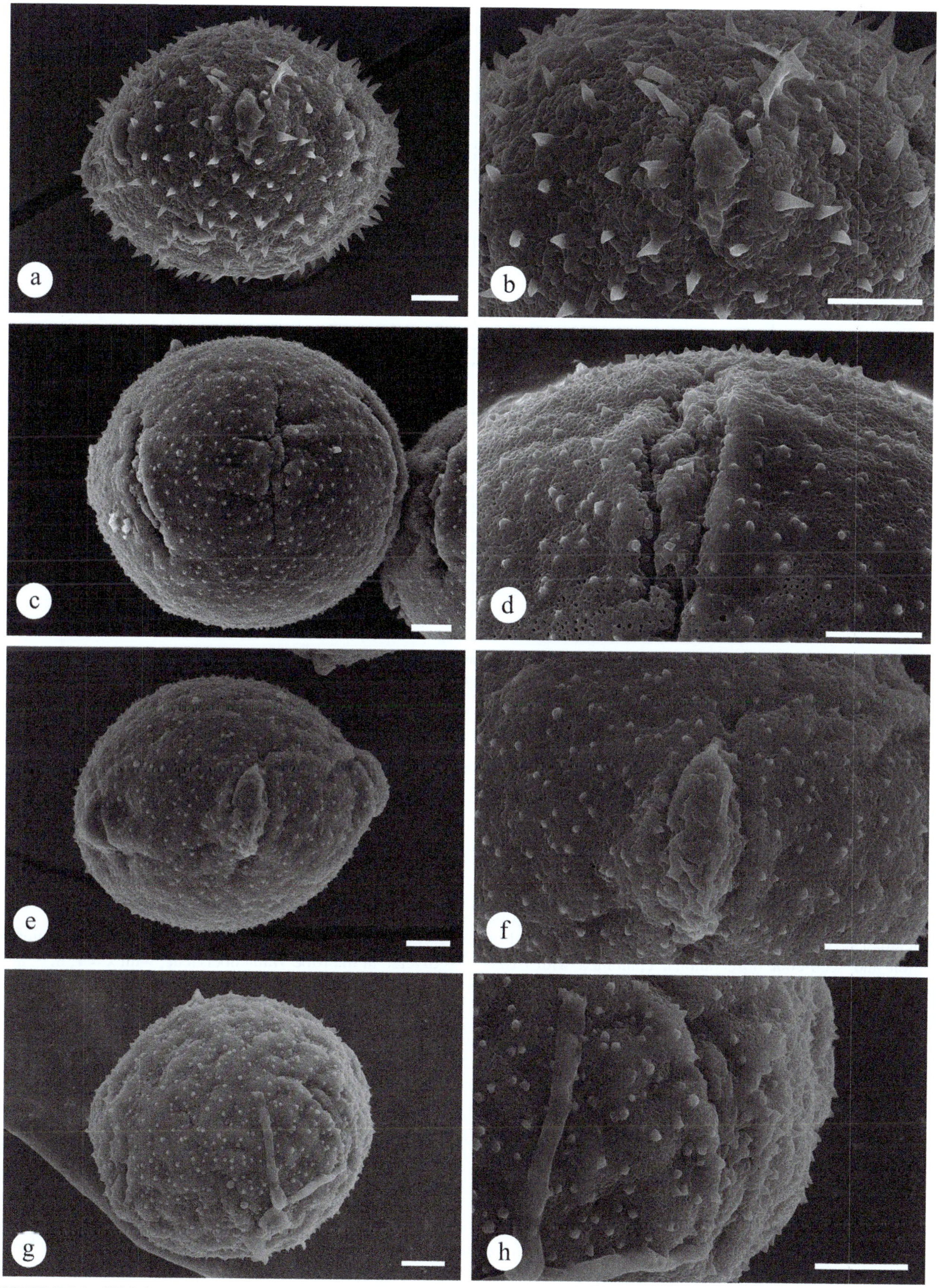

FIGURE II 3:4 SEM photographs of pollen grains of **a** & **b**, *Pseudocodon rosulatus* (W. W. Smith) D. Y. Hong (*Codonopsis rosulatus* W. W. Smith); **c** & **d**, *Himalacodon dicentrifolius* (C. B. Clarke) D. Y. Hong & Q. Wang [*Codonopsis dicentrifolia* (C. B. Clarke) W. W. Smith] ; **e** & **f**, *Pankycodon purpureus* (Wall.) D. Y. Hong & X. T. Ma (*Codonopsis purpurea* Wall.); **g** & **h**, *Codonopsis chimiliensis* J. Anthony.

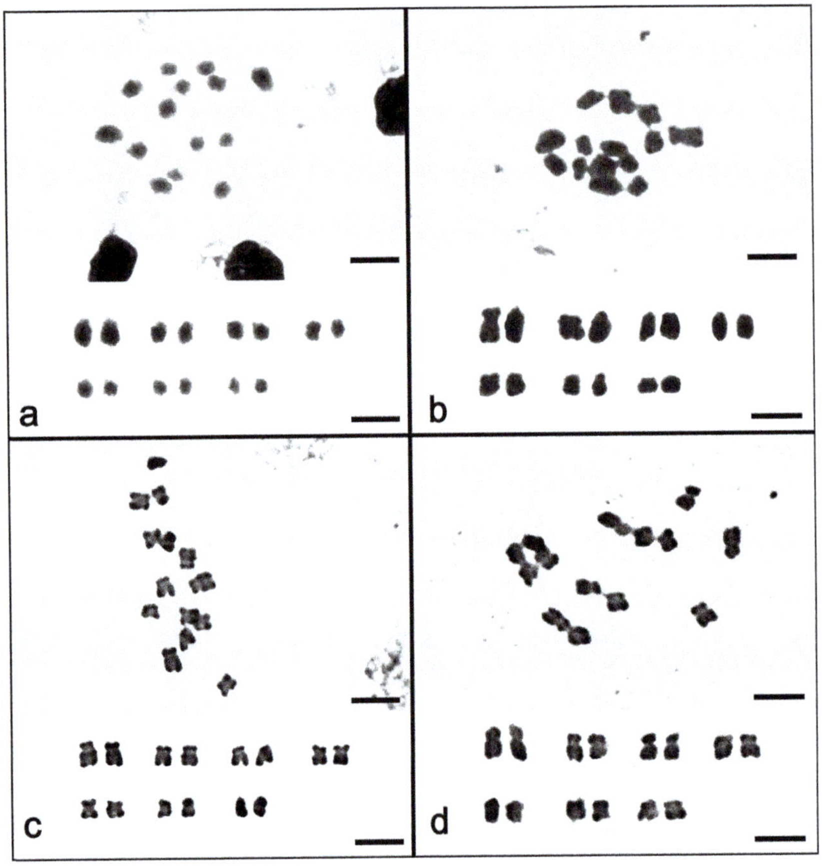

FIGURE II 4:1 Chromosomes of **a**, *Himalacodon dicentrifolius* (C. B. Clarke) D. Y. Hong & Q. Wang [*Codonopsis dicentrifolia* (C. B. Clarke) W. W. Smith]; **b**, *Pankycodon prupureus* (Wall.) D. Y. Hong & X. T. Ma (*Codonopsis purpurea* Wall.); **c**, *Pseudocodon convolvulaceus* (Kurz) D. Y. Hong & H. Sun (*Codonopsis convolvulacea* Kurz) subsp. *forrestii* (Diels) D. Y. Hong; **d**, *Pseudocodon hirsutus* (Hand. -Mazz.) D. Y. Hong.

number of *Codonopsis purpurea* has been counted as 2n = 14, instead of 2n = 16 in the rest species of *Codonopsis* subg. *Codonopsis*. It is reasonable to consider the independence of this species from *Codonopsis*.

The genus *Cyclocodon* Griff. ex Hook. f. & Thomson had been controversial among authors with its taxonomic status, as an independent genus, as a section in *Campanumoea*, or totally merged into *Codonopsis* (see Chapter I for a detailed review). We restored its generic status based on pollen and seed-coat observations, combined with external morphology (Hong & Pan, 1998). The action was soon accepted by Lammers (1998, 1999). Recent report on chromosome number of this group indicates that two species both have 2n = 18 (Zhang et al., 2010), also different from that of *Codonopsis*, 2n = 16.

Evolution of Chromosomes in the Platycodonoid Group (Tribe Cyanantheae)

The basic chromosome number (x) is greatly diverse in the group (Table II 4:1), from x = 5 and 6 in *Cyananthus*, x = 7 in *Cyananthus*, *Himalacodon*, *Pankycodon* and *Pseudocodon*, x = 8 in *Codonopsis* and *Echinocodon*, x = 9 in *Cyclocodon* and *Platycodon*, to x = 17 in *Canarina* and *Ostrowskia*. Facing these diverse basic chromosome numbers one would naturally raises two questions. Which is primary and how is the evolutionary trend ? 2) How has the basic number x = 17 in *Canarina* and *Ostrowskia* originated ? We show chromosome numbers and infer basic numbers of the other two tribes (Table II 4:2 and 3) for our consideration on this tribe in a broad view.

It can be seen from the two tables that the basic number is diverse in the tribe Wahlenbergieae, with x = 7, 8, 9, and 17, and even more diverse in the tribe Campanuleae, with x going up from 6 continuously up to x = 17. It is rather hard to infer which is primary one in the tribe Cyanantheae. However, x = 7 is the most frequent one in the group, and it is also shared by the other two tribes. Furthermore, in the Lobeliaceae (= Campanulaceae subf. Lobelioideae), the closest relative of the Campanulaceae s.s., x = 7 is the most frequently found. Therefore, it is justifiable to infer x = 7 as the ancient basic number of chromosomes of the Cyanantheae. Hong (1990) made a general survey on chromosome numbers of angiosperms and found that all the ten subclasses in angiosperms have x = 7 as their primary basic chromosome number except Caryophyllidae, which is of x

TABLE II 4:1 Chromosome Numbers in the Platycodonoid Group (= Tribe Cyanantheae) of the Campanulaceae s. s.

Taxa	n	2n	References	Basic Number (x) Inferred
***Canarina* L.**				
C. canariensis	17	34, 68±	Hedberg, 1961	17
		34	Michaelis, 1964	
C. abyssinica		34±	Hedberg, 1961	
C. eminii		34	Hedberg, 1961	
***Codonopsis* Wall.**				
C. clematidea	8		Khatoon & Ali, 1993	8
C. lanceolata		16	Nishikawa, 1985; Yoo & Lee, 1989; Ge & Wang, 1990; Probatova et al., 2011	
C. obtusa	8	16	Krasnoborov et al., 1980	
C. ovata	8	16	Gadella, 1966; Khatoon & Ali, 1993	
C. minima (= *C. ussuriensis*)	8		Jee et al., 1989	
		16	Yoo & Lee, 1989	
C. pilosula		16	Gadella, 1966; Sui et al., 1985; Yoo & Lee, 1989; Chen et al., 1991; Cheng et al., 1991	
C. rotundifolia var. *angustifolia* (= *C. bomiensis*)		16	Gadella, 1966	
C. ussuriensis		16	Yoo & Lee, 1989	
***Cyananthus* Wall. ex Benth.**				
C. cordifolius		14	Chen et al., 2014	5, 6, 7
C. delavayi		12	Chen et al., 2014	
C. hookeri		12, 24	Chen et al., 2014	
C. flavus		12	Chen et al., 2014	
C. formosus		12	Chen et al., 2014	
C. inflatus		14, 28	Hong & Ma, 1991; Chen et al., 2014	
C. lichiangensis		14	Chen et al., 2014	
C. incanus		10	Chen et al., 2014	
C. lobatus		14	Chen et al., 2014	
C. longiflorus		12	Chen et al., 2014	
C. macrocalyx		10	Chen et al., 2014	
C. microphyllus		14	Kumar & Chauhan, 1975; Chen et al., 2014	
C. pedunculatus		14	Chen et al., 2014	
***Cyclocodon* Griff. ex Hook. f. & Thomson**				
C. axillaris (*C. lancifolius*)		18	Zhang et al., 2010	9
C. parviflorus		18	Zhang et al., 2010	
***Echinocodon* D. Y. Hong**				
E. draco (Pamp.) D.Y. Hong (*E. lobophyllus*)	8		Hong, 1984	8

TABLE II 4:1 Chromosome Numbers in the Platycodonoid Group (= Tribe Cyanantheae) of the Campanulaceae s. s.—cont'd

Taxa	n	2n	References	Basic Number (x) Inferred
***Himalacodon* D. Y. Hong & Q. Wang (*Codonopsis* subg. *Obconicicapsula* D. Y. Hong)**				
H. dicentrifolius		14	Wang et al., 2014a	7
***Ostrowskia* Regel**				
O. magnifica		34	Zakharyeva & Astanova, 1969	17
***Pankycodon* D. Y. Hong & X. T. Ma (*Codonopsis* subg. *Codonopsis*, p. p.)**				
P. purpureus		14	Wang et al., 2014a	7
***Platycodon* (Jacq.) A. DC.**				
P. grandiflorus	9	18	Chen et al., 2013; Gadella, 1966 ; Ge et al., 1987; Ge & Li, 1989 ; Kihara et al., 1931; Li, 1986a & 1986b; Ma et al., 1984; Nishikawa, 1986; Pandey & Pal, 1980a; Pandey & Pal, 1980b; Probatova et al., 2008 & 2011; Sugiura, 1942; Suzuka & Koriba, 1949	9
***Pseudocodon* D. Y. Hong & H. Sun (*Codonopsis* subg. *Pseudocodonopsis* Kom.)**				
P. convolvulaceus subsp. *forrestii*		14	Wang et al., 2014a	7
P. hirsutus		14	Wang et al., 2014a	

TABLE II 4:2 Chromosome Numbers of the Tribe Campanuleae

Genera	n	2n	Basic Number (x) Inferred
Adenophora Fisch.	17, 18, 34	34, 36, 68, 102	17
Asyneuma Griseb. ex Schenk	16, 24	20, 24, 30, 32, 34, 48, 56,	10, 12, 15, 16, 17
Astrocodon Fed.		30, 34	15, 17
Azorina Feer		56	28
Campanula L.	8, 9, 10, 12, 13, 14, 15, 16, 17, 24, 25, 26, 34, 40, 45, 51	14, 16, 18, 20, 22, 24, 26, 28, 30, 32, 34, 40, 46, 48, 50, 52, 56, 58, 68, 80, 90, 102	7, 8, 9, 10, 11, 12, 13, 15, 17
Diosphaera Buser	16	34	16, 17
Edraianthus (A. DC.) A. DC.	16	32	16
Gadellia Shulkina		34	17
Githopsis Nutt.	9, 10, 18, 19, 20		9, 10, 19
Hanabusaya Nakai		34	17
Jasione L.	6, 9, 12, 18, 24	12, 14, 18, 24, 28, 36, 60	6, 7, 9
Legousia Durande		20	10
Michauxia L'Hér.		28, 30, 34	14, 15, 17
Musschia Dumort.	16	32	16
Peracarpa Hook. f. & Thomson		28, 30	14, 15
Petromarula Vent. ex R. Hedw.	15	30	15
Physoplexis (Endl.) Schur	17	34	17
Phyteuma L.	11, 12, 17	16, 18, 20, 22, 24, 26, 28	8, 9, 10, 11, 12, 13, 14, 17
Symphyandra A. DC.		34	17
Trachelium L.	16, 17	32, 34	16, 17
Triodanis Raf.		26, 30, 56	13, 15

TABLE II 4:3 Chromosome Numbers in the Tribe Wahlenbergieae

Genera	n	2n	Basic Number (x) Inferred
Feeria Buser		34	17
Gunillaea Thulin		18	9
Nesocodon Thulin		34	17
Wahlenbergia Schrad. ex Roth	8, 9, 11, 18, 27, 36	14, 16, 18, 36, 42, 54, 72, 90	7, 8, 9

= 9, confirming the inference by Raven (1975). He (Hong, 1990) also indicates that the ancient basic number of the Campanulales (sense of Cronquist, 1981) is x = 7.

Therefore, x = 7 is the ancient basic chromosome number for the Campanulaceae s.s., and of course also for the Cyanantheae including *Codonopsis* and its allies. Our recent study on evolution of characters in the platycodonoid group (= the tribe Cyanantheae) achieved a result that chromosome number 2n = 14 is favored as primitive (Wang et al. 2014b). In the family the general evolutionary trend of the chromosome number is increase, from x = 7 going upwards to x = 17. The only two exceptions are the genus *Jasione*, where the basic chromosome number has gone down to x = 6, and *Cyananthus*, where it has gone down to x = 5 via x = 6. Chen et al. (2014, fig. 34) shows us a molecular tree indicating that the species with 2n = 14 in *Cyananthus* are at basal position.

Now I am addressing the remaining question, i. e. how has x = 17 originated in the Cyanantheae. The chromosome number 2n = 34 in *Canarina* and *Ostrowskia* is not primitive one, since these two genera are nested in a clade consisting of *Cyclocodon*, *Echinocodon* and *Platycodon*, where the chromosome number of *Cyclocodon* and *Platycodon* is 2n = 18, while that in *Echinocodon* is 2n = 16 (Wang et al., 2014b: fig. 6). Therefore, it is highly possible that 2n = 34 is derived from an ancestral hybridization between the ancestor of *Echinocodon* (x = 8) and that of *Platycodon* (x = 9) or *Cyclocodon* (x = 9).

Chapter 3

Systematics and Evolution

CIRCUMSCRIPTION OF *CODONOPSIS* AND ITS RELATIONSHIPS WITH ALLIED GENERA

In the section "Circumscription of *Codonopsis*" of Chapter I we have reviewed the history of frequent changes in the circumscription of *Codonopsis*. We have indicated that 10 generic names have been involved in this issue. For the history of change of the circumscription of *Codonopsis* we have to mention the following references. Hooker and Thomson (1857) recognized *Codonopsis*, *Campanumoea*, *Leptocodon* and *Cyclocodon*, and pulled back the specific names once placed in *Glossocomia*, *Campanula*, and *Wahlenbergia*, into *Codonopsis*. Clarke's (1881) circumscription is nearly the same as Hooker and Thomson's (1857), but he placed *Cyclocodon* in *Campanumoea* as a section. Komarov (1908) gave a list of names and synonyms of *Codonopsis* and allied genera; he placed back all three species once placed in *Campanula* by Sprengel (1825) into *Codonopsis*, treated all specific names in *Glossocomia* as synonyms in *Codonopsis*, transferred six species once placed in *Wahlenbergia* into *Codonopsis* but leaving *Wahlenbergia dicentrifolia* (?) C. B. Clarke (1881) untouched, which was placed in *Codonopsis* by Smith (1912). He (Komorov, 1908) recognized *Campanumoea* and *Leptocodon*, but not *Cyclocodon* Griff. ex Hook. f. & Thomson, since he just enumerated *Campanumoea parviflora* and *Campanumoea celebica*. Moeliono (1960) merged *Campanumoea* s. l. (including *Cyclocodon*) with *Codonopsis*. Hong (1983) still recognized *Campanumoea* s. l. as an independent genus. Shen and Hong (in Hong, 1983) divided *Codonopsis* into three subgenera, subgs. *Codonopsis*, *Pseudocodonopsis* Kom. (1908) and *Obconicicapsula* D. Y. Hong (1980). Hong and Pan (1998) restored *Cyclocodon* at generic state based on observations on pollen and seed-coat morphology, combined with external morphology. Grey-Wilson (1990) merged *Leptocodon* Lem. with *Codonopsis*. The controversial views are clearly expressed in Hong et al. (2011, in notes), where Hong recognizes *Campanumoea* and *Leptocodon*, while the two co-authors, Lammers and Klein, argue for a broader sense of *Codonopsis*.

Recently I found that Pampanini (1910) described a new species from Hubei Province of China under *Codonopsis*, *C. draco* Pamp. Due to his fine figure it is easily recognized as conspecific with *Echinocodon lobophylla* D. Y. Hong (1984), also from Hubei. The genus *Echinocodon* D. Y. Hong differs distinctly from *Codonopsis* in gross morphology, and pollen morphology, and is not closely related to the latter according to our phylogenetic analysis (Figure III 1:1).

Eddie et al.'s (2003) molecular work is heuristic. They used ITS sequences for analysing phylogenetic relationships in the Campanulaceae s. s. with a result showing that *Codonopsis dicentrifolia* (C. B. Clarke) W. W. Smith, the only species of *Codonopsis* subg. *Obconicicapsula* D. Y. Hong (1980), forms a clade with *Cyananthus lobatus* Wall. ex Benth. Their work raises a question: *Codonopsis* even in Hong's (1983; Hong in Hong et al., 2011) narrow sense may well be polyphyletic.

Therefore, the circumscription of *Codonopsis* involves three questions: 1) Should *Campanumoea* and *Leptocodon* be merged with *Codonopsis* or kept separate? 2) Is the restoration of *Cyclocodon* as generic status justifiable or not? 3) *Codonopsis* in the strict sense (Hong 1983; Hong in Hong et al., 2011) is polyphyletic or monophyletic?

We (Wang et al., 2013) made phylogenetic analyses of the platycodonoid group (Eddie et al., 2003; Haberle et al., 2009) based on 41 taxa representing all genera of the group except *Ostrowskia*, using sequences of five DNA fragments from the chloroplast and nuclear genomes. The result shows that our molecular tree had very high resolution (Figure III 1:1). In the tree the in-group forms four highly supported clades, *Codonopsis* in the broad sense forms two clades with strong support, while the two species of *Cyclocodon* and monospecific genus, *Echinocodon* fall in the *Platycodon* clade mostly possessing colporate pollen. From the tree it is clear that *Cyclocodon* is phylogenetically far from *Codonopsis*. Therefore, the genus *Cyclocodon* differs distinctly from the genus *Codonopsis* not only in pollen morphology, seed-coat, and external morphology. The genus *Echinocodon* is also not closely related to *Codonopsis*. *Campanumoea* and *Leptocodon* nest in the core *Codonopsis* clade, and thus they should be merged with *Codonopsis*, and that *Codonopsis* subg. *Pseudocodonopsis* Kom. (1908), subg. *Obconicicapsula* D. Y. Hong (1980) and *Codonopsis purpurea* Wall. forms a clade with full support. All the members of this short-colpus clade have very short apertures and have no disc, and thus they should be separated from the *Codonopsis* clade.

A Monograph of *Codonopsis* and Allied Genera (Campanulaceae). http://dx.doi.org/10.1016/B978-0-12-801933-7.00003-9
Copyright © 2015 China Science Publishing & Media Ltd. Published by Elsevier Inc. All rights reserved.

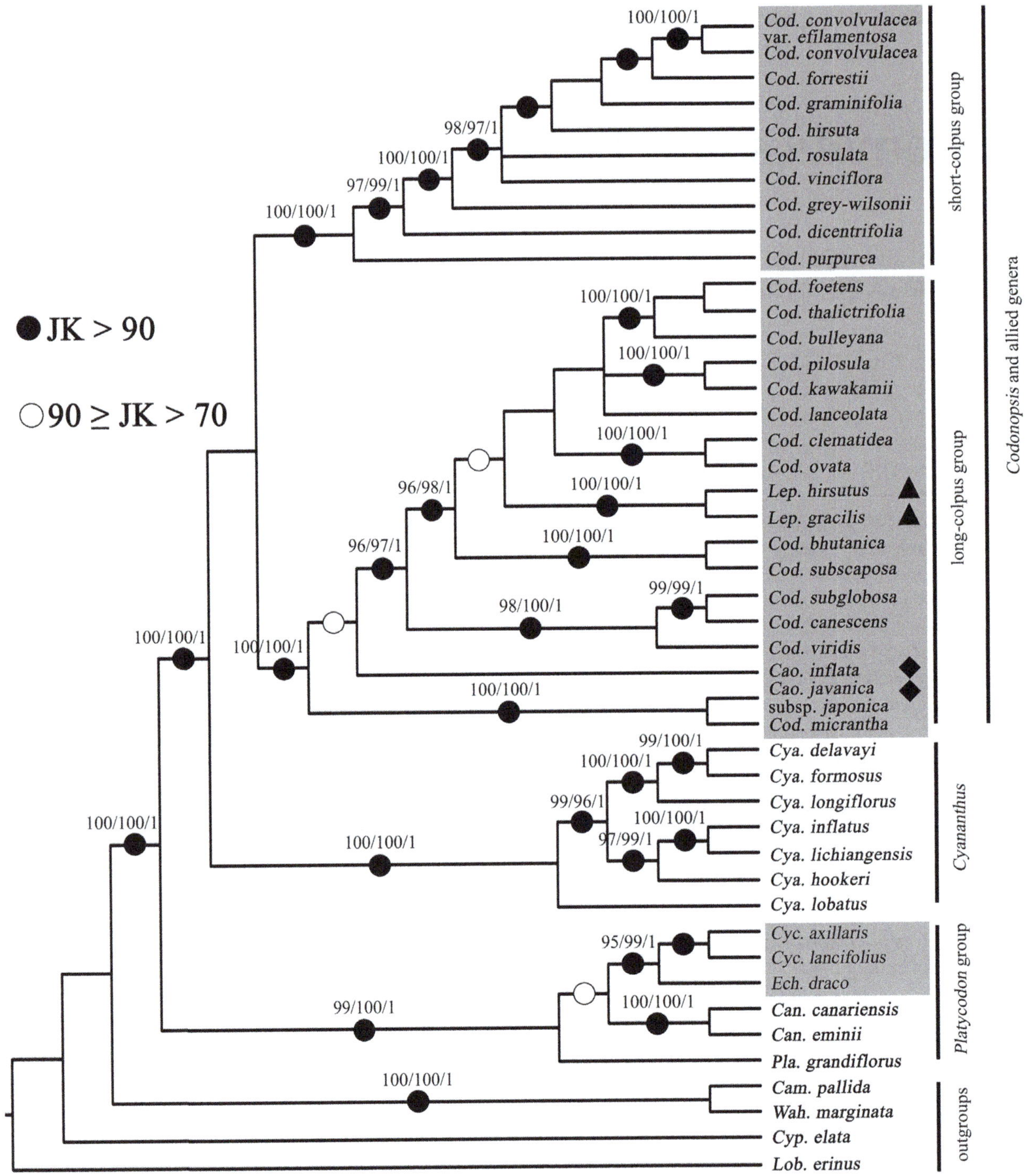

FIGURE III 1:1 The phylogenetic relationship within the Cyanantheae (= platycodonoid group). The 50% majority rule consensus tree combining ITS and four chloroplast DNA markers and its phylogram derived from the Bayesian analysis. Ranges of parsimony jackknife support (JK) above 70 are indicated with dots at the branches. JK of parsimony (> 90%), bootstrap value of likelihood (> 95%), and Bayesian posterior probability values (PP) (= 1.00) are reported above branches (MP/ML/BI). *Cam* = *Campanula*; *Can* = *Canarina*; *Cao* = *Campanumoea*; *Cod* = *Codonopsis*; *Cya* = *Cyananthus*; *Cyc* = *Cyclocodon*; *Cyp* = *Cyphia*; *Ech* = *Echinocodon*; *Lep* = *Leptocodon*; *Lob* = *Lobelia*; *Pla* = *Platycodon*; *Wah* = *Wahlenbergia*. The shaded is *Codonopsis* and allied genera *(from Wang et al., 2013, with minor modification)*.

To address relationships within the short-colpate pollen clade we constructed another molecular tree based on expanded sampling (from 10 to 24 for the clade) (Figure III 1:2). In the tree all the members of *Codonopsis* subg. *Pseudocodonopsis* form a clade with full support. Within the pseudocodonopsis clade (= short-colpus clade), our results also show that subg. *Pseudocodonopsis*, subg. *Obconicicapsula*, and *C. purpurea* should not be circumscribed in a single genus. Subg. *Pseudocodonopsis* is a highly supported monophyletic group (MP/ML/BI = 100/100/1). The subgenus possesses several particular features: roots mostly globose; corolla cleft nearly to the base and rotate; base of filaments dilated into a semiorbicular lamella with lanate hairs at the periphery, and pollen with long spinules on the exine (≥ 2 μm). All of these characteristics are distinctly different from those of subg. *Obconicicapsula*, *C. purpurea*, and the rest of subg. *Codonopsis*. Moreover, subg. *Pseudocodonopsis*, subg. *Obconicicapsula*, and *C. purpurea* are polyphyletic in the ITS tree (Wang et al., 2013), where subg. *Pseudocodonopsis* is sister to the genus *Cyananthus*. Thus, it is better to raise subg. *Pseudocodonopsis* to generic status.

As shown in our molecular tree, subg. *Obconicicapsula* and *C. purpurea* are paraphyletic with strong support (MP/ML/BI = 96/100/1, Figure III 1:2). They also differ from each other in a number of morphological features. In subg. *Obconicicapsula*, the leaves on the main stem are alternate (vs. all opposite in *C. purpurea*); the hypanthium is entirely adnate to ovary (vs. free from the ovary at the top); the ovary/capsule is obconical (vs. semiglobose); the seeds are ovoid, slightly compressed and wingless (vs. lunate, strongly compressed, and wide-winged), and the pollen colpus membrane is granulose (vs. smooth). Therefore, to lump these two phylogenetically paraphyletic and morphologically distinct taxa into a single genus is quite unjustifiable.

Therefore, we separated subg. *Pseudocodonopsis*, subg. *Obconicicapsula*, and *C. purpurea* from *Codonopsis* as three independent genera, i.e. *Pseudocodon* D. Y. Hong & H. Sun, *Himalacodon* D. Y. Hong & Q. Wang, and *Pankycodon* D. Y. Hong & X. T. Ma (Wang et al., 2014a). As a result of our studies (Figures III 1:1 & 2), *Codonopsis* in the broadest sense (Moeliono, 1960; Lammers & Klein in Hong et al., 2011) comprises the following six genera, *Codonopsis*, *Cyclocodon*, *Echinocodon*, *Himalacodon*, *Pankycodon*, and *Pseudocodon*. The following key demonstrates the distinctness of the new genera from *Codonopsis* and between the three new genera in morphology, palynology, and cytology.

Key to Genera of *Codonopsis* and Its Allies

1a. Flowers mostly 6-merous; fruit a berry; calyx lobes always dentate or ramiform; pollen 3-colporate; chromosome number 2n = 18..5. *Cyclocodon* [*Codonopsis* (Moeliono, 1960)]

1b. Flowers mostly 5-merous, less often 4-merous, ovary 3-locular; fruit mostly a capsule, rarely a berry; calyx lobes entire, rarely with one obscure tooth on each side, or with spinose lobules; pollen 5–9- colpate; chromosome number 2n = 14 or 16.

- **2a.** Flowers 3–5- but mostly 4-merous; calyx lobes lobed, with 2–4 spinose lobules; spinules of pollen exine divided at base; chromosome number 2n = 16..6. *Echinocodon*
- **2b.** Flowers 5-merous; calyx lobes entire or denticulate; spinules of pollen exine entire; chromosome number 2n = 14 or 16.
  - **3a.** Disc present; pollen apertures long or medium-long (colpus length/polar axis length ≥ 0.54); chromosome number 2n = 16..1. *Codonopsis*
  - **3b.** Disc absent; pollen apertures short (colpus length/polar axis length ≤ 0.4); chromosome number 2n = 14.
    - **4a.** Corolla rotate, cleft to near base; roots globose, rarely elongated and constricted at middle part; filaments strongly dilated into a semi-orbicular lamella and lanate at base; pollen exine spinules ≥ 2 μm long...4. *Pseudocodon* (*Codonopsis* subg. *Pseudocodonopsis* Kom.)
    - **4b.** Corolla campanulate, shallowly lobed up to middle; roots elongated and attenuate; filaments slightly dilated and narrowly cuneate at base, glabrous; pollen exine spinules ≤ 1 μm long.
      - **5a.** Leaves on main stem alternate; hypanthium entirely adnate to ovary; capsule base obconical; seeds ovoid, slightly compressed, wingless...........................2. *Himalacodon* (*Codonopsis* subg. *Obconicicapsula* D. Y. Hong)
      - **5b.** Leaves all opposite; hypanthium free from top of ovary; capsule base semiglobose; seeds lunate, strongly compressed, widely winged...3. *Pankycodon* (*Codonopsis* *purpurea* Wall.)

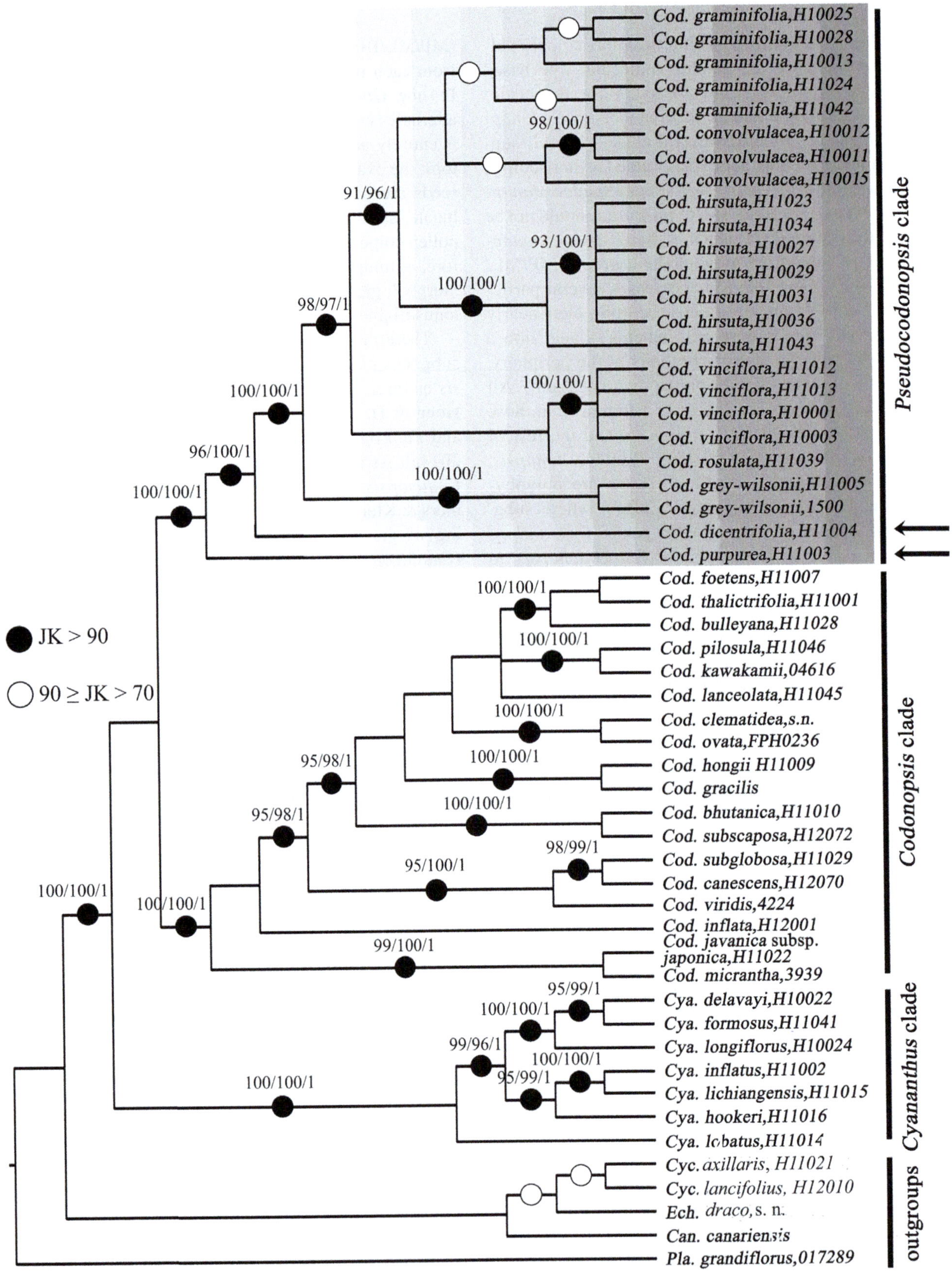

FIGURE III 1:2 Molecular phylogenetic analysis (50% majority rule consensus tree) of *Codonopsis* s. l. (Lammers & Klein in Hong et al., 2011) derived from the Bayesian analysis combining four chloroplast DNA fragments. Ranges of parsimony jackknife support (JK) above 70 are indicated with dots at the branches. JK of parsimony, bootstrap value of likelihood, and Bayesian posterior probability values (MP / ML / BI) are reported only above branches with black dot. The shaded is *Codonopsis* with short-colpate pollen. The two arrows indicate paraphyletic clades (see text). *Pla = Platycodon; Can = Canarina; Ech = Echinocodon; Cyc = Cyclocodon; Cya = Cyananthus; Cod = Codonopsis (from Wang et al., 2014a, with modification).*

A THREE-TRIBAL CLASSIFICATION SYSTEM OF THE CAMPANULACEAE S. S. AND THE SYSTEMATIC POSITION OF *CODONOPSIS* AND ALLIED GENERA

The circumscription of the Campanulaceae has been changed greatly and is still controversial among authors. Schönland's (1889) concept is the broadest one, including Pentaphragmataceae and Sphenocleaceae. It was prevailing until 1981, when Cronquist (1981) excluded these two families from the Campanulaceae. Cosner et al. (1994) constructed a *rbc*L tree, which shows that *Pentaphragma* and *Sphenoclea* are not members of the Campanulaceae s. l. (Cronquist's sense), and that the Campanulaceae (= Campanulales) consists of five monophyletic clades. Gustafsson and Bremer (1995) made an intensive cladistic analysis based on morphological data. Their conclusion is that the campanulaceous clade includes five monophyletic families: Campanulaceae, Cyphiaceae, Cyphocarpaceae, Lobeliaceae, and Nemacladaceae. However, on how to treat these five entities, botanists are divided into two groups. Takhtajan (1987, 1997), Lammers (1992), Cosner et al. (1994), Gustafsson and Bremer (1995), Gustafsson et al. (1996), Lundberg and Bremer (2003), and Roquet et al. (2009) hold a narrow sense, recognizing the Campanulaceae s. s. as an independent family. However, Cronquist (1981), Thorne (1998), Lammers (1998, 2007), Brummit (2007), and APG III (2009) hold a broad sense, recognizing the Campanuloideae within the Campanulaceae s. l. As these five groups are all monophyletic, we think that how to treat them, in a narrow sense or in a broad sense, is just a matter of taste. Furthermore, the Campanulaceae s. s. clearly differs morphologically from the remaining four in having the corolla actinomorphic (vs. zygomorphic, rarely slightly zygomorphic) and the ovary/fruit (2-)3-5(-10)-locular, vs. (1-)2-locular (Lammers, 2007 a). Also, pollen grains of the family always have secondary sculpture, spinules or rarely verrucae, and thus they are distinctly different from those in the other four families which lack secondary sculpturing (Dunbar 1975 a, b & 1984). Therefore, we like to use the Campanulaceae in the narrow sense.

However, intrafamilial subdivision and arrangement of genera have been controversial among authors. De Candolle (1830) first divided the family into two subtribes based on dehiscent pattern of fruit (apically or laterally dehiscent), but nine years later he (de Candolle, 1839) divided it into three tribes based on number of ovules and the dehiscent pattern of fruit, i. e. Wahlenbergieae, Campanuleae, and Merciereae. He placed *Codonopsis* and allied genera, *Platycodon*, *Campanumoea*, together with *Wahlenbergia*. Bentham & Hooker (1876) divided the Campanuleae (= Campanulaceae s. s.) into five groups based on the fruit features. They placed *Platycodon*, *Codonopsis*, *Cyananthus* and *Leptocodon* possessing a loculicidal capsule together with *Wahlenbergia* in the first group, whilst *Campanumoea* and *Canarina* with baccate fruit were placed together with *Peracarpa* and *Pentaphragma* in the second group. Schönland (1889) divided the tribe Campanulieae (= Campanulaceae s. s.) into three subtribes based on features of ovary and fruit, Campanulinae, Wahlenberginae and Platycodinae. He placed *Codonopsis*, *Campanumoea*, *Cyananthus*, *Leptocodon* together with *Wahlenbergia* in the Wahlenberginae, while placed *Canarina* and *Ostrowskia* together with *Adenophora* in the Campanulinae. However, this arrangement of genera is unjustified. For example, *Platycodon*, *Microcodon* and *Musschia* in the Platycodinae are distinctly different from each other in external morphology, and particularly in pollen morphology. Thulin (1975) also pointed out that all three subtribes of Schönland's system are heterogeneous in pollen morphology. However, Schönland's (1889) system is one of the most influential systems in the taxonomic history of the Campanulaceae s. s. Fedorov (1957) grouped the genera of the Campanulaceae s. s. in Soviet Union territory directly into eight tribes, of which six were new and five monogeneric. He did not accept Schönland's Platycodinae, but instead he placed *Platycodon* together with *Codonopsis* in the tribe Wahlenbergieae. Unfortunately, as Erdtman (1952) had already indicated, *Wahlenbergia* has porate pollen with poroid apertures, whereas *Platycodon* and *Codonopsis* possess colporate or colpate pollen with elongate apertures. Avetisjan (1969) and Dunbar (1975) made extensive observations on pollen grains of the Campanulaceae s. l. and related families. Their works confirmed Erdtman's (1952) observation on pollen morphology and revealed the great significance of pollen morphology in the systematics of the Campanulaceae. Thulin (1975) also strongly emphasized the significant value of pollen morphology for the taxonomy of the Campanulaceae, and suggested that it should be the first consideration for rearrangement of genera in the Campanulaceae. However, "based on the morphology of ovary and capsule", Kovanda (1978) still ignored the palynological data and divided the Campanulaceae s. s. into three subtribes, adopting Schönland's (1889) names of subtribes. Again, he placed *Codonopsis*, *Cyananthus*, *Campanumoea* possessing colpate pollen together with *Wahlenbergia* possessing porate pollen in the Wahlenberginae, while separated *Platycodon* from *Canarina* (both with colporate pollen). Essentially his system is the same as that of Schönland (1889).

Rosatti (1986) recognized two tribes, the Campanuleae and Wahlenbergieae Endl. (1858), as two subdivisions of the Campanuloideae (= Campanulaceae s. s.) in the Southeastern United States, and Yeo (1993) added his new tribe, Platycodoneae, as the third tribe of the Campanulaceae s. s. Therefore, according to Yeo the Campanulaceae s. s. consists of three subgroups, the Campanuleae, Wahlenbergieae and Platycodoneae, but his Platycodoneae includes only one genus, *Platycodon*.

Ignoring the achievement of palynological studies and overemphasizing the significance of fruit characters in the taxonomy of the family, Kolakovsky (1987, 1994) published his classification system of the Campanulaceae s. s. from the Old World. He divided the family into 22 tribes in four subfamilies, including Campanuloideae, two new subfamilies (Prismatocarpoideae and Canarinoideae), and Wahlenbergioideae as a new status. We should indicate that Kolakovsky established the new subfamily, Canarinoideae Kolak. with two genera, *Canarina* and *Campanumoea*, just because these two genera share baccate fruit. Again, he placed *Codonopsis*, *Platycodon*, and *Cyananthus* with colpate/coporate pollen in his tribe Wahlenbergieae (all the other genera have porate pollen). In Takhtajan's system (1997) of Campanulaceae s. s., pollen morphology was taken into consideration for his classification. Takhtajan divided the family also into four subfamilies, the Campanuloideae, Canarinoideae, Cyananthoideae, and Ostrowskioideae. Kolakovsky's (1994) system and Takhtajan's (1997) one are different in subfamily names and content, though they both recognized four subfamilies. Takhtajan (1997) maintained only *Canarina* in the Canarinoideae and established a new subfamily, Ostrowskioideae. He separated for the first time *Platycodon*, *Codonopsis*, *Leptocodon* from *Wahlenbergia*, and placed them together with *Cyananthus* in the Cyananthoideae. Therefore, he made a considerable progress toward a natural classification of the Campanulaceae s. s. He placed *Wahlenbergia* and its relatives, however, together with *Campanula* and its allies in the Campanuloideae. Although they both have porate pollen, the former has a loculicidal capsule, which is dehiscent by apical valves, or pores or opercula (above the insertion of calyx lobes), whereas the latter has a poricidal capsule, which is dehiscent by lateral valves or pores (below the insertion of the calyx lobes), or dry and indehiscent fruit. Brummit (2007) adopted Takhtajan's four-subdivisions system for the Campanulaceae s. s. Lammers (2007a) did not make subdivision or adopt any classification system of the Campanuloideae (= Campanulaceae s. s.).

From the above review it is clear that the subdivision of the Campanulaceae s. s. is very controversial, as Kovanda (1978) has already clearly expressed this problem, "The family Campanulaceae...... is rather natural and homogeneous but its subdivision presents serious problems......".

Recent studies on molecular systematics have obtained some insights into phylogenetic relationships within the Campanulaceae s. s. Eddie et al. (2003) constructed a molecular tree using ITS sequences of nuclear ribosomal DNA, in which three major clades could be clearly recognized: the patycodonoids with colpate or colporate pollen, the wahlenbergioids with porate pollen and an apically loculicidal capsule, and the campanuloids also with porate pollen but with a laterally poricidal capsule. Haberle et al. (2009) used DNA sequences of three chloroplast genes, *atp*B, *mat*K and *rbc*L, for constructing phylogenetic relationships of the Campanulaceae s.s., giving the same result that the family comprises three major groups, the platycodonoids, wahlenbergioids and campanuloids. Their platycodonoids comprises *Platycodon*, *Codonopsis*, *Leptocodon*, *Cyananthus* and *Canarina*. Although the names of the three groups are the same as those of Schönland's (1889) and those proposed by Rosatti (1986) and Yeo (1993), the contents of the three groups are different, particularly the relationship between *Platycodon*, *Codonopsis* and their allies on one side and *Wahlenbergia* on the other.

Therefore, despite some new insights into the subdivision of the Campanulaceae s. s. from molecular data, there is no stable classification system of the family. Recently Hong and Wang (2015) eastablished a new classification system based on molecular phylogenetic analysis, combining with the data from external morphology, pollen morphology, and cytology. As a result, the following three points should be emphasized in particular.

Confirmation of Three Primary Subdivisions of the Campanulaceae s. s.

Our result shows that the family comprises three clades with high support, i.e. campanuloids (MP/ML/BI = 77/95/0.97), wahlenbergioids (MP/ML/BI = 99/100/1), and platycodonoids (MP/ML/BI = 100/100/1) (Figure III 2:1). The subdivision of three groups of the family is also highly supported by pollen morphology and external morphology. The pollen in the platycodonoids is colpate or colporate with elongate apertures (Erdtman 1952; Avetisja, 1969; Dunbar, 1975; Hong & Pan, 2012), while that in the other two groups is porate, with poroid apertures. The fruit of the campanuloids is poricidal, dehiscent by lateral pores (below the insertion of calyx lobes), while that in the other two groups is loculicidal, dehiscent by apical valves (above the insertion of calyx lobes). Therefore, three subdivisions in the Campanulaceae s. s. is natural, whereas the treatments of four subfamilies by Kolakovsky (1987, 1994) and by Takhtajan (1997), or of four tribes by Brummit (2007), can not be justified.

Our conclusion of the three subdivisions in the Campanulaceae s. s. seems to be consistent with Schönland's (1889) one, and Rosatti's (1986) and Yeo's (1993) joint one, but ours is different from theirs in content. For example, they placed *Platycodon*, *Codonopsis*, etc., possessing colpate or colporate pollen together with *Wahlenbergia* possessing porate pollen. That is to say, they mixed the platycodonoids and wahlenbergioids.

On Baccate Fruit and the Canarinoideae Kolak.

Kolakovsky (1987, 1994) established a new subfamily in the Campanulaceae s. s., Canarinoideae, which was followed by Takhtajan (1997). The former placed *Canarina* and *Campanumoea* together in the subfamily, while Takhtajan included only *Canarina*. They two recognized the subfamily just because the two genera have baccate fruit. However, we (Wang et al., 2014b) reached the conclusion that the baccate fruit is a derived character that has evolved

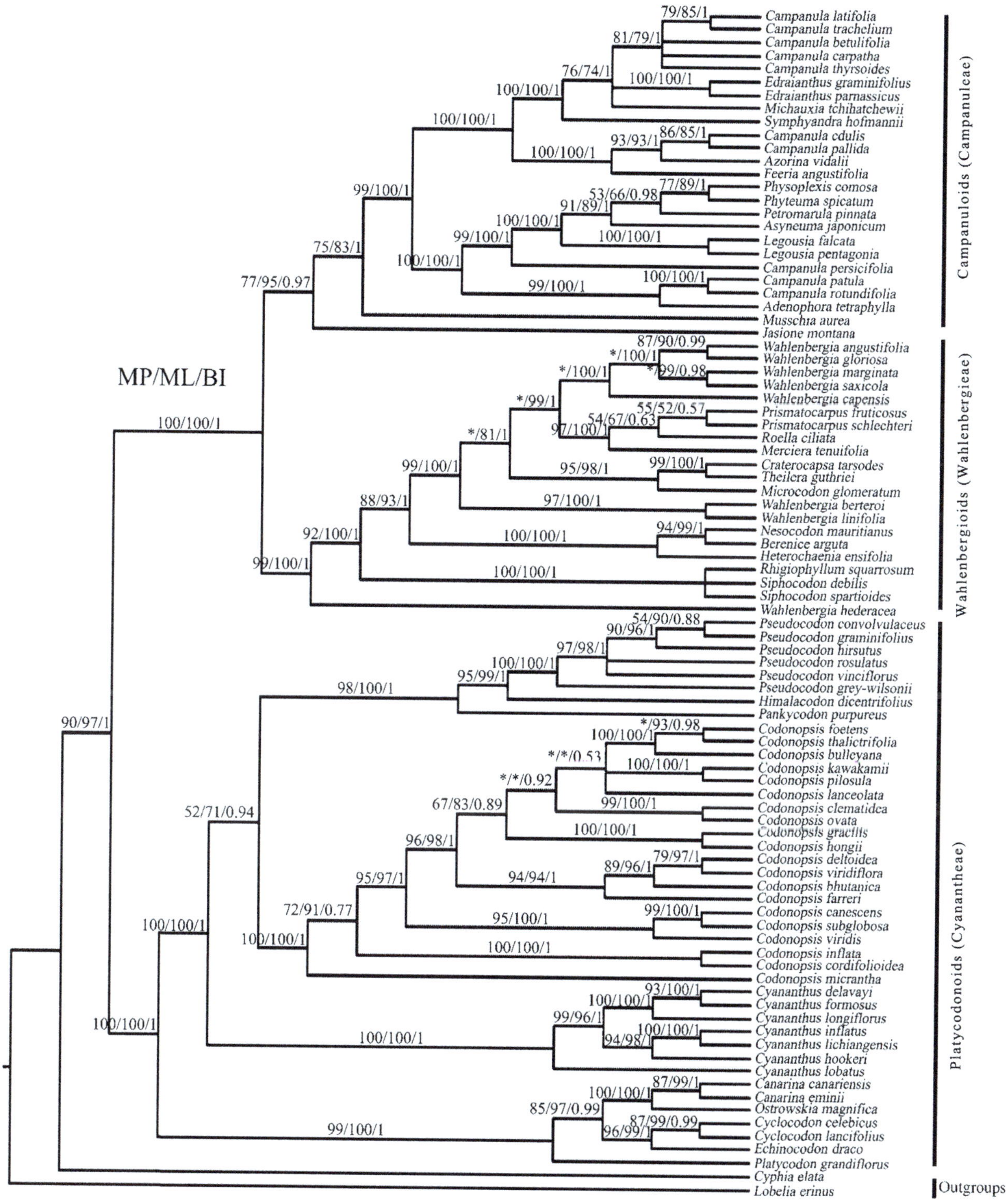

FIGURE III 2:1 50% majority rule consensus tree derived from the Bayesian analysis of four chloroplast DNA fragments. Jackknife support of parsimony/bootstrap value of likelihood/Bayesian posterior probability values are reported above branches (maximum parsimony/maximum likelihood/Bayesian inference (MP/ML/BI)). * Support values lower than 50 of parsimony and likelihood. *From Hong & Wang, 2015, with minor modification.*

independently in *Canarina*, *Cyclocodon* and *Campanumoea* s.s. (= *Codonopsis*), and the appearance time of the berry is dated as no more than 4.59 mya. Our another work (Wang et al. 2013) showed that the two species of *Campanumoea* s. s. with a berry (*C. javanica* and *C. inflata*) even nested within the *Codonopsis* clade in the molecular tree with high support, indicating that *Campanumoea* s. s. with a berry and *Codonopsis* with a capsule are congeneric. It is apparent that there is no reason to accept the Canarinoideae, because the establishment of the subfamily is not based on solid phylogenetic evidence.

On *Ostrowskia* and the Ostrowskioideae Takht.

The genus *Ostrowskia* is monospecific and endemic to a small area in Middle Asia. Due to its unique characteristics, it was treated as a monogeneric tribe by Fedorov (1957), Ostrowskieae Fed., which was raised to subfamilial rank by Takhtajan (1997), Ostrowskioideae Takht. Its flowers are in a terminal pyramidal panicle, (5–) 7 (–9) - merous, fruit capsular, laterally dehiscent by several parallel slits (Lammers, 2007a), seeds with a large air sac (Chapter II 2), and chromosome number 2n = 34 (Zakharyeva & Astanova, 1968). However, it has 6–7-colpate pollen, similar to that in the *Cyananthus* / *Codonopsis* group, and its chromosome number is the same as that in *Canarina*. Mansion and his co-author's work (Mansion et al., 2012) revealed that this genus is sister to the genus *Canarina*, and should belong to the platycodonoid group. The molecular phylogenetic analysis in Hong & Wang's (2015) work (Figure III 2:1) confirms Mansion et al.'s work, i.e., *Ostrowskia* is sister to *Canarina* and belongs to the tribe Cyanantheae (= the platycodonoid group).

Key to Tribes of the Campanulaceae s. s.

1a. Pollen colpate or colporate, with elongate apertures; fruit a loculicidal capsule dehiscent by apical valves above insertion of calyx lobes (exceptionally laterally dehiscent in *Ostrowskia*) or a berry ....Trib. 3. Cyanantheae.

1b. Pollen porate, with poroid apertures; fruit a loculicidal or poricidal capsule

- **2a.** Fruit a loculicidal capsule, dehiscent by 2–5 apical valves, less frequently dehiscent by pores or opercula, rarely by fissures or slits, very rarely dry and indehiscent ....Trib. 2. Wahlenbergieae
- **2b.** Fruit a poricidal capsule, variously dehiscent laterally below insertion of calyx lobes by pores or valves ....Trib. 1. Campanuleae

Tribe 1. **Campanuleae** Durmort., Fl. Belg. 58. 1827. Type: *Campanula* L.
Jasioneae Dumort., Fl. Belg.: 59 (1827). Type: *Jasione* L.
Phyteumeae Dumort., Fl. Belg.: 59 (1827). Type: *Phyteuma* L.
Peracarpeae Fed. in Komarov, Fl. URSS 24: 471 (1957). Type: *Peracarpa* Hook. f. & Thomson.
Michauxieae Fed. in Komarov, Fl. URSS 24: 472 (1957). Type: *Michauxia* L'Hér.
Edraiantheae Fed. in Komarov, Fl. URSS 24: 475 (1957). Type: *Edraianthus* A. DC.
Azorineae Kolak., Bot. Zhurn. 12: 1575 (1987). Type: *Azorina* Feer.
Muehlbergelleae Kolak., Bot. Zhurn. 12: 1575 (1987). Type: *Muehlbergella* Feer.
Musschieae Kolak., Bot. Zhurn. 12: 1575 (1987). Type: *Musschia* Dumort.

Pollen porate, with poroid apertures; fruit capsular and dehiscent by lateral pores or valves (below the insertion of calyx lobes), or dry, indehiscent. Chromosomes x = 6, 7, 8, 9, 10, 11, 13, 15, 17.

Genera included: *Adenophora* Fisch., *Asyneuma* Griseb. & Schenk, *Azorina* Feer, *Campanulastrum* Small, *Cryptocodon* Fed., *Cylindrocarpa* Regel, *Edraianthus* A. DC., *Favratia* Feer, *Githopsis* Nutt., *Hanabusaya* Nakai, *Heterocodon* Nutt., *Homocodon* D.Y. Hong, *Jasione* L., *Legousia* Durande, *Michauxia* L' Hér., *Muehlbergella* Feer, *Musschia* Dumort., *Peracarpa* Hook. f. & Thomson, *Petromarula* Vent. ex R. Hedw., *Phyteuma* L., *Physoplexis* (Endl.) Schur, *Sergia* Fed., *Trachelium* L., *Triodanis* Raf., *Zeugandra* P. H. *Davis*

Tribe 2. **Wahlenbergieae** Endl., Gen. Pl. 513. 1838. Wahlenbergioideae (Endl.) Kolak., Bot. Zhurn. 72(12): 1574. 1987, p. p. Type: *Wahlenbergia* Schrad. ex Roth.
Merciereae Mesin., Pl. Vasc. Gen.: Tab. Diagn. 238. 242. comm.. 150. 1839. Type: *Merciera* A. DC.
Prismatocarpoideae Kolak., Bot. Zhurn. 72(12): 1574. 1987. Type: *Prismatocarpus* L'Hér.

Pollen porate, with poroid apertures; fruit capsular and dehiscent by apical (above the insertion of calyx lobes)valves, pores or opercula. Chromosomes x = 7, 8, 9, 17.

Genera included: *Berenice* Tul., *Craterocapsa* Hilliard & B. L. Burtt, *Feeria* Buser, *Gunillaea* Thulin, *Heterochaenia* A. DC., *Merciera* A. DC., *Microcodon* A. DC., *Nemacodon* Thulin, *Nesocodon* Thulin, *Prismatocarpus* L' Hér., *Rhigiophyllum* Hochst., *Roella* L., *Siphocodon* Turcz., *Theilera* E. Phillips, *Traichelia* Vatke

Tribe 3. **Cyanantheae** Mesin., Pl. Vass. Gen.: Tab. Diagn. 273. comm.. 180. 1840. Type: *Cyananthus* Wall. ex Benth.
Canarineae Webb. & Berth., Hist. Nat. Iles Canaries 2(3): 1. 1844. Type: *Canarina* L.
Ostrowskieae Fed. in Komarov, Fl. URSS 24: 471. 1957. Type: *Ostrowskia* Regel.
Platycodoneae Yeo, Taxon 42: 109. 1993. Type : *Platycodon* A. DC.
Campanumoeae Kolak., Bot. Zhurn. 79: 111. 1994. Type: *Campanumoea* Bl.

Pollen colpate or colporate, with elongate apertures; fruit a loculicidal capsule dehiscent by apical valves (above the insertion of calyx lobes) (with an exception in *Ostrowskia*), or a berry. Chromosomes x = 5, 6, 7, 8, 9, 17.

Genera included: *Canarina* L., *Codonopsis* Wall., *Cyananthus* Wall. ex Benth. *Cyclocodon* Griff. ex Hook. f. & Thomson, *Echinocodon* D. Y. Hong, *Himalacodon* D. Y. Hong & Q. Wang, *Ostrowskia* Regel, *Pankycodon* D. Y. Hong & X.T.Ma, *Platycodon* A. DC., and *Pseudocodon* D. Y. Hong & H. Sun.

Subtribe 1. **Platycodoninae** Schönland in Engler & Prantl, Nat. Pflanzenfam. 4, 5: 59. 1889, p. p., quoad *Platycodon.* Type: *Platycodon* A. DC.

Pollen 5–6-colporate; fruit a 5-loculicidal capsule, with valves opposite to calyx lobes. Chromosomes 2n = 18.

Monogeneric subtribe.

Subtribe 2. **Canarininae** D. Y. Hong, J. Syst. Evol. 53(3): 208. 2015. *Canarineae* Webb. & Berth., Hist. Nat. Iles Canaries 2(3): 1. 1844. *Canarinoideae* Kolak., Bot. Zhurn. 72(12): 1574. 1987, p. p., quoad *Canarina* L. Type: *Canarina* L.

Pollen 3-colporate; fruit baccate. Chromosomes 2n = 34.

Monogeneric subtribe.

Subtribe 3. **Cyclocodoninae** D. Y. Hong, J. Syst. Evol. 53(3): 208. 2015. Type: *Cyclocodon* Griff. ex Hook. f. & Thomson.

Pollen 3-colporate; fruit baccate; ovary inferior to corolla but superior or half inferior to calyx; flowers (4–5) 6-merous; calyx lobes dentate or ramiform. Chromosomes 2n = 18.

Monogeneric subtribe.

Subtribe 4. **Echinocodoninae** D. Y. Hong, J. Syst. Evol. 53(3): 208. 2015. Type: *Echinocodon* D. Y. Hong (1984).

Pollen 4–5-colpate, exine spinules divided at base; flowers (3–) 4 (–5)-merous; fruit capsular, 3-loculicidal; calyx lobes ramiform with 2-4 spinose lobules. Chromosomes 2n = 16.

Monogeneric subtribe.

Subtribe 5. **Ostrowskiinae** D. Y. Hong, J. Syst. Evol. 53(3): 208. 2015. *Ostrowskieae* Fed. in Komarov, Fl. URSS 24: 126. 1957. *Ostrowskioideae* Takht., Divers. Class. Fl. Pl. 408. 1997. Type: *Ostrowskia* Regel

Pollen 6–7-colopate, exine verrucose, verrucae rounded, ca. 2 μm high; flowers (5–) 7 (–9)-merous; fruit capsular, dehiscent by lateral oblong slits below the insertion of calyx lobes. Chromosomes 2n = 34.

Monogeneric.

Subtribe 6. **Cyananthinae** D. Y. Hong, J. Syst. Evol. 53(3): 208. 2015. *Cyanantheae* D. Y. Hong & Y. S. Lian in Hong, Fl. Reipubl. Popularis Sin. 73(2): 183. 1983. Cyananthoideae Takht., Divers. Class. Fl. Pl. 408. 1997, nom. invalid. Type: *Cyananthus* Wall. ex Benth.

Campanumoeae Kolak., Bot. Zhurn. 79(1): 111. 1994. Type: *Campanumoea* Bl.

Pollen 5–10-colpate; ovary superior or inferior; fruit a 3-loculicidal capsule, or a berry. Chromosomes 2n = 10, 12, 14, 16.

Genera included: *Cyananthus*, *Codonopsis*, *Himalacodon*, *Pankycodon*, and *Pseudocodon*.

EVOLUTION AND BIOGEOGRAPHY OF THE CYANANTHEAE

The tribe Cyanantheae (= platycodonoid group) is strongly supported as a monophyletic group (Eddie et al., 2003; Borsch et al., 2009; Haberle et al., 2009; Wang et al., 2013). Recent studies (Wang et al., 2013; Wang et al., 2014a & b; Hong & Wang, 2015) indicate that the tribe consists of ten genera: *Canarina*, *Codonopsis*, *Cyananthus*, *Cyclocodon*, *Echinocodon*, *Himalacodon* (= *Codonopsis* subg. *Obconicicapsula* D. Y. Hong), *Ostrowskia*, *Pankycodon* (*Codonopsis purpurea* Wall.), *Platycodon*, and *Pseudocodon* (*Codonopsis* subg. *Pseudocodonopsis* Kom.) (Figure III 3:1). The distribution patterns of these ten genera are very interesting (Table III 3:1). *Canarina* with three species is widely disjunct across the Sahara Desert with two species in East Africa and one in the Canary Islands. It is widely separated geographically from the other nine genera, which occur in Middle and East Asia, and New Guinea. *Cyclocodon* also with three species is distributed in Southeast Asia and New Guinea. The monospecific genus *Platycodon* is found in East Asia. *Echinocodon*, also with only one species, is confined to a small area of Central China. *Ostrowskia* is also monospecific and confined to the Pamir-Alai in Middle Asia. These five monospecific or oligospecific genera form a strongly supported monophyletic clade in our recent molecular trees (Wang et al., 2013; Wang et al. 2014b). They grow in lowlands from sea level to an altitude of 2600 m (the two species of *Canarina* in E Africa are found at up to 3000 m). The remaining five genera, i.e. *Codonopsis*,

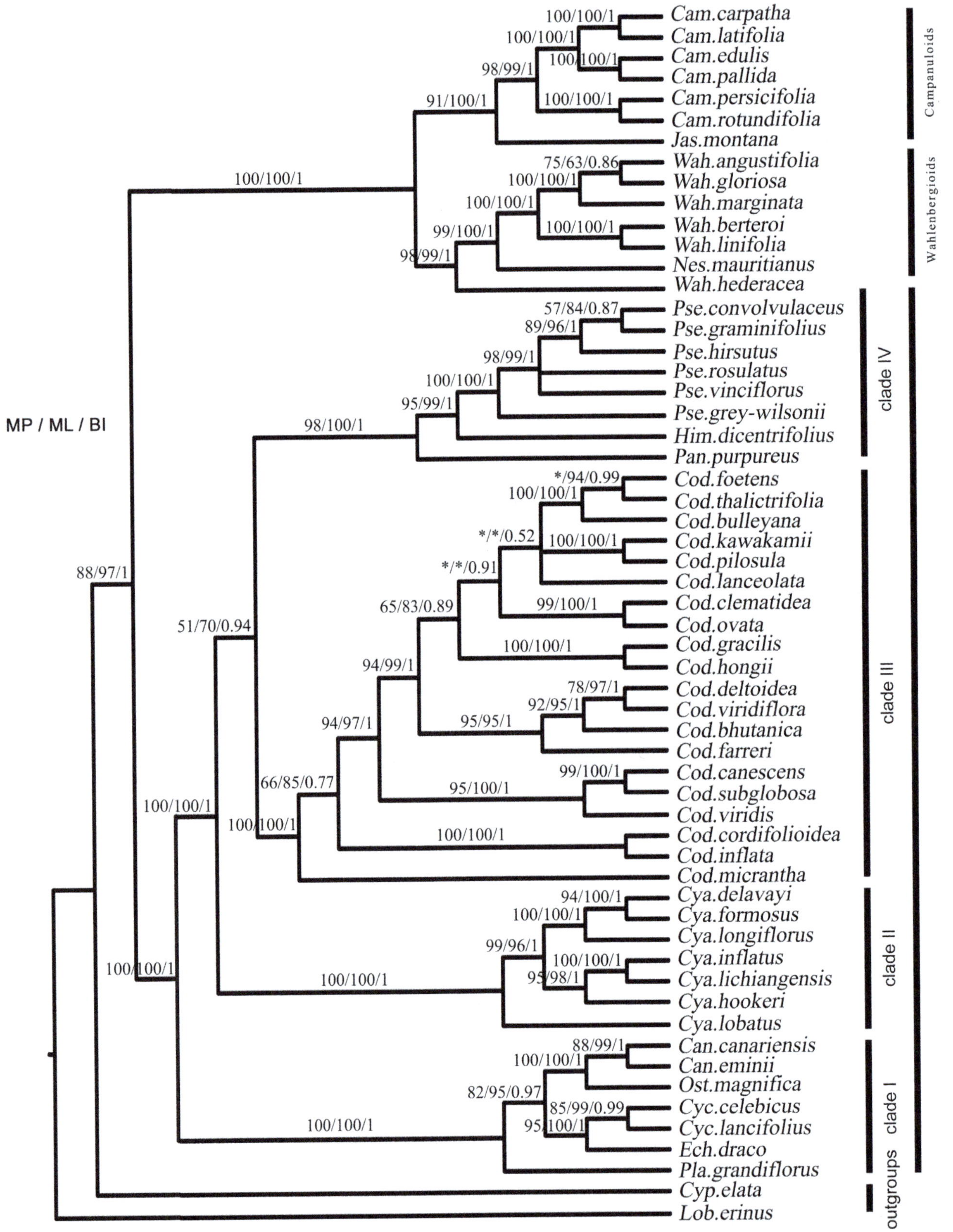

FIGURE III 3:1 Phylogram derived from the Bayesian analysis of four chloroplast DNA fragments. Jackknife support (JK)of Maximum parsimony/ Bootstrap value of likelihood/ Bayesian posterior probability values are reported above branches (MP/ML/BI). The stars are for the support values lower than 50 % parsimony and likelihood *(from Wang et al., 2014b, with minor modification).*

TABLE III 3:1 The Distribution and Character Variation Patterns of Genera in the Cyanantheae (from Wang et al., 2014b, with revision)

Genera	Total Species/ Sampled Species	Distribution Area	Ovary Position	Fruit Type	Pollen Type	Chromosome Number
Canarina	3/2	E Africa and Canary Islands	Inferior	Berry	Colporate	2n = 34
Codonopsis	46/20	QTP* and adjacent regions	Inferior & subsuperior	Capsule & berry	Long- colpate	2n = 16
Cyananthus	18/7	QTP and adjacent regions	Superior	Capsule	Long-colpate	2n = 10, 12, 14, 24, 28
Cyclocodon	3/2	SE Asia and New Guinea	Inferior	Berry	Colporate	2n = 18
Echinocodon	1/1	Central China	Inferior	Capsule	Short-colpate	2n = 16
Himalacodon	1/1	QTP and adjacent regions	Inferior	Capsule	Short-colpate	2n = 14
Ostrowskia	1/1	Pamir-Alai	Inferior	Capsule	Long-colpate	2n = 34
Pankycodon	1/1	QTP and adjacent regions	Inferior	Capsule	Short-colpate	2n = 14
Platycodon	1/1	East Asia	Inferior	Capsule	Colporate	2n = 18
Pseudocodon	8/6	QTP and adjacent regions	Inferior	Capsule	Short-colpate	2n = 14

**QTP: Qinghai-Tibet Plateau.*

Cyananthus, *Himalacodon*, *Pankycodon*, and *Pseudocodon*, are found in the centre of diversity of the tribe, the Himalaya and its adjacent regions, except for a few widely distributed species of *Codonopsis*. Among these five genera, *Codonopsis* has 46 species, *Cyananthus* has 19, and *Pseudocodon* is with eight. They all grow in alpine regions at altitudes up to 5560 m, with very few species found below 2000 m.

The Qinghai-Tibet Plateau (QTP) is an important region for plant biogeographical studies and has attracted much attention (e.g. Chen et al., 2012; Cun & Wang, 2010; Qiao et al., 2007; Qin et al., 2013; Ran et al., 2006; Song et al., 2003; Vetaas & Grytnes, 2002; Wang et al., 2009; Yang et al., 2008; Zhao et al., 2013; Zheng et al., 2008). Recent studies (Axelrod et al., 1996; Kadereit et al., 2008; Wu, 1988; Wu & Wu, 1996; Zhang et al., 2001, 2004) reveal that many temperate genera might have originated in the QTP. The hypothesis 'out-of-QTP' even points out that the majority of temperate plants now distributed across Eurasia might have first originated in the QTP and adjacent regions, and then migrated from there. This hypothesis was supported by phylogenetic analyses of some genera (Zhang et al., 2007, 2009; Mao et al., 2010; Tu et al., 2010; Xu et al., 2010) and species (Jia et al., 2012). The Cyanantheae is a fascinating group of the Campanulaceae. It does not have a typical Northern Hemisphere temperate distribution. However, species in this group are distributed in the QTP and adjacent areas. Very interesting questions are whether this group originated in the QTP and how its present distribution pattern has formed?

In addition, the evolution patterns of four characters emphasized in previous taxonomic studies (Moeliono & Tuyn, 1960; Kolakovsky, 1987, 1994; Hong & Ma, 1991; Takhtajan, 1997; Hong et al., 2011; Hong & Pan, 2012; Wang et al., 2013), i.e. ovary position, fruit type, pollen type, and chromosome number, are still puzzling. Diverse patterns of these four characters have been observed (Table III 3:1), but no agreement has been achieved on their evolution, which has led to many controversial issues within the group.

Of these four characters, the position of ovary (superior & inferior) is generally considered to be a character very significant for taxonomy, and is often used in family diagnoses. Many families, like Cornaceae, Cucurbitaceae, Rubiaceae, are largely defined on the basis of their inferior ovaries. In the Campanulaceae s. s., the ovary is generally inferior and it is also always inferior in the other four closely related families which form the

Campanulaceae s. l. *Cyananthus* is the only genus always with a superior ovary. This genus was once considered to be the most primitive in the Campanulaceae (Hong & Ma, 1991). All members of *Cyananthus* are alpine herbs with a number of specialized characters, and are confined to the Himalaya and adjacent regions. One would ask whether such a genus is really the most primitive in the family, and whether the evolutionary direction here has been from superior to inferior?

The fruit in the Cyanantheae has two types: the capsule and the berry. Plants with baccate fruit have been treated variously in taxonomy. The taxonomic treatment of *Campanumoea* is representative. This genus possesses baccate fruit and was merged into *Codonopsis* with capsular fruit by Moeliono and Tuyn (1960). Our recent molecular phylogenetic analysis supported this treatment (Wang et al., 2013). On the other hand, Kolakovsky (1987, 1994) separated *Campanumoea* and *Canarina* with baccate fruit from the other genera with capsular fruit as a subfamily, Canarinoideae. Takhtajan (1997) recognized the Canarioideae, but he excluded *Campanumoea* from this subfamily. Therefore, it is important to evaluate the significance of this fruit feature in systematics, and to make a reasonable taxonomic treatment accordingly. For this purpose we should gain an insight into the relationship between these two fruit types and how they have evolved.

The pollen in the Cyanantheae is colpate or colporate with elongate apertures, which differs distinctly from that in the other two groups of Campanulaceae s. s., that is, Wahlenbergieae and Campanuleae, which have porate (poroid) apertures. Pollen is polytypic in the Cyanantheae. Hong and Pan (2012) recognize nine types of pollen morphology in the tribe. Major differences between pollen types are that pollen is colporate in *Canarina*, *Cyclocodon* and *Platycodon*, but colpate in all the other seven genera. Among the seven colpate genera, *Cyananthus*, *Codonopsis* and *Ostrowskia* have long pollen apertures (colpus length/polar axis length ≥ 0.54), while *Echinocodon*, *Pseudocodon*, *Himalacodon* and *Pankycodon* have short pollen apertures (colpus length/polar axis length ≤ 0.4). One would ask: in which direction has pollen evolved, from colporate to colpate or vice versa, and from long apertures to short apertures or vice versa?

Six chromosome numbers in the Cyanantheae have been reported: 2n = 10 and 12 in *Cyananthus*; 2n = 14 in *Cyananthus*, *Himalacodon*, *Pankycodon*, and *Pseudocodon*; 2n = 16in *Codonopsis* and *Echinocodon*; 2n = 18 in *Cyclocodon* and *Platycodon*; and 2n = 34 in *Canarina* and *Ostrowskia* (see Chapter II 4). The chromosome number 2n = 34 is rather common in the Campanulaceae s. l., accounting for about 42% of the published counts (Lammers, 2007 a). However, x = 7 (Raven, 1975; Kumar & Chauhan, 1975; Hong & Ma, 1991) and x = 8 (Böcher, 1964; Contandriopoulos, 1984) each has been considered as the ancestral basic number for the Campanulaceae. Which is the primitive basic chromosome number in the Cyanantheae? In addition, as has been pointed out, there is no absolute rule that a high chromosome number is associated with fitness to extreme habitats (Böcher, 1936). What is the implication of the highest chromosome number 2n = 34, found in *Canarina* and *Ostrowskia*, and how has it been derived?

In a recent work, we attempted to infer the ancestral state of these four characters and subsequently to address the questions on biogeography and character evolution of the tribe Cyanantheae (Wang et al., 2014b) based on phylogenetic and dating analyses. We used DNA sequences of four chloroplast DNA regions for 42 samples representing all 10 genera of the Cyanantheae. Readers may refer that article for material and methodology used. We here try to address the questions mentioned above.

Origin and Diversification of the Cyanantheae

In the phylogenetic tree (Figure III 3:1), all the 10 genera of the tribe Cyanantheae form a monophyletic clade with strong support (MP/ML/BI = 100/100/1). Within the tribe, four clades are readily recognizable, confirming the results of Wang et al. (2013): (1) Clade I (MP/ML/BI = 100/100/1), which comprises *Platycodon*, *Ostrowskia*, *Canarina*, *Echinocodon* and *Cyclocodon*; (2) Clade II (MP/ML/BI = 100/100/1), which comprises seven species of *Cyananthus* representing all three sections of the genus; (3) Clade III (MP/ML/BI = 100/100/1), which comprises 20 species of *Codonopsis* (including *Campanumoea* s. s. and *Leptocodon*, which have been merged into *Codonopsis*); and (4) Clade IV (MP/ML/BI = 98/100/1), which includes *Pseudocodon* D. Y. Hong & H. Sun, *Himalacodon* D. Y. Hong & Q. Wang, and *Pankycodon* D. Y. Hong & X. T. Ma.

The optimal reconstruction is illustrated in Figure III 3:2. The reconstruction reveals 12 dispersals and five vicariances to explain the present distribution pattern of the Cyanantheae. The results indicate that the ancestor of the Cyanantheae should have originated in Himalaya-Hengduan Mountains and East-Asian region (node 41; A+B: 100%). Vicariance events were also detected at node 41. East Asia was also indicated as the ancestral area of clade I (node 40; B: 100%), while Himalaya-Hengduan Mountain region was the common ancestral area (node 34; A: 100%) of the other taxa. In addition, frequent dispersal and vicariance events were indicated within clade III.

Except for *Canarina*, restricted to East Africa and Canary Islands, the genera are mainly distributed in the QTP and adjacent region, eastwards to East Asia and southeastwards to New Guinea. Roquet et al. (2009) has pointed out that the ancestor of the Campanulaceae s. s. could have originated and diverged first in Asia, and dispersed later

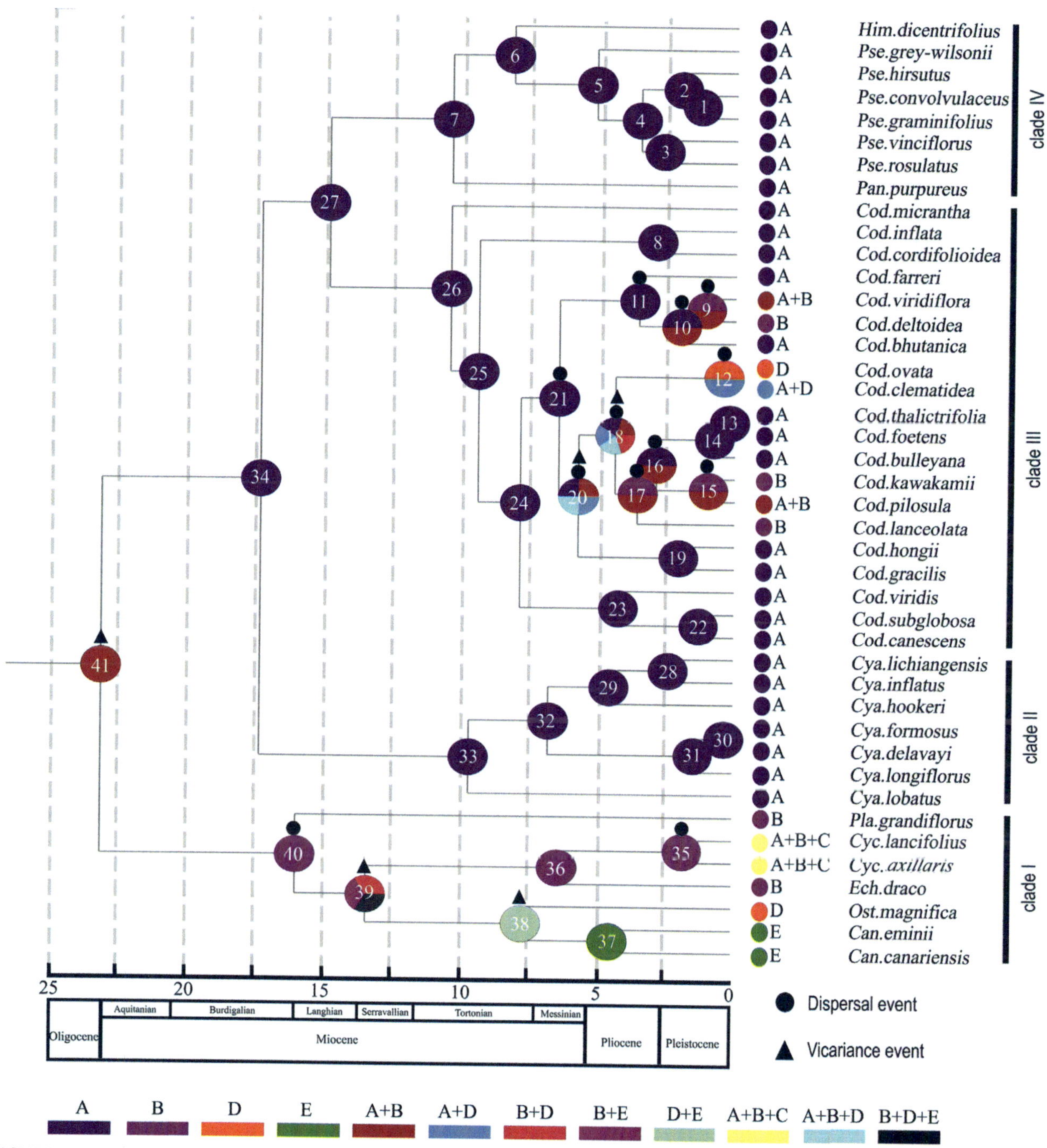

FIGURE III 3:2 Reconstruction of ancestral state in phylogenies and chronogram obtained with exponental relaxed clock (uncorrelated) dating of the combined data. The pie charts represent the relative probability of ancestral areas reconstructed for each node. Biogeographic events are highlighted above each node. (A) The Himalaya-Hengduan mountain region. (B) East Asia (east of Hengduan mountain region). (C) Southeast Asia and New Guinea. (D) Central Asia. (E) East Africa and Canary Islands *(from Wang et al. 2014b).*

to Africa and the Mediterranean region. Furthermore, the Tethyan Tertiary flora was considered to be important for the origin of the Sino-Himalayan flora (Sun, 2002), and was taken into account for the origin of the Cyanantheae. Zhou et al. (2013) also suggested that this group should have descended from widely distributed Tethyan ancestor.

Based on molecular phylogeny (Figure III 3:1) and previous studies of palynology (Hong & Pan, 2012), cytology (Gadella, 1966; Hedberg, 1961; Hong, 1984; Yoo & Lee, 1989; Zhang et al., 2010; Wang et al., 2014a; Chen et al., 2014), and morphological data, it can be summarized that: (1) central taxa (mainly in the QTP and adjacent region) form a clade which is sister to clade I which includes peripheral taxa (in East Asia, Southeast Asia and New Guinea, Middle Asia, and Africa); (2) central taxa all possess lower chromosome numbers (2n = 10 and 12 in *Cyananthus*; 2n = 14 in *Cyananthus, Himalacodon, Pankycodon,* and *Pseudocodon*; 2n = 16 in *Codonopsis*), and colpate pollen (short- colpate in *Pseudocodon*, *Pankycodon*, and *Himalacodon*; long-colpate in *Cyananthus* and *Codonopsis*); (3) peripheral taxa possess colporate pollen and comparatively higher chromosome numbers (2n = 18 in *Platycodon* and *Cyclocodon*; 2n = 34 in *Ostrowskia* and *Canarina*), except for 2n = 16 in *Echinocodon*, and colpate pollen in *Echinocodon* and *Ostrowskia*. Furthermore, the basal genus *Platycodon* of the peripheral group possesses a comparatively lower chromosome number than the derived genera, such as *Canarina* and *Ostrowskia*; and (4) position of ovary and fruit type might not be significant for considering biogeography and evolution of the Cyanantheae. An inferior ovary is present both in the central and peripheral taxa, whilst a superior ovary is present only in *Cyananthus*. Capsules and berries are present both in the central and peripheral taxa.

Combining the data from palynology, cytology, morphology, with the results of our dating analyses and reconstruction of ancestral states in phylogeny, we can infer that the ancestors of the Cyanantheae should have originated in Sino-Himalayan region (node 41, A+B: 100%, Figure III 3:2), and possess the primitive inferior ovary, capsule, long colpate pollen and chromosome number 2n = 14. Since the Tethyan Tertiary flora has been important in the origin of the Sino-Himalayan flora (Sun, 2002), the ancestor of the Cyanantheae might have been derived from widely distributed Tethyan ancestors, which is in accordance with the results of Zhou et al. (2013). In the gradual uplifting phase of QTP before 20 mya (Chung et al., 1998; Royden et al., 2008; Wang et al., 2008), the ancestors of the Cyanantheae had split into two major clades. This separation was dated as about 23.18 mya (Figure III 3:2), approximate to the separation time (26.3 mya) estimated by Roquet (2009). The vicariance event deduced by S-DIVA analysis (node 41, Figure III 3:2) also supports the split within the ancestor of the Cyanantheae. Subsequently, the ancestors of the central taxa continuously diversified in the Himalaya-Hengduan Mountain region, driven by the uplifting of QTP. Following the diversification, characters, such as short colpate pollen, superior ovary, and berry, were derived. However, lower chromosome numbers (2n = 14 and = 16) were preserved in these taxa. Ancestors of the peripheral group (= clade I), might have dispersed to Africa through the Mediterranean region, as well as to Southeast Asia, which was dated as beginning at ca. 16.03 mya (Figure III 3:2), and confirms Roquet et al.'s (2009) hypothesis. Following this dispersal event (node 40, B: 100%, Figure III 3:2), *Ostrowskia* and *Canarina* descended respectively in Middle Asia and North Africa, and evolved derived characters, i.e. berry (in *Canarina*), colporate pollen (in *Canarina*), and chromosome number 2n = 34. However, the remaining three genera in East and Southeast Asia still maintained some primitive characters, such as capsule (in *Echinocodon* and *Platycodon*), and lower chromosome numbers (x = 8 in *Echinocodon*; x = 9 in *Platycodon* and *Cyclocodon*).

In addition, the split of *Cyananthus* and the ancestor of *Codonopsis-Pseudocodon* (17.37 mya, Figure III 3:2), the separation of *Codonopsis* and *Pseudocodon* (14.91 mya, Figure III 3:2), as well as the diversification within clade IV (*Pankycodon*, 10.29 mya; *Himalacodon*, 8.05 mya; Figure III 3:2), all took place in the Himalaya-Hengduan Mountain region. These separation events all occurred in the rapid uplifting of eastern Tibet (Royden et al., 2008; Wang et al., 2008). Strong orogeny of the Himalaya in these phases might be the dynamic driving these separations. Moreover, all of these genera have adapted well to alpine environments, and thus were able to disperse easily in the Himalaya-Hengduan Mountain region to reoccupy the whole area. As a result, the subsequent sympatric distribution pattern of these genera formed.

Formation of the Distribution Pattern of the African Genus *Canarina*: Isolation and Disjunction

As revealed by molecular phylogeny (Figure III 3:1), the genus *Canarina*, which is endemic to East Africa and Canary Islands, is sister to the Middle Asian genus *Ostrowskia*. Since our results have indicated that the ancestor of the Cyanantheae should have originated in Himalaya-Hengduan Mountains and East-Asian region, the ancestor of the African genus might have migrated from Asia. A dispersal event deduced by S-DIVA (node 40, Figure III 3:2) analyses supports this migration of the ancestor of *Canarina*. The node 40 was also dated as 16.03 mya (Figure III 3:2), inosculating the Mid-Miocene collision (16 mya) of the Arabian Plate with Eurasia (Sanmartín, 2003; Oberprieler, 2005). Thus, the ancestor of *Canarina* and *Ostrowskia* could have dispersed to Middle Asia and Africa through the route between Eurasia and Africa across the Levante region. Drought and desertification of the Sahara and Sahel regions occurring about 5500 years ago (Foley et al., 2003) might have led to the isolation of *Canarina canariensis* in Canary Islands from *C. abyssinica* and *C. eminii* in East Africa, while the widespread drought ca. 4500 years ago (Davis & Thompson, 2006) might have led to fragmentation in habitat of *Canarina* and *Ostrowskia*.

Diversification within *Codonopsis*

Codonopsis is the largest genus of the Cyanantheae. In the present study, nearly half (20 species) of *Codonopsis* were sampled for molecular phylogenetic analysis and dating analysis. The results (Figures III 3:1 and III 3:2) reveal that the rapid uplifting phase of QTP (Royden et al., 2008; Wang et al., 2008) and especially the more recent abrupt uplifting beginning at 7 mya (Sun et al., 2000) might be the dynamic driving force of the diversification within *Codonopsis*. The split of ancestral *Codonopsis* was dated to 10.48 mya. Frequent dispersal and vicariance in East Asia beginning at about 5 mya, also greatly drove the diversification of *Codonopsis*, such as the dispersal toward East Asia and the Far East occurring in the ancestor of *C. bulleyana, C. thalictrifolia, C. pilosula* and *C. kawakamii*, and the dispersal toward Middle Asia occurring in the ancestor of *C. clematidea* and *C. ovata*. Dispersal toward East Asia also occurred in the ancestor of *C. bhutanica, C. viridiflora* and *C. deltoidea*.

Evolution of Four Characters

The ancestral states of ovary position, fruit type, pollen type, and chromosome number, are reconstructed and illustrated in Figure III 3:3. The results show that: (1) three types of ovary position can be recognized in the evolutionary history of the Cyanantheae, of which the inferior ovary is fairly certainly primitive with full possibility (Figure III 3:3A), and the superior ovary in *Cyananthus* and subsuperior ovary of *Cyclocodon* respectively evolved only once; (2) the capsule is the primitive and predominant fruit type of the ancestors of the Cyanantheae, whilst the berry evolved twice within clade I (Figure III 3:3B); (3) long-colpate pollen is primitive (root node; long-colpate, 86.6%, Figure III 3:3C), while short-colpate pollen evolved once within clade IV, and colporate pollen evolved once withn clade I. There are some interesting reversals from colporate to long-colpate in clade I; and (4) chromosome number 2n = 14 is favored as primitive (root node, 2n = 14, 47.25%, Figure III 3:3D). The chromosome numbers 2n = 34 and 2n = 18 both evolved only once within clade I, whilst the chromosome number 2n = 16 evolved twice: the first from 2n = 14 in clade III, the second from 2n = 18 in clade I.

The superior ovary is generally considered as primitive within flowering plants as a whole. In Campanulaceae s. l., *Cyananthus* is the only genus entirely possessing superior ovary. However, *Cyananthus* could not be a primitive genus, since molecular phylogeny (Wang et al., 2013; Zhou et al., 2013) has revealed that it is nested within the Cyanantheae, and is not situated at the basal position of the Campanulaceae. Our ancestral state reconstruction also supports that the ovary of the basal ancestor of the Cyanantheae should be inferior with full probability, while the superior and subsuperior ovary (present in *Cyclocodon* and a few species of *Codonopsis*) must have been derived independently during the evolution history of the group. In addition, the appearance of the superior ovary is dated as ca. 9.73 mya, during the rapid uplifting of eastern Tibet (Royden et al., 2008; Wang et al., 2008). Strong orogeny of the Himalaya in these phases might have driven the ancestor of *Cyananthus* to occupy and adapt itself to cold and dry alpine regions. We suggest that the superior ovary might be of advantage to capsule splitting and fast dispersal of seeds in these regions.

Although the fruit type is strongly emphasized in previous classification system of the Campanulaceae s. s. (de Candolle, 1830, 1839; Bentham and Hooker, 1876; Schönland, 1889; Kolakovsky, 1987, 1994), our analyses reveal that the character is not so diagnostic and valuable in taxonomy, especially for any intrafamilial subdivision. The evolution pattern of the fruit type is even clearer than ovary position. The dry capsule is the primary state in the ancestors of the Cyanantheae, whilst the fleshy berry is a derived character which has evolved independently several times in *Canarina*, *Cyclocodon*, and a few species of *Codonopsis* (such as *C. inflata*, *C. javanica*). The appearance time of the berry is dated as no more than 4.59 mya. Interestingly, it is observed that species with berry are always distributed in lowlands, mostly from 1000m to 2000m except for *Canarina*, which can be found up to 3000m in East Africa. We suggest that in the tribe, the appearance of berry might be correlated with the warm and wet environment.

Pollen is an important character in the classification of the Campanulaceae s. s. Thulin (1975) has pointed out that "for the arrangement of genera in the Campanulaceae s. s., the first consideration should be pollen morphology". Our ancestral state reconstruction reveals three main states in the ancestors of the Cyanantheae, in which the long-colpate pollen is most likely the primitive state (Figure III 3:3C). Our results also show that the short-colpate pollen might have been derived from the long-colpate type, and was dated as ca. 10.45 mya, still during the rapid uplifting phase of eastern Tibet (Royden et al., 2008; Wang et al., 2008). There are also two distinct types within the short-colpate pollen: the ancestors of *Pseudocodon* have pollen with long exine spinules (≥2 μm), appearing at ca. 5.20 mya, whilst the basal type has short exine spinules (≤1μm). In addition, colpate and colporate pollen grains both are present in the ancestral state of clade I. Within this clade, long-colpate pollen might have independently evolved again from colporate pollen. The appearance of colporate pollen is dated from 23.18 mya to 16.03 mya, about the beginning of the rapid uplifting phase of eastern Tibet (Royden et al., 2008; Wang et al., 2008). Thus, we favor the following two independent pathways in pollen morphology: long-colpate → colporate → colpate (long / short); and long-colpate → short-colpate. In addition, the evolutionary pathway of pollen exine spinules should be: short spinules → long spinules.

The evolution of chromosome numbers within the Cyanantheae is a little complicated. Based on our results (Figure III 3:3D), the basic chromosome number x = 7 should be primitive, which confirms the opinion of Raven (1975), Kumar and Chauhan (1975), and Hong and Ma

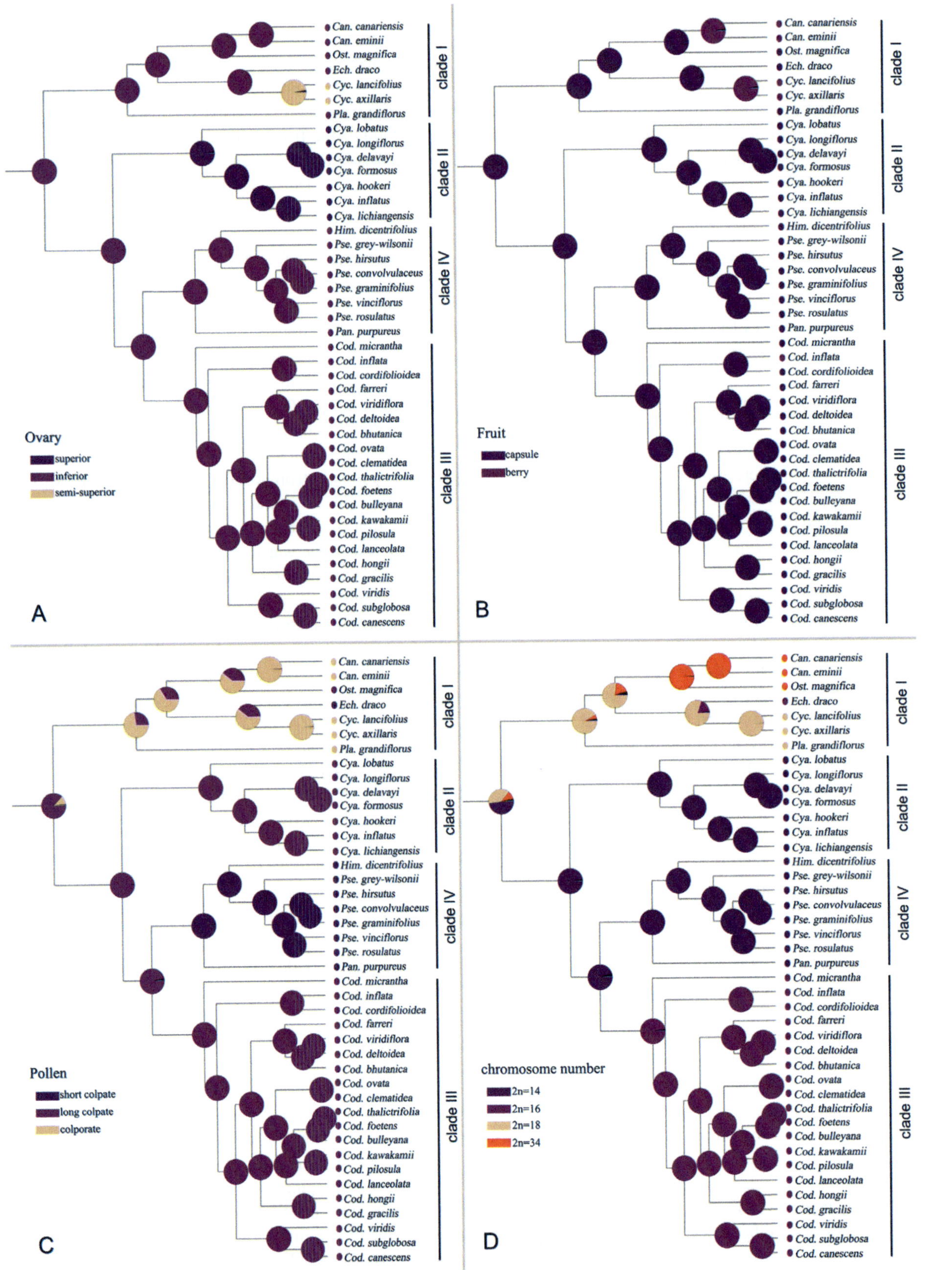

FIGURE III 3:3 Reconstruction of ancestral states of diagnostic characters in phylogenies of the Cyanantheae. The pie charts represent the relative probability of ancestral states reconstructed for each node; (A) ovary positions; (B) fruit types; (C) pollen aperture condition; and (D) chromosome numbers *(form Wang et al., 2014b with modofication).*

(1991). The basic chromosome numbers x = 8 (dated as ca. 10.48 mya) and x = 9 (dated from 23.18 mya to 16.03 mya) should be derived from the primitive x = 7. In addition, x = 8 of *Echinocodon* might have been derived not from primitive x = 7, but from x = 9 of the ancestor of clade I. The second independently evolutionary event of x = 8 is dated as before 6.45 mya. As has been pointed out, the chromosome number 2n = 34 accounts for about 42% of the published counts of the family Campanulaceae s. l. (Lammers, 2007a). However, the chromosome number 2n = 34 of *Canarina* and *Ostrowskia* could not be an ancestral number, since these two genera are firmly nested within clade I, of which the ancestral chromosome number is favored as x = 8. Thus, 2n = 34 of these two genera might have been derived from an ancestral hybridization event. We also suggest that ancestors of *Platycodon* (x = 9) and of *Echinocodon* (x = 8) should be the parents of this hybrid offspring with x = 17. Thus, we suggest that within the Cyanantheae, the pathways of chromosome number evolution should be x = 7→8; 7→9; 7→6→5; and 9→8, while x = 17 should be produced by hybridization between x = 8 and x = 9.

Chapter 4

Taxonomic Revision of *Codonopsis* and Allied Genera

CODONOPSIS WALL.

Wallich in Roxburgh, Fl. Ind. 2: 103. 1824, p. p.; de Candolle, Monogr. Camp. 120. 1830, p.p., quoad sect.1; Hooker & Thomson, J. Proc. Linn. Soc. London 2: 11. 1857, p. p. excl. *Codonopsis purpurea*; Clarke in Hooker, Fl. Brit. Ind. 3: 430. 1881, p. p.; Komarov, Trudy Imp. S.-Peterburgsk. Bot. Sada 29(1): 102. 1908, p. p.; Chipp, J. Linn. Soc. Bot. 38: 374. 1908, p. p.; Anthony, Notes Roy. Bot. Gard. Edinburgh 15: 173. 1926, p. p.; Nannfeldt, Notes Roy. Bot. Gard. Edinburgh 16: 149. 1931, p. p.: Tsoong, Contr. Inst. Bot. Nat. Acad. Peiping 3(3): 90. 1935, p. p.; Shen & Hong in Hong, Fl. Reipubl. Popularis Sin. 73(2): 32. 1983, p. p.; Hong, Lammers & Klein in Wu, Raven & Hong, Fl. China 19: 513. 2011, p. p. *Campanula* sect. *Codonopsis* (Wall.) D. Dietr., Syn. Pl. 1: 757, 1839. *Wahlenbergia* sect. *Megasanthes* (G. Don) A. DC. in de Candolle, Prodr. 7: 424. 1839. *Codonopsis* sect. *Megasanthes* G. Don, Gen. Syst. Gard. 3: 736. 1834. Type: *Codonopsis viridis* Wall. [=*Wahlenbergia viridis* (Wall.) A. DC].

Campanumoea Bl., Bijdr. Fl. Ned. Ind. 13: 726. 1826. Type: *Campanumoea javanica* Bl. [= *Codonopsis javanica* (Bl.) Hook. f.]

Glossocomia D. Don, Prodr. Fl. Nepal. 158. 1825. Type: *Glossocomia tenera* D. Don (=*Codonopsis thalictrifolia* Wall.)

Leptocodon (Hook. f. & Thomson) Lem., Ill. Hort. 3. Misc. 49. 1856. *Codonopsis* subg. *Leptocodon* Hook. f. & Thomson, Ill. Himal. Pl. t. 16, A. 1855. Type: *Leptocodon gracilis* (Hook. f. & Thomson) Lem. (=*Codonopsis gracilis* Hook. f. & Thomson).

Perennial herbs. Roots usually thickened, various in shape, mostly fleshy and carrot-shaped, rarely fusiform or tuberous, rarely woody. Caudex often short. Stems erect, ascending, scandent, or twining. Leaves alternate, opposite, or (3-) 4-verticil-like in sect. *Lanceolatae*. Flowers solitary, less frequently in a cyme of 2–4, terminal only on main stem or on both main stem and branches, some times leaf-opposed, rarely axillary. Calyx tube usually adnate to hypanthium and thus calyx epigynous, often 10-ribbed, calyx lobes 5, in a few cases inserted at base of hypanthium and thus appearing as 5 sepals and calyx nearly hypogynous. Corolla always epigynous, broad-campanulate, campanulate, funnel-shaped, tubular-campanulate, tubular, or urceolate, 5-lobed for 1/8–1/2 of length, red-purple, blue-purple, blue-white, yellow-green or green, often with dark purple veins, sometimes with dark rings or spots. Disc present, various in colour, annular or pentagonal, or interrupted and appearing as 5 glands alternate to stamens. Stamens 5; filaments often dilated at base, glabrous or hairy; anthers basifixed, connective usuall glabrous, rarely setose. Ovary at least half-inferior to corolla, usually 3-locular (5-locular in *C. javanica*); ovules numerous; style hairy; stigma (2- or) 3-fid (4- or 5-fid in *C. javanica*), lobes broad, oblong or ovate-deltoid. Fruit usually a capsule, occasionally a berry, with persistent calyx, lower part semi-globose or obconical, apically 3-loculicidal, often 10-ribbed, upper part rostrate. Seeds numerous, usually ellipsoid, oblong or ovoid, wingless, rarely winged, smooth, in species with berries, seeds irregular or ovoid-globose, reticulate; embryo straight, embedded in copious endosperm. Plants mostly fetid. Pollen 5–7-colpate; colpi long, with colpus length/polar axis length ≥ 0.54; sexine with short spinules (≤ 1 μm). Chromosomes 2n=16.

The circumscription of the genus *Codonopsis* has been controversial among previous authors, and it has been changed greatly in the series of studies leading up to this publication (see Chapter III).

Codonopsis as here circumscribed is monophyletic, perhaps except *Codonopsis chimiliensis* which will be discussed under the treatment of that species, but its subdivision is still problematic. Thus, the subdivision into five sections as here used is tentative.

Fourty-six species in East, Southeast and Central Asia: Afghanistan, Bhutan, China, N and NE India, Indonesia, Japan, Kazakhstan, Kyrgyzstan, Korean Peninsula, Laos, Malesia, W Mongolia, Myanmar, Nepal, Pakistan, Philippines, Russia (the Far East), Tajikistan, Thailand, SE Uzbekistan, and Vietnam (Map IV 1: 0).

A Monograph of *Codonopsis* and Allied Genera (Campanulaceae). http://dx.doi.org/10.1016/B978-0-12-801933-7.00004-0
Copyright © 2015 China Science Publishing & Media Ltd. Published by Elsevier Inc. All rights reserved.

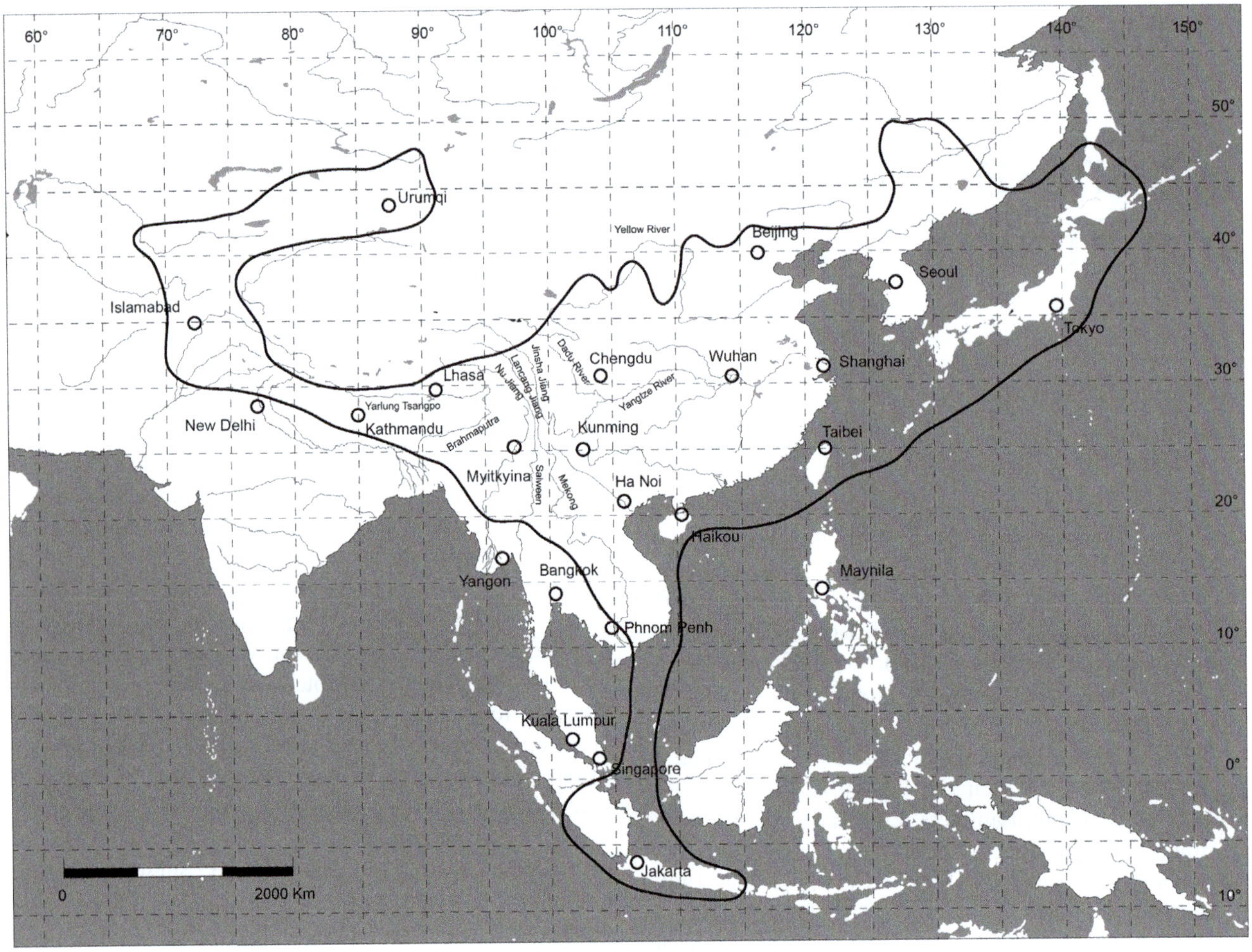

MAP IV 1:0 Distribution of the genus *Codonopsis* Wall.

1a. Disc consisting of 5 lamellar glands, alternating with stamens; calyx almost free from hypanthium; corolla long-tubular; stems twining; plants very fetid (sect. 4. ***Leptocodon***: spp. 43–44).

- **2a.** Plants glabrous throughout except for sparsely villous young leaves; calyx lobes linear-oblong, often reflexed, distant from each other, glabrous, base attenuated into stipe ....43. *C. gracilis*
- **2b.** Plants more or less hairy, rarely villous only on leaf margin; calyx lobes ovate to ovate-lanceolate, not reflexed, overlapping each other, villous at least on margin, base obtuse, truncate, or cordate ....44. *C. hongii*

1b. Disc continuous, circular or pentagonal; calyx usually adnate to hypanthium; corolla usually campanulate, rarely tubular; stems erect or twining; plants fetid or not.

- **3a.** Fruit a fleshy berry; leaf blade cordate (sect. 5. ***Campanumoea***: spp. 45–46).
  - **4a.** Leaves opposite; calyx hypogynous or nealy hypogynous...45. *C. javanica*
  - **4b.** Leaves alternate; calyx epigynous ....46. *C. inflata*
- **3b.** Fruit a dry capsule; leaf blade various, less frequently cordate.
  - **5a.** Leaves 3 or 4 together at top of branches in pseudo-verticillate whorl; seeds winged (sect. 3. ***Lanceolatae***: spp. 41–42).
    - **6a.** Roots carrot-shaped or fusiform; seeds broadly winged; corolla (2.1–) 2.7–3.9 cm long; leaf blade (3–)4.5–7(–10) cm long.. ....41. *C. lanceolata*
    - **6b.** Roots ellipsoid or nearly globose; seeds narrowly winged; corolla 1.6–2.4 cm long; leaf blade (2.3–)3–4.5 (–5.5) cm long...42. *C. ussuriensis*
  - **5b.** Leaves at top of branches alternate or opposite, never whorled.

7a. Stems erect, ascending or scandent, very occasionally twining (sect. 1. ***Erectae***: spp, 1–22).

8a. Stems simple, without many usually sterile pinnate-leaflike branches from lower part.

9a. Stems simple and erect; flowers solitary and terminal; leaves uniform and evenly distributed; corolla purple or dark purple, without markings.

10a. Corolla nearly funnel-shaped, tubular below, nearly rotate above; stems sparsely to densely villous........ 2. *C. elliptica*

10b. Corolla campanulate; stems glabrous .... 1. *C. lixianica*

9b. Stems often branched; leaves uniformly distributed or mostly basal and stem somewhat scapiform; corolla white, yellowish, blue, pale purple, often with dark markings on veins or yellow spots.

11a. Corolla tubular or tubular-campanulate; stems mostly ascending or scandent, rarely twining; leaves evenly distributed on main stem, main stem not scapiform.

12a. Leaf blade 0.6–1.7 cm wide; calyx lobes 5–7 mm long, 1.4–2.6 mm broad; corolla gray or pale purple, lobed to ca. 1/6, with a dark purple ring just below throat........ 5. *C. gongshanica*

12b. Leaf blade 2–7 cm wide; calyx lobes 13–16 mm long, 6–8.5 mm broad; corolla whitish or pale blue, lobed to ca.1/4, without a coloured ring.

13a. Corolla 1.8–4 cm long; fruiting calyx lobes enlarged, up to 2.5 cm long, 1.5 cm wide; stamens villous........ 34. *C. benthamii*

13b. Corolla 1.2–2 cm long; fruiting calyx lobes not enlarged, up to 1.6 cm long, 0.9 cm wide; stamens glabrous .... 11. *C. subsimplex*

11b. Corolla campanulate; stems mostly erect, less frequently ascending; leaves located at lower part of stem, main stem scapiform.

14a. Calyx entirely free from hypanthium; calyx lobes 2–3 cm long, 0.6–1 cm wide........ 3. *C. chimiliensis*

14b. Calyx adnate to whole hypanthium; calyx lobes less than 1.2 cm long, 0.6 cm wide.

15a. Stems often branched at lower and middle parts; petiole 1–10 cm long; calyx lobes 5–7 mm long, 2–3 mm wide, with sinus between them truncate........ 6. *C. subscaposa*

15b. Stems usually without branches at lower and middle parts; petiole less than 1 cm long; calyx lobes 8–12 mm long, 4 6 mm wide, with sinus between them narrow and pointed........ 4. *C. meleagris*

8b. Stems with many usually sterile branches, such branches with many nearly opposite leaves and thus resembling a pinnate leaf.

16a. Stems, lower surface of leaves and calyx densely white hairy; stems mostly with several flowers.

17a. Adaxial leaf surface gray-white; corolla broadly campanulate, wider than long, 1.5–1.8 cm long .... 17. *C. canescens*

17b. Adaxial leaf surface green, hispidulous; corolla campanulate, 1.8–2.8 cm long .... 16. *C. argentea*

16b. Stems, leaves and calyx all green, glabrous or sparsely hairy, not white; stems single-flowered or with several flowers.

18a. Stems and leaves glabrous or nearly glabrous; calyx entirely glabrous or calyx lobes rarely hispidulous.

19a. Petiole 4–6 mm long; leaf blade 0.8–1.5cm long, 0.8–1cm broad; calyx lobes verrucose-denticulate .... 13. *C. alpina*

19b. Petiole less than 3 mm long; leaf blade 1–3 cm long, 0.5–2.5 cm broad; calyx lobes ent ire........ 7. *C. cardiophylla*

18b. Stems and leaves always hairy; calyx hispidulous at least in upper part of lobes, very rarely entirely glabrous.

20a. Calyx lobes usually denticulate above, 4–6 mm long, 2–3 mm broad, sinus between calyx lobes broad and truncate........ 14. *C. chlorocodon*

20b. Calyx lobes entire, rarely denticulate, much larger, sinus between calyx lobes narrow, acute or obtuse.

21a. Leaf blade usually less than 1.5, rarely up to 3 cm long, 1.5 cm broad; sterile branches aggregated at lower part of main stem; main stem scapiform, with usually one, less frequently 2–3 flowers

22a. Corolla tubular-campanulate, 2.5–4.8 cm long, shallowly lobed to 1/6–1/8; anthers villous at connective..20. *C. thalictrifolia*

22b. Corolla ca mpanulate or funnel-shaped, 1.8–3 cm long, lobed to ca. 1/3; anthers glabrous.

23a. Sterile branches and leaves nearly evenly distributed at lower and middle parts of stem; leaf blade 1.5–3 cm long; corolla with a red-purple ring below throat...12. *C. campanulata*

23b. Sterile branches and leaves mostly aggregated towards lower part of stem; leaf blade 0.5–1.8 cm long; corolla without a ring below throat.

24a. Corolla broadly campanulate, pale purple or pink; calyx lobes 3.5–6.5 mm broad...18. *C. foetens*

24b. Corolla tubular below, suddenly dilated from middle and thus infundibular, pale blue above, purplish below; calyx lobes 5–7 mm broad.. 19. *C. bulleyana*

21b. Leaves usually bigger; sterile branches mostly evenly distributed along main stem, main stem not scapiform, usually with several flowers.

25a. Corolla tubular, 1.2–1.5 cm long; capsule 0.8–1 cm in diameter...15. *C. bhutanica*

25b. Corolla campanulate, 1.5–3 cm long; capsule usually more than 1 cm in diameter.

26a. Calyx lobes subulate, widest at base, 2–4 mm wide, always reflexed; leaf blade ovate-lanceolate to lanceolate, with length/width ratio mostly over 2.. 10. *C. obtusa*

26b. Calyx lobes ovate to ovate-lanceolate, widest above base, 4–8.5 mm broad, reflexed or not; leaf blade ovate to ovate-lanceolate, with length/width ratio mostly less than 2.

27a. Calyx lobes always hispidulous entirely or at upper part; corolla yellow-green, bluish or purple.

28a. Calyx tube glabrous, calyx lobes hispidulous at apex and margin; flowers on both main stem and branches; corolla glabrous inside, yellow-green, purplish at base, white, or white with purple veins inside... 21. *C. viridiflora*

28b. Calyx tube hirsute entirely or along ribs, rarely glabrous; flowers usually terminal only on main stem; corolla villous inside, pale to dark purple...22. *C. tsinlingensis*

27b. Calyx lobes glabrous or ciliate, or rarely hispidulous only at apex; corolla pale blue, purple inside.

29a. Calyx lobes revolute, 0.5–1(–1.2) cm long, ciliate; corolla infundibular..8. *C. ovata*

29b. Calyx lobes reflexed, 1–1.8 cm long, totally glabrous; corolla campanulate...9. *C. clematidea*

7b. Stems twining, sometimes scandent (sect. 2. ***Codonopsis***: spp. 23–40).

30a. Corolla tubular; leaf blade dentate or even lobed, very rarely nearly entire; calyx lobes toothed or crenulate, less often entire.

31a. corolla 1–1.3 cm long, 0.3–0.4 cm broad; leaf blade cordate-ovate, cordate at base, 2–3.5 cm long; plants pilose except corolla...36. *C. microtubulosa*

31b. Corolla much larger; leaf blade rounded to cuneate, or cordate at base, 3–10 cm long; plants puberulent, less often hirsute.

32a. Leaves on main stem mostly opposite, less frequeutly alternate; petiole less than 1 cm long; filaments villous...35. *C. tubulosa*

32b. Leaves on main stem mostly alternate; petiole longer than 1.5 cm; filaments glabrous or villous.

33a. Base of leaf blade mostly cordate; corolla broadly tubular, evenly broad, lobed to 1/5–1/4; filaments glabrous.. 27. *C. deltoidea*

33b. Base of leaf blade mostly rounded or cuneate, less frequently cordate; corolla narrowly tubular, slightly dilated above, lobed to 1/7–1/6; filaments villous.......................................34. *C. benthamii*

30b. Corolla campanulate or tubular-campanulate; leaf blade entire or shallowly incised; calyx lobes entire.

34a. Calyx lobes linear to linear-lanceolate, 10–20 mm long, 2–3 mm broad, with length/width ratio more than 5, distant from each other.. 32. *C. viridis*

34b. Calyx lobes wider, with length/width ratio less than 3, close to each other.

35a. Calyx more or less hairy; leaf blade cordate at base.

36a. Leaf blade (2.5–)3–16 cm long, (1–)3–9 cm broad; calyx tube scaberulose; calyx lobes triangular-lanceolate, densely puberulent; corolla broad-campanulate...31. *C. affinis*

36b. Leaf blade 0.5–3 cm long, 0.5–2.5 cm broad; calyx tube hirsute entirely or along ribs; calyx lobes ovate-orbicular or rhomboid-ovate, hirsute; corolla globose-campanulate.

37a. Petiole 0.5–2 cm long; corolla glabrous outside; stigma segments 2-lobed..37. *C. subglobosa*
37b. Petiole less than 2 mm; corolla hirsute outside; stigma segments entire..38. *C. bragaensis*
35b. Calyx glabrous; leaf blade cuneate, rounded or cordate at base.
38a. Leaf blade deeply cordate at base; corolla 0.7–1.8 cm long.
39a. Corolla 7–9 mm long, white; fruit 8–10 mm in diameter, superior part 2.5–3.5 mm long..40. *C. micrantha*
39b. Corolla 17–18 mm long, purple, dark blue, or white but purple-red inside; fruit 13–18 mm in diameter, superior part much longer..39. *C. cordifolioidea*
38b. Leaf blade cuneate to shallowly cordate at base; corolla 1.8–4.5 cm long, rarely shorter.
40a. Calyx lobes strongly reflexed even since anthesis; corolla lobed to 1/2–3/5 of length.
41a. Leaf blade densely white-scaberulose on lower surface; corolla white; filaments glabrous..24. *C. hemisphaerica*
41b. Leaf blade glabrous; corolla dark blue, with a reddish ring near base; filaments villose-ciliate at dilated part..29. *C. reflexa*
40b. Calyx lobes upright; corolla lobed up to 1/3 of length, rarely more deeply.
42a. Leaf blade cordate, rarely truncate or rounded at base; stems and leaves usually hairy, less often glabrous.
43a. Stems and leaves always hairy; calyx hypogynous to perigynous, calyx lobes inserted below corolla..23. *C. pilosula*
43b. Stems and leaves glabrous or sparsely hirsute; calyx epigynous, calyx lobes and corolla inserted at same level..21. *C. viridiflora*
42b. Leaf blade cuneate or rounded, occasionally cordate at base; stems and leaves usually glabrous or sparsely hairy.
44a. Stems and leaves glabrous; leaf blade entire.
45a. Leaf blade 0.5–4 cm long, 0.5–1.5 cm broad, apex acute or obtuse..........................33. *C. farreri*
45b. Leaf blade 6–7 cm long, 2–2.5 cm broad, apex shortly acuminate.....................28. *C. rotundifolia*
44b. Stems and leaves more or less hairy; leaf blade crenulate or serrate.
46a. Calyx lobes ciliate, reflexed...30. *C. henryi*
46b. Calyx lobes glabrous, upright.
47a. Leaf blade 3–15 cm long, apex acuminate...26. *C. bomiensis*
47b. Leaf blade less than 2.3 cm long, apex obtuse or acute...........................25. *C. kawakamii*

Section 1. Erectae (Kom.) D. Y. Hong, Fl. Reipubl. Popularis Sin. 73(2): 48. 1983. *Codonopsis* ser. *Erectae* Kom., Trudy Imp. S. Peterburgsk. Bot. Sada 29(1): 113. 1908.

Roots carrot-shaped. Stems erect, ascending, less frequently scandent, very occasionally twining. Leaves alternate, those on lower branches nearly opposite, branches appearing a pinnate leaf. Calyx always epigynous. Seeds oblong, wingless.

Twenty-two species, distributed from Middle Asia and NE Afghanistan eastwards to central China (W Hubei Province). It reaches the highest altitude among the five sections, up to 5490 m (*C. bhutanica*), and even to 5560 m (*C. thalictrifolia* subsp. *thalictrifolia*).

1. **Codonopsis lixianica** D. Y. Hong, Plant Divers. Resour. 36(3): 292. fig. 5. 2014. Type: China, Sichuan (NW), Lixian, Mt Huangtuliang, 3400 m, 24. Jul. 1957, *W. K. Hu & D. Z. Deng 1988* (holotype: SZ).

Roots carrot-shaped, ca. 5 mm in diameter. Stems erect, simple, glabrous, ca. 35 cm tall. Lower leaves nearly opposite, middle and upper ones alternate; petiole 4–10 mm long, sparsely villous; leaf blade of lower leaves ovate, those of middle and upper ones narrowly ovate to ovate-lanceolate, 1.6–6.3 cm long, 1.8–3.2 cm broad, adaxially glabrous, abaxially very sparsely white-hirsute, base truncate to rounded, margin crenate, apex obtuse to acute. Flowers solitary, terminal only on main stem, or also on small branches; small branches with a pair of very small bracteole-like leaves. Calyx tube adnate to whole hypanthium, 10-ribbed, glabrous; calyx lobes inserted at same level as corolla, ovate-lanceolate, broadest at middle-lower part, 11 mm long, 7 mm broad, adaxially sparsely white-hispidulous, apex acute. Corolla campanulate, dark purple, glabrous, 1.8 cm long, 1.5 cm wide; corolla lobes deltoid-ovate, 8 mm long, 10 mm broad. Disc annular, yellow. Filaments glabrous, 7 mm long, slightly dilated at lower part; anthers 4 mm long. Fruit not seen. Figure IV 1:1A.

Phenology: flowering in July.
Habitats: ca. 3400 m.

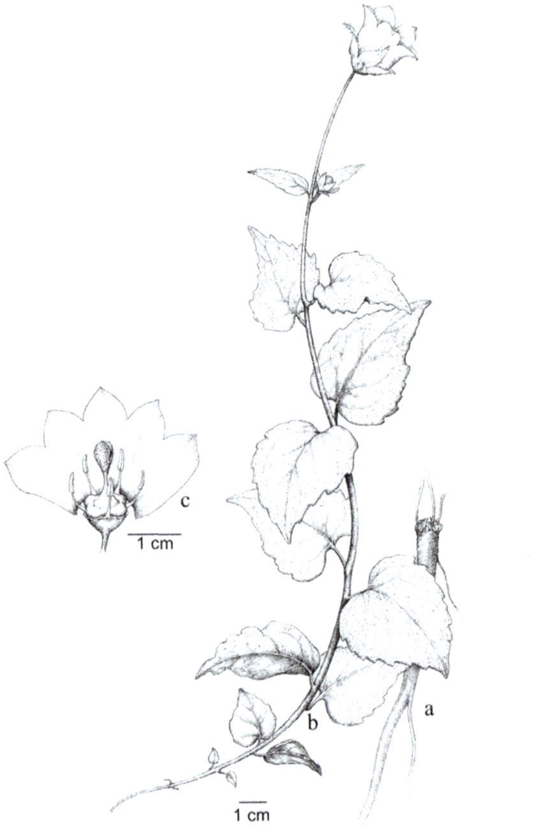

FIGURE IV 1:1A *Codonopsis lixianica* D. Y. Hong. **a**, root, caudex and stem base; **b**, a stem with a terminal flower and a flower bud; **c**, upper part of a flower with corolla opened, showing stamens, disc, style, and stigma.

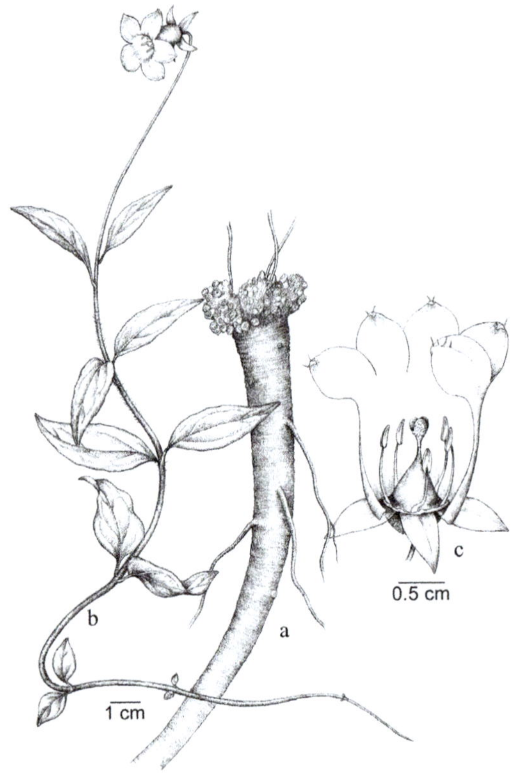

FIGURE IV 1:2A *Codonopsis elliptica* D. Y. Hong. **a**, root and caudex; **b**, a stem; **c**, a flower with corolla opened, showing calyx, stamens, style, and stigma.

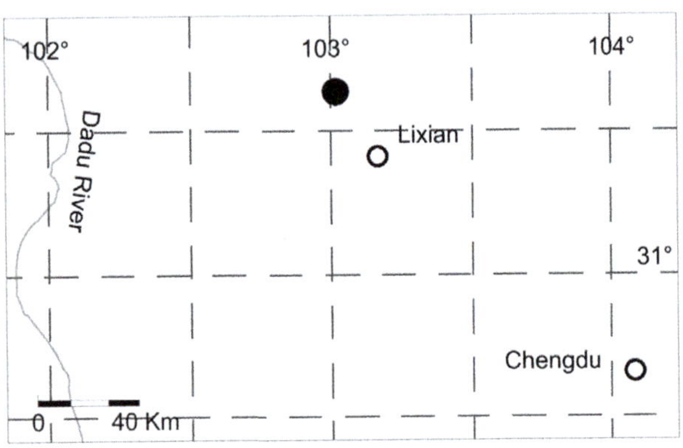

MAP IV 1:1 Distribution of *Codonopsis lixianica* D. Y. Hong.

Distribution: **endemic** to China: confined to Sichuan (Map IV 1:1).

Notes: Only one specimen with two stems was found, so we describe this as a new species with reservation. However, it is so distinct that it does not resemble any other species.

2. Codonopsis elliptica D. Y. Hong, Plant Divers. Resour. 36(3): 292. fig. 6. 2014. Type: China, Sichuan, Beichuan, Jianshe Commune, Heshangtou, 3800 m, sunny slopes, fl. purple, 09. Aug. 1978, *Beichuan Exped. 0383* (holotype: SM; isotype: PE).

Root carrot-shaped, over 15 cm long, ca. 1.3 cm in diameter. Stems several from one caudex, erect, simple, without any branch, 20–35 cm tall, sparsely to rather densely villous. Basal leaves much smaller than those at other parts, sometimes scale-like, those on lower and middle stem opposite or nearly opposite, upper ones alternate, sessile or with a short petiole up to 5 mm long; petiole glabrous or sparsely villous; leaf blade elliptic, less frequently ovate-lanceolate, uppermost ones sometimes linear-elliptic, 3–6 cm long, 0.8–2 cm broad, adaxially glabrous, abaxially sparsely hirsute along veins or totally glabrous, base cuneate, margin entire or slightly crenulate, apex obtuse or acute. Flowers solitary and terminal. Calyx tube adnate to hypanthium up to middle part, hemispheric, glabrous; calyx lobes inserted at same level as corolla, narrowly triangular, or lanceolate, 5–7 mm long, 3–4 mm broad, glabrous or very sparsely hispidulous at apex. Corolla infundibular, tubular below, nearly rotate above, 2–3.4 cm long, lobed to 1/3–2/5, purple, glabrous; corolla lobes orbicular, 1.1 cm long, 1.2 cm broad. Stamens glabrous, filaments ca. 6 mm long, slightly dilated at base,; anthers ca. 4 mm long. Stigma 3-fid; stigma lobes orbicular. Fruit not seen. Figure IV 1:2A.

Phenology: flowering in July and August.
Habitats: meadows or forests; ca. 3800 m.

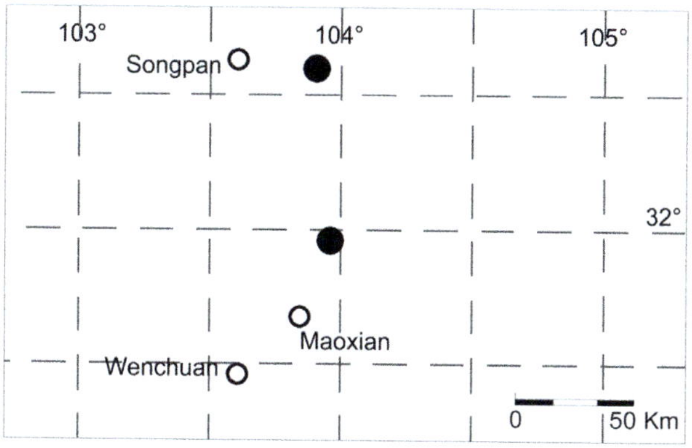

MAP IV 1:2 Distribution of *Codonopsis elliptica* D. Y. Hong.

Distribution: **endemic** to China: confined to Sichuan (Map IV 1:2).

Additional specimens examined, **CHINA**, **SICHUAN**, **Pingwu**: Huya, Dugai, forests, fl. blue, 24. Aug. 1978, *Miandi Exped. 401* (SM); Huya, Damachang, meadows, fl. blue, 16. Jul. 1978, *Pingwu Exped. 458* (SM).

3. **Codonopsis chimiliensis** J. Anthony, Notes Roy. Bot. Gard. Edinburgh 15: 184. pl. 217. 1926; Shen & Hong in Hong, Fl. Reipubl. Popularis Sin. 73(2): 54. 1983; Hong, Lammers & Klein in Wu, Raven & Hong, Fl. China 19: 520. 2011. Type: "N. E. Burma, western flank of the Chimi-li, N'Maikha-Salween Divide", 26°23'N, 98°48'E, 3350–3660 m, Aug. 1924, *G. Forrest 24887* (lectotype here designated: E; isolectotypes: K, P, W); loc. eodem, 26°16'N, 98°48'E, 4270 m, Sept. 1925, *G. Forrest 27223* (paratypes: E, K, P, W); loc. eodem, 26°24'N, 98°48'E, 3350 m, Jul. 1925, *G. Forrest 26998* (paratypes: E, K).

Plants glabrous throughout. Roots carrot-shaped. Stems erect, multi-branched, 60–100 cm tall. Leaves alternate; petiole 0.5–3 cm long; leaf blade ovate to ovate-lanceolate, 5–14 cm long, 3–7 cm broad, base rounded or decurrent, margin crenate or dentate, apex acute to short-acuminate. Flowers solitary, terminal on main stem and branches; pedicel 3–10 cm long. Hypanthium 1–2 mm free from ovary at top; calyx entirely free from hypanthium; calyx lobes oblong-lanceolate, 2–3 cm long, 0.6–1 cm broad, entire, apex acute. Corolla epigynous, campanulate, lobed to middle, white, greenish-white or pale creamy-yellow, with pale purple veins; corolla tube1.2–1.7 cm long, 1–1.2 cm wide; corolla lobes triangular, 1.2–1.5 cm long, 0.8–1.2 cm broad, apex acute. Stamens glabrous; filaments filiform, ca. 8 mm long. Ovary superior to calyx, but inferior to corolla; stigma 3-fid, stigma segments lanceolate. Fruit not seen. Figure IV 1:3A.

Phenology: flowering from July to September.

Habitats: grassy slopes or meadows at the edges of scrubs; 2800–4270 m.

Distribution: found so far only in the border area between China and Myanmar (Map IV 1:3).

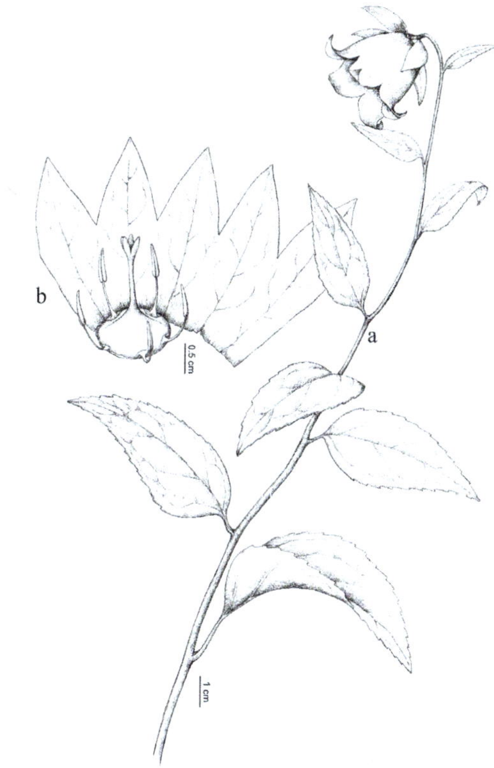

FIGURE IV 1:3A *Codonopsis chimiliensis* J. Anthony. **a**, upper part of a stem; **b**, upper part of a flower with corolla opened, showing stamens, style, and stigma.

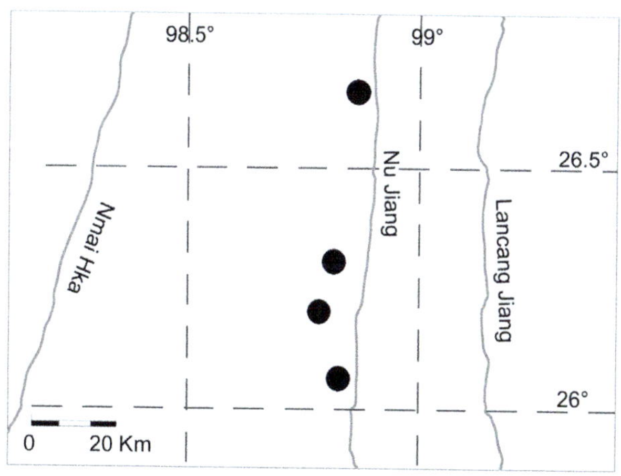

MAP IV 1:3 Distribution of *Codonopsis chimiliensis* J. Anthony.

Notes: All the species observed in *Codonopsis* as here circumscribed have long-colpate pollen with colpus length / polar axis length ≥ 0.54 and with granulose colpus membrane, but the pollen of this species has shorter colpi (colpus length / polar axis length ≤ 0.4) plus a smooth colpus membrane (Hong & Pan, 2012). In the Campanulaceae s. s. pollen morphology is closely correlated with external morphology and molecular data (Erdtman, 1950; Thulin, 1975; Hong & Pan, 2012; Wang et al., 2013). Unfortunately, we have not obtained materials for observations on seeds and chromosomes, nor for molecular analysis. From pollen morphology, hypanthium, and stigma segments, it resembles *Pankycodon* rather than *Codonopsis*.

Additional specimens examined**, CHINA, YUNNAN, Bijiang**: W slope of Gaoligong Shan, 2900 m, broad-leaved forests, 02. Aug. 1978, *Bijiang Exped. 1584* (KUN). **Lushui:** Pianma, 2800 m, among herbs, 14. Aug. 1964, *S. K. Wu 8361* (KUN).

MYANMAR, **Myitkyina**: Hpimaw Pass, 3200 m, 02. Sept. 1919, *RA 1279* (E); Jungle Camp above Hpimaw, 19. Aug. 1938, *Nam Mu Pa 17443* (K); s. loc., s. d.,*G. Forrest 29890* (E, PE).

4. **Codonopsis meleagris** Diels, Notes Roy. Bot. Gard. Edinburgh 5: 172. 1912; Tsoong, Contr. Inst. Bot. Nat. Acad. Peiping 3(3): 97. fig. 9. 1935; Stapf, Bot. Mag. 154: t. 9237. 1931; Shen & Hong in Hong, Fl. Reipubl. Popularis Sin. 73(2): 50. pl. 9. fig. 1. 1983; Hong, Lammers & Klein in Wu, Raven & Hong, Fl. China 19: 519. 2011. Type: China, NW Yunnan, Lichiang (Lijiang), 27°20′ N. 3350–3660 m, Aug. 1906, *G. Forrest 2674* (lectotype here designated: E; isolectotypes: BM, CAL).

Roots carrot-shaped, 15–25 cm long, 0.5–1.8 cm in diameter. Stems erect, 20–90 cm tall, simple, very rarely with long branches from lower part, sparsely hairy, later glabrescent. Leaves on main stem alternate, those on branches opposite, sessile or with petiole less than 1 cm long; leaf blade ovate, oblong, elliptic, or lanceolate, 3–11 cm long, 1.5–3.8 cm broad, adaxially green, glabrous or sparsely hispidulous, abaxially gray-green, densely hirsute, base cuneate, margin crenate or nearly entire, apex acute or obtuse. Flowers solitary, terminal or axillary, long-pedicellate. Calyx tube adnate to whole hypanthium, semiglobose, 10-ribbed, glabrous or sparsely scaberulose; calyx lobes ovate or ovate-triangular, 8–12 mm long, 4–6 mm broad, glabrous, apex acute, margin nearly entire or serrate. Corolla broadly campanulate, 3–4 cm long, 3–4 cm wide, lobed to middle, yellow with red-purple veins, or red-purple with yellow spots, or pale purple with dark purple veins, both surfaces glabrous; corolla lobes often sparsely villous at apex. Stamens glabrous; filaments ca. 7 mm long, slightly dilated at base; anthers ca. 5 mm long. Capsule semiglobose below, conical above, 1–2.2 cm long, 0.8–1 cm in diameter. Seeds, ovoid, brown-yellow. Figures IV 1:4A & B.

Phenology: flowering and fruiting from July to October.

Habitats: grassy slopes or open woods at altitudes from 2600 to 4000 m.

Distribution: **endemic** to China; confined to NW Yunnan (Map IV 1:4).

Additional specimens examined**, CHINA**, **YUNNAN, Jianchuan**: Jianchuan-Mekong Divide, 26°30'N, 99°40'E, Aug.1922, *G. Forrest 23198* (E). **Lijiang** (Yulong): Ganghaizi, 08. Aug. 1939, *R. C. Ching 21124* (KUN, PE); loc. eodem, 3000–3550 m, 12. Sept. 1987, *H. Ohba et al. 733* (KUN); loc. eodem, 2800 m, among herbs, 03. Aug. 1961, *R. L. Xiong 610462* (KUN); loc. eodem, open pastures,

FIGURE IV 1:4A *Codonopsis meleagris* Diels.

FIGURE IV 1:4B *Codonopsis meleagris* Diels. From Yunnan: Lijiang, 08. Aug. 2011, by You-Sheng CHEN.

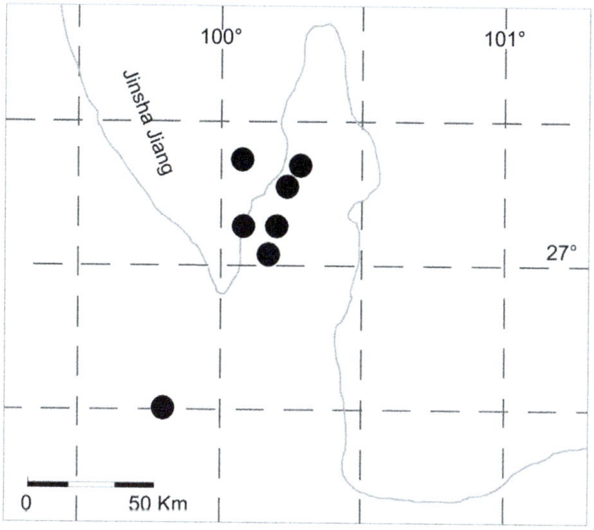

MAP IV 1:4 Distribution of *Codonopsis meleagris* Diels.

R. C. Ching 30503 (KUN, PE); loc. eodem, 10. Aug. 1940, *R. C. Ching 30936* (KUN, PE); loc. eodem, Oct. 1940, *R. C. Ching 31143* (KUN, PE); Gangheba Valley, 3300 m, 05. Oct. 1990, *CLD-90 No 851* (E); loc. eodem, 3170 m, alluvial lands, 07. Sept. 1963, *C. Z. Bao 63–20439* (KUN); near Wutuodi, 3780 m, alpine meadows, limestones, 23. Aug. 2000, *E. von Raab-Straube, et al. 151* (B); Yulong Snow Range, *Handel-Mazzetti 3714* (W, WU); loc. eodem, 3500 m, rocky crevices, 08. Aug. 1959, *K. M. Feng 22567* (HNWP, KUN, PE); loc. eodem, 22. Sept. 1986, *Z. T. Wang 869-397* (PE); loc. eodem, E slopes, 3050–3350 m, Jul. 1922, *J. F. Rock 5453* (PE); loc. eodem, 2600 m, 01. Sept. 1937, *T. T. Yü 15518* (E, PE); loc. eodem, 27°20'N, 3350 3660 m, Aug. 1910, *G. Forrest 6522* (E, K, W); loc. eodem, 3350–3660 m, Aug. 1913. *G.Forrest 10865* (E, K, PE); loc. eodem, 27°21'N, 100°11'E, 3660–3960 m, Sept. 1922, *G.Forrest 22315* (E, K, PE, W); loc. eodem, 01. Aug. 1922, *J. F. Rock 5271* (K) & 20. Sept. 1923, *J. F. Rock 10830* (E, PE); Yulong Snow Range, Liu'ogu Valley, 3660 m, 15. Aug. 1942, *K. M. Feng 9036* (KUN, PE); Yulong Snow Range, near Xianjiyan, 3400–4000 m, 12. Aug. 1942, *K. M. Feng 8933* (KUN, PE); s.loc., 3400–3600 m, 30. Jul. 1914. *C. Schneider 2068* (G, K,PE); s.loc., 3100 m, Oct. 1914, *C. Schneider 2861* (G); s. loc., Jun. 1933, *McLaren's Collectors 162* (E, K). **Shangri-la** (Zhongdian): s. loc., 3400 m, Aug. 1914, *C. Schneider 3252* (G, K, PE).

5. **Codonopsis gongshanica** Q. Wang & D. Y. Hong, Phytotaxa 188 (3): 147. 2014. Type: China, Yunnan, Gongshan, Bingzhongluo, Gawagapu Shan, alpine meadows, 3400–3800 m, 4. Aug. 2013, *X. H. Jin et al. ST1992* (holotype: PE; isotype: PE).

Roots carrot-shaped, 9–25 cm long, 0.4–1.3 cm in diameter. Stems erect, simple, very rarely with branches in lower and middle parts, 20–45 cm long, glabrous or sparsely villous at base. Leaves alternate, subrosulate; petiole 2–5 mm long, glabrous; leaf blade broad-ovate to lanceolate, 1.7–2.9 cm long, 0.6–1.7 cm broad, glabrous, base cuneate, margin entire to slightly sinuate, apex slightly obtuse. Flowers solitary, terminal on main stem. Calyx tube entirely adnate to hypanthium; calyx lobes lanceolate, 5–7 mm long, 1.4–2.6 mm broad, margin entire or slightly sinuate, apex obtuse. Corolla short-tubular, 0.9–1.3 cm long, 0.6–0.9 cm wide, shallowly lobed to ca. 1/6, gray or pale purple, with a dark purple ring just below throat, glabrous; corolla lobes ovate-deltoid. Stamens glabrous; filaments 4–6 mm long, slightly dilated at base; anthers 1–2 mm long. Fruit not seen. Figures IV 1:5A & B.

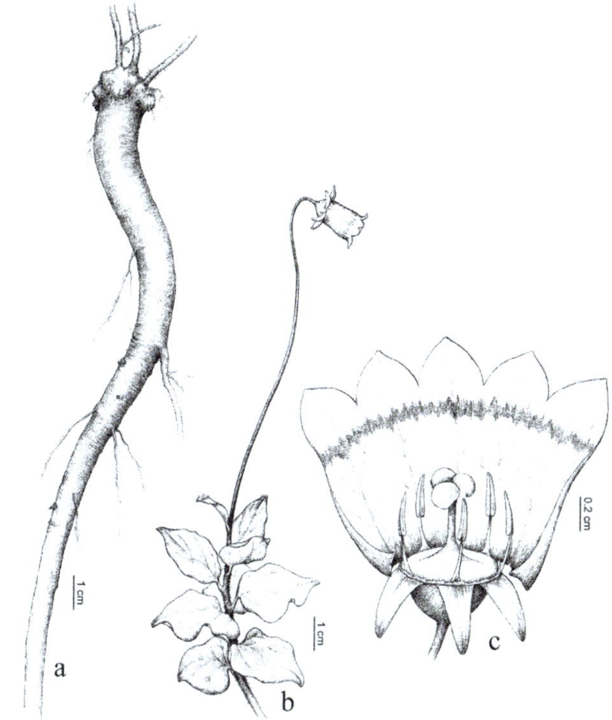

FIGURE IV 1:5A *Codonopsis gongshanica* Q. Wang & D. Y. Hong. **a**, root, caudex, and base of stems; **b**, a stem; **c**, a flower with corolla opened, showing calyx, stamens, disc, style, and stigma.

Phenology: flowering in July and August.

Habitats: alpine scrubs or meadows at an altitude of 3400–4600 m.

Distribution: **endemic** to China: confined to northern Gaoligong Shan in Yunnan (Map IV 1:5).

This species is related to *C. subsimplex* and *C. bhutanica* according to both gross morphology and molecular phylogenetic analysis (Wang & Hong, 2014). Howerer, it differs from them both in having a single flower on each stem, smaller calyx lobes, and a gray or pale purple corolla which has a dark purple ring just below the throat. Furthermore, it differs from the former in having a much smaller leaf blade and shorter petiole, while from the latter in having usually simple stems.

Additional specimens examined, **CHINA, YUNNAN, Fugong**: Lumadeng Xiang, 27°12'N, 98°41'E, 3700 m,

FIGURE IV 1:5B *Codonopsis gongshanica* Q. Wang & D. Y. Hong. **a** & **b**, from Yunnan: Gongshan, 04. Aug. 2013, by Qiang WANG.

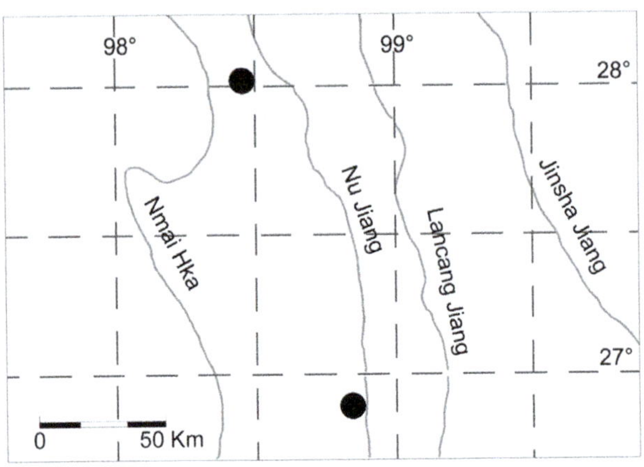

MAP IV 1:5 Distribution of *Codonopsis gongshanica* Q. Wang & D. Y. Hong.

meadows, 16. Aug. 2005, *Gaoligong Shan Biodiversity Survey 28529* (A, CAS).

6. **Codonopsis subscaposa** Kom., Trudy Imp. S.-Peterburgsk. Bot. Sada 29(1): 114. 1908; Chipp, J. Linn. Soc. Bot. 38: 388. 1908; Tsoong, Contr. Inst. Bot. Nat. Acad. Peiping 3(3): 101. fig. 11B. 1935; Anonymous, Icon. Cormophyt. Sin. 4: 380. fig. 6173. 1975; Shen & Hong in Hong, Fl. Reipubl. Popularis Sin. 73(2): 49. 1983; Hong, Lammers & Klein in Wu, Raven & Hong, Fl. China 19: 519. 2011. *Codonopsis pedunculosa* Franch. in sched. Type: China, W Sichuan, near Ta-tsien-lu (Kangding), 2740–3960 m, *A. E. Pratt 474* (lectotype here designated: K; isolectotypes: BM, E, G, LE, P, PE); China, W Sichuan, Schabagou, 08. Aug. 1893, *G. N. Potanin* s. n.(paratype: LE).

Roots carrot-shaped, 15–20 cm long, 0.5–1.5 cm in diameter. Stems erect, 40–100 cm tall, simple or branched below, sparsely pubescent, later glabrescent. Leaves on main stem alternate, those on branches opposite; petiole 1–10 cm long, sparsely villous; leaf blade ovate, oblong, elliptic or lanceolate, 2–13 cm long, 1.5–7.5 cm broad, adaxially green, sparsely hispidulous, abaxially gray-green, glabrous or sparsely villous along veins, base cuneate, margin sparsely dentate or crenate, or nearly entire, apex acute or obtuse. Flowers often solitary and terminal on main stem and branches, or in a terminal cyme of 2–4, long-pedicellate. Calyx tube adnate to whole hypanthium, semiglobose, 10–ribbed, sparsely villous; calyx lobes narrow-deltoid, 5–7 mm long, 2–3 mm broad, glabrous, margin serrate, rarely entire, apex acute; sinus between lobes broad and truncate. Corolla broad-campanulate, 1.5–3 cm long, 2–4 cm wide, lobed to near middle, variously coloured: yellow, creamy yellow, green but with red-purple veins or red-purple with yellow spots, mostly pale blue with dark purple veins, or white above and brown-purple below, glabrous on both sides

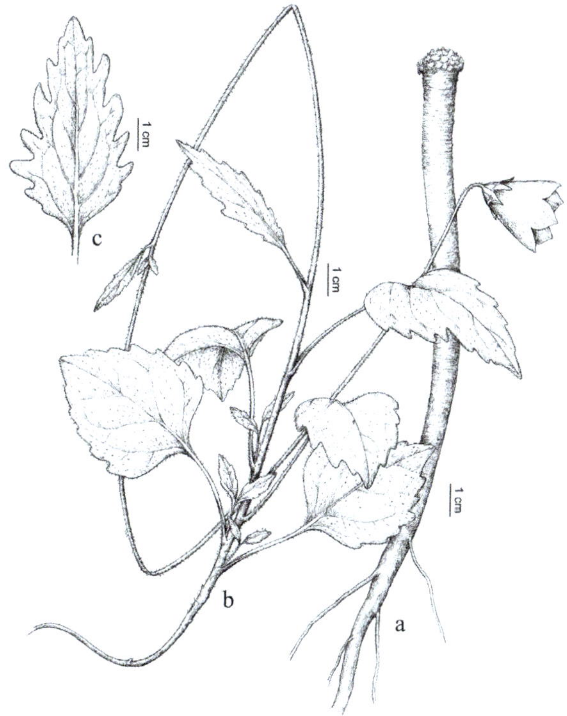

FIGURE IV 1:6A *Codonopsis subscaposa* Kom. **a**, root and caudex; **b**, a stem; **c**, another form of leaves.

or sparsely villous at apex of lobes. Stamens glabrous; filaments ca. 7 mm long, slightly dilated at base; anthers ca. 5 mm long. Capsule semiglobose below, conical above. Seeds ovoid, brown-yellow, smooth. Figures IV 1:6A & B.

Phenology: flowering from July to September; fruiting in September and October

Habitats: grassy slopes, thickets, or open woods at an altitude from 2000 to 5000m.

Distribution: **endemic** to China: W Sichuan and NW Yunnan (Map IV 1:6).

Additional specimens examined, **CHINA, SICHUAN,** s. loc., Jul. 1908, *E. H. Wilson 1086* (K, PE); s. loc., Jul. 1903, *E. H. Wilson 3985* (K, PE). **Barkam**: Ermu Shan, 3600 m, 09. Sept. 1957, *X. Li 71796* (PE, SZ); s. loc., 3000 m, 22. Jul. 1957, *SW China Forestry Lab 150* (CDBI); s. loc., 08. Jul. 1958, *C. L. Wu 32754* (PE) & *32759* (PE). **Danba**: Shachong, 3300 m, 16. Aug. 1958, *No. 8 Forest Managing Team 5622* (SZ). **Daocheng**: Daocheng to Julong road, E of pass, 28°51′N, 100°23′E, 3609 m, 12. Sept. 2010, *Kunming & Edinburgh Exped. to Sichuan 281* (E); Julong Township, 3600 m, forests, 17. Jul. 1973, *Sichuan Veget. Surv. 2143* (CDBI, KUN); Mula Township, 4000 m, edges of *Abies* forests, 17. Aug. 1971, *Sichuan Veget. Surv. 0386* (CDBI); Riwa, 3400 m, 12. Aug. 1982, *X. H. Hu 29341* (CDBI); loc. eodem, 3950 m, forests or edges of *Laris* forests, 11. Aug. 1982, *X. H. Hu 29185* (CDBI); Riwa to Gonggan, *Picea* forests, 25. Aug. 1981, *Qinghai-Xizang Exped. 4328* (CDBI, KUN, PE); Riwa to Julong Township, 3500–3600 m, *Picea* forests, 31. Jul. 1983, *K. Y. Lang et al. 2569* (KUN); Shengmu Township, 3500 m, edges of forests, 01. Aug. 1964, *J. D. Li & Z. M. Xu (Daocheng) 64–52* (PE,SM). **Dawu**: Fangdashen Shan, 3400 m, 16. Jul. 1958, *s. coll. 146* (SZ); Gaka, 3200 m, 20. Aug. 1960, *Medic. Plants Surv. 15322* (CDBI, SM); Gaka, Changlong Gou, 3700 m, 23. Jul. 1961, *S. Jiang 9662* (PE); Gaka, *Quercus* forests, 3700 m, 23. Jul. 1961, *S. Jiang 9667* (PE); from the town to Maoniu 4000–4500 m, forests, 05. Aug. 1940, *K. L. Chu 7517* (PE); Wenshen, 4000 m, forests, 08. Aug. 1974, *Sichuan Plant Exped. 5797* (CDBI, PE); Zheduo Shan Pass, NE of Bamei, on road to Danba, 3750 m, sparse forests, 31. Aug. 2010, *Kunming & Edinburgh Exped. to Sichuan 78* (E); 30°31′N, 101°38′E, 3550–3600 m, 25. Jul. 2007, *D. E. Boufford et al. 38157* (PE); s. loc., 3700 m, 25. Aug. 1974, *s. coll. 5896* (CDBI). **Hongyuan**: Shuajingsi, 3200 m, 08. Aug. 1957, *Z. Y. Zhang & H. F. Zhou 23561* (SZ); loc. eodem, 3500 m, 09. Aug. 1957, *H. F. Zhou & Z. Y. Zhang 23589* (SZ); Zhegu Shan 3300 m, roadsides, 31. Jul. 2007, *X. J. He et al. 146764* (SZ). **Huili**: Shandi Township, 2200 m, 29. Jun. 1960, *S. D. Li 60–1004* (SZ). **Jinchuan**: Chuosijia (Guanyinqiao), 3600 m, forests, 01. Aug. 1958, *P. X. Li 10491* (PE); loc. eodem, 3600 m, *Picea* forests, 20. Aug. 1958, *P. X. Li 10023* (PE); Taiyanghe, 2900 m, 25. Jun. 1979, *s. coll. 531* (SM); Wanli Township, 3400 m, forests, 23. Jul. 1958, *X. Li 78178* (PE, SZ). **Jiulong**: Riluku, 3800 m, oak forests, 20. Sept. 1991, *s. coll. 024* (CDBI); Sanyanlong, 3500 m, woods by stream, 05. Aug. 1979, *Jiulong Exped. 380* (SM); Tanggu, 3800 m, oak forests, 25. Aug. 1980, *Q. Q. Wang 22989* (CDBI); Wuxuhai, 3830 m, 06. Aug. 1979, *X. H. Hu 20539* (CDBI); Xieka, 15. Aug. 1980, *Z. A. Liu 22877* (CDBI). **Kangding:** Cheto La, 3500 m, meadows with shrubs, 03. Aug. 1934, *H. Smith 10924* (PE); Dalongbugou, 3500 m, 14. Jul. 1981, *Z. J. Zhao et al. 114935* (SZ); Dapao Shan, 3740–4210 m, among bushes, 20. Aug. 1934, *C. S. Liu 1226* (PE); Lugba, 3600 m, oak forests, 23. Jul. 1974, *Z. G. Liu 20801* (CDBI); Sandaoqiao, 28. Jul. 1951, *W. K. Hu & Z. He 10613* (PE, SZ); Tongolo, 1893, *J. A. Soulié 426* (G, K, P); loc. eodem, 1893, *J. A. Soulie 2769* (P); Yulingkong, Gomba La, 3400 m, 22. Jul. 1934, *H. Smith 10679* (PE); loc. eodem, 24. Jul. 1951, *W. K. Hu & Z. He 10465* (PE, SZ); loc. eodem, 26. Aug. 1939, *Y. C. Yang 3763* (PE); Zhedo Shan, 4000 m, 05. Aug. 1953, *H. L. Tsiang & J. H. Xiong 36503* (PE, SZ); loc. eodem, 14. Aug. 1986, *Z. T. Wang 868-353* (PE); Zhedotan, 3400 m, oak thickets, 04. Sept. 2012. *D. Y. Hong & Q. Wang H12072* (PE); Zhedotan, Kuangou, 3400 m, thickets, 01. Aug. 1963, *K. C. Kuan, et al. 1302* (PE); Zhedotan, Maojiagou, 3600 m, thickets, 16. Jul. 1953, *H. L. Tsiang 36321* (PE, SZ); Zhonggu Village, 3500 m, grassy slopes, 23. Jul. 1963, *K. C. Kuan, et al. 264* (PE); loc. eodem, 3200 m, edges of forests, 24. Jul. 1963, *K. C. Kuan et al. 327* (PE); s. loc. 2900 m, 23. Jul. 1934, *C. S. Liu 628* (PE); s. loc., 1934, *C. S. Liu 626* (PE); s. loc., Jul. 08. 1930, *Z. P. Huang et al. 1674* (KUN, PE) ; s. loc., 1893, *J. A. Soulié 1060* (G, P) & *2077* (P); s. loc., 1924, *R. Cunningham 484 A* (E) & *484B* (E); s. loc., Sept. 1938, *McLaren AC207* (E). **Litang**: Labo

FIGURE IV 1:6B *Codonopsis subscaposa* Kom. **a**, from Sichuan: Kangding, 19. Jul. 2010, by Yan YU; **b**, from Sichuan: Xiaojin, 09. Aug. 2007, by You-Sheng CHEN; **c** & **e**, from Sichuan: Xiangcheng, 31. Jul. 2012, by Fu-Sheng YANG; **d**, from Sichuan: Muli, 18. Jul. 2007, by You-Sheng CHEN.

Township, 4100 m, forests, 03. Aug. 1973, *Sichuan Veget. Surv. 3718* (CDBI, KUN, PE); Litang-Yalong Divide, 3350 m, 21. Jul. 1910, *F. Kingdon-Ward 4446* (E); between Litang and Shou-Chu River, between Wa-Erh-Dje and Garu, 3900 m, Jul.–Aug. 1928, *J. F. Rock 16786* (PE). **Lixian**: Dabanzhao, 3200 m, 24. Jul. 1957, *D. Z. Deng 1945* (SZ); Miyaluo, 3700 m, 18. Aug. 1956, *D. P. He 45862* (SZ); s. loc., *D. Z. Deng 1933* (SZ); s. loc., 3000 m, 14. Jul. 1956, *W. K. Hu 185* (SZ). **Muli**: Bar Pasture, 5000 m, shrubs, 18. Aug. 1978, *C. S. Cao 6566* (CDBI, SZ); on the way to Donglang Township, 11. Jul. 1966, *s. coll. 66–64* (SM); Gibboh Shan, 3960 m, 04. 05. 1932, *J. F. Rock 24168* (E,K, PE); Gibboh Shan, 3960 m, 10–11, 1932, *J. F. Rock 24596* (PE); Gibo Shan, 2800 m, woods, 07. Aug. 1937, *T. T. Yü 7656* (KUN, PE); Mêri Pasture, 4800 m, 07. Aug. 1978, *C. S. Cao 6511* (CDBI, SZ); Shaoxiangliangzi, 3600 m, grassy slopes, 28. Jul. 1978, *C. S. Cao et al. 6830*

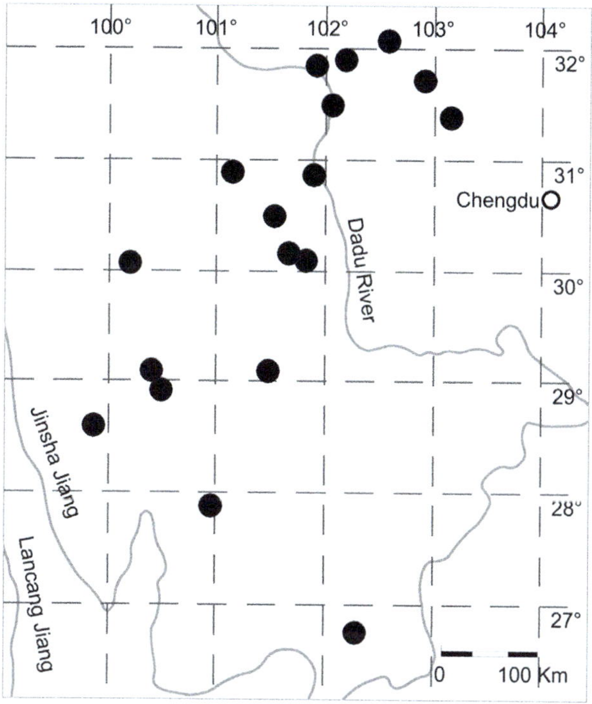

MAP IV 1:6 Distribution of *Codonopsis subscaposa* Kom.

(CDBI, SZ); loc., eodem, 3600 m, 19. Oct. 1982, *Y. B. Yang & G. Yao 294* (CDBI); Wachin, Jinechang, 3800 m, grassy slopes, 30. Oct. 1937, *T. T. Yü 14685* (KUN, PE); Yazui Forest Farm, 3400 m, *Abies* forests, 19. Aug. 1983, *Qinghai-Xizang Exped. 13011* (KUN, PE); Yungning-Muli, 3900–4000 m, 25. Jul. 1915, *Handel-Mazzetti 7234* (E, K, WU); Distr. No.3, near Buchang, 3200 m, pine forests, 12. Sept. 1983, *Qinghai-Xizang Exped. 13978* (KUN); Distr. No.3, Mabanya to Donglang, 3100 m, edges of *Picea* forests, 08. Sept. 1983, *Qinghai-Xizang Exped. 14543* (KUN, PE). **Xiangcheng**: Chitu Township, 28°35'N, 100°18'E, 3200 m, 03. Jul. 2011, *X. F. Gao et al. 13301* (CDBI); Maxionggou, 29°7'26″N, 99°59'48″E, 4000–4300 m, 09. Jul. 1998, *D. E. Boufford et al. 28482*(A, E); loc. eodem, 3926 m, meadows in forests, 31. Jul. 2012, *F. S. Yang et al. 045* (PE); Ra'o, 4500 m, forests, 08. Jul. 1973, *Sichuan Plant Exped. (Xiang) 2894* (CDBI, KUN, PE); Rewu, 3850–3900 m, forests, 09. Aug. 1981, *Qinghai-Xizang Exped. 3981* (CDBI, KUN, PE); No.4 Distr., 3500–3700 m, *Picea* forests, 11. Aug. 1981, *Qinghai-Xizang Exped. 4611* (CDBI, KUN, PE); loc.eodem, to Mt Shizhao, *Picea* forests, 12. Aug. 1981, *Qinghai-Xizang Exped.4726* (CDBI, KUN, PE); s. loc., 3500 m, pine forests, 24. Jul. 1964, *E. P. Tang & Z. Ou 007* (PE). **Xiaojin**: Lianghekou, 3500 m, 24. Jul. 1958, *S. S. Chang & Y. X. Re 6158* (CDBI, PE, SZ); loc. eodem, Mozhigou, 3000 m, 20. Jul. 1958, *S. S. Chang & Y. X. Re 6035* (PE, SZ); Sannonggou, 3000 m, 13. Aug. 1973, *Sichuan Plant Exped. 9720* (CDBI, PE). **Yajiang** : Jianziwan Shan, 3200–3500 m, *Picea* forests, 07. Aug. 1983, *K. Y. Lang et al. 2905* (KUN); Tiemenggou, 3700 m, N slopes, 01. Aug. 1959, *S. Jiang & C. L. Jin 2640* (KUN, PE, SZ); Xiolo, 3800 m, alpine meadows. 09. Aug. 1979, *Yajiang Exped. 300* (SM) ; No.4 Distr., 3750 m, 28. Sept. 1974, *Q. Q. Wang 5033* (CDBI, PE). **Zamtang**: Hongxin, *Picea* openings, 3600 m, 20. Jul. 1975, *Sichuan Plant Surv. 9363* (CDBI, PE). **YUNNAN, Shangri-la** (Zhongdian): Dongwang, 28°37'8″N, 99°49'17″E, 3700–3800 m, 26. Jul.1998, *D. E. Boufford et al.29265* (E); loc. eodem, 3580 m, edges of *Picea* forests, 17. Jul. 2010, *Kham Exped. 10–0185* (PE); Rengtsa, 3700 m, forest margins, 17. Aug. 1937, *T. T. Yü. 12772* (KUN, PE) ; below Rengtsa, 3800 m, forest margins, 17. Aug. 1937, *T. T. Yü 12768* (KUN, PE).

7. Codonopsis cardiophylla Diels ex Kom., Trudy Imp. S.-Peterburgsk. Bot. Sada 29(1): 117. 1908; Chipp, J. Linn. Soc. Bot. 38: 385. 1908; Anthony, Notes Roy. Bot. Gard. Edinburgh 15: 187. 1926, p. p., excl. specim. Tibet.; Nannfeldt, Notes Roy. Bot. Gard. Edinburgh 16: 155. pl. 228. 1931; Anonymous, Icon. Cormophyt. Sin. 4: 380. fig. 6174. 1975; Shen & Hong in Hong, Fl. Reipubl. Popularis Sin. 73(2): 60. 1983; Hong, Lammers & Klein in Wu, Raven & Hong, Fl. China 19: 522. 2011. Type: China, W Hupeh (Hubei), Jul. 1901, *E. H. Wilson 2381* (lectotype here designated: K; isolectotypes: E, P, W).

Roots carrot-shaped, 10–15 cm long, 1–1.5 cm in diameter. Stems several from one caudex, nearly erect, ascending, rarely scandent , 20–100 cm tall; branches 10–25 cm long, very sparsely white-hispidulous, glabrescent, lower branches slender, sterile or fertile, upper branches fertile. Leaves on main stem usually altermate, those on branches mostly opposite; petiole absent or extremely short, less than 3 mm long, glabrous or hirsute; leaf blade ovate to lanceolate, 1–4.7 cm long, 0.5–4.5 cm broad, adaxially nearly glabrous, abaxially sparsely hispidulous, base shallowly cordate or rounded, margin entire, rarely crenulate, recurved, apex obtuse. Flowers terminal on main stem and upper branches, occasionally also on lower branches; pedicel sparsely hirsute, gradually glabrescent. Calyx tube adnate to whole hypanthium, semiglobose, 10-ribbed, glabrous; calyx lobes lanceolate to nearly deltoid, 9–12 mm long, 4–5.5 mm broad, glabrous or sparsely hispidulous, revolute, margin entire, apex obtuse; sinus between calyx lobes narrow and pointed. Corolla campanulate, 2–3.2 cm long, 2.5–3 cm wide, shallowly lobed, purple-red or pale blue with red-purple or brown-red spots; corolla lobes ovate, 1–1.1 cm long, 1–1.2 cm broad, glabrous or villous inside, apex acute. Stamens glabrous; filaments 6–7 mm long, slightly dilated at base; anthers ca. 5 mm long. Capsule 8–10 mm in diameter, inferior part semiglobose, superior part conical, valves 5–7 mm long. Seeds ellipsoid, 1–1.5 mm long, ca. 0.5 mm wide, brown-yellow.

Distribution: **endemic** to China: Chongqing, W Hubei, S Shaanxi, Shanxi, and Sichuan (Map IV 1:7).

Notes: Komarov designated the holotype in B (*Wilson 2381*). However, I visited the herbarium B in 2011, but no any specimen of this species was found there. Apparently, it was destroyed in the second world war, but fortunately, at least four isotypes are preserved. Thus, I designate here the specimen in Kew as lectotype, while the other three as isolectotypes.

1a. Stems erect or ascending, 20–70 cm tall; leaf blade 1–3.5 cm long, 0.5–3 cm broad; corolla lobes villous inside...7a. subsp. *cardiophylla*

1b. Stems ascending or scandent, 50–100 cm long; leaf blade 3–4.7 cm long, 2–4.5 cm broad; corolla lobes glabrous...7b. subsp. *megaphylla*

7a. Codonopsis cardiophylla subsp. **cardiophylla**

Stems usually nearly erect or ascending, 20–70 cm tall. Leaf blade 1–3.5 cm long, 0.5–3 cm broad. Flowers terminal only on main stem and upper branches. Corolla pale blue, with red-purple or brown-red spots; corolla lobes villous inside . Figures IV 1:7A & B.

Phenology: flowering and fruiting from July to October.
Habitats: grassy slopes or cliffs; 2300–3200 m.

Additional specimens examined, **CHINA, CHONGQING, Chengkou**: s.loc., 2500 m, 29. Jul. 1895, *R. P. Farges 1344* (P); s. loc., 1895–1897, *R. P. Farges* s. n. (P). **SHAANXI, Foping:** Dajiangou, mountain summit, 2900 m, 17. Jul. 1952, *P. C. Kuo 1733* (CDBI, PE). **Huaxian**: Erlong Shan, 2400 m, grassy slopes, 21. Jul. 1952, *T. P. Wang 15875* (KUN, PE). **Weinan**: Jianyuling, sunny slopes, 2400 m, 08. Jul. 1982, *T. P. Wang 15720* (PE). **SHANXI, Jiexiu:** Shiwo Shan, forests, 17. 07. 1959, *T. W. Liu 5398* (HNWP). **Yuanqu**: Shunwangping, 2400 m, 24. Jul. 1924, *H. Smith 6762* (E, PE). **SICHUAN, Nanjiang**: Dawu Shan, 2300 m, cliffs, 13. Jul. 2011, *Bashan Exped. 5640* (PE). **Qingchuan**: Qinghua, s. loc., 3200 m, 11. Aug. 1978, *Qingchuan Exped. 194* (SM).

7b. Codonopsis cardiophylla subsp. **megaphylla** D.Y. Hong, Plant Divers. Resour. 36(3): 295. 2014. Type : China, Sichuan, Hongya, Forest Farm, Tuanbao Shan, 2800 m, meadows, 20, Jul. 1994, *W. K. Bao et al. 3165* [holotype: CDBI (CDBI0141276); isotype: CDBI (CDBI0141275)].

Stems usually scandent or ascending, 50–100 cm long. Leaf blade 3–4.7 cm long, 2–4.5 cm broad. Flowers terminal not only on main stem and upper branches, but also on lower branches. Corolla purple-red, entirely glabrous.

Phenology: flowering in July and August.
Habilats: forests or meadows; 2300–2800 m.
Distribution: confined to Sichuan (Hongya).

Notes: The subspecies *megaphylla* differs from the typical one in having larger leaves, stems longer and scandent, nearly twining at the upper part, and the corolla entirely glabrous. It may merit specific rank.

Additional specimens examined, **CHINA, SICHUAN, Hongya**: Hongya Forest Farm, Dazhonggan, 2300 m, 21. Jul. 1995, *W. K. Bao et al. 3308* (CDBI); loc. eodem, 2400 m, 21. jul. 1995, *W. K. Bao et al. 3309* (CDBI).

8. Codonopsis ovata Benth. in Royle, Ill. Bot. Himal. 253. t. 69. fig. 3. 1834; Hooker & Thomson, J. Proc. Linn. Soc. Bot. 2: 15. 1857, p. p., excl. pl. Soongar.; Clarke in Hooker, Fl. Brit. Ind. 3: 433. 1881; Komarov, Trudy Imp. S. -Peterburgsk. Bot. Sada 29(1): 115. 1908, p. p., excl. specim. Chinen.; Chipp, J. Linn. Soc. Bot. 38: 384. 1908. p. p., excl. vars. *Wahlenbergia roylei* A. DC. in de Candolle, Prodr. 7: 425. 1839. Type: India, Kashmir, *Royle* s. n.(holotype: LIV, n. v.)

Roots carrot-shaped, 15 cm long, ca. 1.5 cm in diameter. Stems erect, or sometimes ascending, sparsely villous, 30–80 cm tall, branches mostly or only aggregated at lower part, lower branches leafy, mostly sterile, upper ones more sparsely leafy, fertile. Leaves on main stem alternate; petiole absent to 8 mm long; leaf blade orbicular to ovate-lanceolate, 0.6–2.8 cm long, 0.6–2.2 cm broad, both adaxially and abaxially white-puberulent, sometimes very densely, base shallow-cordate, rounded to broad-cuneate, margin entire, apex rounded to acute; those on branches nearly opposite, similar to those on main stem. Flowers solitary, terminal only on main stem or also on upper branches, rarely also on lower branches; pedicel 1–15 cm long, mostly sparsely villous, rarely glabrous. Calyx tube adnate to whole hypanthium, rounded at base, 10-ribbed, glabrous, rarely sparsely villous along ribs; calyx

FIGURE IV 1:7A *Codonopsis cardiophylla* Diels ex Kom. subsp. *cardiophylla*. **a**, a plant; **b**, abaxial side of leaf, showing revolute margin.

FIGURE IV 1:7B *Codonopsis cardiophylla* Diels ex Kom. subsp. *cardiophylla*. **a–c**, from Sichuan : Nanjiang, 13. Jul. 2011, by Yang-Jun LAI.

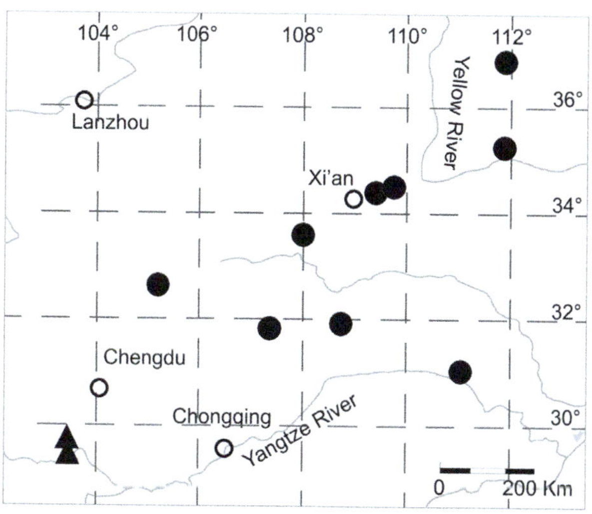

MAP IV 1:7 Distribution of *Codonopsis cardiophylla* Diels ex Kom. Dots: subsp. *cardiophylla*; triangles: subsp. *megaphylla* D. Y. Hong.

lobes ovate to ovate-lanceolate, 5–10(–12) mm long, 4–8 mm broad, glabrous or hispidulous at apex, margin mostly entire, rarely serrate, more or less ciliate, revolute. Corolla funnel-shaped, lobed to 1/3, 1.6–3 cm long, 2–3 cm wide, light blue or purple-blue, purple inside, glabrous. Stamens totally glabrous; filaments slightly dilated at base. Capsule with inferior part semiglobose, superior part conical, ca. 1.7 cm long, 1.2 cm in diameter. Seeds not seen. Figures IV 1:8A.

Chromosomes: 2n=16 (Khatoon & Ali, 1993).

Phenology: flowering in July and August; fruiting in September.

Habitats: bushes among rocks or alpine meadows; 2130–4370 m.

Distribution: NW India, Kyrgyzstan, N Pakistan, Tajikistan, and Uzbekistan (Map IV 1:8).

Notes: 1) *Codonopsis ovata* was often confused with *C. clematidea*, *C. obtusa*, and other species. As a result of

FIGURE IV 1:8A *Codonopsis ovata* Benth. **a**, root, caudex, and basal part of stem; **b**, a stem; **c**, upper part of a flower with corolla opened, showing stamens, disc, style, and stigma; **d**, a calyx lobe.

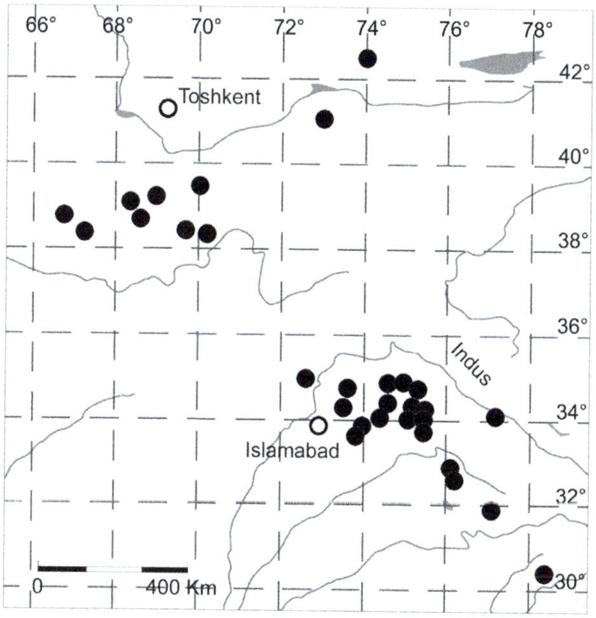

MAP IV 1:8 Distribution of *Codonopsis ovata* Benth.

critical examination of a large quantity of specimens of this group in BM, E, K, LE, etc., I found that these three are all distinct species. The most diagnostic characters of *C. ovata* are calyx lobes which are revolute, ciliate, shorter, 5–10 (–12) mm long, 5–8 mm broad. With reflexed and glabrous calyx lobes and ovate leaf blade, *C. clematidea*, is easily distinguished from the other two. *C. obtusa* is characterized with lanceolate leaf blade (length / width over 2), reflexed, subulate, and nearly always hispidulous calyx lobes. 2) Nasir (1984: Fl. Pakist. 155: 2) failed to record *Codonopsis ovata* Benth. in Pakistan, because he identified specimens of this species as *C. clematidea*.

Additional specimens examined**, INDIA, Himachal Pradesh:** Pangs, Dunai to Bindrabani, 2900 m, 14. Aug. 1897, *J. H. Lace 1620* (E); Pangs Valley, Chamba, Dec. 1880, *R. Ellis* s. n.(K); Rahla Kulu, 3660 m, 3. Aug. 1916 *R. E. Cooper 5242* (E); Sural Valley, 2440–3660 m, 17. Jul. 1899, *Harsuku* s.n. (K); s. loc., *G. Watt 2493* (E). **Jammu-Kashmir**: Aphawab, 3350 m, 06. Aug. 1933, *Jvenning K45* (K); Baltal, 2900 m, 28. Sept. 1848, *T. Thomson* s. n. (K); Butin Pantsar, Wangat Valley, 3350 m, 13. Aug. 1940. *P. M. Pinfold 270* (BM, E); Gagongiv, Sim Valley, 3350 m, 22. Aug. 1940, F. *Ludlow & G. Sherriff 7977* (BM); ca. 3350 m, Aug. 1905, *A. Meebald 580* (G); above Gulmarg, 3540 m, fl. light blue, *G. L. B. Fuller 128* (K); Keilan, above Gulmarg, 3660 m, Aug. 1875, *Aitcheson 05* (K); Khilanmarg, ca. 3050 m, 10. Aug. 1956, *O. Polunin 56/138* (B, BM, E); Kolahoi Valley, Earthy, ca. 3660 m, 27. Aug. 1956, *O. Polunin 56/570* (B, BM, E); Kolahoi & Zojpal, over 3350 m, *M. A. Evershed* s.n. (BM); Ladakh, 3420 m, 17. Jul. 1976, *Hans Hartmann 2061* (G); Liddar Valley, 34°10′ N, 75°15′ E, 3650 m, 02. Aug. 1965, *J. D. A. Stainton 5017* (BM); Margan Pass, 3960–4270 m,, Aug. 1936, *M. K. Timins 189* (BM); Sonsal Nala, Liddar Valley, 3960–4270 m, 31. Jul. 1893, *J. F. Duthie 13355* (BM, E); Tilpatra Forest, Nr. Rampur, Jhelum Valley, 2930 m, 6. Jul. 1940, *F. Ludlow & G. Sherriff 7737* (BM, E); Tragbal, 3350 m, 20. Jul. 1876, *C. B. Clarke 29251A* (BM) & *C. B. Clarke 29251 B* (K); Upper Chenab Valley, 20. Jun. 1879, *R. Ellis 265* (K); Zojpal, Kolahoi Glacier, 07. Sept. 1913, *M. A. Evershed* s. n. (BM); 2290 m, 27. Jun. 1939, *Desu 145* (BM). **Uttar Pradesh**: Tehri-Garhwal, 3660–3960 m, 19. Jul. 1883, *J. F. Duthie 609* (BM).

KYRGYZSTAN, **Alexandrovsk**, 42°30′N, 74°E, 19. Jul. 1930, *M. M. Iljin 119* (LE); Djalalabad Distr., Kirovsky Nutwood Farm, 41°N, 73°E, *O. E. Neustrueva-Knorring 226* (LE).

PAKISTAN, Baltistan: Deosai, Mir Panzil Pass, 28. Jul. 1940, *R. R. Stewart 19955* (PE); Deosai Plains, 3660 m, 19. Jul. 1940, *R. R. Stewart 19500* (PE); Gieuwal, 4570 m, *Slo-livzka* s. n.(CAL); Kashmir, Panchtami, Aug. 1938. *M. Nath* s. n. (KUH); Kashmir, Surgun et au-dessus, 3200–4000 m, 28. Jul. 1953, *F. Schmid 533* (BM, G); Kishenganga, Barai Pass, 3510 m, 22. Jul. 1935, *F. Ludlow & G. Sherriff 1454* (E); Kishenganga Valley, Folowai, 2130 m, 17. Jul. 1935, *F. Ludlow & G. Sherriff 1431* (E); Kishenganga Valley and the road to Nanga Parbat, 3350–3960 m, *R. R. & I. D. Stewart 19909* (KUH); Kishenganga Valley, Kel-Taubat, 1980 m, 24–25. Jul. 1939, *R. R. Stewart 17859* (PE); Nanga Parbat, 3300 m, 19. Aug. 1993, *Nüsser 357* (B); loc. eodem, 3200 m, 21. Jul. 1993, *Nüsser 121* (B); above Tarakhal, 3050 m, alpine meadows, fl. purple blue, 31. Jul. 1936, *W. Koelz 9155* (PE). **Chitral**: Barir Kafristan, 23. Jul. 1976, *M. Shah et al. 2063* (ISL); Molikhoo, Joormoli /Sahat top, 2845 m, 16. Jun. 2007, *H. Ali 5723* (KUH). **Gilgit**: Nattar Sokes, 3960 m, 21. Jul. 1954, *R. R. Stewart 26406* (BM); s. loc., 1885, *G. M. Giles Fl. of Gilgit* (CAL). **Hazára**: Bantara Gali, Poonch, 2740–3050 m, 05. Jul. 1952, *R. R.Stewart & E. Nasir 24112* (KUH, W) ; Kaghan, Silán, 12. Jul. 1899, *Inayat* s. n.(CAL); Kaghan Valley, 3110 m, 19. Jul. 1896, *Inayat 19896a* (W); Kaghan Valley, Paya, ca. 3050 m, 25. Jul.1984, *Muqarrab Shah & Jamohed 41* (E); Kaghan Valley, Sarnl, 13. Aug. 1897, *Inayat 21946* (CAL); Siam Valley, 30. Jun. 1896, *Inayat 19896* (CAL); s. loc., 14. Jul. 1978, *M. Shah et al. 2314* (ISL). **Swat**: Bishigram, 3350 m, 9. Jul. 1953, *R. R. Stewart & A. Rahman 24928* (BM); loc. eodem, 3050–3350 m, 10. Jul. 1953, *R. R. Stewart & A.Rahman 24969* (BM); s. loc., 09. Aug. 1992, *M. Shah et al.* s. n.(PMNH).

TAJIKISTAN, Buhara, Gissar, 38°35′ N, 68°35′ E, 30. Jun. 1896, *V. I. Lipsky 2303* (LE); Mt. Childuhtaron, 38°20′ N, 70°10′ E, 30. Jul. 1969, *R. Kamelin* s. n. (LE); Gissarsky Ridge, 39°10′ N, 68°25′ E, meadows, 08. Jul. 1949, *O. E. Neustrueva-Knorring 61* (LE); Sank Gardan Ravine, between Sarich Saglyk and Bahcha, 04. Aug. 1878, *s. coll.* s. n. (LE); Vahshsky Ridge, near Tagicomar Village, 38°30′ N, 69°40′ E, 17. Jun. 1960, *V. P. Bochantsev & T.*

V. Egorova 1424 (LE); Zeravshansky Ridge, Nofin Village, 39°25′ N, 70° E, 24. Aug. 1925, *V. A. Dubyansky & N. A. Basilevskaya* s. n. (LE); Zeravshansky Ridge, near Hadnel-Makmet Village, 39°20′ N, 69°E, *P. Ovehinniko & A. Slobodov 1007* (LE).

UZBEKISTAN, Mt. Baisun, 38°15′ N, 67°10′ E, 1500 m, 07. Aug. 1958, *Sabirov & Shmatova 47* (LE); West Pamir-Alai, near Tash-Kurgan Village, 38°55′ N, 66°45′ E, 12. Jul. 1936, *B. Bochantsev & A. Butkov 849* (LE).

9. **Codonopsis clematidea** (Schrenk) C. B. Clarke in Hooker, Fl. Brit. Ind. 3: 433. 1881; Komarov, Trudy Imp. S. - Peterburgsk. Bot. Sada 29(1): 115, 1908; Nannfeldt, Acta Horti Gothob. 5: 27. 1930; Tsoong, Contr. Inst. Bot. Nat. Acad. Peiping 3(3): 93. 1935; Rechinger & Schiman-Czeika in Rechinger, Fl. Iran. 13:3. 1965; Anonymous, Icon. Cormophyt. Sin. 4: 381. fig. 6175. 1975; Shen & Hong in Hong, Fl. Reipubl. Popularis Sin. 73(2): 52. 1983; Hong, Lammers & Klein in Wu, Raven & Hong, Fl. China 19: 520. 2011. ≡*Wahlenbergia clematidea* Schrenk in Fischer & Meyer, Enum. Pl. Nov. Songar. 1: 38. 1841. ≡*Glossocomia clematidea* (Schrenk) Fisch. in Regel, Gartenflora 5: 226. t. 167. fig. 2. 1856. Type: "Songarei, Alatau, Dschillkaragai", *Schrenk* s. n.(holotype: LE).

Codonopsis ovata Benth. var. *cuspidata* Chipp, J. Linn. Soc. Bot. 38: 385. 1908, p.p., excl. specim. "China: Tachielu (Kangding), *Soulie*, 593" (= *Codonopsis pilosula* subsp. *handeliana*). Type: " Tibet Occ." (India: Kashmir), Yulchung, Zanskar River, 2750–3350 m, 02. Jul. 1848, *T. Thomson* s. n. [lectotype here designated: K (K000814194)]; "Tibet Occ." (India: Kashmir), Wandla, Zanskar, 2750–3350 m, 05. Jul. 1848, *T. Thomson* s. n.[paratypes: K (K000814192 & K000814195)]; "Tibet Occ." (India: Kashmir), Jadah, 3050 m, 05. Jul. 1848, *T. Thomson* s. n.[paratype: K (K000814197)]; "Tibet Occ." (India: Kashmir), s. loc., 2750–3350 m, 02. Jul. 1848, *T. Thomson* s. n. [paratype: K (K000814196)]; "Tibet Occ." (India: Kashmir), s. loc., 2750–3350 m, Sept. 1849, *T. Thomson* s. n. [paratype: K (K000814193)].

Roots carrot-shaped, 25–45 cm long, 1–3 cm in diameter. Stems single or caespitose, erect or ascending, rarely trailing, 50–100 cm long, branches upright or slightly spreading, mostly at lower part and sterile but those at upper part usually fertile, hispidulous when young. Leaves on main stem alternate, those on branches opposite; petiole up to 2.5 cm long, sparsely hispidulous; leaf blade ovate to lanceolate, 1–2.8(–5.2) cm long, 0.8–1.5(–3.2) cm broad, adaxially sparsely or densely hispidulous, abaxially densely hispidulous, base slightly cordate or rounded, margin entire, apex acute. Flowers solitary, terminal on main stem and branches, long-pedicellate; pedicel sparsely white hispidulous. Calyx tube adnate to whole hypanthium, semiglobose, 10-ribbed, glabrous or rarely sparsely white-scaberulose; calyx lobes ovate, elliptic or ovate-lanceolate, 10–18 mm long, 5–8.5 mm broad, glabrous or apex sparsely hispidulous, strongly reflexed from anthesis, margin entire, apex acute; sinus between calyx lobes narrow and pointed. Corolla campanulate, 1.8–2.8 cm long, purplish or bluish with dark blue veins and purple spots inside, or purple, glabrous; corolla lobes 1/3–1/4 as long as corolla length. Stamens glabrous; filaments 5–6 mm long, slightly dilated at base; anthers oblong, 5–6 mm long. Capsule semiglobose below, conical above, 1.5–2 cm long, 1.2–1.5 cm in diameter, apex acute; persistent calyx lobes much enlarged. Seeds narrowly ellipsoid, slightly compressed, pale brown-yellow, smooth. Figures IV 1:9A & B.

Chromosomes: 2n=16 (Khatoon & Ali, 1993).

Phenology: flowering and fruiting from July to October.

Habitats: forests, edges of forests, or meadows ; 1500–4600 m.

Distribution: widely distributed in the W Himalaya, Karakoram, NE Hindu-Kush, Pamir, Tianshan, and Altai mountains (Map IV 1:9).

Additional specimens examined, **AFGHANISTAN, Nuristan:** Pashki, 2300 m, 24. 07. 1948, *L. Edelberg 1339* (W). **Paktia**: Kurram Valley, at the base of Sikaram, near Kaiwas, 3050 m, Jul.–Aug. 1879, *J. E. T. Aitcheson 748* (p. p.) (K).

CHINA, XINJIANG, s. loc., 1841, *Karelin & Kiriloff* s. n.(G, K, W, WU); Songar, Mt Alatau, Jul. 1841, *C. A. Meyer* s. n.(G); Alatau, *A. de Bunge* s. n.(K); Alatau, 1841. *Karelin & Kiriloff 1702* (B, BM, K). **Aksu**: Aksu to Tagelake, 1960 m, 18. Jun. 1978, *Xinjiang Collection Team 533* (PE). **Bogda Shan**: Jun. 1962, *s. coll. 342* (PE); Aug. 1930, *T. H. Ting* s. n.(PE); Bogda Mts, ca. 2290 m, Aug. 1931, *R. H. S. Ridley* s. n.(K); near Wuji Palace, 18. Aug. 1931, *T. N. Liou 3378* (PE). **Fuhai**: Daqiao Forest Farm, meadows, 02. Aug. 1964, *G. L. Chu 5773* (PE); Ershesi, 1530 m, rocky crevices near river banks, 02. Aug. 1957, *J. Z. He & Y. L. Li* s. n.(PE). **Fuyun**: near the town, 1200 m, *Larix* forests, 18. Aug. 1956, *R. C. Ching 2153* (KUN, PE); s. loc., flood lands, 1200 m, *Betula* forests, 13. Aug. 1956, *R. C. Ching 1804* (PE). **Gongliu**: Hetaogou, 800 m, forests, 19. Jul. 1974, *Xinjiang Exped. 708* (HNWP); Mohe, Kurdeng, 1330m, 21. Aug. 1978, *NW Inst. Bot. Xinjiang Exped. 2629* (PE). **Hejing**: near Balguntay, 1770 m, 13. Aug. 1978, *NW Inst. Bot. Xinjiang Exped. 2322* (PE); near Bayanbulak, 2700 m, meadows, 16. Aug. 1978, *NW Inst. Bot. Xinjiang Exped. 2454* (PE). **Hoxud**: Bort Pasture, rocky slopes, 1835 m, 03. Aug. 1958, *Acad. Sin. Xinjiang Compreh. Exped. 6947* (KUN, PE). **Huocheng**: Daxigou, 1200 m, forests, 15. May. 1959, *A. R. Li & J. N. Zhu 10382* (PE); Guozigou, 1500 m, edges of forests, 14. Jul. 1974, *Xinjiang Exped. 451* (HNWP); Xin'ertai, Guozigou, 1500 m, *Picea* forests, 14. Jul. 1974, *Y. R. Lin 74187* (PE). **Jinghe**: 40 km S of Jinghe Town, 02. Sept. 1957. *K. C. Kuan 4124* (PE). **Kashgar Mts.**, Jul. 1924, *C. P. Skrine* s. n.(K). **Kizil**: Kutan

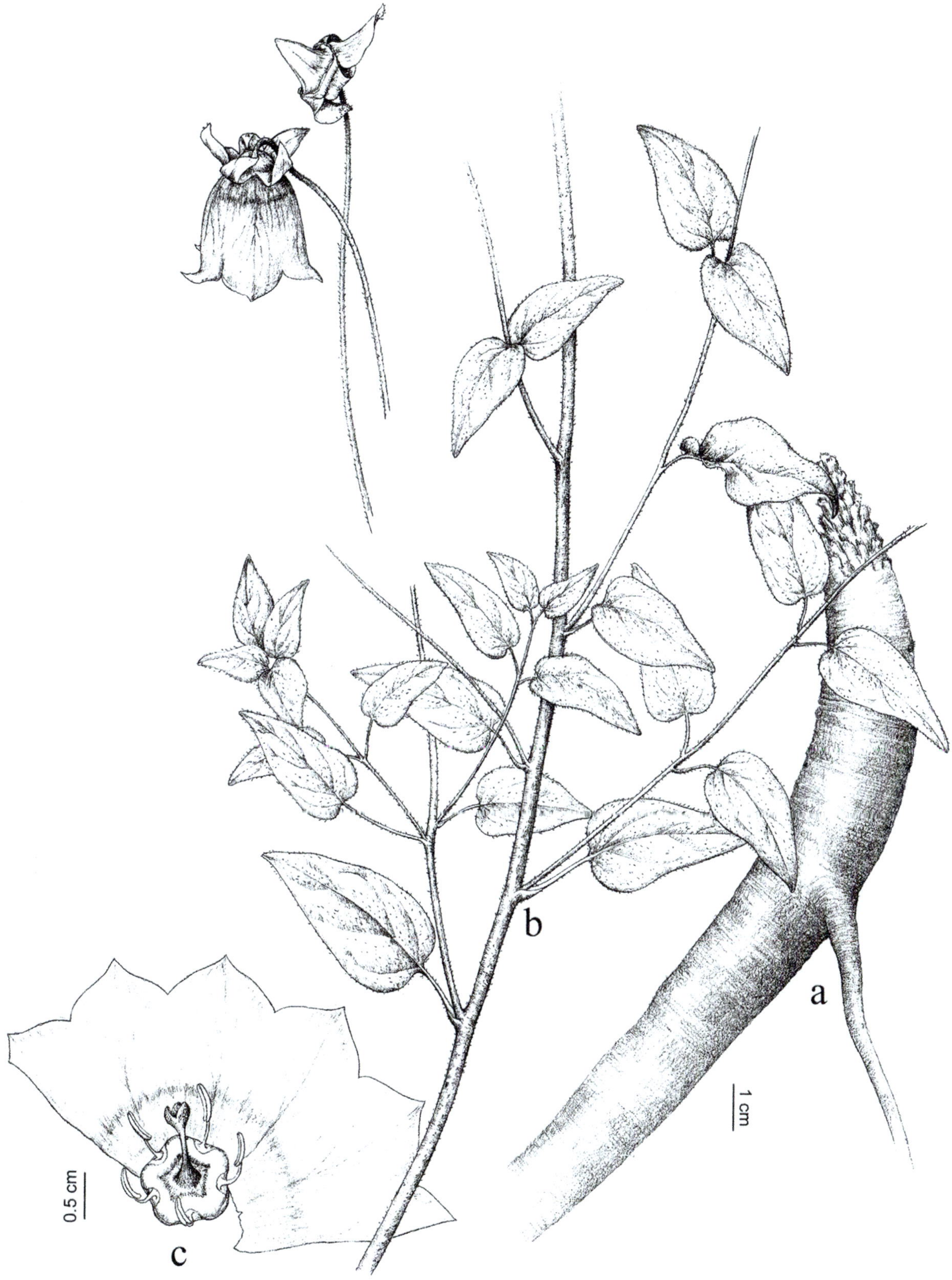

FIGURE IV 1:9A *Codonopsis clematidea* (Schrenk) C. B. Clarke. **a**, root and caudex; **b**, middle part of a stem; **c**, upper part of a flower with corolla opened, showing stamens, disc, style, and stigma.

FIGURE IV 1:9B *Codonopsis clematidea* (Schrenk) C. B. Clarke. **a–c**, from Xinjiang: Ürümqi CO, 21. Jul. 2007, by Dun-Yan TAN; **d**, from Xinjiang: Xinyuan CO, 31. Jul. 2012, by Lian-Zhong FU.

Valley, 2250 m, 25. Jul. 1954, *Acad. Sin. Xinjiang Compreh. Exped. 9975* (PE). **Mori**: Nangou, 2065 m, *Picea* forests, 30. Jun. 1956, *s. coll. (Tian) 0130* (PE). **Qapqal**: Jakesi, Dayuhai, 08. Aug. 1957, *K. C. Kuan 3166* (PE); Nanshan, 2000 m, meadows, *s. coll.*s. n.(PE). **Qitai**: Maguoliang, meadows, 22. Sept. 1957, *K. C. Kuan 4433* (PE). **Shache**: Kelatuzi Mine Region, 2800 m, meadows, 22. Jul. 1987, *Y. H. Wu 870693* (HNWP). **Shawan**: Ziniquan, shady slopes, 1750 m, 18. Jul.

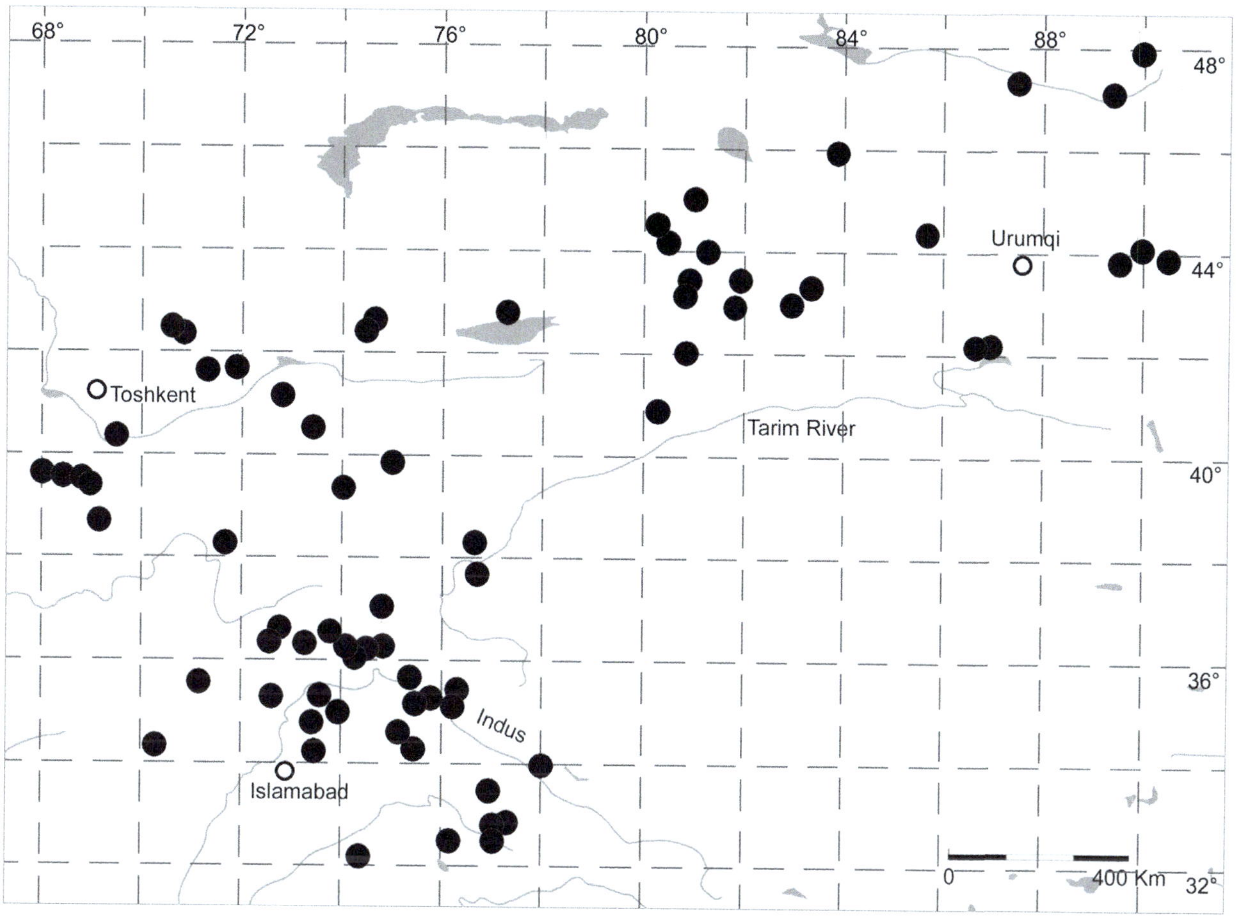

MAP IV 1:9 Distribution of *Codonopsis clematidea* (Schrenk) C. B. Clarke.

1965, *T. Y. Cheo 651052* (HNWP, KUN, PE, SZ). **Tekes**: No. 4 Pasture, Xihetan, flood lands, 1200 m, 02. Jun. 1965, *T. Y. Cheo 650659* (HNWP, KUN, PE, SZ). **Tianshan**: Daniuhe, 2500 m, edges of forests, 22. Jul. 1957, *K. C. Kuan 2177* (PE); loc. eodem, 16. Jul. 1957, *K. C. Kuan 1049* (PE); Kelisu, 1700 m, 16. Jul. 1957. *K. C. Kuan 1821* (PE); Kensu Valley, Tekkes, 09. 07. 1930, *F. Ludlow 709* (BM, E); Shiquanzi, 1700 m, 19. Jun. 1953, *Z. N. Feng 203* (PE); Tianci, 1800 m, forests, 02. Jul. 1974, *Xinjiang Exped. 140* (HNWP); s. loc., *Picea* forests or meadows, Aug. 1956, *Forestry Survey* s. n.(PE); s. loc., 1906, *R. E. Appleton* s. n.(K); s. loc., 20. 07. 1903, *G. Merzbacher 552* (W). **Toli**: Albakzine Shan, 07. Aug. 1957, *K. C. Kuan 2698* (PE); loc. eodem, 06. Aug. 1957, *K. C. Kuan 2539* (PE); Barluke Shan Forest Farm, 1800 m, meadows, 04. Jul. 1993, *S. Li H93205* (PE); s. loc., 1820 m, meadows, 29. Jun. 1993, *K. Deng A93010* (PE). **Tuomuer Mt**: Base Camp, 2300 m, mixed forests, 23. Jun. 1977, *s. coll. 770069* (PE). **Ürümqi**: Baiyanggou, 2500 m, forests, 30. Jul. 1931, *T. N. Liou 2921*(PE); Nanshan, 2000 m, 21. Jul. 1956, *R. C. Ching 527* (PE); 80 km SE of Ürümqi, 1800 m, edges of *Picea* forests, 02. Jul. 1974, *Y. R. Lin 74078* (PE); Nanshan, Xiaoquzi, forests, 14. Jul. 1956, *R. C. Ching 208* (KUN, PE); Nanshan, Xuejiagou, 1640 m, glades, 20. Jun. 1987, *T. Q. Zheng 1-036* (PE); Ürümqi to Yanqi, 2300 m, moist slopes, 22. Jul. 1957, *A. R. Li & J. N. Zhu 6219* (HNWP, KUN, PE). **Wenquan**: Hexia Forest Farm, 2400 m, 07. Jul. 1990, *M. Ma B90-281* (PE). **Wuqia**: Jigenximu Hana Border Station, 3200 m, meadows, 29, Jul. 1978, *NW Inst. Bot. Xinjiang Exped, 2140* (PE). **Xinyuan**: Kok-su Valley, ca. 43° N, 83°E, 15. Aug. –15. Oct. 1901, *G. R. Littledale* s. n.(K); Nalati, alpine meadows by forest, 31. Jul. 2012, *Z. Y. Li et al. 12478* (PE); Nanshan, ca. 900 m, meadows, 18. May 1965, *T. Y. Cheo et al. 650391* (HNWP, KUN, PE, SZ); 70 km from Xinyuan to Bayanbulak, 1600 m, meadows, 26. Jun. 1982, *C. Y. Wu et al. 303* (KUN); Yeguolin Improv. Farm, 1000 m, sparse woods, 18. Aug. 1978, *NW Inst. Bot. Xinjiang Exped. 2590* (PE). **Yecheng**: Sukemaya, meadows by *Picea* forests, 3000 m, 14. Aug. 1987, *Qinghai-Xizang Exped. 1051* (PE); Yangkeya Township, Jiaoshasi Valley, 2600 m, 07. Aug. 1987, *Qinghai-Xizang Exped.870938* (HNWP, PE). **Yining**: Yining Forest Area, 1700 m, Aug. 1974, *Forestry 721 Group* s. n.(PE). **Zhaosu**: Army Horse Farm, Xiaokalagan Pass, 2580 m, sunny slopes, 29. Jul. 1974, *Y. R. Lin 74643* (PE); Hongqi No 2 Farm, Timber Ground, meadows, 2100 m, 09. Jun. 1965, *T. Y. Cheo et al. 650714* (KUN, PE); N of the city, 2000 m, 26. Jul. 1974, *Xinjiang Exped. 1060* (HNWP); Wenquan to Xiaota, 2100–2400 m, 15. Aug. 1957, *K. C. Kuan 3566* (PE); Zhaosu to Tekes, 16. Aug. 1957, *K. C. Kuan 3575* (PE).

INDIA, Himachal Pradesh: Kyelang, 24. Jun. 1888, *J. R. Drummond 25267*(E, K); Lahul, 3050 m, 1865, *H. Jaeschke* s. n.(K, WU); Lahul, Darcha 3350 m, 13. Jul. 1976, *J. D. A. Stainton 7664* (E); Lahul, above Kyelang, Kulu-Lahul, 06. Jul. 1888, *J. R. Drummond 22508* (K); Lahul, Kyelang, 3200 m, 27. Jun. 1936, *W. Koelz 8533* (PE); loc. eodem, 3050 m, 3–4. Jul. 1933, *W. Koelz 5198* (PE); Lahul, Zingzingbar, 4570 m, 23. Jul. 1941, *N. L. Bor 16373* (E, K) ; Randaug, 09. Jul. 1888, *J. R. Drummond 25266* (K); s. loc., 02. Jul. 1888, *J. R. Drummond 25268* (E, K). **Jammu-Kashmir**: Chamba (Pauri), 4270 m, 11. Jul. 1899, *J. F. Duthie* s. n.(K); Ladakh, between Zoji Pass and Sonamarg, 3250 m, 13. Jul. 1976, *F. Billiet & J. Leonard 6901* (K); Ladakh, 1 km from Kanji, 3960 m, 02. Jul. 1984, *A. R. Brown & M. A. Rothera 79* (K); Ladakh, Panichar, 2900 m, 1980, *Southampton Univ. 20* (K); Zanskar Distr., Karsha, 3510 m, 18. Jul. 1981, *Grace, Bailey, Bull & Carr 81* (BM, K); loc. eodem, 3660 m, 19-21. Jul. 1933, *Walter Koelz 5785* (G). **Punjab**: Kangra, 7–10, Sept. 1933, *W. Koelz 7101* (PE).

KAZAKHSTAN, **Almaty** Oblast: **Alatau Mts**, Djarkent Distr., 44°30′N, 80°19′ E, 29. Jul. 1928, *N. L. Desyatkin* s. n. (LE); Kopal, 10 km S of Kopal Town, 45°03′N, 79°04′ E, 10. Aug. 1930, *E. P. Matveeva 333* (LE); Sarkand, 36 km from Pokatilovka, 45°12′N, 80°10′ E, 24. Aug. 1930, *E. P. Matveeva 522* (LE); Karatala River Valley, 44°46′N, 79°26′ E, 21. Aug. 1930, *N. V. Shipchincky 426* (LE). **Alexander** Mts, 2130 m, 06. Jul. 1883, *Fetissow* s. n.(K); loc. eodem, 30. Jun. 1914, *V. Titor 1658* (PE); Zagan-tunge, 1500–1830 m, 08. Jun. 1879, *A. Regel* s. n.(K). 10 km of **Almaty** center, in lower Tian-Shan Mts, 1350–1540 m, 19. Jul. 1985, *T. S. Elias et al. 9661* (A); 1750 m, 17. May. 1971, *F. Yamazaki* s. n.(TI). **Ketmenj**, 2000 m, 09. Jul. 1964, *I. Roldugin & V. Fissjun* s. n.(G, K, PE, TI).

KYRGYZSTAN, Chuy Oblast, Alamüdün, 42°33′19″N, 74°28′ 58″E, 2225 m, 02. Jul. 2008, *R. Borosova, KYRG 2* (K); Alamedin, 42°39′N, 74°30′ E, 2100 m, 06. Aug. 2008, *S. Mosolova & S. Prikhodko 87* (K). **Jalal - Abad** Oblast Aksy Rayon, 41°47′N, 71°59′ 08″E, 1483 m, 07. Jul. 2008, S. Landrein (KYRG) 142 (K). **Djalalabad** Distr., Fergana Ridge, Togus-Bulak Village, 41°22′N, 72°54′ E, 19. Aug. 1945, *A. A. Fedorov. & E. M. Iljina 421* (LE). **Asu-Djabagly**, 42°23′N, 70°34′ E, 08. Jul. 1950, *A. I. Poyarkova 84* (LE); loc. eodem, 42°13′N, 70°42′ E, 24. Jul. 1958, *L. P. Markova & L. I. Medvedeva* s. n. (LE). **Kapka** Valley, Kapka Village, 2300 m, 27. Sept. 1944, E. M. Iljina 113 (LE). **Uzgen** Distr., Besh-Terek Village, 40°47′N, 73°27′ E, 05. Aug. 1959, *S. Sagatov* s. n. (LE).

MONGOLIA, W Mongolia, Bayan-Olgiy Prov., Mongolia Altai, near the border with China, 09. Aug. 1979, *I. A. Gubanov 7617* (LE); loc. eodem, 22. Jul. 1988, *R. V. Kamelin et al. 2073* (LE); loc. eodem, 25 km to the border with China, 16–17. Jul. 1988, *R. V. Kamelin 1518* (LE).

PAKISTAN, Baltistan: Astor, 3050 m, 07. Jul. 1946, *R. R. Stewart & E. Nasir* s. n.(K); Deosai, ca. 3660 m, 26. Aug. 1987, *Muqarrab Sha & Jamohed 273* (BM, E); loc. eodem, 4570 m, 16. Aug. 1966, *Siddigi, Nasir & Ali 4128* (K); Deosai Plains, 4270 m, 16. Aug. 1966, *M. A. Giddeji* s. n.(BM); Doghani, Shyoh River, 2440 m, 13. Jun. 1928, *F. Ludlow 345* (BM,E); Haramosh, Khaharew Valley, Sharal, 05. Jul. 2001, *S. W. Uhaw 21* (KUH); near Hushe Village, ca. 3050 m, 03. Jul. 1955, *G. L. Webster & E. Nasir 5955* (PE); Karakoram, between Turmun and Makerum, 3350–3400 m, 18. Jul. 1939, *R. S. Russell 1224* (BM); Karakoram, Barpu Glacier, 3050–3350 m, 07. Jul. 1939, *R. S. Russell 1093* (BM); Karakoram, upper end of Hushe Valley, 3810–3960 m, 15. Jul. 1955, *G. L. Webster & E. Nasir 6137* (PE); Kishenganga Valley, road to Nanga Parbat, below Kun Patthaw, 3050–3350 m, 14. Aug. 1939, *R. R. Stewart 18486* (KUH, PE); Kishenganga Valley, Badwan, 2440 m, 21. Jul. 1940, *R. R. Stewart 19534* (PE); Paskyum, 2830 m, 08. Jun. 1928, *B. B. Osmaston 130* (K); Rama Valley, 35°20′ N, 75°50′ E, 3170 m, 25. Jul. 1967, *T. E. Lankester & T. A. S. Pearson 1222* (BM); Samsam, near Umasi La and Sappline Mines, 3050 m, 13. Jul. 1943, *F. Ludlow & G. Sherriff 9188* (BM); Satpur, above Skardu, 2740–3050 m, 13. Jul. 1892, *J. F. Duthie 11989* (BM, E); Shuko Gah-Nala Glacier, 3510 m, 05. Sept. 1950, *J. W. Thornley 52* (BM); Skardu, Satpara Nullah, 3400 m, 31. Aug. 2002, *J. Alam et al. JA-11* (KUH); Solu Glacier, 3410 m, 26. Aug. 1939, *R. S. Russell 1752* (BM); s. loc., 3000 m, *Österreichische Karakorum Exped. (1970) 1026* (K); NW Himalaya, s. loc., 2100 m, *T. Thomson* s. n. (BM). **Chitral**: Chumarkhan Pass, 3660 m, 15. Jul. 1958, *J. D. A. Stainton 2907* (BM, E, W); Mastooj, Shandoor top, 4223 m, 27, Jul. 2007, *H. Ali 6751* (KUH); Mastooj, Sor Laspur, 26. Jul. 2007, *H. Ali 6682* (KUH); Yarkhun, between 36°30′ N, 72°41′ E and 36°48′ N, 73°5′ E, 2440 m, 15. Jun. 1958, *S. A. Bowes Lyon 915* (BM, E, W); s. loc., 3500 m, 10. Jul. 1974, *E. J. Ecker* s. n. (W); s. loc., *S. M. Toppin 471* (K). **Gilgit**: Derachil, Ultar Nullah, Karimabad, ca. 3600 m, 22. Jul. 2005, *J. & N. Alam 3296* (KUH); Hunza, Mt. Home, above Karimabad, 3500 m, 23. Jun. 2001, *J. Alam 50* (KUH); Imit, 2900 m, 2-3. Aug. 1954, *F. Schmid 2089* (G); Lower Naltar, ca. 2800 m, 28. Jun. 2003, *J. Alam & Q. Ali 1867* (KUH); loc. eodem, ca. 2600 m, 24. Jun. 2003. *J. Alam & K. Madad 1834* (KUH); Minapir Glacier, 3050 m, 01. Aug. 1961, *P. S. Lloyd & S. Megaw 26* (BM, E, PE); 36°22′ N, 74°09′ E, 3200–3900 m, *F. Lobbichler 106* (W); Nagar Village Area, 2380 m, 28. Aug. 1960, *O. Polunin 6429* (B. BM, E, G); Naltar, 3050 m, 03. Aug. 1986, *S. Malik et al. 378* (PMNH); Naltar, 3050 m, Sept. 1960, *D. McVean* s. n.(E); loc. eodem, 31. Jul. 1983, *s. coll. 155* (ISL); ca. 9 km from Yasin, along Nazber nullah, 17. Jun. 1982, *Saood Omer 230* (BM, K, KUH); s. loc., ca. 1280 m, 03. Aug. 1986, *S. Malik et al. 339* (PMNH). **Hazara**: Besal, 3800 m, 08–09. Jul. 1953, *F. Schmid 361* (BM, E, G); Kagan Valley, between Balakot and Babusar Pass, 34°35′–35°10′N, 73°20′–74°2′E, Jul–Sept. 1954, *J. Abei 113* (BM); Kagan Valley, between Battakundi and

Burawai, 26. Jun. 1958, *B. L. Burtt & Roshan Din 925* (E); Naran, 2393 m, 12. Jul. 1975, *I. D. A. Amin et al. 896* (PE); Saif-ul-Maluk-Sar, 3570 m, 1–4. Jul. 1953, *F. Schmid 279* (G). **Swat**: Sho Nala, 3350 m, 22. Aug.. 1955, *A. Rahman 70* (BM); Ushu to Batani, 2440–2740 m, 27. Jul. 1953, *R. R. Stewart & A. Rahman 25295* (BM) s. loc., 3400 m, 05. Aug. 1973, *J. Renz* s. n.(W).

TAJIKISTAN, **Dushanbe:** Preredovoy Range, along Siama River, 1700–1800 m, 12. Jul. 1975, *H. Ohashi* s. n.(TI). **Pamir**: Jersil Valley, 2800 m, 01. Jul. 1930, *C. Persson 023* (LD); near Oikulya Lakes, 39°35′ N, 68°51′ E, 3200 m, 02. Aug. 1959, *S. Yunusov 715* (LE); Leninabad, Shahristan Distr., 39°40′ N, 68°45′ E, 2350 m, 03. Aug. 1968, *Roslov 68387* (LE); Samarkand, Hodjent Distr., 40°20′ N, 69°30′ E, 18. Jun. 1914, *O. E. Knorring 349* (LE); s. loc., 2300 m, 07. Aug. 1964, *A. G. Qikabuka 10650* (PE).

UZBEKISTAN, Samarkand, Jizzak Distr., 39°40′ N, 68° E, 1914, *A. I. Mihelson* s. n. (LE); N Fergana, 1850 m, 41°35′ N, 71°40′ E, 03 Jul. 1908, *s. coll. 115* (LE); Pamiro-Alai, 39°43′ N, 68°25′ E, 18. Jul. 1935, *B. Zakrzhevsky 971* (LE).

10. Codonopsis obtusa (Chipp) Nannf., Acta Horti Gothob. 5: 28. 1930; Rechinger & Schiman-Gzeika in Rechinger, Fl. Iran. 13: 3. 1965; Nasir in Nasir & Ali, Fl. Pakistan 155: 3. 1984. ≡*Codonopsis ovata* Benth. var. *obtusa* Chipp, J. Linn. Soc. Bot. 38: 385. 1908. ≡*Codonopsis clematidea* (Schrenk) C. B. Clarke var. *obtusa* (Chipp) Kitam., Fl. Afgh. 383. 1960. Type: Afghanistan, Kuram Valley, 1879, 3050 m, *J. E. T. Aitchison 748* (lectotype here designated: BM; isolectotypes: CAL, G, K , PE); Afghanistan, s. loc., *Griffith 3435* (paratypes: K, PE).

Codonopsis ovata Benth. var. *ramosissima* Hook. f. & Thomson, J. Proc. Linn. Soc. Bot. 2: 15. 1857. Type: Afghanistan, s. loc., *Griffith 3435* p. p. (holotype: K).

Roots carrot-shaped, up to 2 cm in diameter. Stems erect, 20–100 cm tall, glabrous or villous, multi-branched, lower branches sterile, middle and upper ones often fertile. Leaves on main stem mostly evenly distributed, alternate; petiole 2–8 mm long, hispidulous; leaf blade ovate to lanceolate, 1–5.2 cm long, 1–2.3 cm broad, hispidulous on both surfaces, base rounded to broad-cuneate, margin entire, apex acute to acuminate; leaves on branches nearly opposite, similar to those on main stem but smaller. Flowers terminal on main stem and upper branches, flowering from top; pedicel 1.5–30 cm long, hispidulous or villous. Calyx tube adnate to whole hypanthium, rounded to obconical-rounded at base, glabrous or hispidulous, 10-ribbed; calyx lobes subulate, broadest at base, very rarely slightly broader above, 6–15 mm long, 2–4 mm broad, entire, always ciliate, reflexed at or after anthesis, both surfaces hispidulous, rarely nearly glabrous. Corolla campanulate, 1.5–3 cm long, lobed to 1/3 or 2/5, pale blue outside, deep blue-purple inside or just with 2 purple rings inside; corolla lobes deltoid-orbicular. Stamens totally glabrous; filaments 3–4 mm long, slightly dilated at base; anthers 5–6 mm long. Capsule obconical to globose-obconical in inferior part, long-conical in superior part, 1.8–2.5 cm long, 1–1.5 cm in diameter. Seeds unknown. Figure IV 1:10A.

Chromosomes: 2n=16 (Krasnoborov et al., 1980).

Phenology: flowering from June to August; fruiting from August to September.

Habitats: damp meadow-ground, open calcareous slopes, by streams; 1550–3400 m.

Distribution: NE Afghanistan, NW Pakistan, Tajikistan, and SE Uzbekistan (Map IV 1:10).

Additional specimens examined**, AFGHANISTAN, Badakhshan:** Arakhi, 2550 m, 29. Jun. 1970, *J. E. Carter 1113* (K). **Bamian**: between Bamian and Band-i-Amir, ca. 34°52′ N, 67°32′ E, 2800–3000 m, 13. Jul. 1962, *K. H. Rechinger 18177* (B, K,W); Band -i-Amir, 2900 m, 12. Aug. 1935, *W. R. Hay 400* (K); Band-i-Amir, ca. 2900 m. 27. May. 1971, *C. Grey-Wilson & T. F. Hewer 951* (K); between Behsud & Panjaw, 2710 m, 17. Jun. 1971, *C. Grey-Wilson & T. F. Hewer 1119* (E,K,W); 2750 m, 17. Jul. 1972. *I. Kukkonen 7226* (W); Berg ober Band-i Amir, ca. 3000 m, 29. Aug. 1951, *R. Dückelman* s. n.(W); 34°52′ N, 68°05′ E, 2200 m, 13. Jun. 1962, *K. H. Rechinger 16751* (G,W); 34°23′ N, 67°17′ E, ca. 2800 m, 14. Jul. 1962, *K. H. Rechinger 18355* (E,G,K,LD,W); 2800 m, 27. Jun. 1962, *O. H. Volk (Würzburg) 2720* (W); 2930 m, 14. Jun. 1970, *A. Dieterle 548* (G); 2800 m, 17. Jun. 1971, *A. Dieterle 1315* (G); 34°38′ N, 66°45′ E, 2600–2900 m, 23. Jun. 1967, *K. H. Rechinger 36414* (B, E,G,K,W); 34°30′ N, 66°58′ E, 2800–3000 m, 23-28. Jun, 1967, *K. H. Rechinger 36379* (G,W); 2680 m, 27. Jul. 1970, *D. Podlech 18852* (G); Bulula, damp ground, 2440 m, 9. Aug. 1936, *W. R. Hay 545* (K); Kuh-I Hisar, Inter Sanzak Sumai et Serdak versus Jugum Sad Bark, 34°38′ N, 66°45′ E. 2600–2900 m, 23. Jun. 1967, *K. H. Rechinger 36414* (E,K); 2900 m, 30. Jul. 1965, *D. Podlech 12098* (E, G); 2000 m, 13. Jun. 1962, *I. Hedge & P. Wendelbo 4169* (E); 2701 m, 17. Jul. 1971, *C. Grey-Wilson & T. F. Hewer 1119* (E,K,W). **Ghazni**: Malestam Distr., 33°25′ N, 66°45′ E, 2700–3200 m, 02. Jul. 1962, *K. H. Rechinger 17580* (E,G,LD,W); 34°10′ N, 68°00′ E, 2800 m, 07. Jul. 1962, *K. H. Rechinger 18592* (G,LD,W); 33°13′ N, 67°24′ E, 2980 m, 28. Jun. 1978, *D. Podlech 31871* (G); Waras, 34°07′ N, 66°48′ E, 2900 m, 01. Jul. 1967, *K. H. Rechinger 36679* (G, LD, W). **Kabul**: N of Hajigak Pass, 3100 m, 25. Jun. 1962, *I. Hedge & P. Wendelbo 4624* (E,W); Hajigak Pass, Koh-i-Baba, 2440–3350 m, 11. Jul. 1966, *P. Furse 8474* (K); loc. eodem, 2660 m, 17. Aug. 1954, *W. Thesiger 29* (BM); Hajigak Pass, 3660–4270 m, 31. Aug. 1839, *Griffith* s. n.(K); Paghman, 17 miles W of Kabul, 2440–2740 m, 23. Jun. 1939, *J. L. Chaworth-Muster*

FIGURE IV 1:10A *Codonopsis obtusa* (Chipp) Nannf. **a**, a part of plant, showing root, caudex, and caespitose stems; **b**, a flower with corolla opened, showing calyx, stamens, disc, style, and stigma; **c**, a 3-loculicidal capsule.

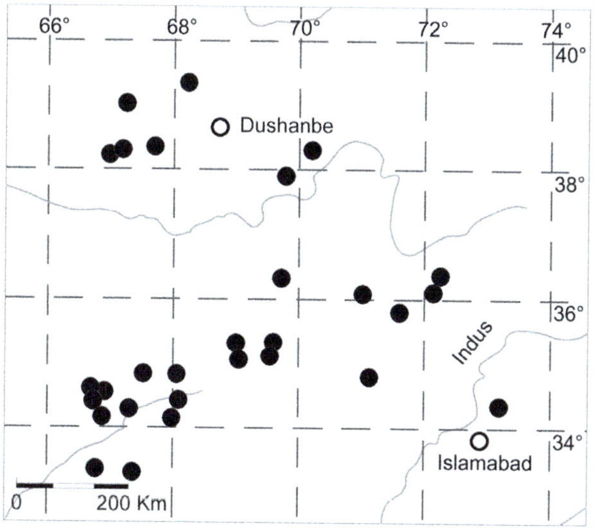

MAP IV 1:10 Distribution of *Codonopsis obtusa* (Chipp) Nannf.

s. n.(BM); loc. eodem, 2300 m, 25. Jun. 1952, *Mme G. Frumkin 286* (G); loc. eodem, 2500 m, 21. Jun. 1962, *I. Hedge & P. Wendelbo 4352* (E). **Kapisa**: Panjir-Tal. 2600 m, 20. Aug. 1965, *D. Podlech 12508* (E). **Khor**: 35 km W of Chakcharan, on road to Herat, 2425 m, 20. Jun. 1969, *T. F. Hewer 1312* (K). **Konar**: 1750 m, 10. Sept. 1972, *J. E. Carter 1285* (G,K); ca. 1540 m, 06. Jun. 1969, *T. F. Hewer 1392* (E); Barg-i-Natal, 1858 m, 05. Jun. 1969 *T. F. Hewer 1391* (K). **Kunar**: 25. Aug. 1969, *M. Rasoul* s. n.(W); 2200 m, 22. Aug. 1969, *D. Podlech 16190* (G); 3400 m, 05. Sept. 1969, *D. Podlech 16613* (G); 3250–3400 m, 28. Aug. 1969, *D. Podlech 16368* (G,K); Linar Valley, 3660 m, 14. Aug. 1956, *W. Thesiger 1531* (BM); Rangul Valley, 3350 m, 07. Aug. 1956, *W. Thesiger 1482* (BM); loc. eodem, 3350 m, 07. Aug. 1956, *W. Thesiger 1478* (BM). **Lugar**: 2500 m, 25. Jun. 1969, *L. Ekberg W9204* (E); Logar, Cacobay River, 3100 m, 05. Sept. 1967, *P. Lalande* s. n.(W). **Maidan**: 34°28′ N, 68°06′ E, 2800 m, 24. Jun. 1970, *O. Anders 4238* (G). **Parwan**: Sheba Pass, c. 2800 m, 08. Jul. 1969, *J. E. Carter 344* (K); 35°22′ N, 69°39′ E, 2050 m, 24. May. 1973, *O. Anders 9820* (G); on road to Shibar Pass, 2600 m, 14. Jun. 1969, *T. F. Hewer 1277* (E,K); 2750 m, 19. Jul. 1962, *I. Hedge & P. Wendelbo 5268* (E). **Qataghan**, 2200 m, 01. Jul. 1965, *J. Lamond 2244* (E). **Salang**: 2200 m, 01. Jul. 1965, *K. H. Rechinger 31658* (B, G, K, W); 15 km N of Salang Pass, 35°20′ N, 69°E, 2450 m, 02. Jul. 1976, *J. F. Veldkamp 7263* (W); Salang Pass, south side of Hindu Kush, 2740–3350 m, 22. Jun. 1966, *P. Furse 8003* (K); south side of Salang, Panjshir Valley, 2440 m, 22. Jul. 1966, *P. Furse 8643* (K); loc. eodem, 2290 m, *R. B. & L. Gibbons 389* (K). **Takhar**: 1600 m, 22. Jun. 1965, *D. Podlech 11480* (E). **Puri Valley**: 2650 m, 29. Jul. 1939, *E. Bacow 104* (PE).

PAKISTAN, **Chitral:** Barum, SE of Tirich Mir, 2440 m, 27. Jun. 1958, *J. D. A. Stainton 2772* (BM); Bashgalian, 2440 m, 17. Jun. 1958, *J. D. A. Stainton 2702* (BM); Chitral, Kasaweer, 2206 m, 23. Jun. 2007, *H. Ali 6175* (KUH); Tirich Gol, 36°25′ N, 72°15′ E, 2740 m, 04. Jul. 1888, *S. A. Bowes Lyon 1064* (BM,E). **Hazara**: Mansehra, Saif-ul-Maluk, 27. Aug. 1990, *S. Malik & J. Saqib 117* (PMNH).

TAJIKISTAN, Samarkand, Kshtut Valley, 39°25′ N, 68°10′ E, 1680 m, 24. Jul. 1913, *G. Kjukental 157* (LE).

UZBEKISTAN, **Baisun** Prov., Buhara acres, 38°10′N, 67°E, 13. Jun. 1916, *A. P. Fedchenko et al. 566* (LE). **Samarkand**: Alpes Sarawschan, 1880 m, 20. Jul. 1913, *J. Bornmüller 206* (B); loc. eodem, 19. Jul. 1913, *J. Bornmüller 162* (B); Samarkand, Buchara, Hissar, alpes Hissar, ca. 2300 m, 03. Aug. 1913. *J. Bornmüller 829* (B). **Sangardak** Valley, Sangardak Village, 38°27′N, 67°45′ E, 25. Aug. 1938, *N. F. Goncharov 61* (LE). **Sherabad** Distr., near Kemnyr-Tjube Village, 38°20′N, 70°10′ E, 1800 m, 01. Jul. 1931, *A. V. Jarmolenko & I. A. Firsv 164* (LE). **Shuro-Abad** Distr., Mt. Imamaskar, 37°55′N, 69°58′ E, 2300 m, 25. Aug. 1935, *I. A. Linchevsky & T. I. Maslennikova 1187* (LE). **Zapadny Gissar**, Kashka-Darya-Aksu Valley, 39°N, 67°20′ E, 13. Jun. 1982, *R. V. Kamelin 1233* (LE).

11. Codonopsis subsimplex Hook. f. & Thomson, J. Proc. Linn. Soc. Bot. 2: 16. 1857; Clarke in Hooker, Fl. Brit. Ind. 3: 432. 1881; Komarov, Trudy Imp. S.-Peterburgsk. Bot. Sada 29(1): 113. 1908; Chipp, J. Linn. Soc. Bot. 38: 388. 1908; Shen & Hong in Hong, Fl. Reipubl. Popularis Sin. 73(2): 49. pl. 8. fig. 2. 1983; Hong, Lammers & Klein in Wu, Raven & Hong, Fl. China 19: 519. 2011. Type: India, Sikkim, Lachen, 4270 m, 15. Jul. 1849, *J. D. Hooker* s. n.(lectotype here designated: K); Sikkim, Lachen, 3660–3960 m, 30. Jul.. 1849, *J. D. Hooker* s. n.(paratypes: CAL, G, K, W).

Roots carrot-shaped, 15–35 cm long, 0.5–1.5 cm in diameter. Stems erect, ascending or trailing, few-branched, 35–70 cm long, nearly glabrous or sparsely white-villous. Leaves on main stem alternate, those on branches nearly opposite and much smaller; petiole 0.5–4.5 cm long, sparsely white-villous; leaf blade broadly ovate, ovate or elliptic, 2–12 cm long, 2–6 cm broad, adaxially hispidulous, abaxially hispidulous but hirsute along veins, base cuneate or rounded, margin crenate or dentate, apex acute or slightly obtuse. Flowers solitary or in a cyme of 2–4, terminal on main stem and upper branches. Calyx tube adnate to whole hypanthium; calyx lobes ovate-lanceolate or ovate, 13–16 mm long, 6–8.5 mm broad, abaxially hirsute, margin usually denticulate, sparsely ciliate, apex acute or obtuse; sinus between calyx lobes narrow and pointed. Corolla shortly tubular, 1.2–2 cm long, 1.1–2 cm wide, shallowly lobed to ca. 1/4, whitish, pale yellow, yellow-green or pale blue, glabrous; corolla lobes ovate-deltoid. Stamens glabrous; filaments ca. 2–3 mm long, slightly dilated at base; anthers ca. 5 mm long. Capsule semi-globose or obovoid-conical below, short-conical above, 8–10 mm in diameter, rostrum 4–6 mm long. Seeds ellipsoid, ca. 1.5 mm long, yellow. Figures IV 1:11A.

Chromosomes: 2n=16 (Lammers, 2007b).

Phenology: flowering in July and August; fruiting in September and October.

Habitats: mountain forests or scrubs; 1220–4270 m.

Distribution: Bhutan, China (S Tibet), N India (Darjiling and Sikkim) and Nepal (Map IV 1:11).

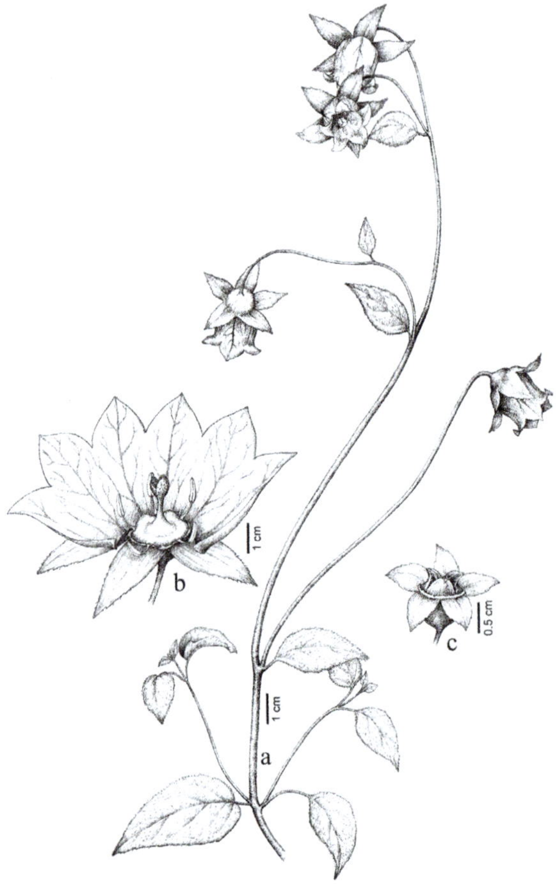

FIGURE IV 1:11A *Codonopsis subsimplex* Hook. f. & Thomson. **a**, upper part of a stem; **b**, a flower with corolla opened, showing calyx, stamens, style, and stigma; **c**, a 3-loculicidal capsule.

Additional specimens examined, **BHUTAN**, **Ha:** Chelai La, 3660 m, 13. 08. 1949, *F. Ludlow*, et al. *19612* (BM, E); Ha to Chile La, 2750–3650 m, 22. Jul. 1938, *B. J. Gould 1247* (K). **Mangde Chu**: Dur Chuteen, 19. Jul. 1949, *F. Ludlow et al. 19456* (BM); Mangde Chu, Goktang La, 3660 m, *F. Ludlow et al. 16891* (BM, E). **Shingbe**: Me La (Cho La Valley), 3050 m, 20. Aug. 1949, *F. Ludlow et al. 21052* (BM, PE) ; loc. eodem,27°58'N, 91°37'E, 3810 m, 01. Jul. 1949, *F. Ludlow et al. 20451* (BM, E); loc. eodem, 27°58'N, 91°37'E, 4110 m, 22. Aug. 1934, *F. Ludlow & G. Sherriff 876* (BM); loc. eodem, 4270 m, 06. Aug. 1933, *F. Ludlow & G. Sherriff 426* (BM). **Thimphu:** Parshang, 3660 m, 27. Jul.1914, *R. E. Cooper 1948* (E); Phajudin, 3660m, 05. Aug. 1914, R. E. *Cooper 2740* (BM, E). **Rinchen Chu** (central Bhutan), 3350-3660 m, 06. Aug. 1937, *F. Ludlow* & G. *Sherriff 3530* (BM). **Yuto La** (central Bhutan), 3510 m, 21. Jul. 1937, *F. Ludlow & G. Sherriff 3469* (BM).

CHINA, **TIBET**, **Gyirong**: Gyirong-Gyirong Town, 28°33'N, 85°14'E, 3700 m, grasses in thickets, fl. blue, 22 Jul. 2013, *PE-Tibet Exped. 3840* (PE). **Yadong**: Champitang to Yadong, 3050–3650 m, 02. Aug. 1936, *F. S. Chapman 301* (K); Chumbi, 15. Jul. 1882, *Dungboo* s. n.(K); Chumbi & Phari, 1879, *Dungboo* s. n.(K); Chumbi Tang, meadows in forests, 27. Jul. 1953, *P. C. Tsoong 5931* (PE); loc. eodem, thickets, 3860 m, 14. Sept. 1974, *Qinghai-Xizang Exped. 74–2509* (PE); loc. eodem, 3660 m, 01. Aug. 1936, *F. S. Chapman 1168* (K); Tanggapu Gou , 3800 m, openings of *Abies* forests, 09. Aug. 1975, *s. coll. 75–914* (PE).

INDIA, **Darjiling**, Singalila, Phalut, 3800 m, 16. Jul. 1969, *H. Hara et al. 69713* (TI); loc. eodem, 18. Jul. 1969, *H. Hara et al. 69714* (TI); loc. eodem, 3600 m, 03. Aug. 1972, *H. Kanai et al. 721414* (TI); loc. eodem, 3350 m, 07. Oct. 1870, *C. B. Clarke 12542* (CAL, K); loc. eodem, 3350 m, 08. Oct. 1870, *C. B. Clarke 13450* (CAL) & *13450A* (K). **Sikkim**: Gnatong, 3660–4270 m,

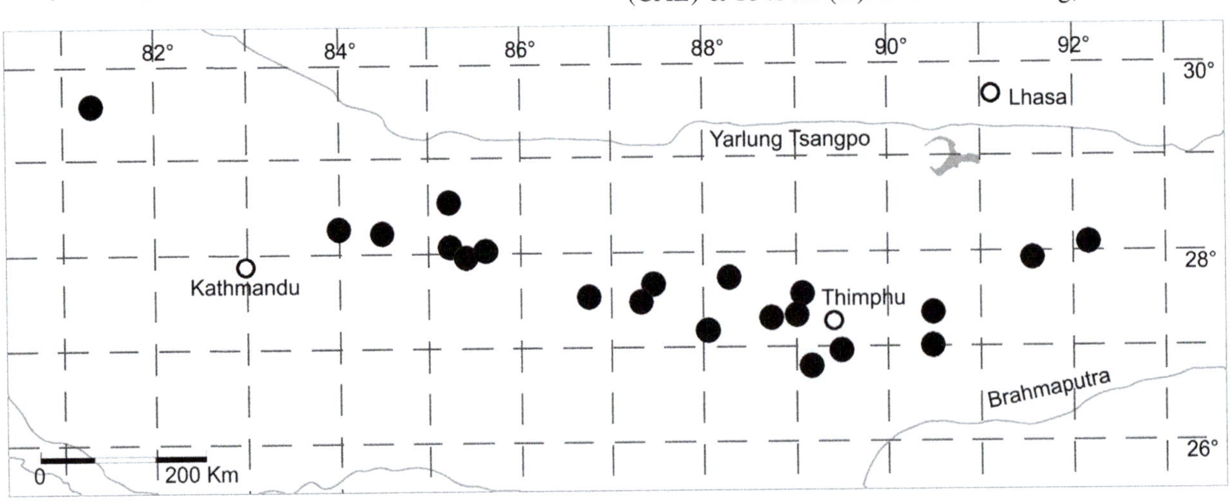

MAP IV 1:11 Distribution of *Codonopsis subsimplex* Hook. f. & Thomson.

30. Aug. 1913, *R. E. Cooper 750* (E); S of Imgri, 07. 1887, *King's Collector* s. n.(CAL); Quatung, 1891, *Cummins* s. n.(CAL); Rathangoong, 10. 1908, *Ribu 111* (CAL); loc. eodem, 08. Dec. 1903, *Prain's Collector 267* (CAL); Shaitanpokrikapûp, 3810 m, 03. Sept. 1913, *R. E. Cooper 768* (E); Sherabthang, 3960 m, 13. Aug. 1910, *W. W. Smith 4298* (B, CAL, E); Zemu Valley, 3960 m, 20. Jul. 1909, *W. W. Smith & G. H. Cave 1653* (B, CAL); s. loc., 12. Aug. 1910, *W. W. Smith 4250* (CAL, G); s. loc., 3660 m, 15. Aug. 1913, *R. E. Cooper 507* (E); s. loc., *G. H. Cave* s. n.(E, G, PE); s. loc., 4750 m, 04. Aug. 1893, *Cummins* s. n.(CAL); s. loc., 3660–3960 m, 1861, *J. D. Hooker* s. n.(G, P, W).

NEPAL, **Bagmati**: Nuwakot Distr., N of Gulu Bhanjyang, 27°55'N, 85°30'E, 3300 m, 09. Aug. 1974, *J. H. de Haas 2156* (BM); Nuwakot Distr., Kalanki, 3660–3960 m, *L. Dhevoj 0467* (BM, E); Rasuwa Distr., Chilime Kharka-Thangjet, 4410 m, 26-28. Jul. 1949, *O. Polunin 1331* (BM); Chilime Kharka, 3962 m, Jul. 1949, *O. Polunin 1244* (BM); Rasuwa Distr., E of Mailung Khola River, 28°10'N, 85°13'E, 3700 m, 20. Aug. 1974, *J. H. de Haas 2466* (BM); Pach Pokhri, 3960–4720 m, 1927, *Clive Wigram* s. n.(E); Rasuwa Distr., Taglung (S. of Tukucha), Kali Gandaki, 3660 m, 12. Jul. 1954, *Stainton, Sykes & Williams 1764* (BM); between Gopte (3500 m) and Surjakund (4400 m), 25. Aug. 1972, *H. Kanai et al. 723168* (TI); Sindhupalchok Distr., Gopte-Thale Patil, 3400 m, 26. 08. 1972, *H. Kanai et al. 723169* (BM, TI). **Gandaki**: Kaski Distr., Annapurna Himal., Seti Khola, 3800 m, 03. Aug. 1954. *Stainton, Sykes & William 6653* (BM, E); Lamjung Distr., Lamjung Himal., 3960 m, 14. Jul. 1954, *Stainton, Sykes & Williams 6352* (BM, E). **Kosi**: Chhurchathange (3600 m) – Piling La (4350 m) – Thudam (3400 m), 3800 m, 11.Aug. 1977, *H. Ohashi et al. 772283* (BM, E, PE, SZ, TI) & *773705* (TI); Lamrak, 3050–4270 m, 1929, *L. Dhevoj 169* (BM); Sankhuwasabha Distr.,Arun Valley, Thudam, E of Chyamtang, 3660 m, 16. Jul. 1966, *J. D. A. Stainton 1010* (BM); Sankhuwasabha Distr., Jaljale Himal., between Singo Kharka (4140 m) and Pahakhola 27°35'N, 87°25'E, (2400 m), 11. Aug. 1991, *H. Ohba et al. 9110464* (BM, TI) & *9120394* (BM, TI); Thudam, 3750 m, 19. Jul. 1971, *L. W. Beer et al. 8391* (BM). **Sagarmatha**: Solukhumbu Distr., 27°37'N, 86°46'E, 3900 m, 28. Aug. 1997, *M. Wakabayashi & al. 9715261* (KUN, E); Solukhumbu Distr., Khumbu, 3960 m, 01. Aug. 1966, *A. D. Schilling 999* (K). **Seti**: Bajhang Distr., 3274 m, 12. Jul. 2009, *H. Ikeda et al. 20915092* (E).

12. Codonopsis campanulata D. Y. Hong, Plant Divers. Resour. 36(3): 290. fig. 4. 2014. Type: Nepal, Gandaki Zone, Marsyandi Valley, Manang Distr., Annapurna, slopes opposite to SW of Braga, Jul. 1989, *C. Grey-Wilson* s. n.(holotype: K).

Roots unknown. Stems ascending or procumbent, 50–80 cm long, slender, sparsely villous, with very slender, sterile branches from lower and middle parts. Leaves evenly distributed along most parts of stems, those on main stem alternate, those on branches nearly opposite, basal leaves very small, scale-like, middle and upper leaves normal: petiole 2–5 mm long; leaf blade hispidulous on both surfaces, base more or less cordate, margin entire or shallowly crenulate, apex obtuse or acute. Flowers solitary, terminal on main stem; pedicel over 10 cm long, villous. Calyx tube adnate to whole hypanthium, and thus calyx lobes and corolla inserted at same level; calyx tube densely villous; calyx lobes lanceolate, 1–1.4 cm long, 0.4–0.5 cm broad, villous on abaxial side, entire, revolute. Corolla campanulate, 1.8–2.2 cm long, lobed to ca. 1/3, with a red-purple ring below throat; corolla lobes deltoid, 6–8 mm long, sparsely hispidulous outside. Stamens glabrous; filaments ca. 7 mm long, slightly dilated at base; anthers ca. 8 mm long. Stigma 3-fid, stigma segments broad-oblong. Fruit not seen. Figure IV 1:12A & B.

Phenology: flowering in July.

Habitats: meadows with shrubs; ca. 3700 m.

Distribution: China (Tibet: Gyirong) and C Nepal (Gandaki Zone) (Map IV 1:12).

Notes: This newly described species resembles *Codonopsis foetens* Hook. f. & Thomson mostly in having a campanulate corolla and sterile branches, but it differs from that species in having more evenly distributed and larger leaves (1.5–3 cm vs. 0.5–1.5 cm long), much longer stems (50–80 vs. 20–40 cm long), densely villous calyx tube (vs. sparsely hispidulous), and by a red-purple ring below the corolla throat (vs. lacking such a ring).

Additional specimens examined, **CHINA**, **TIBET**, **Gyirong**; Gyirong to Gyirong Town, 28°33'N, 85°14'E, 3700 m, shrubby meadows, 22. Jul. 2013, *PE-Tibet Exped. 3840* (PE).

13. Codonopsis alpina Nannf., Notes Roy. Bot. Gard. Edinburgh 16: 154. pl. 227. & fig. 1g. 1931; Shen & Hong in Hong, Fl. Reipubl. Popularis Sin. 73(2): 59. pl. 10. fig. 2 & 3. 1983; Hong & Ma in Wang & Wu, Vasc. Pl. Hengduan Mountains 2: 1959. 1994; Hong, Lammers & Klein in Wu, Raven & Hong, Fl. China 19: 522. 2011. Type: China, "Tibet austr.-orient., Sarong, on Doker-la, Mekong-Salween divide" (Dêqên, Yunnan), 28°20′ N, 3660 m, Aug. 1917, *G. Forrest 14659* (holotype: E; isotypes: K, W)

Codonopsis foetens Hook. f. & Thomson var. *major* Hand.-Mazz., Akad. Wiss. Wien Math. -Naturwiss. Kl. Anz. 61(20): 169. 1924. Type: China, NW Yunnan, Mekong-Salween, Mt Maya, 28°4′ N, 4050–4300 m, 03. Aug. 1916, *Handel-Mazzetti 9638* (lectotype here designated: WU; isolectotypes: K, W)

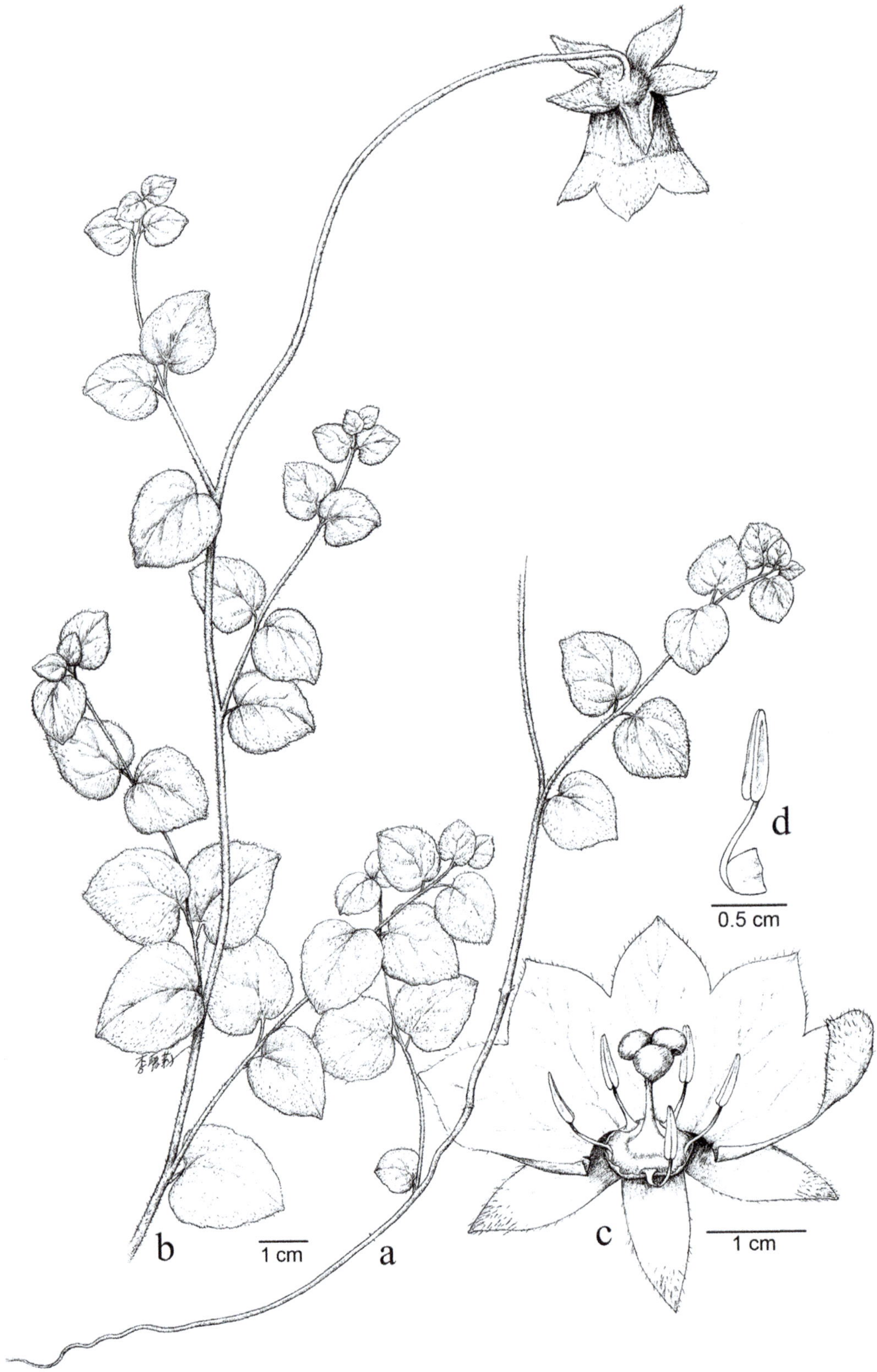

FIGURE IV 1:12A *Codonopsis campanulata* D. Y. Hong. **a** & **b**, lower and upper parts of a stem; **c**, a flower with corolla opened, showing calyx, stamens, disc, style and stigma; **d**, a stamen.

FIGURE IV 1:12B *Codonopsis campanulata* D. Y. Hong. **a** & **b**, from Tibet: Gyirong, 22 Jul. 2013, by Bin-Bin LIU; **c** & **d**, from the same locality as above, 22. Jul. 2013, by Xin-Tang MA.

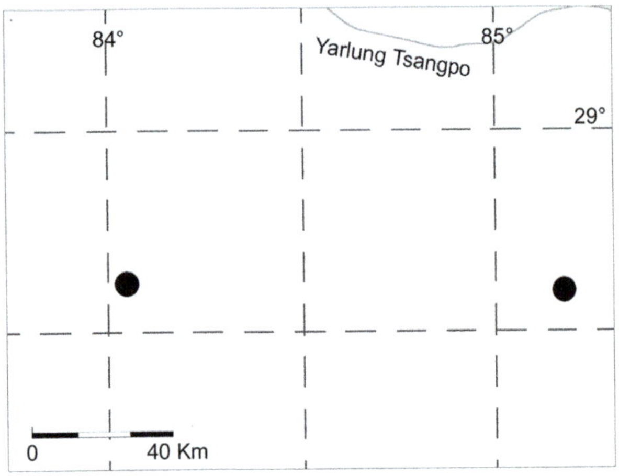

MAP IV 1:12 Distribution of *Codonopsis campanulata* D. Y. Hong.

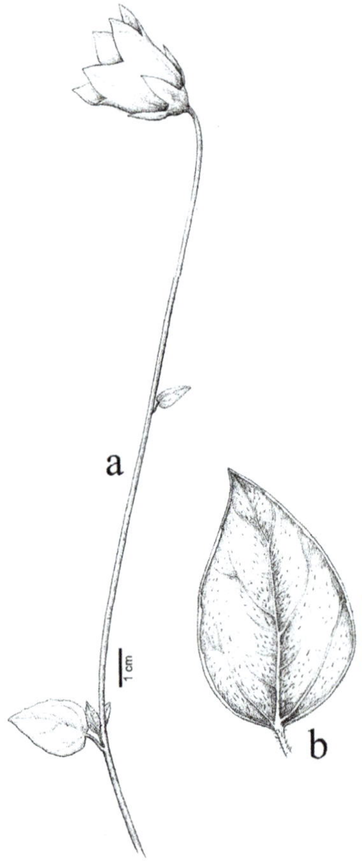

FIGURE IV 1:13A *Codonopsis alpina* Nannf. **a**, upper part of a stem; **b**, abaxial side of leaf blade, showing revolute margin.

Roots carrot-shaped, up to 30 cm long, 1.5 cm in diameter. Stems several from one caudex, simple or branched, main stem erect or ascending, 15–45 cm long, or longer, glabrous or sparsely white-hirsute; branches aggregated at lower part, leafy, sterile, sparsely white-hirsute. Leaves on main stem sparse and alternate, those on branches nearly opposite, rarely alternate; petiole 4–6 mm long, glabrous or sparsely hirsute; leaf blade ovate-cordate, 0.8–2.8 cm long, 0.8–2 cm broad, adaxially glabrous or sparsely white-hispidulous, abaxially sparsely white-hispidulous, base shallowly cordate or rounded, margin sinuous-crenate or entire, recurved, apex obtuse. Flowers solitary, terminal, usually only on main stem, less frequently also on upper branches; pedicel sparsely hirsute. Calyx tube adnate to whole hypanthium, obconical or semiglobose, glabrous, 10-ribbed; calyx lobes triangular-oblong, 7–11 mm long, 3.5–6.5 mm broad, clearly 3–5-nerved, both surfaces glabrous or abaxially very sparsely hispidulous at apex, sometimes revolute, margin entire, dentate or sparsely crenate, apex acute or obtuse; sinus between calyx lobes broad and obtuse. Corolla campanulate, 2.5–3 cm long, ca. 2.5 cm wide, shallowly lobed, variously coloured: purple-green with purple veins, red, violet-blue, but mostly purple, glabrous; corolla lobes deltoid-orbicular. Stamens glabrous; filaments 7–8 mm long, slightly dilated at base; anthers ca. 4 mm long. Capsule 1.3–1.6 cm long, ca. 1.2 cm in diameter, obovoid-conical in inferior part. Figures IV 1:13A & B.

Phenology: flowering from July to October; fruiting in October.

Habitats: alpine meadows or rocky grassy slopes; 3600–4420 m.

Distribution: China (W Sichuan, SE Tibet, and NW Yunnan) and N Myanmar (Map IV 1:13).

Notes: 1) In the protologue of *C. foetens* var. *major* Hand.-Mazz., Handel-Mazzetti designates *Handel-Mazzetti 9636* as the type, but the three specimens with the correct collection information in K, W and WU all show "9638".

2) Anthony (1926, Notes Roy. Bot. Gard. Edinburgh 15: 187) included material of this species from "Tibet" (Tibet and Yunnan) under *Codonopsis cardiophylla* Diels ex Kom.

Additional specimens examined, **CHINA, TIBET (SE), Markam**: s. loc., 3660 m, alpine meadows, 05. Jul. 1932, *J. F. Rock 22266* (E); s. loc. 4420 m, alpine scree, 08. Oct. 1932, *J. F. Rock 22680* (PE). **YUNNAN, Dêqên**: Baima Shan, 3660 m, 22. Aug. 1912, *F. Kingdon-Ward 5330* (E); by Mekong River, Yongzi Shan, 3800–4200 m, grassy slopes, 09. Aug. 1940, *K. M. Feng 6451* (KUN, PE); loc. eodem, 3800–3900 m, rocky slopes, 18. Aug. 1940, *K. M. Feng 6739* (KUN, PE); Mekong-Salween Divide, Sewalongba, 4000 m, rocky places, 30. Aug. 1938, *T. T. Yü 22576* (KUN, PE); loc. eodem, 4000 m, alpine medows, 02. Sept. 1938, *T. T. Yü 22650* (KUN, PE). **Gongshan**: Salween–Kiukiang (Drungjiang) Divide, Netahtsu, 3600 m, grassy slopes, 20. Oct. 1938, *T. T. Yü 20792* (KUN, PE).

MYANMAR, Irrawaddy, Adung Valley, 28°20′ N, 97°40′ E, 31. Jul. 1931. *F. Kingdon-Ward 9884* (BM); Nam Tamai Valley (Adung Wang-Gamlang Wang), 28°15′ N, 97°30′ E, 3660 m, 03. Oct. 1937. *F. Kingdon-Ward 13357* (BM); Senighku Valley, 28°10′ N, 97°20′ E, 3660–3960 m, 10. Aug. 1926, *F. Kingdon-Ward 7563* (K); Senighku Wang, 28°8′ N, 97°24′ E, 3960 m, 06. Jul. 1926, *F. Kingdon-Ward 7056* (K).

FIGURE IV 1:13B *Codonopsis alpina* Nannf. **a** & **b**, from Tibet: Zayü, 10. Aug. 2013, by Qiang WANG; **b**, inside of flower, showing the disc.

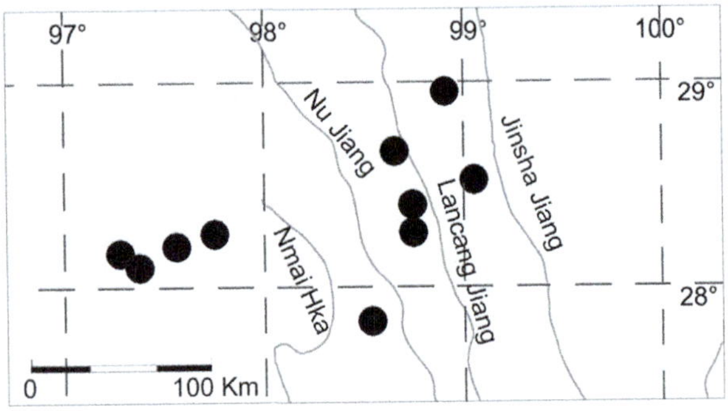

MAP IV 1:13 Distribution of *Codonopsis alpina* Nannf.

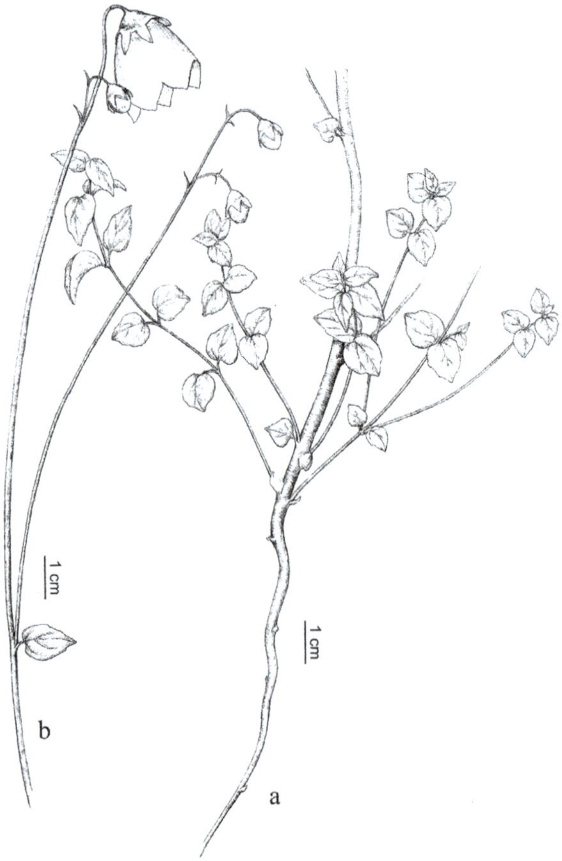

FIGURE IV 1:14A *Codonopsis chlorocodon* C. Y. Wu. **a**, lower part of a stem; **b**, upper part of a stem.

14. Codonopsis chlorocodon C. Y. Wu in Wu & Li, Yunnan Trop. Subtrop. Fl. Res. Rep. 1: 82. pl. 30. fig. 2.1965; Shen & Hong in Hong, Fl. Reipubl. Popularis Sin. 73(2): 63. 1983; Hong, Lammers & Klein in Wu, Raven & Hong, Fl. China 19: 523. 2011. ≡*Codonopsis viridiflora* Maxim. var. *chlorocodon* (C. Y. Wu) S. H. Huang, in Wu & Chen, Fl. Yunnan. 5: 490. 1991. Type: China, Yunnan, Dêqên, Baima Shan, Gujianong, 3400 m, 09. Aug. 1937, *T. T. Yü 9461* (holotype: KUN; isotypes: PE).

Roots carrot-shaped, up to 20 cm or even longer, up to 2.5 cm in diameter. Stems often single from caudex, erect, 60–100 cm tall, lower branches many, leafy, sterile, rarely fertile, middle ones often absent, upper ones, if present, fertile but leafless, or with leaves linear and much smaller. Leaves on main stem alternate, those on branches nearly opposite, sessile or with petiole up to 1 cm long; leaf blade triangular-ovate to lanceolate, 1–2.5 cm long, 0.5–2 cm broad, both surfaces sparsely hispidulous, base shallowly cordate to rounded, apex acute, margin entire or crenate, recurved. Flowers solitary, terminal on main stem, upper branches often also with 1 or 2 flowers. Calyx tube adnate to whole hypanthium, semiglobose, 10-ribbed; calyx lobes distant from each other, narrowly triangular-lanceolate, 4–6 mm long, 2–3 mm broad, glabrous or rarely very sparsely hispidulous apically, margin denticulate above, less frequently entire; sinus between calyx lobes truncate. Corolla tubular-campanulate, 1.5–1.8 cm long, lobed to 1/3, green, pale yellow-green, or pale yellow. Capsule ca. 1 cm in diameter, inferior part hemispheric, slightly obtuse below. Seeds ellipsoid, ca. 1.5 mm long, brown-yellow. Figures IV 1:14A & B.

Phenology: flowering from June to August; fruiting in late August and September.

Habitats: sunny grassy slopes or open thickets; 2700–4000 m.

Distribution: **endemic** to China: W Sichuan and NW Yunnan (Dêqên) (Map IV 1:14).

Additional specimens examined, **CHINA, SICHUAN, Batang**, s. loc. 1903, *J. A. Soulie 3703* (P). **Daocheng**: Sumdo, 03. Aug. 1964, *J. D. Li et al. 64–54* (PE, SM). **Dawu**: near Baimai (Qianning), 3300 m, sunny rocky slopes, 22. Jul. 1964, *T. L. Dai et al. 64–16* (PE, SM); near Baimai, 30°26′N, 101°32′E, 3300 m, thickets, 04. Sept. 2012, *D. Y. Hong & Q. Wang H 12071* (PE). **Dêrong**: Way to Riyu, pine forests, 3400 m, 05. Aug. 1981, *Qinghai-Xizang Exped. 3299* (CDBI, KUN, PE). **Kangding**: Xinduqiao, thickets, 3500 m, 20. Jul. 1963, *K. C. Kuan et al. 824* (PE). **Litang**: between Maoya Qusangzhi and Sumdo of Daocheng, 3680 m, sunny grassy slopes, 06. Aug. 1964, *J. D. Li et al. 64–54* (PE, SM). **Xiangcheng**: Ma'an Shan, 28°57′ N, 99°46′E, 3873 m, edges of forests, 20. Jul. 2010, *Kham Exped. 10–0227* (PE); Reda, 29°6′11″ N, 99°37′55″E, 3450–3650 m, *D. E. Boufford & al. 28739* (A, E, P, W); No 4 Distr., sunny grassy slopes, 4000 m, 16. Jul. 1972, *Z. G. Liu 1439* (CDBI, PE). **Yajiang**: No 4 Distr., 21. Aug. 1986, *Z. T. Wang 868-362* (PE); loc. eodem, 3720 m, moist slopes, 16. Sept. 1974, *D. Q. Wang 5004* (CDBI, PE). **YUNNAN, Dêqên**: Baima Shan, 3600 m, *Betula* forests, 23. Jul. 1981, *Qinghai-Xizang Exped. 3650* (CDBI, KUN, PE); Doker-la, Mekong-Salween Divide, 3960 m, 13. Jul. 1913, *F. Kingdon-Ward 762* (E); Mekong-Salween Divide, 28°25′ N, Jul. 1917, *G. Forrest 14394* (E, K); s. loc., 2700 m, Sept. 1935, *C. W. Wang 69913* (paratypes: KUN, PE).

15. Codonopsis bhutanica Ludlow, J. Roy. Hort. Soc. 97:127. 1972; Grey-Wilson, Plantsman 12(2): 71. 1990; Hong, Lammers & Klein in Wu, Raven & Hong, Fl. China 19: 523. 2011. Type: Bhutan, Shingbe, Me La,

FIGURE IV 1:14B *Codonopsis chlorocodon* C. Y. Wu. **a** & **c**, from Sichuan : Garzê, 20. Jul. 2011, by You-Sheng CHEN ; **b** & **d**, from Sichuan: Dawu, 04. Sept. 2012, by Qiang WANG & De-Yuan HONG.

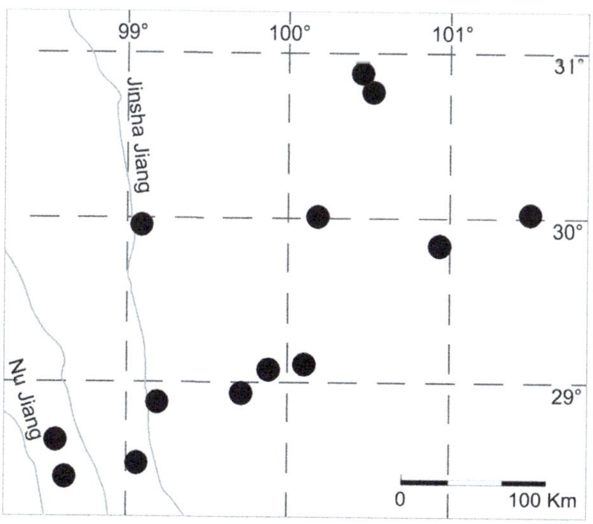

MAP IV 1:14 Distribution of *Codonopsis chlorocodon* C. Y. Wu.

27°58′N, 91°37′E, 3810 m, 02. Jul. 1949, *F. Ludlow, G. Sherriff & J. H. Hicks 20786* (holotype: BM; isotypes: E, K, TI).

Codonopsis xizangensis D.Y. Hong, Acta Phytotax. Sin. 18: 246. fig. 1, 1–3. 1980. Type: China, Tibet, Cona, 1975, *C. Y. Wu & S. K. Chen 75-907* (holotype: PE; isotype: KUN).

Roots carrot-shaped, up to 20 cm long, 1.8 cm in diameter. Stems erect, ascending or procumbent, 20 50 cm long, branches several to many from lower and middle parts, slender, usually sterile. Leaves on main stem alternate, those on branches opposite or nearly so; all leaves sessile or with petiole up to 5 mm long; leaf blade shallowly cordate, 1–2.8 cm long, 0.8–2.4 cm broad, glabrous or villous, or sparsely hispidulous, margin recurved. Flowers solitary, terminal on main stem, sometimes also on upper branches, rarely also on some slender, lower branches. Calyx tube adnate to whole hypanthium, glabrous or sparsely villous; calyx lobes long-ovate, 5–15 mm long, 3–5 mm broad, hispidulous entirely or only at upper part, or entirely glabrous, revolute; sinus between calyx lobes narrow and pointed. Corolla tubular-campanulate,1.2–1.8 cm long, lobed to 1/3, pale blue with dark blue at base. Stamens glabrous. Capsule globose in inferior part, conical in superior part, 1.4–1.6 cm long, 0.8–1 cm in diameter. Seeds ellipsoid, ca. 1 mm long. Figures IV 1:15A & B.

Phenology: flowering from July to September; fruiting in September and October.

Habitat: grassy slopes or thickets; 3100–5490m.

Distribution: Bhutan, China (S Tibet), and Nepal (Map IV 1:15).

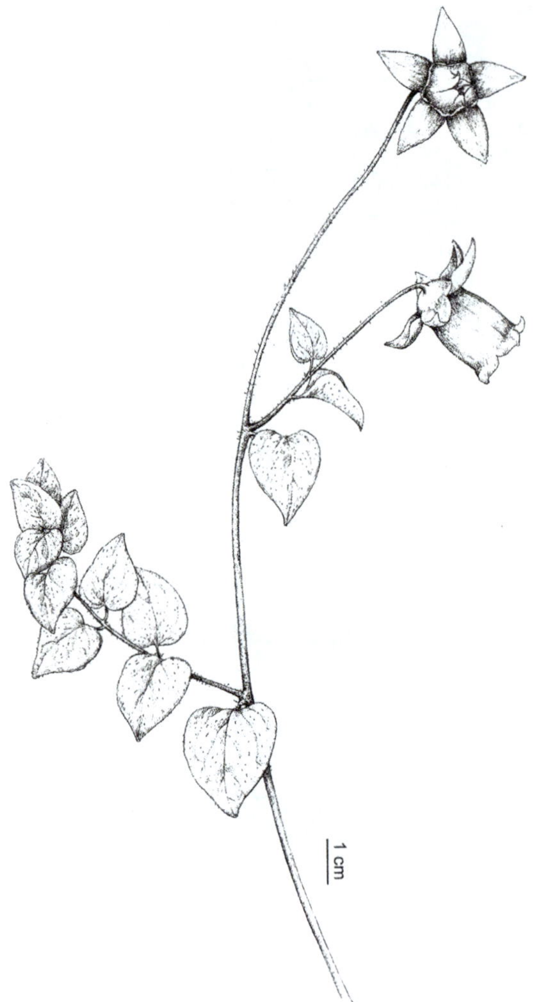

FIGURE IV 1:15A *Codonopsis bhutanica* Ludlow. **a**, upper part of a stem, showing a sterile branch appearing a pinnate leaf.

Additional specimens examined: **BHUTAN**, **Mangde Chu**, Khoa La, 27°46′ N, 90°24′ E, 4420 m, 25. Aug. 1949. *F. Ludlow, G. Sherriff & J. H. Hicks 17202* (paratypes of *C. bhutanica:* BM, TI); upper Mangde Chu, 27°54′ N, 90°26′ E, 4110 m, 14. Jul. 1949, *F.Ludlow, G. Sherriff & J. H. Hicks 16820* (paratypes of *C. bhutanica*: BM, TI). **Shingbe**: Me La, 27°58′ N, 91°35′ E, 4270 m, 06. Aug. 1933. *F. Ludlow & G. Sherriff 417* (BM); loc. eodem, 4110 m, 27°58′ N, 91°37′ E, 08. Sept. 1949, *F. Ludlow & G. Sherriff 21167* (BM); loc. eodem, 27°58′ N, 91°37′ E, 4270 m, 21. Aug. 1949, *F. Ludlow, G. Sherriff & J. H. Hicks 21408* (paratypes of *C. bhutanica*: BM, PE). **Tang Chu**, 4270–4570 m, 07. Aug. 1937, *F.Ludlow & G. Sherriff 3539* (paratype of *C. bhutanica*: BM). **Tashigang** Distr.: Shingphel, Trashi Yangtsi, 3900 m, very wet *Abies* forest clearings, 01. Aug. 2000, *S. A. Bowes Lyon 13198* (E). **Thampe La** : 27°40′ N, 90°17′ E, 4720 m,17. Oct. 1949, *F. Ludlow & G. Sherriff 19838* (BM). Thampe La, 27°40′ N, 90°17′ E, 4570 m, 14. Aug. 1949, *F. Ludlow, G. Sherriff & J. H. Hicks 17132* (paratype of *C. bhutanica*: BM). **Tsampa**, Marlung, 27°56′ N, 90°38′ E, 4420 m, 07. Jul. 1949, *F. Ludlow, G. Sherriff & J. H. Hicks 19372* (paratypes of *C. bhutanica*: BM, K); Tsampa, Pangotang, 27°50′ N, 90°42′ E, 4270–4570 m, 10. Jul. 1949, *F. Ludlow, G. Sherriff & J. H. Hicks 19311* (paratype of *C. bhutanica*: BM). **Upper Mo Chu** Distr.: below Chhew La, above Chebesa, 27°57′ N, 89°29′ E, among shrubs on river banks, 4130 m, 30. Sept. 1984. *I. W. J. Sinclair & D. G. Long 5439* (BM, K); loc. eodem, Lingshi – Chebesa, 4020 m, rocky shrubs, 24. Jul. 2003, *S. A. Bowes Lyon 15076* (BM); near Chhew La, 09. Aug. 1973, *S. Tshering & R. Nawang* s. n.(BM). **Yale La**: E of Chamolhari, 4570 m, 15. Jul. 1938, *B. J. Gould 1134* (BM).

CHINA, TIBET, **Cona**: S of Cona Town, 27°55′ N, 91°51′ E, 4000 m, 30. Aug. 2011, *D. Y. Hong et al. H11010* (PE); Jeba Township, 27°55′ N, 91°50′ E, 3840 m, meadows with shrubs, 18. Jul. 2011, *X. F. Gao et al. 13902* (CDBI); Phile' La, 4270 m, 23. Jul. 1914, *R. E. Cooper 1723* (BM); Cona, S slope of Mt Poshan, 3800–3900 m, thickets, 10. Sept. 1975, *Qinghai-Xizang Suppl. Exped. 75-1947* (paratypes of *C. xizangensis*: E, KUN, PE); Cona, 4000 m, thickets, 21. Aug. 1960, *K. X. Fu 605* (paratype *C. xizangensis*: PE); s. loc., 4570 m, 18. Jul. 1934, *F. Ludlow & G. Sherriff 682* (E); s. loc., *R. E. Cooper 1927* (BM). **Lhünzê**: Lhozhag, Lakang, 28°5′ N, 91°9′ E, 4450 m, meadows, fl. blue-brown, 18. Aug. 2013, *Y. S. Chen et al. 13–1591* (PE). **Nangxian**: by Yarlung Zangbo, 3100 m, 19. Aug. 1977, *P. C. Kuo 22966* (HNWP, KUN).

NEPAL, **Dhawalagiri**: Tegar (N of Mustang), 5490 m, 05. Aug. 1954, *Stainton, Sykes & Williams 2237* (BM). **Kosi**: Arun Valley (Phungchu), 3350 m, 02. Jul. 1922, *s. coll. 296* (K).

16. Codonopsis argentea P. C. Tsoong, Contr. Inst. Bot. Nat. Acad. Peiping 3(3): 92. pl. 11. 1935; Shen & Hong in Hong, Fl. Reipubl. Popularis Sin. 73(2): 58. 1983; Hong, Lammers & Klein in Wu, Raven & Hong, Fl. China 19: 522. 2011. Type: China, Guizhou, Fanjing Shan (Vanchien Shan), *s. coll. 51436* (holotype: SYS).

Roots not seen. Stems erect or ascending, 25–40 cm tall; branches fertile, densely leafy, pale brown or yellow, lanate. Leaves on main stem alternate or opposite, those on branches nearly opposite; petiole absent to less than 3 mm long; leaf blade ovate, or oblong-ovate, 1–2.5 cm long, 0.5–2 cm broad, adaxially appressed tomentose, abaxially white tomentose, base cordate, margin sinuous or entire, slightly revolute, apex obtuse. Flowers terminal on main stem and branches, slightly pendent; pedicel brownish yellow, tomentose. Calyx tube adnate to whole hypanthium , semiglobose, 10-ribbed, white-tomentose; calyx lobes ovate-lanceolate or lanceolate, broadest at lower-middle part, 7–9 mm long, 3–4 mm broad, brownish yellow or white tomentose, nearly entire, revolute, apex acute or obtuse; sinus between calyx lobes broad and obtuse. Corolla campanulate, 1.8–2.8 cm

FIGURE IV 1:15B *Codonopsis bhutanica* Ludlow. **a–c**, from Tibet: Cona, 30. Aug. 2011, by Xin-Tang MA & De-Yuan HONG.

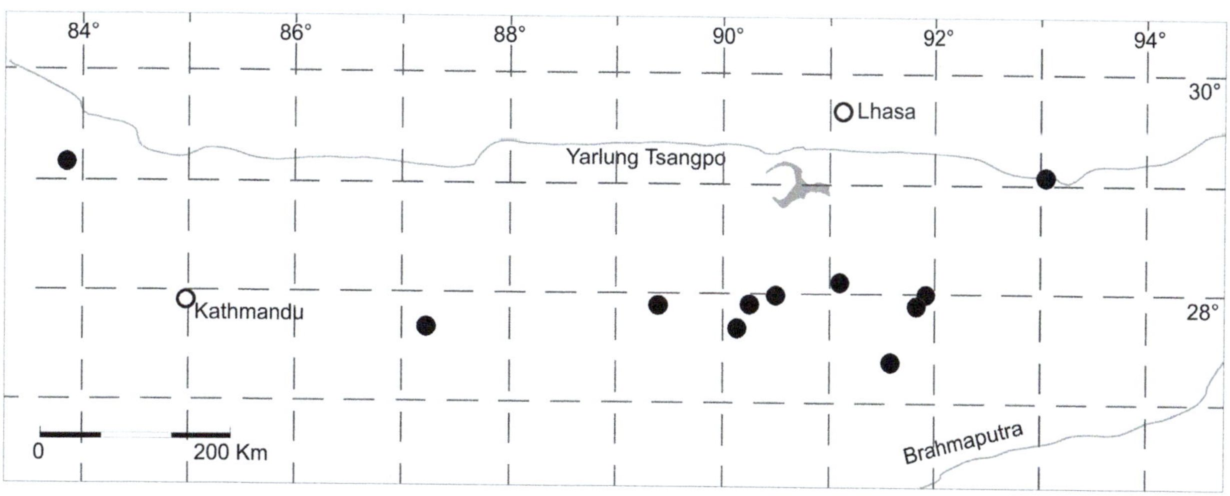

MAP IV 1:15 Distribution of *Codonopsis bhutanica* Ludlow.

long, 2–2.5 cm wide, lobed to 1/3 of corolla length, white with a purplish flush or pale blue; corolla lobes rounded, sparsely setose on both sides above, apex acute. Stamens glabrous; filaments ca. 6 mm long, slightly dilated at base; anthers 4–5 mm long. Fruit ca. 1.1 cm long, 1 cm in diameter, inferior part hemispheric, superior part conical. Figures IV 1:16A & B.

Phenology: flowering in July and August; fruiting in late August to September.

Habitats: exposed rocky slopes or meadows at an altitude of 1800–2400 m.

Distribution: **endemic** to China: Guizhou (Map IV 1:16).

Additional specimens examined**, CHINA, GUIZHOU, Jiangkou**: Fanjing Shan, ca. 2000-2300 m, 28-29. Aug. 1986, *Sino-Amer. Guizhou Bot. Exped. 627* (PE); loc. eodem, 05. Oct. 1963, *C. P. Tsien et al. 32119 B* (PE); Fanjing Shan, Wanbaoyan, 2400 m, meadows, 26. Jul. 1983, *Hunan-Guizhou Exped. 2688* (KUN) *& 2717* (KUN).

FIGURE IV 1:16A *Codonopsis argentea* P. C. Tsoong. **a**, a stem with a terminal flower.

17. Codonopsis canescens Nannf., Svensk Bot. Tidskr. 34: 386. pl. 4. 1940; Anonymous, Icon. Cormophyt. Sin. 4: 382. fig. 6178. 1975; Shen & Hong in Hong, Fl. Reipubl. Popularis Sin. 73(2): 58. 1983; Hong, Lammers & Klein in Wu, Raven & Hong, Fl. China 19: 522. 2011. Type: China, "Sikang: Taofu" (W Sichuan, Dawu), ca. 3100 m, 16. Sept. 1934, *H. Smith 12221* (holotype: UPS, n. v.; isotype: PE).

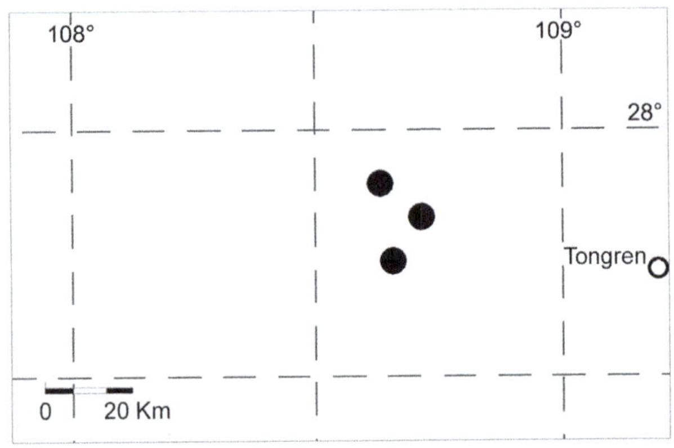

MAP IV 1:16 Distribution of *Codonopsis argentea* P. C. Tsoong.

FIGURE IV 1:16B *Codonopsis argentea* P. C. Tsoong. **a**, **c** & **d**, from Guizhou: Fangjing Shan, 07–08, Oct. 2005, by Qing-Wen SUN; **b**, from the same locality, 20. Aug. 2009, by Qiang WANG.

Roots carrot-shaped or elongated, 20–40 cm long, sometimes even longer, 1–2.5 cm in diameter. Main stems usually many from one caudex, forming a clump, erect or ascending, 25–85 cm tall, leafy in middle part; branches numerous, mostly aggregated on lower or middle part, leafy, fertile or sterile, densely white-villous. Leaves on main stem alternate, those on branches nearly opposite; petiole shorter than 4 mm; leaf blade ovate, broad-ovate or nearly cordate, 0.8–1.8cm long, 0.5–1.2 cm broad, densely white-puberulent on both surfaces, margin entire, apex acute or obtuse. Flowers solitary, terminal on main stem and upper branches; pedicel 2–15 cm long. Calyx tube adnate to whole hypanthium, semiglobose, 10-ribbed, densely white hispidulous; calyx lobes ovate-lanceolate, 5–6 mm long, 2–5 mm broad, widest at middle, densely white-puberulent, margin entire; sinus between calyx lobes broad and obtuse. Corolla broad-campanulate, 1.5–1.8 cm long, 2–2.5 cm wide, shallowly lobed, white, some with blue veins at base inside, some purple at base; corolla lobes deltoid, villous at apex. Stamens glabrous; filaments slightly dilated at base, 2–2.5 mm long; anthers ca. 3 mm long. Capsule semiglobose below, conical above, 1–1.7 cm long, 0.6–1 cm in diameter. Seeds ellipsoid, brown-yellow. Figure IV 1:17A & B.

Phenology: flowering from June to September; fruiting from August to October.

Habitats: grassy slopes, sunny and stony river terraces; 2900–4500 m.

FIGURE IV 1:17A *Codonopsis canescens* Nannf. **a**, lower part of a plant, showing root, caudex, and two stems; **b**, upper part of a stem.

Distribution: **endemic** to China: S Qinghai, NW Sichuan and E Tibet (Map IV 1:17).

Additional specimens examined, **CHINA, QINGHAI, Nangqên:** 32°7′ N, 97°2′ E, 3570 m, 30. Aug. 1996, *T. N. Ho et al. 2717* (E, PE); 32°7′ N, 96°28′ E, 3650 m, 04. Sept. 1996, *T. N. Ho et al. 2887* (E, PE); 32°5′ N, 97°1′ E, 3600 m, 27. Aug. 1996, *T. N. Ho 2405* (E, PE); Ji'nyinsib Township, thickets, 4200 m, 15. Jul. 1965, *Y. C. Yang 1179* (KUN, PE); by Za Qu, 02. Aug. 1978, *P. C. Kuo 345* (HNWP); s. loc., 3700 m, 25. Aug. 1972, *Xizang Chinese Mater. Medic. & Herbs Exped. 1321* (HNWP, KUN, PE). **Yushu**: Jiangxigou Forest Farm, 3800 m, dry slopes, 30. Jun. 1971, *P. C. Kuo & W. Y. Wang 8392* (HNWP); Xiao Surmang, 32°6′ N, 97°16′ E, 24. Aug. 1996, *T. N. Ho et al. 2276* (E, KUN, PE); s. loc., thickets, 3650 m, 17. Jul. 1964, *s. coll. 504* (HNWP, PE). **SICHUAN, Baiyü**: by Jinshajiang, 31°17′ N, 98°49′ E, 3157 m, grassy slopes, 25. Jul. 2010, *Kham Exped. 10–0650* (PE); Shanyan, ca. 4000 m, 10. Jul. 1979, *s. coll. 129* (SM). **Dawu**: mountain near Larmasery 2900 m, rocky crevices or herbs, *T. L. Dai et al. Dawu 64-1* (PE); near the city, Longwang Temple, 3100 m, 04. Jun. 1958, *s. coll. 75* (SZ); mountain near the city, 3400 m, grassy slopes, 29. Jun. 1974, *Y. T. Wu & B. C. Kao 111658* (CDBI, PE, SZ); loc. eodem, 3000 m, on terraces, 22. Jul. 1959, *S. Jiang & C. L. Jin 2292* (PE, SZ); loc. eodem, 04. Sept. 2012, *D. Y. Hong & Q. Wang H12070* (PE); Mazê, Goupu, 3100m, 06. Jul. 1979, *s. coll. 92* (SM); s. loc., 16. Aug. 1986, *Z. T. Wang 868–354* (PE). **Dêgê**: Damayanda Gou, s. d., *S. X. Jia* s. n.(PE); Dongkor, rocky crevices by river, 22. Aug. 1964, *T. L. Dai et al. 64-3* (PE); Jangra, 3200 m, sunny slopes, 20. Jun. 1974. *Sichuan Plant Exped. 7026* (CDBI); loc. eodem, 31°40′ N, 98°36′ E, 3026–3200 m, 22. Jul. 2010, *X. F. Guo et al. 12613* (CDBI); loc. eodem, 3000 m, 11. Jun. 1979, *s. coll. 163* (SM); Mengda Gou, 20. Jul. 1957, *Y. W. Tsui 4513* (PE) ; near the city, 10. Sept. 1951, *W. K. Hu 13088* (PE); s. loc., dry slopes, 14. Jul. 1977, *P. C. Kuo & W. Y. Wang 21353* (HNWP, KUN); s. loc., 3100 m, sunny grassy slopes, 10. Aug. 1960, *Sichuan Mater. Medic. Resources Exped. 15559* (PE); s. loc., grassy slopes, 3700 m, 02. Jul. 1961, *S. Jiang & T. S. Ying 9170* (KUN, PE). **Garzê:** Sênkar, Baili Township, 3380 m, 18. Aug. 1964, *J. D. Li 64–61* (PE, SM); Tuoba, 11. Jul. 1977, *P. C. Kuo & W. Y. Wang 21204* (HNWP); Yongcha Gou, 19. Aug. 1964, *J. D. Li et al. 64-65* (PE). **Luhuo** (Zhaggo): Left bank of Xianshui River, Buddhist nunnery, 3300 m, grassy slopes, 16. Jun. 1961, *J. S. Ying 9402* (KUN, PE); Yiliang Zhaizi, 3100 m, debris, 11. Aug. 1964, *T. L. Dai et al.* (*Luhuo*) *64-3* (PE, SM); Yimu, Ala Gou, 3200–4000 m, 11. Jul. 1979, *General Surv. Exped. 005* (SM); Zhêhor, 3300 m, grassy slopes, 15. Aug. 1964, *T. L. Dai et al.* (*Luhuo*) *64–10* (PE). **Sêrtar**: Wungda, 3400 m, grassy slopes, 14. Jul. 1974, *Sichuan Plant Exped. 6576* (CDBI, PE). **Serxü:** Mabja, Kardê, 3500 m, 14. Jul. 1979, *s. coll. 287* (SM). **Zamtang**: Hongxin, by fields, 3600 m, 17. Jul. 1975, *Sichuan Plant Exped. 9285* (CDBI, PE). **TIBET, Gonjo**: N of the city, grassy slopes, 3650 m, 29. Jun. 1991, *J. S. Yang 91-489* (KUN, PE); near the city, dry slopes, by cliff, 3900 m, 03. Aug. 1976, *Qinghai-Xizang Exped. 12555*

FIGURE IV 1:17B *Codonopsis canescens* Nannf. **a**, from Tibet: Riwoqê, 14. Jul. 2012, by Bo XU; **b** & **c**, from Tibet: Baxoi, 20. Jun. 2009, by Hang SUN; **d**, from Tibet: Riwoqê, 14. Jul. 2012, by Zhuo ZHOU.

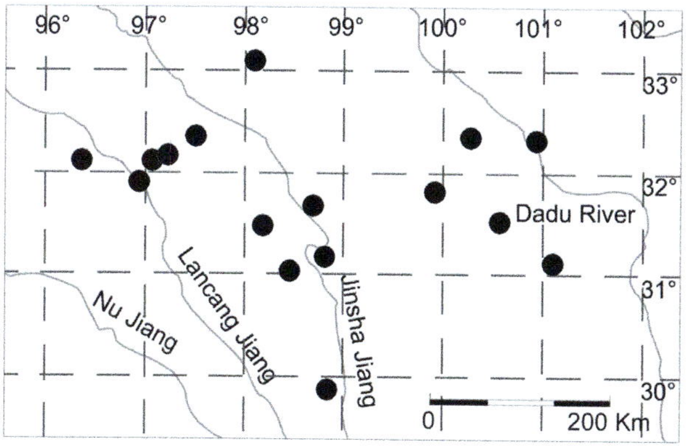

MAP IV 1:17 Distribution of *Codonopsis canescens* Nannf.

(KUN, PE); loc. eodem, 3680 m, dry slopes, 13. Aug. 2010, *Kham Exped. 10–1749* (PE). **Jomda**: Aila Shan, 3400–3700 m, thickets, 21. Jul. 2012, *H. J. Wang et al. JD–001* (PE); W SW of Jomda city, 31° 26′ 54″ N, 98° 9 ′55″ E, 3625 m, 29. Jul. 2004, *D. E. Boufford et al. 31208* (A); Tangpu, 3500 m, herbs at edges of forests, 23. Jul. 1976, *Qinghai-Xizang Exped. 12443 A* (PE); loc. eodem, 3600 m, 22. Jul. 1976, *Qinghai-Xizang Exped. 12442* (HNWP, KUN); loc. eodem, 3400 m, thickets, 25. Aug. 1973, *Xizang Exped. 2218* (HNWP); s. loc., 3700 m, 12. Aug. 1976, *Wu 4504* (KUN). **Markam**: Jiaolong, *Picea* forests, 4500 m, 09. Jun. 1976, *Qinghai-Xizang Exped. 11827* (KUN, PE). **Qamdo**; Garma, 31° 58′ N, 96° 57′E, 3490 m, dry sparse *Cupressus* forests, 02. Aug. 2010, *Kham Exped. 10–1130* (PE).

18. Codonopsis foetens Hook. f. & Thomson, J. Proc. Linn. Soc. Bot. 2: 16. 1857; Clarke in Hooker, Fl. Brit. Ind. 3: 433. 1881; Chipp, J. Linn. Soc. Bot. 38: 383. 1908; Anthony, Notes Roy. Bot. Gard. Edinburgh 15: 185. 1926; Nannfeldt, Acta Horti Gothob. 5: 29. 1930; Shen & Hong in Hong Fl. Reipubl. Popularis Sin. 73(2): 56. pl. 9. fig.2. 1983; Hong, Lammers & Klein in Wu, Raven & Hong, Fl. China 19: 521. 2011. Type: India (Sikkim), Kengna Lanch, 4570 m, 24. Jul. 1849, *J. D. Hooker* s. n.(lectotype here designated: K; isolectotype: BM); India (Sikkim), Lachen, 4270 m, 15. Jul. 1849. *J. D. Hooker* s. n.(paratype: K); India (Sikkim), Samding, 4880 m, 11. Sept. 1849, *J. D. Hooker* s. n.(paratype: K)

Roots slender, 10–15 cm long, 1–1.5 cm in diameter. Caudex up to 15 cm long, abundantly branched. Stems several to many from one caudex, caespitose, ascending, 20–40 cm tall, very sparsely villous; branches many, aggregated in lower part, slender, sterile. Leaves on main stem alternate, mostly yellow, scale-like, only several upper ones green and ordinary, sessile, smaller than those on branches, those on branches opposite or nearly opposite, all ordinary leaves with petiole ca. 2 mm long; leaf blade cordate or cordate-ovate, 0.5–1.5 cm long, 0.5–1.5 cm broad, rather densely white hispidulous on both surfaces, margin entire, apex obtuse. Flowers solitary, rarely 2 or 3 in a cyme, terminal on main stem, more or less pendent. Calyx tube adnate to whole hypanthium, semiglobose, glabrous or very sparsely white-hispidulous; calyx lobes ovate-oblong or ovate-lanceolate, 6–12 mm long, 3.5–6.5 mm broad, rather densely hispidulous, margin entire or verrucose-denticulate, revolute; sinus between calyx lobes mostly rounded, rarely slightly truncate. Corolla campanulate, 1.8–3 cm long, blue or pale purple, rarely white, with dark purple veins; corolla lobes suborbicular, 5–7 mm long, mostly puberulent, apex obtuse or acute. Stamens glabrous; filaments slightly dilated at base. Capsule obconical at inferior part, conical at superior part, 1.5–2 cm long, 1.2–1.5 cm in diameter. Seeds not seen.

Distribution: from Bhutan eastwards to SE Gansu, China (Map IV 1:18).

1a. Leaf blade 5–9 mm long, 5–7 mm broad, rarely up to 15 mm long, 12 mm broad; calyx lobes often verrucose-serrate .... 18a. subsp. *foetens*
1b. Leaf blade 10–15 mm long, 10–15 mm broad; calyx lobes always entire .... 18b. subsp. *nervosa*

18a. Codonopsis foetens subsp. **foetens**

Leaf blade 5–9(–15) mm long, 5–7(–12) mm broad; calyx lobes often verrucose-serrate. Figures IV 1:18A & B.

Phenology: flowering from July to September; fruiting in September and October,

Habitats: meadows, thickets, or alpine debris; 3650–4880 m.

Distribution: Bhutan, N India, China (SE Tibet), and N Myanmar.

Additional specimens examined, **BHUTAN**, **Shingbe**: Me La, 27°56′ N, 91°38′ E, 4270 m, 04. Aug. 1933, *F. Ludlow & G. Sherriff 381* (BM); loc. eodem, 27°56′ N, 91°40′ E, 4110 m, 05. Oct. 1934, *F. Ludlow & G. Sherriff 1027* (BM) & *1028* (BM); loc. eodem, 27°58′ N, 91°48′ E, 4420 m, 29. Jun. 1949, *F. Ludlow et al. 20774* (BM); loc. eodem, 3960 m, 23. Aug. 1949, *F. Ludlow et al. 21077* (BM, G); loc. eodem, 4270 m, 19. Sept. 1949, *F. Ludlow et al. 21200* (BM). **Thimphu**: Yele La, 4270 m, 25. Jul. 1914, *R. E. Cooper 1840* (BM, E). **Upper Bumthang Chu** Distr., Pangotang, Tsampa (Tolegong), 4420 m, 27°54′ N, 90°47′ E, 03. Jul. 1949, *F. Ludlow et al. 19320* (BM). **Upper Kulong Chu** Distr.: Shingphel, Pemaling, near Tashi Yangtsi, 4200 m, 01. Aug. 2000, *S.A. Bowes Lyon 13156* (E). **Upper Kuru Chu** Distr., Singhi Dzong, Lhuntse, 4200 m, 25. Jul. 2000, *S. A.*

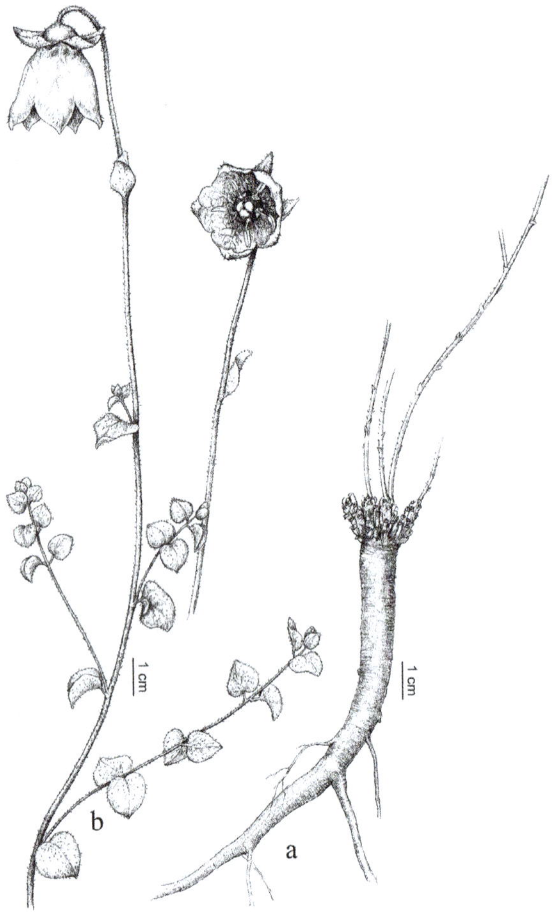

FIGURE IV 1:18A *Codonopsis foetens* Hook. f. & Thomson subsp. *nervosa* (Nannf.) D. Y. Hong. **a**, root, caudex, and base of stems; **b**, upper part of a stem.

Bowes Lyon 13092 (E). **Upper Mo Chu** Distr.: Valley SW of Sinchu La, 28°02′ N, 89°36′ E, 4000 m, 02. Aug. 1983, *C. Sargent 174* (E); Chasa area, 27. Jul. 1983, *C. Sargent 36* (E); Cekha, 3750 m, 21. Aug. 1971, *R. Bedi 906* (K); Shinja La, Upper Mo Chu, 27°56′ N, 89°30′ E, 05. Oct. 1949, *F. Ludlow et al. 17369* (BM); Chum La-Chenchena, 4400 m, 05. Sept. 1966, *K. Nishioka 1107* (BM); Dumpu, 4210 m, 18. Sept. 1967, *S.A. Bowes Lyon 5083* (BM); Kapoop, 3960 m, 24. Sept. 1923, *G. H. Cave* s. n.(E); Narimthang, 4110 m, 25. Sept. 1949, *F. Ludlow et al. 21466* (BM, E); Phagudin, 3960 m, 05. Aug. 1914, *R. E. Cooper 3230* (BM, E); Yala, Pyala, Gilela areas, 3350 m, Aug. 1938, *B. J. Gould 1360* (K).

CHINA, TIBET, **Bomi**: Guxiang, 29°55′ N, 95°30′ E, 4200 m, 02. Aug. 1965, *J. S. Ying & D. Y. Hong 650852* (PE); Yi'ong Farm, pass of Mt Bala, 4000 m, alpine meadows with shrubs, 18. Aug. 1983, *B. S. Li et al. 6697* (PE). **Cona**: Baishan Pass, 27°55′ N, 91°52′ E, 4530 m, rocky and grassy slopes, 12. Aug. 2013, *Y. S. Chen et al. 13–0993* (PE); Boshankou, 4500–4600 m, summit thickets, 25. Aug. 1975, *Qinghai-Xizang Suppl. Exped. 751537* (KUN, PE); Boshankou, N slopes, 4500 m, ancient glacier debris, 7. Jul. 1975, *C. Y. Wu et al.* 75–1007 (KUN, PE); Chumolari, 4880 m, 12. Sept. 1912, *R. Lepca 485* (PE); W of Cona Town, 4450 m, *Rhododendron* thickets, 29.Aug. 2011, *D. Y. Hong et al. H11007*(PE); Tawang, 4150 m, 23. Jul. 2002, *S. A. Bowes Lyon 15049* (E); Yumen, 28°41′ N, 93°2′ E, 4600 m, meadows with shrubs, 05. Aug. 2013, *Y. S. Chen et al. 13–0355* (PE); Ze La, ca. 4570 m, 20. Aug. 1938, *F. Kingdon-Ward 14124* (BM). **Damdoi**: Weiyuanliang, Aug. 1985, *J. H. Tong 3025* (PE). **Lhasa**: Natu to Champitang, *F. S. Chapman 1167* (K); Tang La, 4270–4570 m, *F. Kingdon-Ward 5943* (K); Tuna to Khamba, 4880 m, *B. J. Gould 2353* (K); s. loc., May–Jul. 1947, *J. Guthrie* s. n.(K). **Mainling**: near Kaamula Lake, 4300 m, meadows with shrubs, 27. Jul. 1983, *B. S. Li et al. 05913* (PE). **Nyingchi**: Nambu La, 4270–4570 m, 15. Aug. 1924, *F. Kingdon-Ward 6097* (BM, K); Sherqi La, W slope, 4360 m, 02. Aug. 1975, *Qinghai-Xizang Suppl. Exped. 751091* (KUN, PE); Sherqi La, at the pass, 4700 m, meadows with shrubs, 04. Aug. 1983, *B. S. Li & S. Z. Cheng 6133* (PE); loc. eodem, 4700 m, 04. Aug. 1983, *B. S. Li & S. Z. Cheng* 6123 (PE); loc. eodem, meadows, *FLPH Tibet Exped.12–2002* (PE); loc. eodem, 4508 m, alpine meadows, 11. Aug. 2012, *D. Y. Hong et al. H12046* (PE). **Yadong** : Chumbi, Bue-tang, 08. Jul. 1878, *Dungboo* s. n.(K); between Chumbi and Pagri, Jul. 1879, *Dungboo* s. n.(G, K); near Chumbi, Pem La, 29. Jul. 1892, *G. King's Collector* s. n.(B); Chumolari, 4880 m, 12. Sept. 1912, *R. Lepcha 485* (E, G, K); Chumbi Valley, 3660 m, 05. Aug. 1936, *F. S. Chapman 265* (K, PE); Donggo La Pass, alpine debris, 4400 m, 14. Aug. 1975, *s. coll.* 75–1005 (PE) ; Kangbu Spring, 27°50′ N, 88°59′ E, 4471 m, fl. blue, 06. Jul. 2013, *PE-Tibet Exped. 2664* (PE); Lower Yadong, Zelila pass, 4200 m, thickets, 12. Sept. 1974, *Qinghai-Xizang Exped.* 74–2351 (KUN, PE); Pagri, 4300 m, meadows, 28. Jul. 1974, *Qinghai-Xizang Exped.* 2854 (PE); Pagri to Heiwu, 22. Jul. 1953, *P. C. Tsoong 5827* (PE); Yadong to Pagri, 3900–4100 m, 27. Jul. 1960, *G. X. Fu* 455 (PE) ; s. loc., 1897, *H. E. Hobson* s. n.(K). **Zayü**: Mingqi Cattle Farm, 3900 m, meadows with shrubs, 12. Aug. 1973, *Integrated Surv. Team 804* (PE); Sangju Village to Ridong, 4500 m, 17. Sept. 2012, *FLPH Tibet Exped. 12–1425* (PE); Shangzayü, A'za to Rawu, 4100 m, stony herbs, 13. Aug. 1973, *Qinghai-Xizang Exped. 73–1072* (KUN, PE).

INDIA, **Darjiling**: Hahnak, 3960 m, Oct. 1938, *G. Ghose & Co.* s. n.(PE). **Sikkim**: Alpine region, 4260–4870 m, *J. D. Hooker* s. n.(P, PE); Chakung Chu, 3960 m, 21. Sept. 1913, *R. E. Cooper 931* (E); Changu, 3660 m, 18. Aug. 1913, *R. E. Cooper 528* (E); Chum-la-Chenchena, 4400 m, 05. Sept. 1966, *K. Nishioka* s. n.(TI); Ewutso la, 2200 m, 12. Jul. 1909, *W. W. Smith & G. H.Cave 1298* (B); Jamrachen Chu, 3960 m, 22. Jul. 1909, *W. W. Smith & G. H. Cave 1693* (B, K); North Distr., Lasha Chhu, below Phaklung, 27°55′ 22″N, 88°35′ 08″E, 4420 m, 19. Jul. 1996, *Edinburgh Exped. to N Sikkim (1996) 336* (E, TI); 27°51′ N,

FIGURE IV 1:18B *Codonopsis foetens* Hook. f. & Thomson subsp. *foetens*. **a** & **b**, from Tibet: Nyingchi, Serji La, 10. Aug. 2012, by De-Yuan HONG; **b**, with corolla opened, showing disc (arrow); **c**, from Tibet: Zêtang to Cona, 29. Aug. 2011, by Xin-Tang MA & De-Yuan HONG.

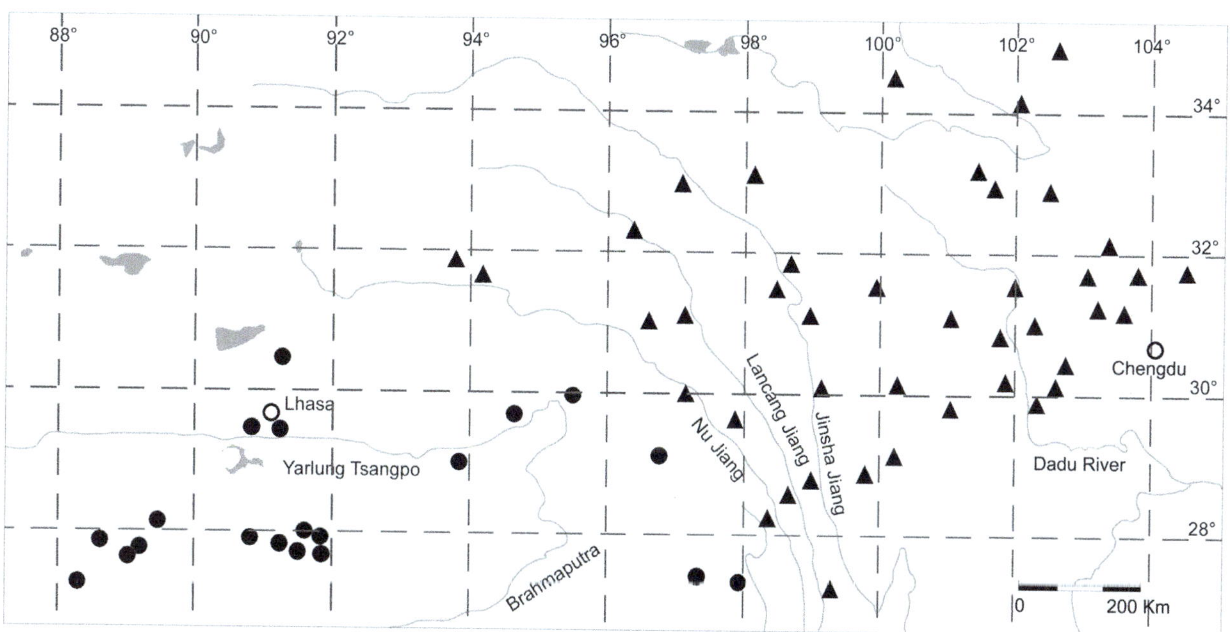

MAP IV 1:18 Distribution of *Codonopsis foetens* Hook. f. & Thomson. Dots: subsp. *foetens*; triangles: subsp. *nervosa* (Nannf.) D. Y. Hong.

88°35′ E, 1897, *H. E. Hobson* s. n.(K); Nathui La, 3960 m, *F. M. Baiby* s. n.(E) ; Shanka La, 3960 m, 30. Jul. 1913, *R. E. Cooper 352* (E); N Sikkim, 23-27. Jul. 1933, *L. R. Wager 325* (K) & *339* (K).

MYANMAR (N): Seinghku Valley, 28°10′ N, 97°20′ E, 3960 m, 09. Oct. 1926, *F. Kingdon-Ward 7568* (K).

18b. Codonopsis foetens subsp. **nervosa** (Chipp) D. Y. Hong, Novon 20(4): 422. 2010; Hong, Lammers & Klein in Wu, Raven & Hong, Fl. China 19: 521, 2011. ≡*Codonopsis ovata* Benth. var. *nervosa* Chipp, J. Linn. Soc. Bot. 38: 385. 1908. ≡*Codonopsis nervosa* (Chipp) Nannf., Acta Horti Gothob. 5: 26. pl. 13b. 1930; Shen et al., Acta Phytotax. Sin. 13(3): 56. pl. 6–16. 1975; Anonymous, Icon. Cormophyt. Sin. 4: 382. fig. 6177. 1975; Shen & Hong in Hong Fl. Reipubl. Popularis Sin. 73(2): 56. 1983. Type: China, Sichuan, Tachienlu (Kangding), 2740–4110 m, *A.E. Pratt 632* (lectotype here designated: K; isolectotype: BM); W China, *E. H. Wilson 3984* (paratype: K)

Codonopsis macrantha Nannf., Notes Roy. Bot. Gard. Edinburgh 16: 157. fig. 1a, pl.229. 1931 & in Handel-Mazzetti, Symb. Sin. 7(4): 1079. 1936. ≡*Codonopsis nervosa* (Chipp) Nannf. var. *macrantha* L. T. Shen in Hong, Fl. Reipubl. Popularis Sin. 73(2): 57. 1983. ≡*Codonopsis nervosa* (Chipp) Nannf. subsp. *macrantha* (Nannf.) D. Y. Hong & L. M. Ma in Li & Zhu, Fl. Sichuan. 10: 541. 1992. Type: China, SE Tibet, Salween-Kiujiang Divide, 28°40′N, 98°15′E, Aug. 1919, *G. Forrest 18949* (holotype: E; isotypes: K, W).

Leaf blade mostly 1–1.5 cm long, 1–1.5 cm broad, rarely up to 2.4 cm long, 2 cm broad; calyx lobes always entire; stems sometimes densely hirsute, multiflowered. Figure IV 1:18C.

Phenology: flowering in July and August; fruiting from August to October.

Habitats: grassy slopes, meadows with shrubs, or edges of forests at altitudes from 3000 to 4700 m.

Distribution: **endemic** to China: SE Gansu, E Qinghai, W Sichuan, SE Tibet, and NW Yunnan.

Additional specimens examined, **CHINA**, **GANSU**, **Maqu**: s. loc., grassy slopes, 26. Jul. 1960, *G. L. Chu 4146* (PE). **Xiahe**: Labrang to Yellow River, Wanchen Nang Valley, 3810 m, 29. Jul. 1926, *J. F. Rock 14522* (BM, E, K,PE); Yaqukarde, 3700 m, 18. Aug. 1937, *K. T. Fu 1472* (PE); loc. eodem, 3700 m, 18. Aug. 1937, *T. P. Wang 7539* (PE). **QINGHAI**, **Jiuzi**: 33°15′ N, 101°29′ E, 3850–3900 m, 09. Aug. 2007, *D. E. Boufford et al. 39313* (PE). **Maqên**: A'nyêmaqên Shan (W of Yellow River), Mt Druggu, 3810 m, Jul. 1926, *J. F. Rock 14436* (BM, E, K,PE); Jungung to

FIGURE IV 1:18C *Codonopsis foetens* Hook. f. & Thomson subsp. *nervosa* (Nannf.) D. Y. Hong. **a**, from Sichuan: Jiuzhaigou, 02. Aug. 2011, by Yan YU; **b**, from Sichuan: Daocheng, 31. Jul. 2012, by Fu-Sheng YANG; **c**, from Yunnan: Dêqên, 01. Aug. 2009, by Yan YU.

Maqên, 34°30′ N, 100°24′ E, 3900 m, 26. Jul. 1993, *Ho, Bartholomew & Gilbert 372* (E). **Nangqên**: Maozhuan Township, 4500 m, meadows with shrubs, 15. Aug. 1965, *Y. C. Yang 01416* (KUN, PE); Xiangda Township, 3800 m, flood lands, 07. Aug. 1965, *Y. C. Yang 01405* (KUN, PE). **Yushu**: Baitang, Reshuigou, 4300 m, meadows, *s. coll. 0596* (PE); Go Qu, 32°45′ N, 97°13′ E, 3950 m, 21. Aug. 1996, *T. N. Ho et al. 2253* (A, E, PE). **SICHUAN**, **Aba** (Ngawa): between upper Aba and Jigzhi, 04. Aug. 1975, *Sichuan Plant Surv. 10498* (CDBI, PE). **Baiyu**: Azi Township, 4300 m, meadows with shrubs, 12. Aug. 1982, *Z. X. Tang et al.728* (PE). **Baoxing**: Daqiangpeng, 4000 m, 17. Jul. 1954, *T. P. Soong 39085* (PE, SZ); Shizi Shan, 3000m, grassy slopes, 15. Jul. 1933, *T. T. Yü 2349* (PE). **Barkam**: Zhegu Shan, 15. Jul. 1958, *C. L. Wu 33001* (PE); Longriba, 3600 m, 20. Aug. 1957, *X. Li 72090* (SZ); Zhuoke Ji, 31°42′N, 102°19′E, 3950–4150 m, meadows, 06. Sept. 1997, *D. E. Boufford et al. 27906* (E) ; s. loc., 03. Jul. 1958, *C. L. Wu 32628* PE . **Batang**: Rongguo Shan (between Zangqênrong & Cêwu), 3800 m, 20. Jul. 1964, *J. D. Li & Z. M. Xu 46* (SM); Yidun, Haizi Shan, 4700 m, 27. Jul. 1982, *Tang & Xia 531* (PE); loc. eodem, by Haizi Lake, 4300 m, 05. Aug. 1959, *S. Jiang & C. L. Jin 02736* (KUN, PE). **Beichuan**: Jianse, 3600 m, 04. Aug. 1978, *Beichuan Exped. 341* (SM). **Danba**: Taiping Township, 3800 m, by streams, 11. Jul. 1964, *T. L. Dai et al. 64-1* (PE, SM); s. loc., 18. Jul. 1930, *Z. P. Huang 1490* (KUN). **Daocheng**: Bowa Shan, 4570 m, *Rhododendron* thickets, 30. Jul. 1982, *Sichuan Veget. Surv. (Daocheng) 28967* (CDBI); Daocheng Town to Julong, 4000 m, thickets, 17. Aug. 1983, *Y. F. Han 83-284* (PE); Haizi Shan, 4420 m, meadows, 31. Jul. 2012, *F. S. Yang et al. 044* (PE); Julong, Bowa Shan, 4600 m, meadows with shrubs, 22. Jul. 1973, *Sichuan Veget. Surv. 3608* (CDBI, KUN, PE); Shengmu Township, Gongba Shan, 4700 m, meadows with shrubs, 27. Jul. 1973, *Sichuan Plant Surv. (Daocheng) 3637* (CDBI, KUN, PE); Sumdo, 3500 m, alpine meadows, 30. Jul. 1964, *J. D. Li & Z. M. Xu 64–51* (PE, SM); Sumdo, 3900 m, thickets, 07. Aug. 1971, *Sichuan Veget. Surv. (Daocheng) 0349* (CDBI). **Dawu**: Baimai, 3500 m, grassy slopes. 07. Aug. 1960, *J. S. Ying 04527* (PE, SZ); Gedaliangzi (pass), 35°5′N, 101°4′E, 3900 m, meadows, 22. Aug. 1995, *D. Y. Hong et al. 95056* (PE); Songlingkou, 4230 m, meadows, 16. Apr. 1960, *Forest Exped. Second Team 6-0233* (PE) ; Yuko Distr., 4300 m, 15. Sept. 1974, *Sichuan Plants Surv. 5963* (CDBI, PE). **Dêgê**: Damayan Gou, s. d., *S. X. Jia 148* (PE); Haizi Shan, 24. Jul. 1951. *Y. W. Tsui 4983* (PE); loc. eodem, 24. Jul. 1951, *Y. W. Tsui 5909* (PE). **Garzê**: Dongge Distr., 3600 m, 08. Aug. 1960, *Medic. Plants Surv. 15503* (CDBI); S of Garzê, 31°29′50″N, 99°59′33″E, 4105–4350 m, 04. Aug. 2005, *D. E. Boufford et al. 34147* (A); 15 km S of Garzê, 4020 m, 24 Aug. 2006, *D. E. Boufford et al. 37221* (A); W of Garzê, 31°52′41″N, 99°45′50″E, 3690 m, 02. Aug. 2005, *D. E. Boufford et al. 34008* (A); Sênkar Distr., Baili Township, 3700–3800 m, 19. Aug. 1964, *J. D. Li 64–63* (PE, SM); Yulonghaizi, 4250 m, stony places, 26. Jul. 1959, *S. Jiang & C. L. Jin 02408* (PE, SZ). **Heishui**: Shidiaoluo Township, 2900 m, among herbs or by fields, 27. May. 1959, *Heishui Group 1183* (KUN); Waboliangzi, 2300 m, by fields, 10 Jun. 1959, *Sichuan Econ. Plant Surv. (Aba) 2622* (KUN); s. loc., 2300 m, by rocks in valley, 08. May. 1959, *Heishui Group 1076* (KUN). **Hongyuan**: Amoikog, 4000 m, wet meadows, 07. Jul. 1964, *Z. Z. Ni 247* (CDBI); Chazhenliangzi, 3700 m, 31. Sept. 1989, *C. S. Cao 1472* (PE); Shuajingsi, 3500 m, 30. Jul. 2007, *X. J. He et al. 146637* (SZ); Zamkou, 3400 m, thickets, 30. Jul. 2007, *X. J. He et al. 145172* (SZ); loc. eodem, 3600 m, 30. Jul. 2007, *X. J. He et al. 133549* (SZ); loc. eodem, 4200 m, 21. Aug. 1957, *X. Li & J. X. Zhou 74103* (PE, SZ); loc. eodem, 3500 m, 30. Jul. 1958, *S. Y. Chen 5693* (SZ); loc. eodem, 4000 m, 15. Aug. 1957, *Z. Y. Zhang & H. F. Zhou 23772* (SZ); Zhegu Shan, 3900 m, 31. Jul. 2007, *X. J. He et al. 132798* (SZ). **Jinchuan**: Anning, 3800 m, 07. Sept. 1964, *J. D. Li et al. 64–68* (PE, SM); Wanli Township, 4600 m, grassy slopes, 21. Jul. 1958, *X. Li 78151* (PE). **Kangding**: Luba, 4030 m, thickets, 29. Jul. 1981, *Q. Q. Wang 28618* (CDBI); near Sandaoqiao, 27. Jul. 1951, *W. K. Hu & Z. He 10567* (PE); loc. eodem, 3200 m, 26. Jul. 1934, *C. S. Liu 905* (PE); Tagong Township, 4380 m, thickets, 18. Aug. 2008, *D. Y. Chen & C. B. Wang 198113* (SZ); Tongolo, 1893, *J. A. Soulié 286* (B, K); loc. eodem, Aug. 1908, *E. H. Wilson 2289* (BM, K); Tsipula, 4000 m, 05. Aug. 1922, *H. Smith 4198* (E); Yulonggong, Gomba La, 3800–4000m, 22. Jul. 1934, *H. Smith 10678* (PE); Zheduo La, 3600 m, 04. Aug. 1934, *H. Smith 11080* (PE); Zheduo Shan, 4000 m, meadows with shrubs, 28. Jul. 1963, *K. C. Kuan et al. 1014* (PE); loc. eodem, 4080 m, shrubs, 03. Sept. 2012, *D. Y. Hong & Q. Wang H12057* (PE); loc. eodem, E of pass, 4020 m, 30. Aug. 2010, *Kunming & Edinburgh Exped. to Sichuan 38* (E); loc. eodem, 14. Aug. 1986, *Z. T. Wang 868–352* (PE); loc. eodem, forests, 3900 m, 05. Aug. 1953, *H. L. Tsiang & J. H. Xiong 36504* (PE); Zheduo Shan, the pass, 4200 m, meadows, 25. Jul. 2007, *X. J. He et al. 144928* (SZ); Zhongjiecun, Dagaigou, 3900 m, meadows, 25. Jul. 1963, *K. C. Kuan et al. 346* (PE); s. loc., 1934, *C. S. Liu 1007* (PE); s. loc., summit, 1938, *McLaren's Collectors AC203* (E, LD); s. loc. 1924, *R. Cunningham 348* (E); s. loc., 1893, *J. A. Soulié 540* (BM, G, K). **Litang**: Cogsum, 4000 m, meadows, 15. Aug. 1964, *L. T. D. S 004* (PE). **Lixian**: s. loc., 1952, *Z. He & Z. L. Zhou 13702* (PE); s. loc., 3600 m, 12. Aug. 1951, *T. C. Lee 2823* (SZ); s. loc., *s. coll. 64-6* (PE). **Luding**: Moxi, grassy slopes, 29. Jul. 1983, *Sichuan Veget. Surv. 31321* (CDBI); loc. eodem, 3200 m, grassy slopes, 08. Sept. 1980, *G. R. Xu 23371* (CDBI); Xinxing, 3700 m, grasses, 19. Sept. 1983, *Sichuan Veget. Surv. 31658* (CDBI). **Maoxian**: Laoren Shan, Machang, 25. Jul. 1955, *Y. Xu et al.004* (PE); Shaba, 4300 m, meadows, 16. Jul. 1979, *Maowen Exped. 1050* (SM). **Sêrxü**: N of Luoxu, 32°39′39″N,

82°2′53″E, 4800–4850 m, 29. Jul. 2005, *D. E. Boufford et al. 3120833745* (A). **Tianquan**: Jinpeng Shan, 4400 m, 03. Aug. 1978, *Tianquan Exped. 78–571* (SM). **Wenchuan**: Balang Shan, 4300 m, meadows, 10. Aug. 2007, *X. J. He et al. 146376* (SZ) & *153347* (SZ); loc. eodem, 3800 m, 10. Aug. 2007, *X. J. He et al. 149928* (SZ); loc. eodem, 3000 m, 09. Aug. 2007, *X. J. He et al. 150892* (SZ); Balang Township, 2100 m, 09. Aug. 2007, *X. J. He et al. 153204* (SZ); Fengyi, 3850 m, meadows, 07. Jul. 1963, *X. Y. Tang & S. J. Yue 63-44* (PE); Fengyi, Machang Shan, 3550 m, meadows, 15. Jun. 1959, *Sichuan Econ. Plant Surv. (Maowen) 2622* (PE); Tsao-Puh, 3960 m, Aug. 1942, *S. Y. Hu 2650* (BM, K). **Xiangcheng**: Distr. No.1, Chaoyangdaoban, 4200 m, meadows at edges of oak forests, 11. Aug. 1973, *Sichuan Veget. Surv. (Xiangcheng) 3103* (CDBI, KUN, PE); Wuming Shan, Maxionggou, 4200 m, meadows with shrubs, 18. Aug. 1981, *Qinghai-Xizang Exped. 4979* (CDBI, KUN, PE); s. loc., 4600 m, 29. Jul. 1972, *Sichuan Veget. Surv. 1148* (CDBI, PE). **Xiaojin**: Balang Shan, 4200 m, among herbs, 15. Jul. 2008, *X. J. He & Q. S. Zhao 188672* (SZ); Lianghekou, Dashuigou, 3600 m, 24. Jul. 1958, *S. S. Chang & Y. X. Ren 6189* (PE); Rilong, 3900 m, thickets at edges of forests, 28. Aug. 1975, *Sichuan Plant Surv. 9700* (PE); loc. eodem, 3200 m, 11. Aug. 2007, *X. J. He et al. 149209* (SZ) & *149222* (SZ); Shepiliangzi, 4300 m, meadows, 14. Aug. 1975, *Sichuan Veget. Surv. 9743* (CDBI). **Yajiang**: Bailuogongka, 3900–4300 m, grassy slopes, 09. Aug. 1964, *Z. M. Xu & J. D. Li 64–59* (SM) ; Jianziwan Pass, 4456 m, oak shrubs, 10. Aug. 2010, *E. D. Liu et al. 2666* (KUN). **TIBET**, **Baxoi**: Demula Shan, N slopes, meadows, 4500 m, 17. Aug. 1980, *C. C. Ni et al. 1217* (PE). **Biru**: Nujiang Commune, 3900 m, sunny thickets, 12. 09. 1976, *D. D. Tao 11366* (KUN, PE). **Gonjo**: Sadong, 30°37′N, 98°47′E, 4395 m, meadows with shrubs, fl. pale blue, 15. Aug. 2010, *Kham Exped. 10–1980* (PE). **Jomda**: Qinglongdong, 3750–4100 m, grassy slopes, 17. Jul. 1961. *s. coll. 1363* (KUN). **Markam**: Cawarong, Himala, 3500 m, grassy slopes, Aug. 1935, *C. W. Wang 65634* (PE); Cawarong, Lajo, 3400 m, Aug. 1935, *C. W. Wang 65843* (PE); Cawarong, Nga La, grassy slopes, 3800 m, Aug. 1935, *C. W. Wang 66061* (KUN, PE) ; Lawula, 3800 m, 12. Aug. 1976, *Wu 4495* (KUN); Yanjing, 3200 m, 07. Jun. 1961, *s. coll. 1100* (KUN). **Qamdo**: Dama La, 31°10′N, 97°36′E, 4400 m, grassy slopes, 03. Aug. 2010, *Kham Exped. 10–1150* (PE); Toba, 3700–4100 m, meadows, 09. Aug. 1976, *Qinghai-Xizang Exped. 12712* (KUN, PE); Zhudou Si, 4000 m, forest margins, 16. Aug. 1976. *Wu 4759* (KUN). **Riwoqê**: Riwoqê Distr., edges of *Picea* forests, 3800 m, 18. Aug. 1976, *Qinghai-Xizang Exped. 12882* (KUN, PE). **Sogxian**: Gaqing Shan, 4100–4300 m, 01. Sept. 1976, *Qinghai-Xizang Exped. 11131* (KUN, PE); Rongbo to Ya'ngamdo, Qala Shan, 3700–3800 m, thickets on rocks, 20. Aug. 1976, *Qinghai-Xizang Exped. 11010* (PE). **Zogang**: Nujiang Valley, 3700 m, 11. Aug. 1975, *M. Zang 007* (KUN). **YUNNAN**, **Dêqên**: Baima Shan, 28°20′ 42″N, 99°04′06″E, 4300 m, *Roy. Bot. Gard. Edinburgh Exped. To Dêqên 258* (E); loc. eodem, 4880 m, 22. Aug. 1916, *F. Kingdon-Ward 5346* (E); Baima Shan, 3500 m, Sept. 1935, *C. W. Wang 69426* (PE); loc. eodem, 4000 m, 24. Jul. 1937, *T. T. Yü 9071* (KUN, PE); loc. eodem, 3900 m, 12. Aug. 1937, *T. T. Yü 9591* (PE); loc. eodem, 4280 m, stony slopes, 13. Aug. 1981, *X. Y. Tian et al. 868* (PE, SZ); loc. eodem, 4600 m, 26. Sept. 1998, *Q. E. Yang & H. Z. Kong 98-407* (PE); Mekong-Salween Divide, Doker-la, 3960–4270 m, 1913, *F. Kingdon-Ward 1100* (paratype of *C. macrantha*: E); loc. eodem, edges of woods, 3000 m, 3–5. Aug. 1935, *C. W. Wang 65123* (KUN, PE); Mekong-Salwen Divide, 28°4′ N, 4050–4300 m, 03. Aug. 1916, *Handel-Mazzetti 5825* (WU); Mekong-Salween Divide, 28°09′ N, 3800–3950 m, 06. Aug. 1916, *Handel-Mazzetti 9703* (E, K, W, WU); Mekong-Salween Divide, Doker-la, 3600–3800 m, grassy slopes, 01. Aug. 1940, *K. M. Feng 6029* (KUN, PE); by Mekong, Yongzi, 3800–4000 m, stony grassy slopes, 13. Aug. 1940, *K. M. Feng 6579* (KUN, PE); loc. eodem, 09. Aug. 1940, *K. M. Feng 6457* (KUN, PE); Yunnan-Tibet, Doker-la, 28°15′ N, 4200–4250 m, 17. Sept. 1915, *Handel-Mazzetti 8071* (W, WU); NW Yunnan-E Tibet, 4420 m, 1913, *F. Kingdon-Ward* s. n.(E). **Gongshan**: Upper Kiukiang Valley, Tsugum, 3800 m, alpine meadows, 11. Aug. 1938, *T. T. Yü 19887* (KUN, PE). **Weixi**: Tungchu-ling, 4100 m, on rocks, 17. Jul. 1937, *T. T. Yü 9051* (KUN, PE); N of Tungchu-ling, 4000 m, rocky slopes, 08. Nov. 1937, *T. T. Yü 10654* (E, PE); Yezhi, 3600 m, grassy slopes, Aug. 1935, *C. W. Wang 68497* (PE).

19. Codonopsis bulleyana Forrest ex Diels, Notes Roy. Bot. Gard. Edinburgh 5: 171. 1912; Nannfeldt, Notes Roy. Bot. Gard. Edinburgh 16 : 155. fig. 1b. 1931; Tsoong, Contr. Inst. Bot. Nat. Acad. Peiping 3(3) : 92. fig. 8A. 1935 ; Shen & Hong in Hong, Fl. Reipubl. Popularis Sin. 73(2) : 55. pl. 9. fig. 3. 1983; Hong, Lammers & Klein in Wu, Raven & Hong, Fl. China 19: 521. 2011. Type: China, Yunnan, "eastern flank of the Lichiang Range " (Lijiang), 27°15′ N, 3300-3660 m, Jul. 1906, *G. Forrest 2731* (holotype: E).

Cyananthus mairei H. Lév., Cat. Pl. Yunnan 25. 1915. Type: China, Zochers du Io-ch'an (Yunnan, Qiaojia), 3300 m, Aug. 1911, *E. E. Maire* s. n.(holotype: E).

Roots carrot-shaped, 15–30 cm long, up to 2 cm in diameter. Main stem erect or ascending, 25–60 cm tall, sparsely white-villous; branches aggregated at lower part of main stem, usually in axils of scale-like leaves, leafy, sterile, 1–10 cm long, densely villous. Leaves on main stem often only one or two at middle and upper parts, alternate, those on branches nearly opposite; petiole 2-4 mm long, densely villous; leaf blade shallowly cordate, broad-ovate or ovate, 1–1.8 cm long, 0.8–1.4 cm broad, both surfaces white-hispidulous, base cordate or rounded, margin sinuous or obscurely crenate, or nearly entire, apex acute or obtuse. Flowers solitary, terminal on main stem, very rarely second flower terminal on an upper branch, slightly pendent; pedicel 4–8 cm long, usually sparsely villous. Calyx tube adnate to whole hypanthium, semiglobose, nearly glabrous; calyx lobes ovate, 8–13 mm long, 5-7 mm broad, strongly reflexed, both surfaces hispidulous at least towards apex, apex acute, margin sinuous, revolute; sinus between calyx lobes narrow and pointed, or broad and obtuse due to revolute margin. Corolla tubular at lower half or 1/3, but suddenly spreading and becoming infundibular above, 1.8–2.8 cm long, shallowly lobed to 1/3, pale blue with purplish tube; tube 0.5–1.2 cm wide; limb 2–2.8 cm wide; corolla lobes ca. 1.5 cm broad. Stamens glabrous; filaments ca. 6 mm long. Capsule hemispheric in inferior part, conical and rostrate in superior part, 2–2.4 cm long, 1–1.5 cm in diameter. Seeds ellipsoid, ca. 1.5 mm long, brown-yellow. Figures IV 1:19A & B.

Phenology: flowering from July to September; fruiting in September and October.

Habitats: grassy slopes or scrubs; 2800–4600 m.

Distribution: **endemic** to China: SW Sichuan and N Yunnan (Map IV 1:19).

Notes: *Codonopsis bulleyana* very much resembles *C. foetens*, particularly *C. foetens* subsp. *nervosa*, in appearance. They differ in corolla shape and colour. The corolla of the latter is broad-campanulate and uniformly pale purple or pink, whereas that of the former is tubular for the lower half to 1/3, but then is abruptly dilated into a funnel-shape and is purplish below, blue above. Thus they are easily distinguished at anthesis, but it is rather difficult to identify them without flowers. However, their distribution ranges are different, with the latter in the north, while the former in the south, overlapping only in Dêqên, NW Yunnan. Furthermore, the former has broader calyx lobes, 5-7 mm broad vs. 3.5-6.5 mm broad in the latter.

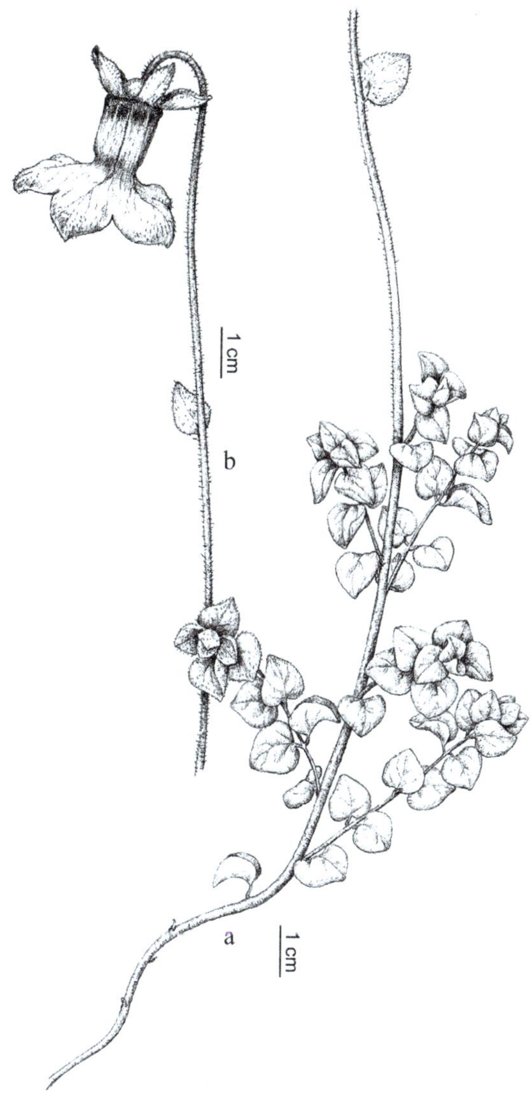

FIGURE IV 1:19A *Codonopsis bulleyana* Forrest ex Diels. **a**, lower part of a stem, showing sterile branches which appear pinnate leaves; **b**, upper part of a stem.

Additional specimens examined, **CHINA, SICHUAN, Daocheng:** Gongga-ling, 3960-5335 m, Aug. 1928, *J. F. Rock* 16804 (PE); near Laocunkou, 4000 m, thickets, 03. Aug. 1983, *K. Y. Lang et al. 2697* (KUN); Xiachonggu of Gongga Shan to Luorong, 4100 m, forests, 29. Aug. 1981, *Qinghai-Xizang Exped. 5461* (CDBI, KUN, PE). **Jiulong**: 43 km N of Jiulong, 29°19′ N, 101°29′ E, 4188 m, limestones, 15. Sept. 2010, *Kunming-Edinburgh Exped. to Sichuan 303* (E); Sanyanlong Commune, pastures, 3700 m, 20. Jul. 1980, *s. coll. 189* (SM); loc. eodem, 3500 m, 29. Jul. 1979, *s. coll. 328* (SM); Tanggo, 4100 m, thickets, 06. Aug. 1980, *F. A. Liu & Q. Q. Wang 22852* (CDBI); loc. eodem, 4250-4600 m, 18. Jul. 2005, *D. E. Boufford et al. 33224* (A) ; Wuxuhai, 4350 m, meadows, 09. Aug. 1979, *Z. G. Liu 20959* (CDBI). **Muli:** Baidiao, 3800 m, 05. Sept. 1978, *C. S. Cao et al. 8442* (CDBI, SZ); Gibboh Shan, 3960 m, Apr. May. 1932. *J. F. Rock 24167* (E, K, PE); loc. eodem, 3960 m, 21. Jul. 1924, *F. Kingdon-Ward 4618* (E); Kangwu, 3500 m, meadows with shrubs, 21. Aug. 1983, *Qinghai-Xizang Exped. 13173* (KUN, PE); Kai Pasture, 3600 m, 01. Aug. 1978, *s. coll. 561* (SM); Kumatian, 3900 m, margins of woods, 28. Jul. 1937, *T. T. Yü 7412* (PE); Litang-Yalong Divide, 28°15′ N, 101°05′ E, 3960–4270 m, 21. Jul. 1914, *F. Kingdon-Ward 4479* (E); Mairi to Ba'er Pasture, 3800 m, meadows with shrubs, 06. Sept. 1983, *Qinghai-Xizang Exped. 14446* (KUN, PE); Mêri, 4600 m, 30 Jul. 1978, *C. S. Cao 6294* (CDBI, SZ); Muli-Taoba, 4100 m, pastures with scattered shrubby *Rhododendron*, 21. Sept. 2011, *D. Y. Hong et al. H11028* (PE); Wachin, 3900 m, margins of thickets, 19. Oct. 1937, *T. T. Yü 14576* (E, KUN, PE); Yungning-Muli, 3900-4000 m, 25. Jul. 1915, *Handel-Mazzetti 7231* (W, WU); No. 3 Distr., Warzhai, 28, Jul. 1966, *T. L. Dai & L. Y. Xu 66–28* (SM); No. 3 Distr., Zhenxi Pasture, meadows with shrubs, 3900 m, 08. Sept. 1983, *Qinghai-Xizang Exped. 13713* (PE). **Xiangcheng**: Daxue Shan, 4200 m, meadows with shrubs, 05. Aug. 1981, *Qinghai-Xizang Exped. 3769* (CDBI,

FIGURE IV 1:19B *Codonopsis bulleyana* Forest ex Diels. **a** & **b**, from Yunnan: Shangri-la, 01. Aug. 2012, by Fu-Sheng YANG; **c**, from Yunnan: Shangri-la, 06. Aug. 2011, by Hang SUN.

KUN, PE); loc. eodem, 4300 m, meadows, 01. Aug. 2012, *F. S. Yang et al. 062* (PE). **Xide**: Lianghekou, 3660 m, Aug. 1938, *T. K. Wang & T. S. Wen 655* (PE) *& 623* (PE). **Yanyuan**: Bailing, 3700 m, meadows with shrubs, 08. Aug. 1983, *Qinghai-Xizang Exped. 12730* (KUN, PE); loc. eodem, 4100 m, 02. Aug. 1978, *Yanyuan Exped. 518* (SM); Maoniujuan, 3300-3400 m, 24. Sept. 1965, *F. G. Zhu 007* (PE) ; Yuanbao Distr., Dalu Shan, 3800 m, meadows with shrubs, 21. Jul. 1982, *Qinghai-Xizang Exped. 12233* (KUN, PE). **YUNNAN, Dêgên:** Baima Shan, 4570–4880 m, 1913, *F. Kingdon-Ward 1095* (E); loc. eodem, 4300 m, *Rhododendron* thickets, 14. Jul. 1981, *Qinghai-Xizang Exped. 2759* (CDBI, KUN); Huann-fu-ping, 3700 m, grassy slopes, Aug. 1935, *C. W. Wang 68777* (KUN, PE); Mekong-Salween Divide, Doker-La, 28°20′ N, 3660-3960 m, 09. 1917, *G. Forrest 14839* (E, K); Mekong-Salween Divide, Naka-La, E of Yuragan, 4270 m, 06. Jul. 1932, *J. F. Rock 23442* (E, K, PE); loc. eodem, 4270 m, Jun.–Jul. 1932, *J. F. Rock 23095* (E, K, PE); Yangtsa Snow Range, Rusa-la, 4120 m, rocky hills, 20. Aug. 1937, *T. T. Yü 9683* (KUN, PE); Yunling, 4150–4300 m, meadows with shrubs, 24. Aug. 1981, *Inst. Bot. Hengduan Mts Exped. 3616* (PE). **Fugong:** Chiuchuan-Mekong Divide, Jun. 1922, *G. Forrest 23177* (E, K, W); Mt. Fu-Chuan, *McLaren 219* (E). **Heqing**: Siangshui River, Ganhaizi, 3800 m, 03. Sept. 1929, *R. C. Ching 24300* (PE). **Huize**: Dahai, 3700 m, *Rhododendron* thickets, 30. Jul. 1964, *NE Yunnan Exped. 432* (KUN); Shekuai Township,

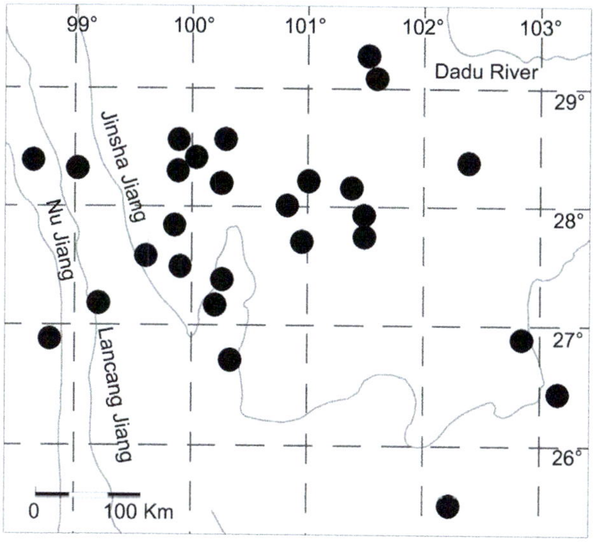

MAP IV 1:19 Distribution of *Codonopsis bulleyana* Forrest ex Diels.

4100 m, grassy slopes, 18. Jul. 2009, *E. D. Liu & W. Fang 2114* (KUN); loc. eodem, 4010 m, rocky meadows, 28. Jul. 2008, *H. Peng et al. 8654* (KUN). **Lijiang** (Yulong): between Lijiang, Tungshan, Tuinaoko & Tsilijiang, 3960-4270 m, dry Yangtze basin, May 1923, *J. F. Rock 9756* (PE); Nguluko, 2800 m, grassy slopes, 11. Aug. 1937, *T. T. Yü 15422* (KUN, PE); Yuhu Township, Xuesongcun, 3800 m, meadows, 08. Sept. 1955, *K. M. Feng 21490* (PE, SZ); Yulong Snow Range, 4000 m, 20. Jul. 1914, *Handel-Mazzett 4302* (G, WU); loc. eodem, 08.Jul. 1922, *J. F. Rock 4939* (K); loc. eodem, 4570 m, Jul. 1922, *J. F. Rock 5444* (PE); loc. eodem, open pasture, 03. Aug. 1939, *R. C. Ching 30459* (KUN, PE); loc. eodem, 29. Jul. 1940, *R. C. Ching 30895* (KUN, PE); loc. eodem, 3500-3900 m, grasses in *Abies* forests, 28. Jul. 1981, *Inst. Bot. Hengduan Mts Exped. 2160* (PE); loc. eodem, alpine meadows, 23. Jul. 1981, *Inst. Bot. Hengduan Mts Exped. 1978* (PE); loc. eodem, edges of *Abies* forests, 27. Jul. 1981, *Inst. Bot. Hengduan Mts Exped. 2069* (PE); loc. eodem, Mahuangba, 3700-3900 m, rocky slopes, 12. Aug. 1942, *K. M. Feng 8965* (KUN, PE); loc. eodem, 3900 m, meadows with shrubs, 23. Jul. 1981, *Y. F. Han et al. 81-1312* (PE); loc. eodem, Wutoudi, 3600 m, meadows, 22. Aug. 1964, *X. W. Yu 64-096* (KUN); loc. eodem, Wutoudi , 3500 m, 10. Oct. 1990, *CLD-90 1065* (E); loc. eodem, 3800-4000 m, 02. Aug. 1914, *C. Schneider 2128* (G, K, PE); 27°15′ N, 100°10′ E, Jul. 1929, *G. Forrest 27843* (E, PE); loc. eodem, 27°25′ N, 3960 m, Sept. 1910, *G. Forrest 6608* (E, K, PE, W); loc. eodem, 4270 m, Aug. 1925, *J. F. Rock 9884* (E); s. loc., 4400 m, meadows, 18. Aug. 1976, *NW Inst. Plateau Biol. & Chengdu Inst. Biol. 033* (CDBI). **Luquan**: Wumong Township, 3900 m, rocky meadows, 03. Aug. 2008, *H. Peng et al. 9379* (KUN); loc. eodem, 4223 m, rocky meadows, 03. Aug. 2008, *H. Peng et al. 9463* (KUN) & *9474* (KUN). **Qiaojia**: Dacun to Qiaoding Shan, 3300 m, 17. Aug. 2003, *H. Wang et al. 03-0856* (KUN) & *03-0866* (KUN); Yaoshan, 3600 m, 16. Jul. 1973, *B. S. Sun 1078* (KUN, PE). **Shangri-la** (Zhongdian): S side of Daxue Shan, 25. Jul. 1998, *D. E. Boufford et al. 29167* (A, E); Haba Snow Range, 3800 m, *Rhododendron* thickets, 11 Aug. 1981, *Inst. Bot. Hengduan Mts Exped. 2976* (PE); loc. eodem, 27°21′ 29″N, 100°3′ 48″E, *D. E. Boufford et al. 35190* (A); loc. eodem, near Haba Hai, glacier lake, 4100 m, open stony pastures, granites, 20. Aug. 2000, *E. von Raab-Straube et al. 242* (B); Hong Shan, 35 km NNE of Zhongdian town, 4400 m, open meadows, 02. Sept. 2000, *E. von Raab-Straube et al. 500* (B); Jisan Forest Farm, 3850 m, meadows with shrubs, *Y. F. Han et al.* 81-1652 (PE); Juabe, 3700 m, rocky hills, 06. Aug. 1937, *T. T. Yü 12671* (KUN, PE); Kaiza, 4550 m, meadows, 15, Jul. 1981, *Z. X. Tang et al. 582* (PE); from the town to Nanshan-ling, 3800 m, thickets at edges of forests, 16. Aug. 1962, *Zhongdian Exped. 885* (KUN, PE); between Pino and Peilun Shan, open pastures, 13. Aug. 1939, *K. M. Feng 2003* (KUN, PE); Wongshiu, 3900 m, margins of thickets, 20. Aug. 1937, *T. T. Yü 12790* (KUN, PE); Xianrendong, 3600 m, thickets, 18. Oct. 1937, *T. T. Yü. 13751* (E); Xiaozhongdian, 3960 m, Apr. –May. 1932, *J. F. Rock 24627* (E, G, K, PE); loc. eodem, meadows with shrubs, 3800 m, 06. Aug. 1981, *S. W. Tian et al. 792* (PE); 28°35′ N, 99°50′ 14″E, 24. Jul. 1998, *D. E. Boufford et al. 29140* (E); s. loc., 3400 m, Sept. 1916, *C. Schneider 3859* (G, K, PE). **Weixi**: Kangpu, 3500 m, grassy slopes, Jul. 1935, *C. W. Wang 64664* (PE); Yezhi, 3600 m, grassy slopes, Aug. 1935, *C. W. Wang 68655* (KUN, PE).

20. Codonopsis thalictrifolia Wall. in Roxburgh, Fl. Ind. 2: 106. 1824; Hooker & Thomson, J. Proc. Linn. Soc. Bot. 2: 16. 1857; Clarke in Hooker, Fl. Brit. Ind. 3: 432. 1881; Komarov, Trudy Imp. S.-Peterburgsk. Bot. Sada 29(1): 114. 1908; Chipp, J. Linn. Soc. Bot. 38: 381. 1908; Nannfeldt, Notes Roy. Bot. Gard. Edinburgh 16: 159. 1931; Shen & Hong in Hong, Fl. Reipubl. Popularis Sin. 73(2):54. 1983; Hong, Lammers & Klein in Wu, Raven & Hong, Fl. China 19: 521. 2011. ≡*Glossocomia tenera* D. Don, Prodr. Fl. Nepal. 158. 1825. ≡*Glossocomia thalictrifolia* (Wall.) Wall., Cat. 1297. 1828. ≡*Wahlenbergia thalictrifolia* (Wall.) A. DC. in de Candolle, Prodr. 7: 425. 1839. ≡*Campanula thalictrifolia* (Wall.) Spreng., Cura Post. 2(2): 77. 1825. Type: Nepal, Emodu ad Gossain Than, *Wallich 1297* [lectotype here designated: K (K000814201); isolectotypes: BM, E, G, K (K000814200)].

Roots carrot-shaped, 15–20 cm long, 0.5–1 cm in diameter. Stems erect or ascending, main stem robust, 30–80 cm tall, glabrous or sparsely villous; branches sterile, slender, leafy, many aggregated at base of main stem, 4–6 cm long, villous. Leaves on main stem alternate, those on branches nearly opposite; petiole 2–4 mm long, white-hirsute; leaf blade nearly orbicular, 3–19 mm long, 3–6 (–14) mm broad, both surfaces hirsute, base cordate or truncate, margin nearly entire or crenate, apex obtuse or acute. Flowers several or single, terminal on main stem and upper branches, slightly pendent. Calyx tube adnate to whole hypanthium, semiglobose, 3–4 mm long, 6–10 mm broad, glabrous, rarely sparsely hispidulous; calyx lobes oblong or linear-lanceolate, broadest at middle, 5–11 mm long, 2–4 mm broad, hispidulous outside, apex obtuse, margin entire, crenulate or denticulate, often revolute; sinus between calyx lobes broad, rounded or truncate. Corolla tubular-campanulate, lower half tubular, upper half infundibular, 2.2–5.5 cm long, 1.5–4.3 cm wide, shallowly lobed, pale blue, totally glabrous; corolla tube 1.8–4.2 cm long, 0.8–1.2 cm wide; corolla lobes triangular, 3–4 mm long, apex obtuse. Filaments ca. 1 cm long, slightly dilated at base, glabrous; anthers ca. 3 mm long, villous at connective. Capsule semi-globose below, conical and rostrate above, 1.5 cm long, ca. 9 mm in diameter. Seeds ellipsoid, brown-yellow, smooth.

Distribution: China (C & S Tibet), N India, and E Nepal (Map IV 1:20).

Two allopatric subspecies are easily recognized

1a. Corolla mostly shorter than 3.3 cm ....20a subsp. *thalictrifolia*
1b. Corolla mostly over 3.5 cm long ....20b. subsp. *mollis*

20a. Codonopsis thalictrifolia subsp. **thalictrifolia**

Corolla 2.2–4 cm long, but mostly shorter than 3.3 cm. Figures IV 1:20B.

Phenology: flowering from June to August; fruiting in September and October.

Habitats: grassy slopes, alpine meadows with shrubs, and forests; 3400–5560 m.

Distribution: China (S and SE Tibet), India (Darjiling and Sikkim) and E Nepal.

Additional specimens examined, **CHINA, TIBET, Dinggyê**: Chengtanggou, 27°54′ N, 87°34′ E, 4151 m, meadows by forests, 15. Jul. 2013, *PE-Tibet Exped. 3516* (PE); Kelabei Shan, 4300–4800 m, meadows with *Sabina* thickets, 16. Aug. 1990, *B. S. Li et al. 13129* (PE); loc. eodem, 15. Jun. 1990, *B. S. Li et al. 13016* (PE); Riwu Township, 27°54′ N, 87°33′ E, 3900 m, 25. Jul. 2011, *X. F. Gao et al. 14470* (CDBI). **Gyirong**: Bahaogou, 4200 m, *Betula* forests, 29. Jun. 1972, *Xizang Chinese Mater. Medic. & Herbs Exped. 699* (HNWP, PE). **Nyalam**: near the city, 4300 m, 20. Aug. 2010, *D. Y. Hong & Q. Wang XZ 002* (PE); near Quxam, 3850 m, 27. Aug. 1972, *Xizang Chinese Mater. Medic. & Herbs Exped. 1578* (PE); Quxam to Nyalam, 3600–3700 m, 31. Aug. 1972, *Xizang Chinese Mater. Medic. & Herbs Exped. 1676* (HNWP, PE); Suannai Lake, 4500 m, alpine meadows with shrubs, 19. Jul. 2012, *F. S. Yang 002* (PE).

INDIA, Darjiling: s. loc., 3960 m, Oct. 1838, *G. Ghose* s.n.(PE). **Sikkim**: Bikbari, 4270m, Sept. 1983, *S. Macpherson 76* (E); Dongkung, 5560 m, 16. Aug. 1972, *Pradhan, Norbu & Naku 23* (BM); West Distr., Dzongri Pass, 27°29′ N, 88°09′ E, 4400 m, 16. Jul. 1992, *D. G. Long et al.* 424 (E); Dzongri, 3960 m, 14. Sept. 1983, *AGSES 429*(K); Jongri, 4270 m, Aug. 1881, *G. Watt 5767* (E); loc. eodem, 4270 m, 12. Aug. 1913, *Rohmoo Lepcha 944* (E); Leonak, 3660 m, 26. Jul. 1909, *W. W. Smith & G. H. Cave 1759* (BM); loc. eodem, 4480 m, 05. Aug. 1907, *W. W. Smith & G. H. Cave 2124* (B); Thangshing to Lambi, Prek Chhu Valley, 27°31′ N, 88°11′ E, 4050 m, 21. Jul. 1992, *D. G. Long et al. 575* (E); Zemu Valley, 3810 m, 15. Jul. 1909, *W. W. Smith & G. H.Cave 145* (E), loc. eodem, 3810 m, 21. Jul. 1909, *W. W. Smith & G. H. Cave 1670* (B).

NEPAL, Bagmati: Rasuwa Distr., Laurebinayak, 28°06′ 37″N, 85°21′E, 3450–3950 m, 23. Jul. 1995, *T. Hoshino & al. 9536089* (KATH); Rasuwa Distr., Chilime Kharka, 4110 m, Jul. 1949, *O Polunin 1182* (BM); Rasuwa Distr., N of Gosainkund Lake, 28°05′ N, 85°25′ E, 4250 m, 13. Aug. 1974, *J. H. de Haas 2238*(BM); loc. eodem, 4250 m, rocky grasslands, 27. Aug. 1969, *Mella 16227* (KATH); Rasuwa Distr., Kyangchin Gomba, 28°13′ N, 85°44′E, 3750 m, 09. Sept. 1971, *J. F. Dobremez 1053* (G); Rasuwa Distr., Singum Gompa (3200 m)-Gosainkund (4200 m), 23. Aug. 1972, *Kanai, Hara & Ohba 721857* (BM, E, KUN, TI); loc. eodem, 23. Aug. 1972, *Kanai, Hara & Ohba 723170* (TI); Yala Kharka, 28°12′ 37″N, 85°36′ 52″E, –28°12′ 59″N, 85°30′ 32″E, 4100 m, 22. Jul. 1992, *Takayama, Arai & Miyamoto 9220338* (E); Sindhupalchok Distr.,Ganja La, 28°09′ N, 85°34′ E, 4200 m, 05. Aug. 1970, *J. F. Dobremez 487* (BM, G); Bhalike Pahar, 4270 m, 23. Sept. 1958, *Stainton, Sykes & Williams 4574* (BM); Langtang Himal., 28°14′ N, 85°35′E, 4450 m, 01. Aug. 1962, *J. D. A. Stainton 4090* (BM); Chalike Pahar, 4110 m, 03. Aug. 1954, *Stainton, Sykes & Williams 3728* (BM, E, PE). **Gandaki**: Manang Distr., Bimtakothi, 3810 m, 26. Aug. 1950. *D. G. Lowndes 1458 A* (BM, E); loc. eodem, 4570 m, 25. Aug. 1950, *D. G. Lowndes 1435* (BM); Rambrong, Lamjung Himal. 4110 m, 05. Jul. 1954, *Stainton, Sykes & Williams 6169* (BM, E). **Janakpur**: Dolakha Distr., Gaurivinayak, 3810 m, 10. Sept. 1965, *T. B. Shrestha & P. R. Shakya 3788* (G, KATH); Dolakha Distr., Yalung region, 4112 m, 17. Sept. 1978. *Binns, Mason & Wrigh 27* (E); Ramechhap Distr., near Thare, 4150 m, 23. Jul. 1985, *H. Ohba et al. 8570743* (E); loc. eodem, 27°45′ N, 86°28′ E, 23. Jul. 1985, *H. Ohba et al. 8580432* (BM, E, TI); Jadhang, 4110 m, 20. Jul. 1976, *Durham Univ. Himalaya Exped. 24* (BM). **Kosi**: Sankhuwasabha Distr., Jaljale Pokhari, 27°28′ N, 87°27′ E, 4000 m, 22. Aug. 1972, *J. F. Dobremez 1653* (BM, G); Sankhuwasabha Distr., Kipu Pokhari, 3905 m, 27°49′ N, 87°21′ E, 23. Aug. 1998, *S. Noshiro et al. 9830142* (E); loc. eodem, Pakhola Valley, ca. 3700 m, 16. Sept. 1981, *C. Grey-Willson et al. 4456* (K); loc. eodem, Jaljale Himal., 27°30′ N, 87°30′ E, ca. 4100 m, 04. Aug. 1991, *H. Ohba et al. 9120200* (BM, TI); loc.eodem, 10, Aug. 1991, *H. Ohba et al. 9120367* (TI); loc. eodem, 25. Jul. 1991, *H. Ohba &*

FIGURE IV 1:20B *Codonopsis thalictrifolia* Wall. subsp. *thalictrifolia*. All from Tibet: Nyalam. **a**, 20. Aug. 2010, by Xin-Tang MA & De-Yuan HONG; **b**, 20. Aug. 2010, by De-Yuan HONG; **c** & **d**, 19. Jul. 2012, by Fu-Sheng YANG.

al. 9153274 (TI); loc. eodem, 27°45′ N, 87°08′ E, ca. 3700 m, 20. Jul. 1988, *M. Suzuki et al. 8820691* (BM); Sankhuwasabha Distr., Thudam, 3400 m, 23. Jun. 1972, *H. Kanai et al. 720698* (TI); Thudam (3400 m) – Lamni Nama (4200 m), 14. Aug. 1977, *H. Ohashi et al. 772375* (TI); around Lamni Nama, 15 Aug. 1977, *H. Ohashi et al. 772441* (TI); near Thudam, 1070 m, 23. Jul. 1971, *L. W. Beer 9439* (BM). **Mechi**: Taplejung Distr., Kambachen, 27°44′ N, 87°59′ E, 4060 m, 09. Sept. 1989, *KEKE 493* (E, K); Taplejung Distr., Tamur Valley, Mewa Khola, Topke Gola, 4270 m, 10. Jul. 1956, *J. D. A. Stainton 919* (BM, CAL, E, PE); Topke Gola, 27°38′ N, 87°35′ E, 01. Aug. 1971, *G. F. Smith 19* (BM); Topke Gola (3600 m) – Tasagon (4350 m), 17. Aug. 1977. *H. Ohashi et al. 772506* (TI); Topke Gola to Rato Pokhari, 4420 m, grasses, 26. Jul. 1971, *T. B. Shrestha & D. P. Joshi 349* (KATH). **Sagarmatha,** Solukhumbu Distr., Kunde, Khumbu, 3960 m, 27. Jul. 1966, *A. D. Schilling 949* (K); Solukhumbu Distr., Lumding Khola-Raule, 27°30′ N, 86°30′E, 4270 m,

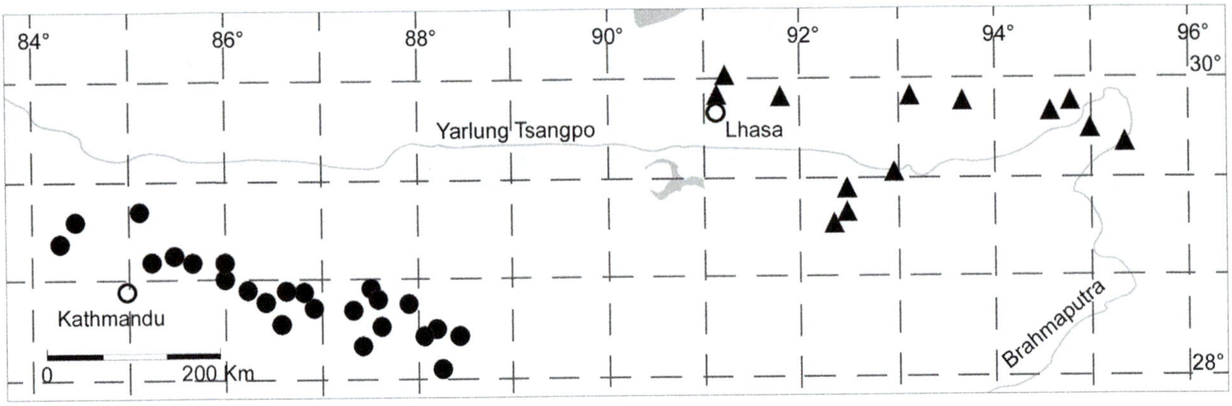

MAP IV 1:20 Distribution of *Codonopsis thalictrifolia* Wall. Dots: subsp. *thalictrifolia*; triangles: subsp. *mollis* (Chipp) D. Y. Hong.

14. Jul. 1964, *D. McCosh 412* (BM, E, KATH); Pale Bhu (Bhce), Mt Everest, 4270–4570 m, 1921, *A. F. R. Wallaston* s. n.(K); Solukhumbu Distr., Pangka, S of Cho Qyo, 4600 m, 17. Sept. 1982, *Miebe 1058* (BM); Solukhumbu Distr., Phulaji Kala, 27°54′ 30″N, 86°48′ 20″E, 4360 m, 20. Sept. 2005, *DNEP 3 AY 108* (E); Solukhumbu, Thami, ca. 3810 m, 03. Aug. 1966, *Banerji* s. n.(PE); Solukhumbu Distr., 27°47′ N, 86°55′ E, 4590 m, 12. Aug. 1995, *F. Miyamoto et al. 9596284* (E); loc. eodem, 4500 m, 12. Aug. 1995, *F. Miyamoto et al. 9592258* (E); loc. eodem, 27°56′ 24″N, 86°36′ 13″E, Sept. 2005, *DNEP 3 BY10* (E); loc. eodem, Jar Kharka, 4125 m, 03. Aug. 1995, *F. Miyamoto et al. 9584098* (E); loc. eodem, Aug. 1995, *F. Miyamoto et al. 9596201* (E); loc. eodem, Khola Kharka, 4110 m, 07. Aug. 1997, *M. Wakabayshi et al. 9715094* (E); loc. eodem, 4110 m, 17–19. Jul. 1949, *O. Polunin 1079* (BM, E); loc. eodem, Horsola Kharka, 3930 m, 11. Sept. 1997, *M. Wakabayashi 9720178* (E); loc. eodem, Thosa Kharka, 27°38′ N, 86°35′ E, 3670 m, 22. Aug. 1985, *H. Ohba et al. 8571933* (E); loc. eodem, Dik Kharka, 27°44′ N, 86°52′ E, 4465 m, 07. Aug. 1985, *F. Miyamoto et al. 9596241* (E); Solukhumbu Distr., Dinjung to Chhule, 27°54′ 10″N, 86°37′ 54″E, 4170 m, 18. Sept. 2005, *DNEP 3 BX 150* (E); Solukhumbu Distr., Tyangbochee: ca. 3960 m, 29. Jul. 1966, *Banerji 1919* (PE).

20b. Codonopsis thalictrifolia subsp. **mollis** (Chipp) D. Y. Hong, comb. et stat. nov. Basionym: *Codonopsis mollis* Chipp, J. Linn. Soc. Bot. 38: 381. 1908; Nannfeldt, Notes Roy. Bot. Gard. Edinburgh 16: 153 & 158. 1931; Fletcher, Bot. Mag. 164: pl. 9677. 1946; Anonymous, Icon. Cormophyt. Sin. 4: 383. fig. 6179. 1975. ≡*Codonopsis thalictrifolia* Wall. var. *mollis* (Chipp) L. T. Shen in Hong, Fl. Reipubl. Popularis Sin. 73(2): 55. 1983. Type: China, Tibet, hills above Lhasa, Aug. 1904, *H. J. Walton* s. n.(lectotype here designated: E; isolectotype: K).

Corolla 3.1–5.5 cm long, but mostly longer than 3.5 cm. Figure IV 1:20C & D.

FIGURE IV 1:20C *Codonopsis thalictrifolia* Wall. subsp. *mollis* (Chipp) D. Y. Hong, redrawn from *Icon. Cormophyt. Sin.* 4. 1975. Lower and upper parts of a stem.

Phenology: flowering from July to September; fruiting in September.

Habitats: thickets and meadows at an altitude of 3400–4600 m.

Distribution: endemic to China: SE Tibet.

Notes: The protologue of *Codonopsis mollis* Chipp indicates that it differs from *C. thalictrifolia* Wall. by"longer corolla, longer filaments, and more hirsute and larger leaves."All the previous authors recognized *C. mollis*

FIGURE IV 1:20D *Codonopsis thalictrifolia* Wall. subsp. *mollis* (Chipp) D. Y. Hong. **a–d**, from Tibet: Gongbo'gyamda, 25. Aug. 2010; **a** & **d**, by Qiang WANG & De-Yuan HONG; **b** & **c**, by De-Yuan HONG; **e**, from Tibet: Maizhokunggar, 25. Aug. 2010, by Qiang WANG & De-Yuan HONG. **b** & **c**, showing disc (arrow) and half-inferior ovary.

Chipp until 1983, when Shen and Hong reduced it to a variety of *C. thalictrifolia* Wall. In "*Flora of China*", Hong and his co-authors (Hong et al., 2011) went further, reducing it as a synonym of *C. thalictrifolia* based on the consideration that the length of corolla varied continuously. However, I made an expedition to Tibet with Dr. Qiang WANG in 2010. We sampled three populations respectively in Nyalam, Gongbo'gyamda and Maizhokunggar. We found that though the variation of corolla length is continuous, the difference in this character between the two entities respectively in the west and the east is remarkable (Map IV 1:20). And furthermore, there is a rather large gap between the two geographic areas. Therefore, I am now considering it more reasonable to treat these two entities as two subspecies.

Additional specimens examined, **CHINA, TIBET, Gongbo'gyamda:** Numari, 18. Aug. 1952, *P. C. Tsoong 7185* (PE); Pasum Lake, 3660 m, 15. Jul. 1947, *F. Ludlow et al. 14140* (E); Sumdo, 4300 m, thickets,25. Aug. 2010, *D. Y. Hong & Q. Wang H10008* (PE); loc. eodem, 07. Aug. 2001, *H. L. Qin et al. 80* (PE); loc. eodem, 4170 m, meadows, 07. Aug. 1965, *Y. T. Chang & K. Y. Lang 1380* (KUN, PE). **Gyaca**: behind the Harbor, 3800–4300 m, 29. Aug. 1972, *Xizang Chinese Mater. Medic. & Herbs Exped. 4535* (HNWP, PE). **Lhasa**, 5 km N of Naiguodong, 4300 m, meadows, 17. Aug. 1965, *Y. T. Chang & K. Y. Lang* 1696 (PE); s. loc., 05. Jul. 1947, *J. Guthrie* s. n.(K). **Lhünzê**: Shirap, Char Chu, 3960–4270 m, 08. Aug. 1936, *F. Ludlow & G. Sherriff 1972* (E); loc. eodem, W of Sangacho-ling, 02. Aug. 1997, *S. A. Bower Lyon 11137* (E); Xinggexi to Nyaimai, 4300 m, *Rhododendron* thickets, 22. Aug. 1960, *G. X. Fu 622* (PE); Xoisar-202 high way, 28°35′ N, 92°29′ E, 4343 m, grassy slopes, 04. Aug. 2013, *Y. S. Chen et al. 13-0270* (PE). **Lhünzhub:** Poindo, 4100–4400 m, thickets, 07. Oct. 1972, *Xizang Chinese Mater. Medic. & Herbs Exped. 1917* (PE); Wulong Commune, Gyironggou, 4550 m, thickets, 08. Oct. 1972, *Xizang Chinese Mater. Medic. & Herbs Exped. 1963* (PE). **Mainling**: Nyima La, 4110 m, 27. Aug. 1947, *F. Ludlow et al. 14446* (E); Pero La, 29°30′ N, 95°0′ E, 3960 m, 09. Jul. 1938, *F. Ludlow et al. 5184* (E); Zêtanggang, 3800 m, forests. 26. Jul. 1972, *Xizang Chinese Mater. Medic. & Herbs Exped. 3896* (HNWP, PE). **Maizhokunggar**: Maizhokunggar to Mi La, 4400 m, alpine thickets, 25. Aug. 2010, *D. Y. Hong & Q. Wang H10010* (PE); Rutog Township, Mila Pass, 4350 m, thickets, 12. Sept. 2012, *FLPH Tibet Exped. 12–0484* (PE). **Mêdog**: Nam La, 4270 m, 25. Jul. 1924, *F. Kingdon-Ward 5995* (K, E). **Nyingchi**: Lunang to Bayi Town, 4000 m, grassy slopes, 13. Aug. 2001, *H. L. Qin et al. 511* (PE); Nambu La, 4420 m, 13. Jul. 1947, *F. Ludlow et al. 15418* (E); Serqi La, 4508 m, alpine meadow, 11. Aug. 2012, *D. Y. Hong et al. H12048* (PE); loc. eodem, 4600 m, alpine meadows with shrubs, 04. Aug. 1983, *B. S. Li & S. Z. Cheng 6062* (PE); loc. eodem, west slope, 4320-4510 m, 02. Aug. 1975, *Qinghai-Xizang Suppl. Exped. 751056* (HNWP, KUN, PE); Valley behind College of Agriculture and Animal Husbandry, 3400-4200 m, *Abies* forests, 08. Aug. 1983, *B. S. Li et al. 6296* (PE).

21. Codonopsis viridiflora Maxim., Bull. Acad. Imp. Sci. Saint-Pétersbourg 27: 496. 1881; Komarov, Trudy Imp. S.-Peterburgsk. Bot. Sada 29(1): 116. 1908; Chipp, J. Linn. Soc. Bot. 38: 383. 1908; Nannfeldt, Acta Horti Gothob. 5: 29. 1930; Shen & Hong in Hong, Fl. Reipubl. Popularis Sin. 73(2): 61. pl. 10. fig. 1. 1983; Hong, Lammers & Klein in Wu, Raven & Hong, Fl. China 19: 523. 2011. Type: China, Kansu, Yussunchatyma Valley, 2740–3050 m, *N. M. Przewalsky, 11/23. Jul. 1880* (holotype: LE).

Codonopsis bicolor Nannf., Acta Horti Gothob. 5: 24, pl. 14. 1930 & Notes Roy. Bot. Gard. Edinburgh 16: 155. 1931. Type: China, Sichuan: Songpan, "Dongrergo, in prato alpino fruticoso, ca. 4200 m", 08. Aug. 1922, *H. Smith 3530* (lectotype here designated: E; isolectotypes: BM, PE).

Roots carrot-shaped, 10–15 cm long, 1–3 cm in diameter. Stems 2 to several from one caudex, nearly erect, ascending, or scandent, sometimes twining apically, 0.3–2 m long, branched at lower part, sparsely hirsute or nearly glabrous. Leaves on main stem alternate, small and bract-like, those on branches opposite or nearly opposite, so branches appearing a pinnate leaf; leaf blade broad-ovate to lanceolate, 1.5–3.5(–5) cm long, 1.3–3 cm broad, both surfaces hispidulous, base slightly cordate or rounded, margin crenate, apex obtuse. Flowers solitary, terminal on both main stem and branches, very rarely only on main stem; pedicel 6–15 cm long, glabrous or sparsely hirsute at top. Calyx tube adnate to whole hypanthium, semiglobose, 10-ribbed, glabrous; calyx lobes ovate to ovate-lanceolate, 10–15 mm long, 6–8 mm broad, sparsely hispidulous, margin ciliate, sparsely crenate or entire, apex obtuse. Corolla campanulate, 1.7–2.5 cm long, 2–3 cm wide, lobed to 2/5–1/2, green or yellow-green, purplish only at base, or blue, white, or whitish blue with purple veins, often purple inside tube at base, totally glabrous; corolla lobes triangular, 8–10 mm long, 9–11 mm broad. Stamens glabrous; filaments ca. 5 mm long, slightly dilated at base; anthers ca. 5 mm long. Capsule ca. 1.5 cm in diameter. Seeds ellipsoid, brown-yellow. Figures IV 1:21A & B.

Phenology: flowering and fruiting from July to October.

Habitats: alpine meadows, thickets, forest margins, or forests; 2200–4130 m.

Distribution: **endemic** to China: SE Gansu, S Ningxia, E & S Qinghai, W Sichuan, E Tibet, and NW Yunnan (Map IV 1:21).

Additional specimens examined, **CHINA**, **GANSU**, Terra Tangutorum, 1872, *N. M. Przewalsky 316* (K, P).

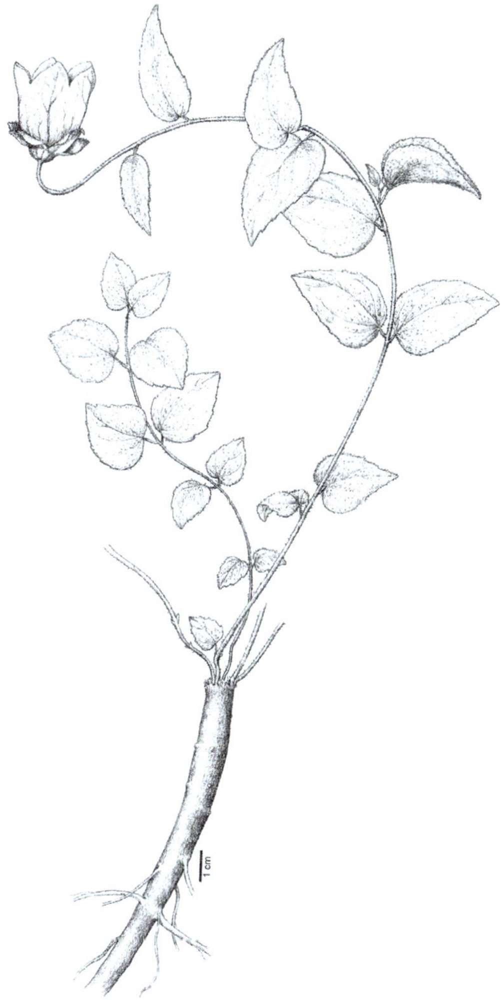

FIGURE IV 1:21A *Codonopsis viridiflora* Maxim., a plant, showing root, caudex, base of 3 stems, and 2 intact stems.

Hezuo: 2900 m, glades, 04. Aug. 1991, *Taohe Exped. 018* (KUN). **Lintan**: Lianhua Shan, stony slopes, 2600 m, 18. Aug. 1996, *Y. S. Lian et al. 96703* (PE); Lianhua Shan, between Taochow and Titao, 3050 m, 14–20. Jul. 1925, *J. F. Rock 12738* (BM, P, PE ,W). **Luqu**: Shuangchashengou, 3500 m, glades, 08. Aug. 1991, *Taohe Exped. 291* (KUN). **Minxian**: Suehloa Shan, 3000 m, 01. Jul. 1936, *T. P. Wang 4773* (PE); Tao River basin, Tsaluku Valley, 2900–3050 m, Aug. 1925, *J. F. Rock 13125* (K, PE). **Yongden**, Liancheng Forest Farm, 2900 m, thickets, 08. Jul. 1991, *T. N. Ho 1801* (HNWP). **NINGXIA** (S), **Jingyuan**: Erlonghe, 1900–2000 m, forests, 08. Jun. 1976, *Q. Du 76–0168* (HNWP); Longde Gou, 2800 m, 31. Jul. 1953, *T. P. Wang 17078* (PE). **Longde**: Huoshanding, 2800–3000 m, 10. Jul. 1942, *T. P. Wang 13086* (PE). **QINGHAI, Baima:** Golog, 3400–3450 m, 27. Jul. 1983, *W. Y. Wang 26814* (HNWP); Valley S of Baima, 32°51′ N, 100°50′ E, 3360 m, 08. Jul. 2000, *Sino-British Qinghai Alpine Garden Society Exped. 758* (E) & *761* (E). **Chindu**: Lochen-la, 3960 m, 20. Aug. 1933, *F. Kingdon-Ward 12239* (BM); Shindu Gemla, 3660–3960 m, 24. Aug. 1933, *F. Kingdon-Ward 10781* (BM); Xiwu Township, 33°0′ N, 97°20′ E, 3620 m, 15. Aug. 1996, *T. N. Ho et al. 1865* (A, BM, E, PE); loc. eodem, 3700 m, 10. Aug. 1977, *S. W. Liu 2476* (HNWP); loc. eodem, 4000 m, 15. Aug. 1996, *T. N. Ho et al. 1844* (BM, E, PE). **Datong**: Baoku, 2680 m, 27. Jul. 1989, *Z. H. Zhang 3711* (PE); loc. eodem, 2600 m, 27. Jul. 1989, *Z. H. Zhang 3712* (PE); loc. eodem, 2700 m, *Z. H. Zhang 3713* (PE); loc. eodem, 2789 m, *Z. H. Zhang 3714* (PE); loc. eodem, 2670 m, *Z. H. Zhang 3717* (PE); loc. eodem, 2800 m, *Z. H. Zhang 3718* (PE); loc. eodem, 2700 m, *Z. H. Zhang 3719* (PE); Xishan Township, 2900 m, 10. Aug. 1989, *L. Z. Mei 156* (HNWP). **Guide**: Jiala Forest Farm, 3000 m, 24. Jul. 1989, *T. N. Ho 1147* (HNWP, PE). **Huangyuan**: Zhugu Temple, *Cupressus* forests, 18. Aug. 1958, *P. C. Tsoong 10083* (PE). **Huangzhong**: Chunjia Forest Farm, thickets, 2910 m, 22. Jul. 1989, *T. N. Ho 1057* (HNWP, PE); loc. eodem, 3900 m, thickets 22. Jul. 1989, *S. W. Liu 3224* (HNWP); Sanhe Forest Farm, thickets, 24, Aug. 1972, *S. W. Liu 2128* (HNWP); loc. eodem, 2750 m, thickets, 06. Aug. 1973, *S. X. Wang 1059* (HNWP). **Huzhu**: Jiading, forests, 10. Aug. 1971, *P. C. Kuo & T. N. Ho 9388* (HNWP). **Ledu**: Shangbei Shan Forest Farm, 3000 m, thickets, 03. Aug. 1986, *Y. H. Liu et al. 4067* (HNWP, PE); loc. eodem, 2800 m, 27. Jul. 1986, *Y. H. Liu et al. 3868* (HNWP); loc. eodem, 3000 m, forests, 03. Aug. 1986, *Y. H. Liu et al. 2252* (HNWP); Yinzhonggou-Xiaoxigou, 3050 m, 10. Sept. 1959, *Qinghai-Gansu Exped. 1467* (HNWP, KUN). **Mengyuan**: Xianmi Township, 2800 m, forests, 17. Jul. 1970, *P. C. Kuo 7280* (HNWP, KUN). **Minghe**: Gushan, 07. Jul. 1973, *L. H. Zhou 2389* (HNWP). **Tongren**: Lancai to Xiazilan Shan, 3200 m, thickets, 28. Jul. 1970, *L. H. Zhou & L. N. Sun 1322* (HNWP); Lancai, 3000 m, thickets, 28. Jul. 1970, *S. W. Liu & D. S. Luo 1533* (HNWP). **Xunhua**: Mengda Forest Farm, 2200 m, edges of forests, 09. Jun. 1981, *P. C. Kuo & W. Y. Wang 25195* (HNWP). **Yushu**: Baitang, SE of Baitang, 3700 m, shrubs, 18. Aug. 1996, *T. N. Ho et al. 2009* (A, KUN). **Zekog**: Maixi forest Farm, 2900 m, meadows, 21. Aug. 1967, *s. coll. 1983* (HNWP). **SICHUAN**, s. loc., Oct. 1908, *E. H. Wilson 1170* (K). **Barkam**: s. loc., 3000–4000 m, grassy slopes, 17. Sept. 1964, *J. D. Li & Z. M. Xu* (*Barkam*) *64–87* (PE, SM). **Dêgê**: Damayanda Gou, s. d., *S. X. Jia* s. n.(PE). **Heishui**: Guagu Shan, meadows, 3550 m, 16. Aug. 1957, *X. Li 74028* (PE, SZ); Luhua Valley, 2500 m, 23. Jul. 1957, *X. Li & J. X. Zhou 73756* (SZ); Sergu, 3300 m, 23. Sept. 1964, *J. D. Li et Z. M. Xu 64-91* (PE, SM). **Hongyuan**: Shuajinsi, 3500 m, 09. Aug. 1957, *T. Y. Chang & H. F. Zhou 23594* (PE, SZ); loc. eodem, 3200 m, 08. Aug. 1957, *T. Y. Chang & H. F. Zhou 23561* (SZ); loc. eodem, 07. 1986, *Z. T. Wang 867–181* (PE). **Maoxian**: Sanlong Township, 17. Aug. 1952, *Z. He 13702* (SZ). **Nanping**: Xowa Distr., 4100 m, 19. Jul. 1979, *s. coll. 881* (SM). **Pingwu**: Huya, 3500 m, 24. Aug. 1978, *s. coll. 853* (SM); loc. eodem, 17. Jul. 1978, *Pingwu Exped. 500*

FIGURE IV 1:21B *Codonopsis viridiflora* Maxim. **a**, from Sichuan: Jiuzhaigou, 18. Aug. 2011, by Yan YU; **b**, **c**, & **g**, from Tibet: Zayü, 10. Aug. 2013, by Qiang WANG; **d**, from Tibet: Qamdo, 10. Aug. 2010, by You-Sheng CHEN; **e** & **f**, from Qinghai: Datong, 07. Sept. 2012, by De-Yuan HONG.

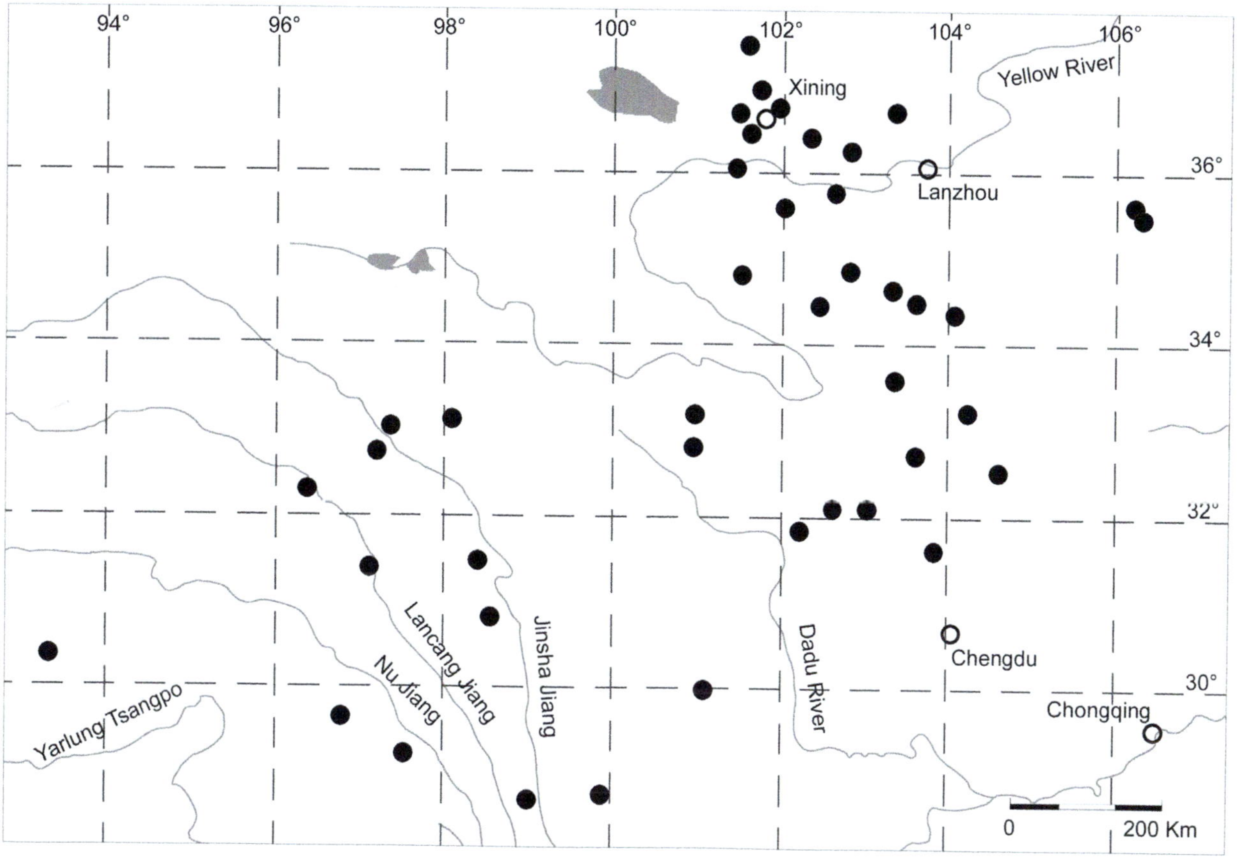

MAP IV 1:21 Distribution of *Codonopsis viridiflora* Maxim.

(SM). **Sêrxü**: N of Luoxu, 32°34′ 13″N, 98°4′55″E, 3780 m, 29. Jul. 2005, *D. E. Boufford et al. 33676* (A). **Songpan**: Yanchong, 3650 m, thickets, 02. Aug. 1983, *Z. X. Tang 1584* (PE); s. loc., 3100 m, sunny grassy slopes, 17. Jul. 1964, *X. Y. Tang & W. X. Wang 64–004* (PE). **Tsipula**: 4000 m, 05. Aug. 1922, *H. Smith 4186* (PE). **Xiangcheng**: Daxue Shan, 4297 m, meadows in forests, 01. Aug. 2012, *F. S. Yang et al. 063* (PE). **Zoigê**: Qammê, meadows with shrubs, 3480 m, 04. Aug. 1975, *Sichuan Plant Surv. 10338* (PE); E of Zoigê, 33°38′ 38″N, 103°30′28″ E, 3975 m, 22. Aug. 2007, *D. E. Boufford et al. 40279* (A); s. loc., 4000 m, meadows, 1970, *s. coll.* s. n.(SM). **TIBET**, **Baxoi**: Rawu, just W of Rawu, 3950 m, 25. Jul. 2000, *D. E. Boufford et al. 29688* (A). **Gongbo'gyamda**: Pasum-la, 30°28′ N, 93°22′ E, 3960 m, 24. Aug. 1938, *F. Ludlow et al. 6922* (BM); Pasum Lake, 3050–3350 m, 24. Aug. 1924, *F. Kingdon-Ward 6102* (K). **Gonjo**: Wabar Township, 3800 m, 21. Aug. 1976, *Qinghai-Xizang Exped. Veget. group 9524* (PE). **Jomda**: Aila Shan, 3641 m, meadows, 27. Jul. 2012, *H. J. Wang et al. JD-002* (PE); on the way to Dêqê, 31°38′48″ N, 98°23′36″ E, 3485–3550 m, 02. Aug. 2004, *D. E. Boufford et al. 31551* (A); S side of Du River, 31°29′17″ N, 98°13′24″ E, 3675 m, 29. Jul. 2004, *D. E. Boufford et al. 31286* (A); WSW of Jomda, 31°19′58″ N, 98°7′43″ E, 3850 m, 30. Jul. 2004, *D. E. Boufford et al. 31305* (A) ; near the town, 3500 m, thickets, 29. Jul. 1976, *Qinghai-Xizang Exped. 12535* (HNWP, KUN, PE); loc. eodem, 3600 m, 22. Jun. 1976, *Qinghai-Xizang Exped. 12417*(KUN, PE); Tangpu, meadows at edges of forests, 3500 m, 23. Jul. 1976, *Qinghai-Xizang Exped. 12443* (HNWP, PE); Tangpu, 31°38′ N, 98°26′ E, 4000 m, grassy slopes by forests, 10. Aug. 2010, *Kham Exped. 10–1557* (PE). **Qamdo**: W of Toba, 31°30′26″ N, 97°20′4″ E, 3480–3550 m, 17. Aug. 2004, *D. E. Boufford et al. 32524* (A). **Zayü**: vicinity of Naxue, rock crevices among herbs, 28. Jun. 1973, *Qinghai-Xizang Exped. 73–457* (KUN). **YUNNAN**, **Dêqên**: Baima Shan, Shelong, 4130 m, on rocks, 26. Jul. 1937, *T. T. Yü 9191* (PE); Baima Shan, 4500 m, 22. Aug. 1999, *N. Fujii et al. F00613* (KUN); Doker-la, 3000 m, woods, *C. W. Wang 64907* (PE).

22. Codonopsis tsinlingensis Pax & K. Hoffm., Repert. Spec. Nov. Regni Veg. Beih. 12: 500. 1922; Tsoong, Contr. Inst. Bot. Nat. Acad. Peiping, 3(3): 102. fig. 12, 1935; Shen & Hong in Hong, Fl. Reipubl. Popularis Sin. 73(2): 61. 1983; Hong, Lammers & Klein in Wu, Raven & Hong, Fl. China 19: 523. 2011. Type: China, "Schenhsi , Tsinling schan……Tai Pai schan" (Shaanxi, Qinling Range, Taibai Shan), 3400 m, 01. Aug. 1916, *W. Limpricht 2730* (holotype: WRSL, n. v.; isotype: WU).

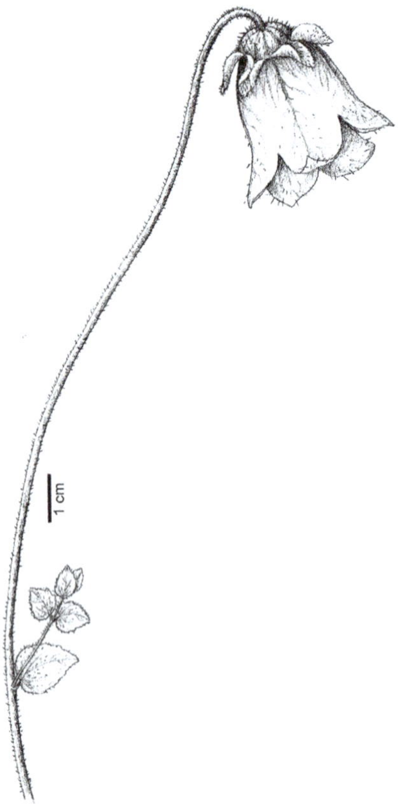

FIGURE IV 1:22A *Codonopsis tsinlingensis* Pax & K. Hoffm., upper part of a stem, showing the terminal flower.

Roots carrot-shaped, 10–20 cm long, 0.6–3.5 cm in diameter. Stems several from one caudex, erect or ascending, 25–60 cm long, white-hirsute; branches slender, with leaves aggregated at top. Leaves on main stem alternate, those on branches nearly opposite; petiole less than 1 cm long, white-hirsute; leaf blade ovate to broad-ovate, 1–2.6 cm long, 0.6–1.8 cm broad, both surfaces usually white-hispidulous but sometimes white-hirsute, rarely adaxial surface glabrous, base cordate to rounded, margin crenulate to nearly entire, slightly revolute, apex acuminate to obtuse. Flowers solitary, terminal, usually only on main stem but rarely also on branches; pedicel 3–15 cm long, densely hirsute. Calyx tube adnate to whole hypanthium, hemispheric, 10-ribbed, hirsute entirely or only along ribs, rarely glabrous; calyx lobes narrowly triangular, 8–10 mm long, 4–6 mm broad, abaxially hispidulous, margin entire, revolute, apex acute. Corolla campanulate, 2–3 cm long, 2–3 cm wide, lobed to near middle, pale or dark purple, villous inside. Stamens glabrous; filaments ca. 5 mm long, slightly dilated at base; anthers ca. 4 mm long. Capsule hemispheric at lower part, conical toward apex. Seeds brown-yellow, ellipsoid. Figures IV 1:22A & B.

Phenology: flowering and fruiting from July to October.
Habitats: alpine scrubs or glassy slopes; 1300–3600 m.
Distribution: **endemic** to China: Shaanxi (the Qinling Range) (Map IV 1:22).

FIGURE IV 1:22B *Codonopsis tsinlingensis* Pax & K. Haffm. All from Shaanxi: Taibai Shan. **a** & **b**, 22. Jul. 2011, by Jun LIU; **c**, 07. Jul. 2007, by You-Sheng CHEN.

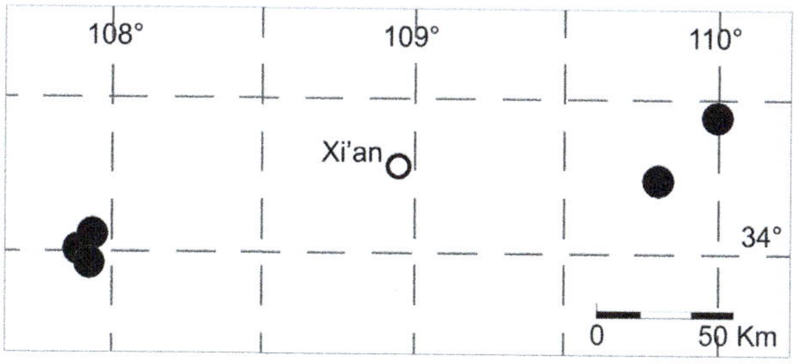

MAP IV 1:22 Distribution of *Codonopsis tsinlingensis* Pax & K. Hoffm.

Notes: This species is very similar to *C. viridiflora* Kom. in appearance, but its corolla is purple (vs. pale green, blue, or whitish), villous inside (vs. entirely glabrous), its calyx tube mostly hirsute, rarely glabrous, and the calyx lobes always hispidulous.

Additional specimens examined, **CHINA, SHAANXI, Taibai Shan**: Dayehai to Nantianmen, 3300–3600 m, meadows, 09. Jul. 1999, *Zhu, Chen, Xu & Wang 1509* (PE); Fangyang-si, 3100–3600 m, 05. Aug. 1933, *T. P. Wang 1871* (PE); loc. eodem, Sept. 1938, *W. Y. Hsia 4708* (PE); near Fangyang-si, Sept. 1938, *W. Y. Hsia 4647* (PE); Fangyang-si to Wenggongmiao, 26. Jul. 1938, *T. N. Liou & P. C. Tsoong 2800* (PE); above Fangyang-si, 3300 m, 07. Aug. 1956, *K. T. Fu 8459* (KUN); Guaili Township, 2730 m,rocky crevices, 26. Jul. 1958, *D. X. Wang 3-0412* (PE); Haopingsi, 1300 m, forests, 10. May. 1959, *J. X. Yang 435* (PE); Wenggongmiao to Dayegou, 3400 m, thickets, 01. Aug. 1991, *s.coll. PB-112* (PE); Wenggongmiao, 3000 m, 20. Jul. 1955, *Taibai Exped. 55090* (PE); loc. eodem, 19. Jul. 1981, *C. G. Ma et al. 225* (PE); ca. 3200 m, 22. Aug. –05. Sept. 1934, *G. Fenzei 834* (W); 1910, *W. Purdom 449* (K). **Weinan**: Jianyuling, 2100 m, sunny slopes, 08. Jul. 1952, *T. P. Wang 15720* (KUN). **Zhouzhi**: S slopes of Taibai Shan, Houzhenzi, 3400 m, 10. Sept. 1999, *Zhu, Chen, Xu & Wang 2031* (KUN, PE).

Section 2. Codonopsis

Codonopsis subg. *Eucodonopis* Kom. ser. *Volubiles* Kom., Trudy Imp. S. - Petersburgsk. Petersburgsk Bot. Sada 29: 104. 1908.

Roots all carrot-shaped. Stems all twining, rarely scandent, very occasionally ascending. Ovary inferior, very rarely half-inferior to both calyx and corolla, very occasionally superior to calyx (*C. pilosula*). Fruit a capsule. Seeds numerous, wingless.

Eighteen species, distributed from the W Himalaya in N Pakistan eastwards to the Korean Peninsula, and Taiwan, China. They grow from sea level in the Korean Peninsula to 4300 m in the Himalaya.

23. Codonopsis pilosula (Franch.) Nannf., Acta Horti Gothob. 5: 29. 1930; Tsoong, Contr. Inst. Bot. Nat. Acad. Peiping 3(3): 99, 1935; Shen et al., Acta Phytotax. Sin. 13(3): 57. 1975; Anonynous, Icon. Cormophyt. Sin. 4: 378. fig. 6169.1975; Shen & Hong in Hong, Fl. Reipubl. Popularis Sin. 73(2): 40. 1983; Hong, Novon 20: 422. 2010; Hong, Lammers & Klein in Wu, Raven & Hong, Fl. China 19: 515. 2011. ≡*Campanumoea pilosula* Franch., Nouv. Arch. Mus. Hist. Nat., ser. 2, 6: 72. 1883. Type: China, "Mont Pekinensis" (Beijing), cult. in "Ta-tchiao-cha" as medicinal plants, Jul. 1863, *A. David 2333* (lectotype designated by Hong, 2010: P).

Codonopsis glaberrima Nannf., Svensk Bot. Tidskr. 34: 387. 1940. ≡*Codonopsis pilosula* Nannf. var. *glaberrima* (Nannf.) P. C. Tsoong ex Anonymous, Icon, Cormophyt. Sin. 4: 774. 1975, comb. invalid. Type: China, " Shansi centr. (central Shanxi): Koan yün miao", 1500–1700m, 24.Aug.1916, *E. Licent 2793* (holotype: W; isotypes: K, PE).

Codonopsis modesta Nannf., Acta Horti Gothob. 5: 26. 1930; Nannfeldt in Handel-Mazzetti, Symb. Sin., 7(4): 1078. 1936; Shen et al., Acta Phytotax. Sin. 13(3): 56. pl. 5. fig.15. 1975. ≡*Codonopsis pilosula* (Franch.) Nannf. var. *modesta* (Nannf.) L. T. Shen in Hong, Fl. Reipubl. Popularis Sin. 73(2): 41. 1983. Type: China, "Sze-ch'uan bor. (N Sichuan): Chún-ché", ca. 3200 m, 24. Aug. 1922, *H. Smith 4304* (lectotype designated by Hong, 2010: BM; isolectotypes: E, PE).

Codonopsis silvestris Kom., Trudy Imp. S. -Peterburgsk. Bot. Sada 18: 425. 1900; ibid, 25 (Fl. Mansh. 3): 573. tab. 9. B. 1907 & ibid. 29: 105. 1908. Type: "Manshuria", 05. Aug. 1896, *Komarov* s. n. (holotype: LE; isotype: LE).

Codonopsis volubilis Nannf., Svensk. Bot. Tidskr. 34: 388. pl. 5. 1940. ≡*Codonopsis pilosula* (Franch.) Nannf. var. *volubilis* (Nannf.) L. T. Shen in Hong, Fl. Reipubl. Popularis Sin. 73(2): 41. 1983. Type: China, Shanxi, Chiao-chéng Distr. (Jiaocheng CO), Pa-shui-ko, 1900 m, 20. Aug. 1924, *H. Smith 7201* (holotype: GH, n. v.; isotypes: BM, PE).

Roots carrot-shaped or fusiform-cylindrical, often branched, 15–50 cm long, 1–3 cm in diameter. Stems twining, glabrous, 1–2 m long, branched, lateral branches 15–30 cm long, branchlets 1–5 cm long, sterile or fertile. Leaves on main stem and branches alternate, those on branchlets opposite; petiole 0.5–4 cm long, sparsely hispidulous; leaf blade ovate or narrowly ovate, 1–8.5 cm long, 0.8–6 cm broad, adaxially green, abaxially gray-green, both surfaces sparsely or densely hispidulous, densely or sparsely appressed hirsute or villous, rarely glabrous; base cordate or nearly cordate to rounded, apex obtuse or acute, margin crenate to entire. Flowers solitary and terminal on branches, pedicellate. Calyx from nearly totally free from hypanthium to calyx tube adnate to hypanthium up to middle; calyx lobes lanceolate or elliptic, 1–2.2 cm long, 0.5–1 cm broad, apex obtuse or acute, margin nearly entire or slightly waved; sinus between calyx lobes narrow and pointed. Corolla campanulate, 1.8–2.3 cm long, lobed to 1/5–1/4, yellow-green, with purple spots inside; corolla lobes deltoid, apex acute. Filaments ca. 5 mm long, glabrous, slightly dilated at base; anthers elongate, 5–6 mm long. Style densely hairy above. Capsule semiglobose in inferior part, conical at superior part. Seeds ovoid, brown-yellow.

Widely distributed in China, Korea Peninsula, and the Far East of Russia. (Map IV 1:23).

It is a challenge how to deal with *Codonopsis pilosula*, *C. handeliana* and *C. tangshen* in taxonomy. For example, the relationship between *C. pilosula* and *C. tangshen* has made me think for a long time. *C. tangshen* (with hypogynous calyx) differs distinctly from the extreme form of *C. pilosula* (with half-epigynous calyx), but there exists a continuous series between these two forms. We made a phylogenetic analysis using four DNA fragments of chlofoplast. The result shows that materials of *C. pilosula* (from Beijing, the type locality), *C. handeliana* (from Shangri-la, the type locality) and *C. tangshen* (from Chongqing, near type locality) form a clade with *C. handeliana* at basal position (Wang & Hong, unpublished data). Therefore, we here treat them as three subspecies, but with some reservation.

1a. Leaves usually hirsute on both surfaces....................23c. subsp. *handeliana*
1b. Leaves hispidulous, rarely glabrous.
  2a. Calyx nearly entirely free from hypanthium, calyx tube 1–3 mm long....................23b. subsp. *tangshen*
  2b. Calyx apparently adnate to hypanthium, calyx tube over 3 mm long....................23a. subsp. *pilosula*

23a. Codonopsis pilosula subsp. pilosula

Leaves 1–6.5 cm long, 0.8–5.5 cm broad, mostly shallowly cordate or truncate, less often rounded, at the base, both surfaces hispidulous, rarely glabrous. Adnate part of calyx ≥ 3 mm long, ca. 1 cm across; calyx lobes 1.4–1.8 cm long. Corolla 2–2.3 cm long. Figures IV 1:23A & B.

Chromosomes: 2n=16 (Chen et al., 1991 ; Cheng et al., 1991 ; Gadella, 1966 ; Sui et al., 1985 ; Yoo & Lee, 1989).

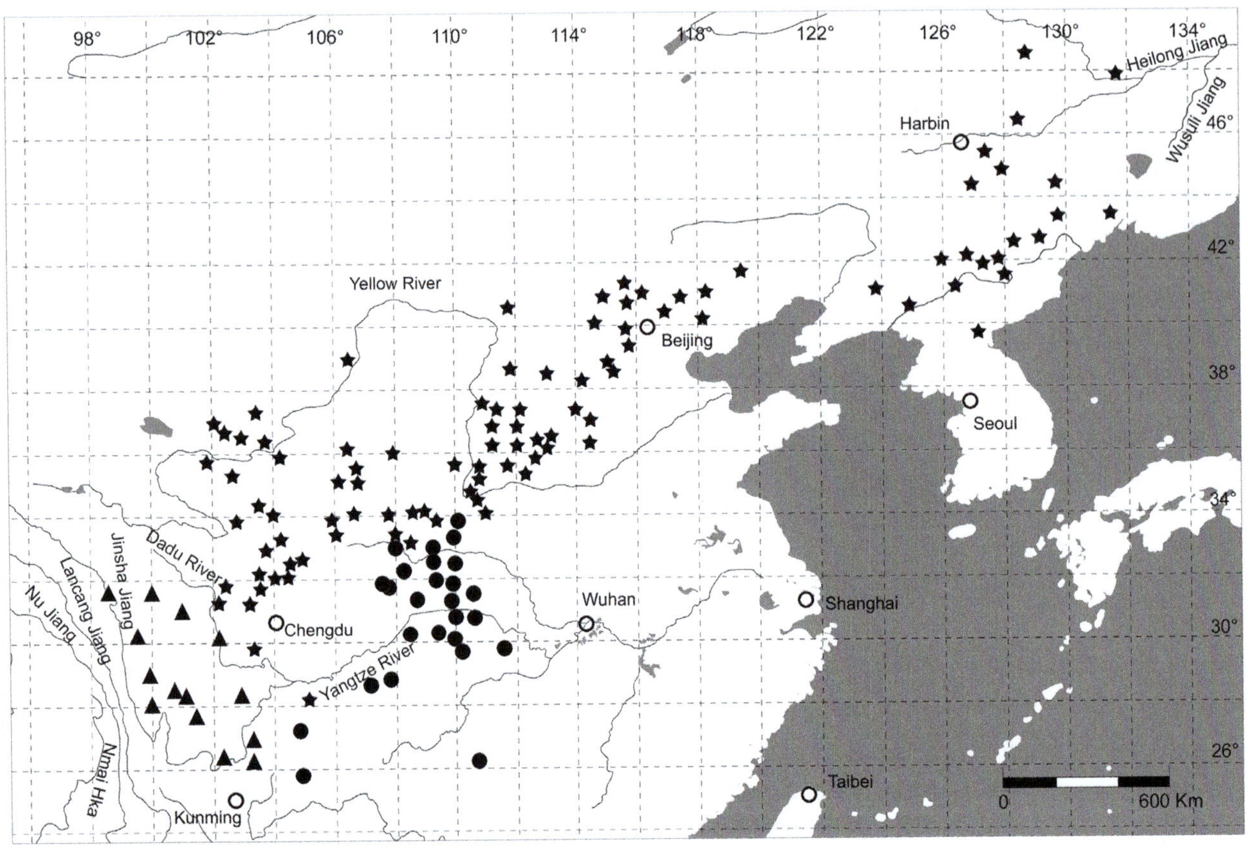

MAP IV 1:23 Distribution of *Codonopsis pilosula* (Franch.) Nannf. Stars: subsp. *pilosula*; dots: subsp. *tangshen* (Oliv.) D. Y. Hong; triangles: subsp. *handeliana* (Nannf.) D. Y. Hong.

Phenology: flowering from June to September; fruiting in September and October.

Habitats: grasses, thickets, edges of woods, or woods; from sea level in Korea Peninsula to 3480 m in Sichuan, China.

Distibution: China (NE China to N Sichuan), Korea Peninsula, and the Far East of Russia.

Additional specimens examined, **CHINA, BEIJING, Baihua Shan:** Sept. 1936, *T. F. King 352* (PE); loc. eodem, 27. May. 1950, *W. T. Wang 288* (PE); loc. eodem, 30. Jul. 1930, *W. Y. Hsia 2586* (PE); loc. eodem, 1400 m, 29. Jul. 1930, *W. Y. Hsia 2316* (PE); Heiduegou, 1400 m, 25. Jul. 1990, *PE Exped. 151* (PE); loc. eodem, 30. Jul. 1930, *W. Y. Hsia H2580* (W). **Fangshan**: Xiayun-ling, 15. May 1971, *Beijing Chinese Mater. Medic. & Herbs Surv. (Fangshan)32* (PE). **Miyun**: Potou Forest Farm, ca. 1000 m, thickets, 03. Oct. 2011, *D. Y. Hong et al. H11046* (PE). **Xiaolongmen**: Forest Farm, 18. Aug. 1998, *Beijing Normal Univ. 9701142* (PE). **GANSU, Chengxian**: Jishan, 1800 m, rocks, 03. Oct. 1951, *T. P. Wang 15031* (PE). **Heshui**: Taibai Town,

FIGURE IV 1:23A *Codonopsis pilosula* (Franch.) Nannf. subsp. *pilosula*. **a**, upper part of a twining stem; **b**, root, caudex, and base of stems; **c**, a flower with corolla opened, showing calyx, stamens, disc, style, and stigma.

FIGURE IV 1:23B *Codonopsis pilosula* (Franch.) Nannf. subsp. *pilosula*. **a**, **b**, **g** & **h**, from Beijing: Miyun, 14. Aug. 2011, by Lian-Zhong FU; **c–f**, from the same locality as above, 03, Oct. 2011, by De-Yuan HONG. **c** & **g**, showing disc; **d** & **f**, showing ovary inferior to corolla, but half-inferior to calyx.

Kuanshang, thickets, 18. Jul. 1954, *Huanghe Exped. 622* (PE). **Huating**: Jindipo to Xiameng, 1700 m, *Populus* forests, 29. Jul. 1953, *S. Q. Zhong & P. L. Yang 119* (PE). **Jonê**: near Jonê, 2800 m, 26. Jul. 1936, *T. P. Wang 5408* (PE); Taohe River Basin, 2440 m, Aug. 1925, *J. F. Rock 13196* (K,PE). **Kangxian**: Yangba, 1200 m, 11. Aug. 1963, *Z. Y. Zhang 16579* (PE). **Minxian**: Luoda Township, 2600 m, open woods, 26. May. 1957, *Taohe Exped. 3133* (KUN). **Pingliang**: Kongdong Shan, 08. Aug. 1956, *Acad. Sin. Huanghe Exped. 2058* (PE). **Tianshui**: Baiyang-ling, Zifanggou, 1900 m, 26. Jul. 1951, *J. M. Liu 10228* (PE); Dongcha Township, Zhangjiagou, 1570 m, grasses, 21. Aug. 1951, *J. M. Liu 10530* (KUN, PE). **Tianzhu**: s. loc., 15. Sept. 1991, *J. S. Yang 91–858* (KUN). **Wenxian**: Zhong'an, 2000 m, sunny slopes, 15. Jul. 1959, *Z. Y. Zhang 6745* (HWNP); Baishui Jiang, 2150–2300 m, 27. Jun. 2006, *Baishui Jiang Exped. 0406* (PE). **Xiahe**: Qingshui, 2300 m, forests, 03. Jul. 1937, *T. P. Wang 7038* (PE); loc. eodem, 2400 m, forests, 05. Jul. 1937, *K. T. Fu 1018* (PE). **Xigu**: Heiyugou, Laogou, 2700 m, 20. Jul. 1951, *T. P. Wang 14481* (PE); Santang Dahaigou, 2500 m, 29. Jul. 1951, *T. P. Wang 14638* (PE). **Yuzhong**: Maxian Shan, 2580–2780 m, grassy slopes, 07. Sept. 1956, *Huanghe Exped. 3055* (PE); loc. eodem, 2600 m, 08. Aug. 1959, *Y. C. Hou 5954* (PE); Wulong Shan, 2060–2100 m, 31. Aug. 1956, *Huanghe Exped. 3215* (PE); Xinglong Shan, 2150 m, deciduous forests, 12. Jun. 1990, *C. Y. Wu et al. 9204* (KUN). **HEBEI, Chengde**: s. loc., 1959, *Nankai Univ. 1088* (PE). **Chicheng**: Dahaituo Shan, Sanjianfang, 1230 m, 16. Sept. 1959, *PE Exped. 7705* (PE). **Cixian**: Lufeng Shan, Nangou, 24. Jul. 1958, *K. C. Kuan 6079* (PE). **Huailai**: s. loc., 1500 m, 19. Sept. 1959, *Hebei Exped. 2nd Team 181* (PE). **Laiyuan**: Baishi Shan, 2200 m, 23. Jul. 1959, *Hebei Agr. Univ. Bot. Teaching Group (4) 3551* (PE); Dianziliang, 1600 m, thickets, 14. Jul. 1959, *Hebei Agr. Univ. Bot. Teaching Group (4) 2139* (PE). **Longguan**: Laopengzi, 1580 m, 12. Sept. 1959, *PE Exped. 4528* (PE). **Neiqiu**: Xiaolingdi Village, 1750 m, upon thickets, 30. Jul. 1950, *Y. Liu 13227* (PE). **Pingshan**: Muchang, 1280 m, 22. Aug. 1959, *Hebei Exped. 5th Team* (V) *1031* (PE). **Shexian**: Wudu, 13. Jun. 1958, *K. C. Kuan 5367* (PE). **Wuling Shan**: Dakeng, grasses, 20. Jun. 1955, *Dongling Exped. 226* (PE); grasslands, Aug. 1935, *Y. Liu 11711* (PE). **Xiaowutai Shan**: 1800 m, 18. Sept. 1921, *H. Smith 48* (LD). **Xuanhua** (Hsüenhwa): 20. Jul. 1930, *H. Serre* s. n.(W). **Yixian**: Dayukou, Nantaigou, 850 m, 25. Jul. 1959, *PE Exped. 3295* (PE). **Zanhuang**: Loudi Village, 04. Aug. 1951, *X. Y. Liu 859* (KUN, PE). **Zhuolu**: Wangjiagou, 03. Aug. 1955, *J. G. Yang 1645* (PE); Yangjiaping, 19. Aug. 1951, *J. G. Yang 398* (PE); loc. eodem, Beigou, 08. Oct. 1950, *W. T. Wang 1049* (PE). **Zunhua**: Dongling, Aug. 1931, *T. Tang 2230* (PE); loc. eodem, grasslands, Aug. 1930, *H. T. Tsai 50432* (PE). **HEILONGJIANG, Acheng**: Xiaoling, fl. blue, 30. Jul. 1951, *G. Z. Wang 652* (IFP). **Dailing**: Dongshan, *Quercus-Lespedeza* forests, 04. Aug. 1959, *W. Wang et al. 788* (PE); loc. eodem, mixed forests, 08. Jul. 1956, *Sino-Germ. Exped. 6981* (PE); loc. eodem, 24. Jul. 1954, *T. N. Liou et al. 5937* (IFP). **Hailin**: fl. pale green, 26. Jul. 1975 *P. Y. Fu 3411* (IFP). **Mudanjiang**: s. loc., 1959, *Mudanjiang Exped. 904* (PE) & *1037* (PE); s. loc., 1959, *Mudanjiang Regional Exped. 162* (PE). **Shangzhi**: Mao'ershan, 23. Jun. 1951, *G. Z. Wang et al. 320* (IFP). **Songhuajiang**: s. loc., 1959, *Songhuajiang Exped. 43* (PE). **Yongsu**: 22. aug. 1963, *T. N. Liou et al. 10307* (IFP). **HENAN, Lingbao**: Funiu Shan, 1240 m, river banks, 16. Aug. 1958, *J. Q. Fu 92* (KUN); s. loc., 1959, *General Exped. 14097* (KUN, PE). **Lushi**: Qihe, 1600 m, 02. Oct. 1958, *J. Q. Fu 1679* (KUN). **JILIN, Antu**: Hualazi, 800 m, sunny places, 19. Aug. 1959, *Yanbian Second Group 178* (PE). **Changbai**: Hengshan to Erdaogou, by forests, 29. Aug. 1963, *Y. X. Li et al. 961* (IFP). **Changbai Shan**: 28 km NW of Changbai, 900–1000 m, 20. Aug. 1986, *D. Y. Hong, W. H. Sauer et al. 34262* (PE) ; Erdaobaihe, edges of *Larix* forests, 12. Jul. 1975, *s. coll. 75243* (CDBI); Erdaobaihe, grasses by forests, 09. Sept. 1959, *Yanbian First Group 673* (PE); Huapi, 900 m, 28. Jul. 1950, *T. N. Liou 1770* (KUN); upper Mai-Ho River, 23. Aug. 1924, *N. Kozlow 14250* (W); Manjiang, forest edges, 09. Aug. 1957, *J. J. Chien 781* (PE); loc. eodem, 1000 m, forest edges, 09. Aug. 1957, *J. J. Chien 791* (KUN); loc. eodem, 31. Jul. 1962, *Temperate Forest Group 413* (PE); Shijiudaogou Valley, 840 m, thickets, 20. Aug. 1986, *D. Y. Hong, W. H. Sauer et al. 34207* (PE); to Tang-Ho-Ko, Sungari River, May–Aug. 1886, *H. E. M. James* s. n.(K); Toudaobaihe, meadows, 23. Jul. 1963, *W. Wang et al. 2190* (PE); near Yimianpo, 1500 m, grasses by forests, 03. Aug. 1957, *J. J. Chien et al. 675* (PE). **Dunhua**: Malugou, meadows by forest, 20. Aug. 1951, *T. N. Liou 3516* (IFP). **Fusong**: between Longtou Shan Monitory Station and old Weidong Forest Farm, 1030–1150 m, virgin mixed forests, 05. Aug. 1986, *D. Y. Hong, W. H. Sauer et al. 33427* (PE) ; Manjiang, Houshangan, broad-leaved forests, 11. Aug. 1959, *Tonghua Team 126* (PE). **Helong**: Bailiping, Xishanpo, mixed forests, 14. Sept. 1959, *Yanbian Second Group 828* (PE). **Shuangshan**: N of Shuangshan slopes of hills, 500 m, 28. Jul. 1950, *Y. C. Ma 10164* (PE). **Tonghua**: s. loc., s. d. *J. Sato 9995* (PE). **Wangqing**: Chunyang to Laomiao, forest edges, 08. Aug. 1959, *P. Y. Fu et al. 992* (IFP, PE). **Oniodo Valley**: 08. Aug. 1896, *V. L. Komarov 1494* (K). **LIAONING, Benxi**: Caohezhang, forests, by river, 14. Aug. 1965, *C. Q. Lin 979* (IFP); Xiejiachun, forests, 02. Sept. 1951, *C. Wang & Y. X. Liou 1465* (IFP, PE). **Fengcheng**: Dongxingbaozi, by forest, Aug. 1965, *C. Q. Lin 1072* (IFP). **Huaren**: Laotudingzi Nature Reserve, by forest, fl. pale green, 07. Sept. 1986, *P. Y. Fu & Z. S. Qing 1555* (IFP); loc. eodem, thickets, 02. Sept. 1959, *C. S. Wang et al. 4239* (IFP). **NEI MONGOL, Daqing Shan**: Pichaigou, forests, 07. Jul. 1987, *Inner Mong. Univ. (85-1) 066* (PE). **Ningcheng**: Dayingzi, forests, fl. pale yellow-white, 27. Jul. 1981, *S. X. Li et al. 4718* (IFP). **NINGXIA**,

Guyuan: Wating, 2100 m, 08. Aug. 1953, *T. P. Wang 17215* (PE). **Helan Shan**: Nuanquan, 19. Jul. 1973, *Ningxia Drug Inspect. (Bei) 228* (PE). **Jinyuan**: Dananchuan, Nantaigou, 2300 m, forests, 28. Jul. 1964, *Ningxia Drug Inspect. (Jin) 0028* (PE). **QINGHAI**, **Datong**: Baoku, 2700 m, 27. Jul. 1989, *Z. H. Zhang 3709* (HNWP). **Minghe**: Gushan, 2500 m, thickets, 13. Jul. 1989, *S. W. Liu 2754* (HNWP); Xigou Forest Farm, 2000 m, thickets at edges of forests, 14. Jul. 1989, *T. N. Ho 620* (HNWP); s. loc., 1960, *C. Y. Xu 86* (HNWP). **Tongren**: Xiao Valley, 2800 m, 12. Sept. 1936, *T. P. Wang 6169* (PE). **Xining**: 2300 m, gullies, 30. Jul. 1985, *Y. H. Wu 1881* (B). **Xunhua**: Muchangci, 2300 m, forests, 13. Aug. 1973, *H. Z. Zhang 238* (HNWP). **SHAANXI**, s. loc., 1897, *R. J. Giraldi* s. n.(K). **Baoji**: Matoutan Forest Park, 1480 m, thickets, 01. Aug. 1996, *Y. S. Lian et al. 96096* (PE). **Chang'an**: Nanwutai, 08. May. 1956, *Huanghe Exped. Gansu Team 164* (PE). **Foping**: Dudemeng, 26. Jun. 1952, *P. C. Kuo 1559* (CDBI, PE); Dudehe to Foping, 11. Aug. 1938, *T. N. Liou & P. C. Tsoong 3222* (PE); Gaojiaba, 1400 m, 18. Jul. 1952, *K. T. Fu 5088* (KUN, PE). **Huanglong Shan**: below Qigan Temple, bushes, 18. Jul. 1939, *K. T. Fu 3357* (PE). **Huxian**: Gaoguanyu, 1700 m, sunny woods, 02. May. 1959, *Medic. Pl. Exped. 2001* (PE). **Liuba**: Ziguan-ling, 1500 m, 06. Sept. 1977, *J. W. Wang & Z. C. Shi 302* (PE). **Meixian**: Yingtou Township, 1247 m, 08. Jul. 2005, *J. Y. Wu et al. 88* (PE). **Ningshaan**: Huoditang, 1650 m, thickets, 26. Jul. 1978, *K. T. Fu 17618* (PE); s. loc., 1300 m, shady slopes, 21. Jul. 1933, *H. W. Kung 3139* (PE). **Shaanxi-Gansu Basin**: s. loc., *T. Y. Le* s. n.(PE). **Taibai Shan**: near Dadian, 2350 m, 05. Sept. 1937, *T. N. Liou & P. C. Tsoong 645* (PE); near Haoping-si, 1300 m, 24. Aug. 1937, *T. N. Liou & P. C. Tsoong 256* (KUN, PE); La Trappe and vicinity, 680–900 m, 24. Jul. 1917, *E. Licent 3191* (K); loc. eodem, 1938, *W. Y. Hsia 33* (PE); Lianbinchang, 09. Jul. 1938, *T. N. Liou & P. C. Tsoong 1972* (PE); near Ping'an-si, 2880 m, 08. Sept. 1937, *T. N. Liou & P. C. Tsoong 710* (PE); loc. eodem, 1910, *W. Purdom* s. n.(K); loc. eodem, 1909, *W. Purdom 675* (K); Ravine to W of Ping'an-si, 20. Jul. 1938, *T. N. Liou & P. C. Tsoong 2540* (PE); s. loc., 1400 m, forests, 27. Jul. 1933, *T. P. Wang 1382* (PE). **Zhongnan Shan**: 1600 m, shady slopes, 15. Jun. 1933, *H. W. Kung 2828* (PE). **SHANXI**, s. loc., 06. Jul. 1916, *E. Licent 2213* (K); SE Shanxi, Tsien-Ou, 25. Aug. 1934, *E. Licent 11440* (W). **Huoxian**: Huoshan, 1600–2100 m, 30. Jul. 1935, *E. Licent 12389* (W); Qiliyu, 1650 m, 09. Jun. 1957, *Huanghe Exped.790* (KUN) & *1637* (PE); loc. eodem, 1600–2100m, 16–17. Jul. 1935, *E. Licent* 12139 (PE). **Jiaocheng**: Kuandi Shan, 1800 m, 24. Jun. 1950, *Huabei Inst. Agr. 46* (PE); Kuandi Shan, Nangou, 1700 m, 27. Jun. 1956, *Huanghe Exped.* 1038 (PE). **Jiexiu**: Xingdi Township, 1325 m, 19. Jun. 1957, *J. M. Liu 1235* (PE). **Jincheng**: Gujiao, Nangetai, 13. Jul. 1959, *S. Y. Bao* 336 (HNWP, PE). **Jishan**: Xishe, 31. Jul. 1959, *s.coll. 2282* (HNWP). **Kuancheng Shan**: 1520 m, 07. Aug. 1929, *W. Y. Hsia 1470* (PE). **Lingchuan**: Taihang Shan, 1200 m, 22. Jul. 1937, *K. M. Liou 7832* (PE); loc. eodem, 20. Jul. 1937, *K. M. Liou 7759* (PE). **Lishi**: Gesi, Shanyuan Village, 1470 m, 28. Aug. 1960, *X. Y. Liu 21259* (HNWP, PE). **Luliang**: Shikou, 06. Jun. 1959, *s. coll. 356* (HNWP). **Luya Shan**: 17. Jul. 1925, *H. Smith 8289* (LD, PE). **Pingshun**: Taihang Shan, 1150 m, 11. Aug. 1937, *K. M. Liou 7965* (PE). **Qinxian**: Yuerquan, 28. Jul. 1959, *K. C. Kuan & Y. L. Chen 962* (HNWP, PE). **Puxian:** Huangtu-ling, forests, 04. Sept. 1959, *Q. Ma 850* (HNWP). **Ruicheng**: Xuehua Shan, 1860 m, 20. Aug. 1959, *S. Y. Bao 518* (HNWP, PE). **Wutai**: Lianzhen Township, 2200 m, 18. Jul. 1959, *K. C. Kuan & Y. L. Chen 2410* (HNWP, PE); Lingjing Township, 2000 m, 09. Jul. 1959, *K. C. Kuan & Y. L. Chen 2706* (PE); Paomaquanchang, 2440 m, 22. Jul. 1959, *K. C. Kuan & Y. L. Chen 1655* (HNWP, PE). **Wuzhai**: Dianping, 29. Sept. 1955, *Huanghe Exped. 1637* (PE). **Xianning**: Kuanwangmiao, 1150 m, 17. Jul. 1966, *X. Y. Liu 20887* (HNWP, PE). **Xixian**: Xiluo Shan, 14. Aug. 1935, *T. P. Wang 3125* (PE). **Xiaxian**: Sijiao, Taikuanhe, 1250 m, thickets, 30. Jun. 1962, *T. W. Liu 381* (HNWP). **Xiaoyi**: Shangting Shan, 28. Aug. 1935, *T. P. Wang 3387* (PE). **Yangcheng**: Yunmong Shan, forests, 29. Jul. 1959, *S. Y. Bao 2168* (HNWP, PE). **Yicheng**: Nanfan, 1640 m, 12. Jun. 1960, *X. Y. Liu 20162* (HNWP); Shihe Township, 1450 m, 31. May. 1960, *X. Y. Liu 20317* (PE); s. loc., 1200 m, 14. May. 1957, *Huanghe Exped. 248* (PE). **Zhongyang**: Liugongdong, forests, 12. Jun. 1955, *Huanghe Exped. Zhongyang B Team 15* (PE); Liujiaping, 07. Jun. 1959, *Q. Ma 5604* (HNWP). **SICHUAN, Barkam**: Somang Township, 2910 m, 06. Jul. 1957, *T. Y. Chang & H. F. Zhou 22578* (SZ); Zhuokeji, 2800 m, 31. Jul. 1957, *T. Y. Chang & H. F. Zhou 23431* (SZ); s. loc., 2500 m, thickets by field, 15. Sept. 1964, *J. D. Li et al. 82* (PE). **Emei**: Emei Shan, 2150 m, 25. Jul. 1965, *L. N. Zhao 2565* (CZ). **Heishui:** near Luhua, 2500 m, 30. Jul. 1957, *X. Li & J. X. Zhou 73872* (SZ); loc. eodem, 2445 m, 18. Jul. 1957, *X. Li & J. X. Zhou 73649* (SZ); opposite Luhua, 2450 m, 21. Jul. 1957, *X. Li & J. X. Zhou 73712* (SZ); Luhuayu, 2500 m, 23. Jul. 1957, *X. Li 73756* (PE); loc. eodem, 2600 m, 28. Sept. 1964, *Z. M. Xu & J. D. Li 64-101* (PE); Naizigou, 3200 m, 25. Jul. 1957, *X. Li 73280* (SZ); Shidiaolou Township, 2900 m, 27. May. 1959, *Sichuan Econ. Plant Exped. Heishui Group 1183* (PE). **Jiangyou**: 2000 m, open thickets, 19. Aug. 1930, *F. T. Wang 22279* (KUN). **Lixian**: Dabanzhao, 24. Jul. 1956, *W. K. Hu 53–952* (SZ); loc. eodem, 19. Jul. 1957, *D. Z. Deng 1813* (SZ); Miyaluo, forests, 2900 m, 11. Sept. 1964, *T. L. Dai et al. 64-1* (PE); loc. eodem, 2700 m, edges of thickets, 22 Jul. 1959, *Maowen Exped. 2978* (CDBI, KUN); loc. eodem, 2350 m, 25. Jul. 1956, *D. P. He 45170* (SZ); loc. eodem, 3000 m, 27. Jul. 1956, *D. P. He 45301* (SZ); Mungtwin-ko, 2440 m, 14. Aug. 1941, *S. Y. Hu 2106* (K); Zhegu Shan, 3300 m, 16. Aug. 1957, *T. Y. Chang & H. F. Zhou 23862* (SZ); s. loc., *T. C. Lee 2504* (SZ). **Maoxian**: Baiqiao Distr., Jiaochang Township, 2800 m, thickets, 16. Oct. 1964, *T. L. Dai* *et al.* (Maowen) *64-31* (PE); Huangcaoping, 2450 m, thickets, 09. Oct. 1964, *T. L. Dai et al.* (Maowen) *64-7* (PE); Sanlong Township, forests, 11. Jul. 1952, *Z. He & Z. L. Zhou 13235* (SZ); loc. eodem, 19. Jul. 1952, *Z. He & Z. L. Zhou 13345* (PE, SZ); Wangzhou Shan, 2000 m, grasses, 07. Oct. 1964, *T. L. Dai et al.* (Maowen) *64-1* (PE). **Nanping** (Jiuzhaigou): Jiuzhaigou, 2200 m, by river, 23. Sept. 1986, *C. Y. Wu et al. 552* (KUN); s. loc., 2860–3200 m, forests, 02. Nov. 1962, *X. Y. Tang 6211* (PE). **Qingchuan**: Qinghua Distr., 2200 m, 31. Jul. 1978, *Qingchuan Exped. 88* (SM). **Songpan**: near Balang, 3300 m, 13. Sept. 1937, *K. T. Fu 1721* (PE); Hongyankuan, 3000 m, thickets, 07. Jul. 1959, *Sichuan Econ. Plant Surv. 1563* (PE); Xiaoxin, 2600 m, forests, 06. Aug. 1964, *X. Y. Tang 64036* (PE); s. loc., 06. Aug. 1928, *W. P. Fang 4010* (G, K, P, PE); s. loc., Jun.–Aug. 1914, *H. Weigold* s. n.(W); s. loc., 3600 m, 19. Sept. 1937, *T. P. Wang 7768* (PE). **Wenchuan**: Longxi, alpine zone, 20. Jul. 1963, *S. J. Yue 63-27* (PE). **Xiaojin**: Shanong, 3000 m, by road, *Sichuan Plant Exped. 9712* (CDBI). **Xuyong**: Fengshui Distr., 1750 m, 08. Aug. 1977, *Xuyong Exped. 77–1011* (SM). **Zoigê**: 2800 m, thicket edges, 13. Aug. 1960, *Sichuan Medic. Plant Resources Surv. 20598* (PE).

KOREA PENINSULA, **Kangwon** Prov.: Kumgangshan, 06. Aug. 1928, *K. Kondo* s. n, (TI); loc. eodem, 18 Aug. 1902, *J. Uchiyama* s. n.(TI); loc. eodem, 20. Aug. 1916, *T. Nakai 5881* (TI). **Shinkeiji**, Kongo, sea level, 1931, *L. Trotter 40* (K).

RUSSIA, **Amursk**, Birobidzan, near Tsarskaya gora, 01. Aug. 1910, *M. E. Semyagin 22* (LE). **Primorsk**, Mongugai Valley, Ovchinnikovo Village, 43°15′N, 131°23′E, 18. Jul. 1913, *N. Dyukina 774* (LE).

FIGURE IV 1:23A1 *Codonopsis pilosula* (Franch.) Nannf. subsp. *tangshen* (Oliv.) D. Y. Hong.: a 3-loculicidal capsule with hypogenous calyx.

FIGURE IV 1:23C *Codonopsis pilosula* (Franch.) Nannf. subsp. *tangshen* (Oliv.) D. Y. Hong. From Chongqing: Nanchuan, 10. Sept. 2012, by Qiang WANG.

23b. Codonopsis pilosula subsp. **tangshen** (Oliv.) D. Y. Hong, Novon 20(4): 423. 2010; Hong, Lammers & Klein in Wu, Raven & Hong, Fl. China 19: 516. 2011. ≡*Codonopsis tangshen* Oliv., Hooker's Icon. Pl. 20: pl. 1966. 1891; Skan, Bot. Mag. 132: t. 8090. 1906; Komarov, Trudy Imp. S. -Peterburgsk. Bot. Sada 29(1): 104, 1908; Chipp, J. Linn. Soc. Bot. 38: 379. 1908; Tsoong, Contr. Inst. Bot. Nat. Acad. Peiping 3(3): 102. 1935; Anonymous, Icon. Cormophyt. Sin. 4: 377. fig. 6167. 1975; Shen & Hong in Hong, Fl. Reipubl. Popularis Sin. 73(2): 43. 1983. Type: China, W Hupeh (Hubei), Hsingshan (Xingshan CO), *A. Henry 6468* [lectotype here designated: K (K000229965); isolectotypes: K (K000229964; K000229966; K000229969), E, G, LE, P, PE, TI]; W Hubei, *A. Henry 6468A* (paratypes: K, PE); W Hubei, So-Patung (S Badong), 1220–2440 m, *A. Henry 6468B* (paratypes: K, P, PE).

The subspecies differs from the other two in having the calyx nearly entirely free from the hypanthium, and the leaf blade mostly truncate or rounded, less frequently cordate, at the base. Figure IV 1: 23A1 & IV 1: 23C.

Chromosomes: 2n=16.

Phenology: flowering from July to September; fruiting in September.

Habitats: grassy slopes, grasses in forests, thickets, and forests; at altitudes from 800 to 2600 m.

Distribution: **endemic** to China: Chongqing, N Guizhou, W Henan, W Hubei, W Hunan and SE Shaanxi.

Additional specimens examined, **CHINA, CHONGQING, Chengkou**: Baici Shan, 1300 m, 12. Sept. 1958, *T. L. Dai 102590* (PE, SZ); Beiping, 1600 m, forests, 26. Jul. 1958, *T. L. Dai 101495* (CDBI); Beiping Shan, 1600 m, grassy slopes, 26. Jul. 1958, *T. L. Dai 101499* (HNWP, PE, SZ); Beiping Shan, Anle, 1460 m, 20. Jul. 2008, *Bashan*

Exped. 2243 (PE); Dong'an Township, 1700 m, 06. Jul. 2008, *Bashan Exped. 1219* (PE); loc. eodem, 1580 m, 07. Jul. 2008, *Bashan Collection Team 0081* (PE); loc. eodem, 1394 m, *Bashan Exped. 0991* (PE); Houping, 1420 m, grasses, 20. Jul. 1958, *T. L. Dai 105712* (KUN, PE, SZ); loc. eodem, 1250 m, grassy slopes, 27. Jul. 1958, *T. L. Dai 105837* (PE, SZ); Miaoba, 950 m, 13. Jul. 2008, *Bashan Exped. 1469* (PE); Qigan Shan, 1650 m, forests, 30. Aug. 1958, *T. L. Dai 102195* (HNWP, PE, SZ); Shuangping, 1500 m, 04. Sept. 1958, *T. L. Dai 107098* (CDBI, KUN); Yizil-iang, 1630 m, grasses, 16. Aug. 1958, *T. L. Dai 101921* (PE, SZ); loc. eodem, 1300 m, 27. Aug. 1958, *T. L. Dai 101977* (CDBI, HNWP, PE, SZ); Yongmong Distr., 1800 m, 24. Sept. 1958, *T. L. Dai 106048* (KUN, PE, SZ); loc. eodem, 2300 m, 26. Oct. 1958, *T. L. Dai 106975* (PE); s. loc., 15. Jul. 1956, *T. L. Dai 101231* (SZ). **Fengjie**: Hanrui Township, 1400 m, 28. Jun. 1958, *M. Y. Fang 24508* (SZ) ; Shuangdian, 900 m, 07. Aug. 1958, *M. Y. Fang 24885* (PE, SZ); loc. eodem, 1100 m, 05. Jul. 1964, *H. F. Zhou & H. Y. Su 109278* (PE, SZ) ; Xinglong, 1300 m, 01. Jul. 1958, *T. Y. Chang et al. 25441* (KUN, PE, SZ); loc. eodem, 1400 m, 28. Jun. 1958, *T. Y. Chang et al. 24508* (SZ) ; Xinhe Township, 1200 m, 12. Jun. 1958, *H. F. Zhou et al. 26243* (SZ). **Kaixian**: Manyue Township, 1300 m, 28. Jul. 2008, *Bashan Exped. 2805* (PE); Xuebao Shan, 2457 m, 28. Jul. 2008, *Bashan Exped. 2460* (PE). **Nanchuan**: Huangcaoping, 1700 m, bamboo thickets, 10. Sept. 2012, *Q.Wang H 12074* (PE); Jinfo Shan, Xinluzi, 2000 m, forests, 08. Oct. 1957, *G. F. Li 64444* (KUN); Jinfo Shan, 1945, *F. T. Wang 10110* (PE); loc. eodem, 1700 m, 21. Jul. 1957, *G. F. Li 62992* (PE, SZ); loc. eodem, 1600 m, 30. Jul. 1957, *G. F. Li 63078* (PE, SZ); loc. eodem, 1520 m, 07. Aug. 1957, *G. F. Li 63272* (SZ); loc. eodem, 1780 m, 18. Jul. 1957, *G. F. Li 62896* (SZ); loc. eodem, 1600 m, 11. Jul. 1957, *S. Y. Chen et al. 95114* (SZ); loc. eodem, 1350–1450 m, thickets by road, 27. Jul. 1978, *Phytogeography Exped. 734* (SZ); loc. eodem, 1750 m, forests, 14. Aug. 1957, *J. H. Xiong & Z. L. Zhou 92689* (PE, SZ); loc. eodem, 1550 m, forests, 01. Aug. 1957, *J. H. Xiong & Z. L. Zhou 92373* (PE, SZ); loc. eodem, 1900 m, forests, 03. Aug. 1986, *Jinfoshan Exped. 2012* (PE); loc. eodem, 1750 m, 02. Jul. 1986, *Jinfoshan Exped. 1731* (PE); loc. eodem, 2000 m, 17. Jul. 1984, *S. F. Wang 286* (SZ); Jinku, Liziba, 1960 m, forests, 09. Oct. 1957, *J. H. Xiong & Z. L. Zhou 64504* (PE, SZ); Xiaohe, Sancha, 1700 m, 19. Jul. 1957, *J. H. Xiong & Z. L. Zhou 92115* (PE, SZ). **Shizhu**: Huangshui Township, Tianwan, 1450 m, thickets, 19. Jul. 1988, *F. D. Fu & Y. L. Cao 1007* (PE); Huangshui, 1600 m, 24. Aug. 1978, *W. H. Wang 3247* (CDBI, SZ); Qiyue Shan, 1680 m, 21. Jul. 1978, *Y. Chen 2871* (CDBI, SZ); Shazi Distr., 2600 m, woods, 28. Jul. 1978, *W. H. Wang 2364* (CDBI). **Wanyuan**: Dazhu, 1710 m, 10. Aug. 1959, *P. Y. Li 6135* (PE); Dazhu, Miaozi Township, 09. Aug. 1959, *P. Y. Li 4581* (PE). **Wulong**: Gonghe, 1750 m, mixed forests, 28. Sept. 2007, *Z. Y. Liu 180391* (PE); near Xiangyang Reservoir, 1200 m, 17. Aug. 1978, *s. coll. 996* (SM). **Wushan**: Dengjia Township, mixed forests, 1710–1830 m, 10. Sept. 2009, *Sanxia Exped. 3669* (PE); Luoping, 1820 m, 01. Aug. 1964, *H. F. Zhou & H. Y. Su 109013* (PE, SZ); Quanziping, 1800–2100 m, forests, 08. Oct. 1958, *G. H. Yang 59775* (KUN); Sandengzi, 1900 m, forest margins, 25. Aug. 1958, *G. H. Yang 58951* (CDBI, KUN, PE); s. loc., 1600 m, 21. Jul. 1964, *C. K. Xie & S. S. Zhen 64002* (PE). **Wuxi**: Baiguo Forest Farm, 1310–1810 m, mixed forests, 15. Sept. 2007, *Bot. Inst. Exped. 0020* (PE); Bailu, 1370 m, 11. May 1959, *X. Z. Wang 1002* (SZ); Hongchiba, 1900 m, grasses in glades, 06. Oct. 1988, *Y. D. Chen et al. 2193* (PE); loc. eodem, 1800 m, mixed forests, 06. Oct. 1988, *Y. D. Chen et al. 2235* (PE); loc. eodem, 1800 m, grassy slopes, 07. Oct. 1988, *Y. D. Chen et al.* s.n. (PE); loc. eodem, 1900 m, grasses, 05. Oct. 1988, *Y. D. Chen et al. 2168* (PE); Shuanghe Township, 2600 m, grassy slopes, 18. Jul. 1958, *G. H. Yang 58874* (PE). **HENAN**, s. loc., 30. Jul. 1960, *K. C. Kuan & T. L. Dai 1355* (PE). **HUBEI,** W Hubei, s. loc., 26. Oct. 1901, *E. H. Wilson 269* (E, K); W Hubei, s. loc., Aug. 1900, *E. H. Wilson 1623* (E, K, W); W Hubei, s. loc., Jul. 1907, *E. H. Wilson* s. n.(W); W Hubei, s. Loc., Sept. 1907, *E. H. Wilson* s. n.(W); s. loc., 07. Aug. 1903, *E. H. Wilson* s. n.(K). **Badong**: Lucongpo, 1650 m, margins of broad-leaved forests, 14. Aug. 2009, *C. M. Zhao et al. Ex 3171* (PE); Luoping, 1840 m, forests, 07. Oct. 1952, *Q. L. Chen et al. 1949* (PE); near Niutongwan, 2100 m, thickets, 21. Jul. 1957, *G. X. Fu & Z. S. Zhang 937* (PE); Xiagu, 1700 m, 25. Jul. 1977, *Shennongjia Plant Exped. 24572* (PE); loc. eodem, 1700 m, grassy slopes, 08. Aug. 1977, *Shennongjia Plant Exped 24457* (PE); Yanduhe, 1000 m, broad-leaved forests, 31. Jul. 2009, *C. M. Zhao et al. Ex 2376* (PE). **Enshi**: Banqiao, 1500 m, 21. Jun. 1958, *H. F. Zhou et al. 24382* (SZ). **Hefeng**: Daping Farm, grassy slopes, 16. Aug. 1958, *H. J. Li 6149* (PE). **Jianshi**: Huaguoping, 1250 m, 05. Jul. 1951, *L. Y. Dai & Z. H. Qian 92* (PE). **Lichuan**: *Metasequoia* Area, 1948, *W. C. Cheng & C. T. Hwa 625* (PE); vicinity of Shui-sa-pa, 25. Jul. 1948, *J. Linsley Gressitt 2424* (PE); Xiaohe, Jiantianba, 1170 m, forests, 19. Sept. 1951, *L. Y. Dai & Z. H. Qian 1121* (PE). **Shennongjia**: Dajiuhu, 1600 m, 27. Aug. 1976, *Shennongjia Plant Exped. 11981* (PE); Honghuaduo Forest Farm, 1600 m, thickets, 06. Jul. 1976, *Shennongjia Plant Exped. 21401* (PE); Hongping, forest margin, 2000 m, 08. Aug. 1976, *Shennongjia Plant Exped. 32273* (PE); Kuanmen Shan, 1480 m, 12. Aug. 1976, *Shennongjia Plant Exped. 11232* (PE); Laojun Shan, 1400 m, 07. Jul. 1976, *Shennongjia Plant Exped. 30555* (PE); loc. eodem, open woods, 18. Jul. 1976, *Shennongjia Plant Exped. 31229* (PE); near Sanyuan, 1550 m, dry forests of *Quercus*, *Castanea*, *Betula*, 26. Sept. 1980, *1980 Sino-Amer. Exped. 1881* (KUN, PE); 31°30′ N, 110°30′ E, 2100 m, 27. Aug. 1980, *1980 Sino-Amer. Exped. 186* (E, PE); loc. eodem, 1800 m, 28. Aug. 1980, *1980 Sino-Amer. Exped. 1057* (PE); loc. eodem, 1800 m, scandent over thickets, 28. Aug. 1980, *1980 Sino-Amer. Exped. 234* (KUN, PE); loc. eodem, 1700 m, dense undergrowth along trail, 05. Aug.

1980, *1980 Sino-Amer. Exped. 97* (KUN, PE); 1600 m, grassy slopes, 27. Aug. 1976, *Shennongjia Plant Exped. 31916* (PE); 2000 m, thickets, 15. Aug. 1976, *Shennongjia Plant Exped. 21993* (PE); *s. coll. 21401* (PE); s. loc., *H. J. Li 191* (PE). **Xingshan**: Muzhuxia, 23. Jul. 1956, *H. J. Li 37* (PE). **Xuan'en**: Yexi Township, 1350–1500 m, 27. Jul. 1958, *H. J. Li 5282* (PE); 1520–1870 m, grasses, 17. Jul. 1988, *Y. M. Wang 5408* (PE) ; s. loc. 800-900 m, grasses, *H. J. Li 3828* (KUN, PE). **Zhuxi**: 1600 m, 29. Sept. 1991, *Z. Zheng 91–336* (KUN); 1680 m, 29. Sept. 1991, *Z. Zheng 91–290* (KUN). **HUNAN, Sangzhi**: Bachangxi, Tianping Shan, 1600 m, thickets, 02. Jul. 1975, *B. G. Li & S. B. Wan 750102* (PE); loc. eodem, 1670 m, 15. Sept. 1990, *T. R. Cao 90340* (KUN); Badagong Shan, Doupeng Shan, 01. Sept. 1988, *Beijing Exped. 4328* (PE); Badagong Shan, Jiazui River, evergreen & deciduous broad-leaved mixed forests, 1400–1500 m, 05. Jul. 1988, *Beijing Exped. 2442* (PE); Badagong Shan, Tianping Shan, forests, 1300 m, 08. Jul. 1988, *Beijing Exped.2706* (PE); Shadiping, 1350 m, forests, 22. Jun. 1958, *L. H. Liu 9041* (PE). **Shimen**: Huping Village to summit of Huping Shan, 1800–1900 m, thickets in evergreen & deciduous broad-leaved mixed forests, 26. Jun. 1987, *Hupingshan Exped. 0838* (PE); Qianping-Xiaoxi, 1000–1400 m, evergreen & deciduous broad-leaved mixed forests, 09. Jul. 1987, *Hupingshan Exped. 1219* (PE). **Xinning**: Dayun Shan, summit, 1920 m, s. d., *Y. B. Luo 2981* (PE). **SHAANXI, Ankang:** Furong, 01. Jul. 1959, *s. coll. 378* (HNWP). **Foping**: s.loc., 1070 m, forests by road, 08. Sept. 1958, *Z. M. Zhang 401* (KUN). **Langao**: Shenheyuan, 2300 m, 19. Jul. 2008, *Bashan Exped. 1888* (PE). **Pingli**: Hongluan Shan, 1300 m, thickets, 01. Jul. 1959, *s. coll. 0121* (HNWP). **Sanyang**: Tianzhu Shan, 1400 m, 05. Sept. 1952, *T. P. Wang 16477* (PE). **Yangxian**: Huayang, Mt Longshan, 1900 m, 29. Jul. 1952, *P. C. Kuo 1915* (PE). **Zhenping**: Shizhaihe, 1440 m, forests, 03. Jul. 1959, *Zhen 073* (HNWP). **Ziyang**: Huishui, thickets, 10. Aug. 1959, *P. Y. Li 4683* (PE); Shuanghe, 1800 m, 12. Aug. 1959, *P. Y. Li 6246* (PE). **SICHUAN**, **Nanjiang**: s. loc., 2300 m, 19. Aug. 1978, *s. coll. 908* (SM). **Pingchang**: Zhenglong, 1200 m, 07. Jul. 1978, *Pingchang Exped. 606* (SM). **Pingwu**: Wangbachu, 2200 m, among herbs, 07. Aug. 1985, *s. coll. 352* (CDBI); Pingwu, Sept. 2014, L. Z. Fu 2346 (PE). **Tongjiang**: on way to Cheng'an Forest Farm, 900–1000 m, Aug. 1978, *Tongjiang Exped. 714* (SM). **Xuanhan**: Sandun, 2500 m, 19. Jul. 1978, *s. coll.*, *Xuan-0809* (SM).

23c. Codonopsis pilosula subsp. **handeliana** (Nannf.) D. Y. Hong & L. M. Ma in Li & Zhu, Fl. Sichuan. 10:532. 1992; Lammers, World Checkl. Bibliogr. Campanulaceae. 269. 2007, pro syn. ≡*Codonopsis handeliana* Nannf. in Handel-Mazzetti, Symb. Sin. 7(4): 1078. 1936 & Svensk. Bot. Tidskr. 34: 383 & 384. 1940. ≡*Codonopsis pilosula* var. *handeliana* (Nannf.) L. T. Shen in Hong, Fl. Reipubl. Popularis Sin. 73(2): 41. 1983. Type: China, Yunnan, "Dshungdien" (Zhongdian), 3500–3600 m, 12. Aug. 1914, *Handel-Mazzetti 4777* (lectotype here designated: WU; isolectotype: W)

Leaf blade 1–3 cm long, 0.8–2.5 cm broad, both surfaces usually hirsute. Adnate part of calyx ≥ 3 mm long; calyx lobes 1.5–2 cm long. Corolla 2–2.8 cm long, ca. 1 cm wide.

Phenology: flowering in July and August; fruiting in September.

Habitats: forests, thickets or margins of woods; at an altitude of 2300–3900 m.

Distribution: **endemic** to China: SW Sichuan and N Yunnan.

Additional specimens examined, **CHINA, SICHUAN**, **Baiyu**, Zhandu Xiang, 31°6′N, 98°58′E, 3250 m, thickets, 23. Aug. 2006, *D. E. Boufford et al. 37192* (PE). **Batang**, NE of Batang, 30°13′N, 99°18′E, 3010 m, meadows, 31. Jul. 2006. *D. E. Boufford et al. 35546* (PE). **Daocheng**: Mengzi Township, 3500 m, thickets, 19. Jul. 1971, *Z. G. Liu 288* (CDBI); Gongga-ling, 3300 m, 10. Aug. 1973, *Sichuan Veget. Surv. 2475* (CDBI). **Dawu**: Gaqag, 3900 m, forests, 23. Jul. 1961, *S. Jiang 9633* (PE); Gaqag, 3350 m, 28. Jul. 1958, *s.coll. 170* (SZ); Wesheng, 3100 m, thickets, 09. Aug. 1974, *Sichuan Plant Surv. 5808* (CDBI, PE) ; by Xianshui River, thickets, 3200 m, 21. Jul. 1961, *J. S. Ying 9500* (PE). **Garzê**, Ringo Xiang, 31°49′N, 99°43′E, 3520–3550 m, 02. Aug. 2005, *D. E. Boufford et al. 33972* (PE). **Hongxi**: Leiwu, 2300 m, thickets, 31. Jul. 1959, *Sichuan Econ. Plant Surv. (Liang) 1509* (KUN). **Huili**: Deyang, thickets or herbs, 27. Jul. 1976, *Sichuan Veget. Surv. 12098* (CDBI). **Kangding**: Chungo Valley, ca. 2900 m, 16. Aug. 1934. *H. Smith 11086* (PE); Yala Township, 06. Aug. 1951, *W. K. Hu & Z. He 10652* (PE); Zhonggu Village, 3220 m, forests, 23. Jul. 1963, *K. C. Kuan, et al. 202* (PE); s. loc., 1893, *J. A. Soulie 593* (G, K); s. loc., 2640–2900 m, thickets, 15. Aug. 1934, *C. S. Liu 1076* (PE, SZ). **Muli**: ; Kumatian, 2900 m, margin of woods, 14. Aug. 1937, *T. T. Yü 7730* (KUN, PE); Yazui Forest Farm, 3100 m, 13. Jul. 1978, *Sichuan Plant Surv. 7356* (CDBI). **Xiangcheng**: No. 4 Distr., Qianjin Township, 4000 m, meadows, 16. Jul. 1972, *Sichuan Plant Exped. 1428* (CDBI). **Yanyuan**: Mianba, 2300–2420 m, *F. G. Zhu* 002 (PE). **TIBET, Jamda**: Jamda Town, 31°34′ N, 98°19′ E, 3380–3400 m, 22, Jul. 2010, *X. F. Gao et al. 12581* (CDBI). **YUNNAN**, **Dongchuan**: Fazhe Forest Farm, 2800 m, upon thickets, 30. Jul. 2008, *H. Peng et al. 8989* (KUN): Fazhe Township, 2700 m, thickets, path sides, 31. Jul. 2008, *H. Peng et al. 8924* (KUN). **Qiaojia**: Mt Yeoshan, 3300 m, 1913, *E. E. Maire 1018* (E). **Shangri-la** (Zhongdian): outside Alpine Bot. Gard., 3300 m, tall thickets by stream, 17. Sept. 2010, *D. Y. Hong et al. H10017* (PE); Zhongdian Valley, 3300 m, on *Aconitum* spp., 10. Aug. 1939, *K. M. Feng 1962* (KUN, PE).

24. Codonopsis hemisphaerica P. C. Tsoong ex D. Y. Hong, Plant Divers. Resour. 36(3): 288. fig. 2. 2014. Type: China, Sichuan, Baoxing, Puxigou, 2500 m, 21, Jul. 1954., *T. P. Soong 39208 A* (holotype: PE; isotype: SZ).

Roots carrot-shaped, up to 2.5 cm in diameter, branched or simple. Stems over 1 m long, twining, glabrous. Leaves on main stem alternate; petiole 1–2.7 cm long, hispidulous; leaf blade ovate or ovate-lanceolate, 2.5–5 cm long, 1.5–3 cm broad, both surfaces densely white-puberulent, base cordate to rounded, margin entire; leaves on branches similar, but smaller. Flowers solitary, terminal on main stem and branches; pedicel 0.5–4.5 cm long, glabrous. Calyx tube hemispheric, entirely adnate to hypanthium, glabrous, 10-ribbed; calyx lobes triangular-lanceolate, broadest at base, 8–10 mm long, 4–6 mm broad, glabrous, reflexed from anthesis, margin entire; sinus between calyx lobes narrow and pointed. Corolla campanulate, white, glabrous, 2.8 cm long, lobed to half; corolla lobes as long as tube, triangular. Stamens glabrous; filaments as long as anthers, slightly dilated at base. Style as high as stamens. Figures IV 1:24A & B.

Phenology: flowering in July and August.

Habitats: thickets or forests; 1000–2500 m.

Distribution: **endemic** to China: W Sichuan (Map IV 1:24).

Notes: This newly published species resembles *Codonopsis pilosula* very closely, particularly in habit and leaves, and many of the specimens cited below were originally identified as that species. However, it differs from the latter in having the corolla white (vs. yellow-green or pale purple), deeply divided with the corolla lobes as long as the tube, the ovary inferior to both calyx and corolla, and the calyx lobes entirely reflexed from anthesis onwards.

Additional specimens examined, **CHINA, SICHUAN, Baoxing**: Ganyang Xiaogou, 2400 m, 01. Sept. 1936, *K. L. Chu 3770* (P, SZ); Tú-pa-kou, 2130 m, Jul. –Aug., 1939, *S. Y. Hu 1260* (A, K, SZ); Yongxing, 2450 m, forests, 01. Aug. 1954, *T. P. Soong 39324* (SZ); s. loc., 1954, *T. P. Soong 39326* (PE); s. loc., 30 Jul. 1954, *T. P. Soong 39324A* (KUN, SZ). **Ebian**: Heizhugou, 1000 m, 03. Aug. 2007, *X. J. He et al. 151494* (SZ). **Ganluo**: Sizu, forests, 05. Aug. 1959, *s. coll. 4433* (PE, SZ). **Hongxi**: Leiwu, 2300 m, thickets, fl. white, 31. Jul. 1959, *Sichuan Econ. Plant Resources Surv. (Liang) 1509* (PE). **Leibo**: Huanglang, 2000 m, bamboo thickets, 09. Aug. 1976, *s. coll. 13424* (PE, SZ). **Pingwu**: Sier Township, 2300 m, thickets, 28. Aug. 1958, *H. L. Tsiang 11051* (PE, SZ). **Wenchuan**: Fengyi, 1800 m. thickets, 12. Jul. 1963, *S. J. Yue & X. Y. Tang 63–13* (PE).

25. Codonopsis kawakamii Hayata, J. Coll. Sci. Imp. Univ. Tokyo 30(1): 165. 1911; Shen & Hong in Hong, Fl. Reipubl. Popularis Sin. 73(2): 59. 1983; Hong, Lammers & Klein in Wu, Raven & Hong, Fl. China 19: 518. 2011; Type: China, Taiwan, Mt Niitaka, Nov. 1905, *Nakahara & Kawakami* s. n.(holotype: TI).

Roots carrot-shaped. Main stem scrambling or twining, fertile, sparsely villous; branches sterile or fertile. Leaves on main stem alternate, those on branches nearly opposite; petiole 5–12 mm long, hirsute; leaf blade ovate or deltoid-ovate, 0.7–2.3 cm long, 0.5–1.7 cm broad, adaxially hispidulous, abaxially hirsute, base usually rounded, rarely slightly cordate, margin mostly entire or sinuous, or very obscurely crenate, apex acute or obtuse. Flowers solitary, terminal on main stem and some branches, slightly pendent; pedicel 1–2.5 cm long, glabrous or sparsely hirsute. Calyx tube adnate to hypanthium up to middle, semiglobose, glabrous or rarely sparsely hairy; calyx lobes ovate-lanceolate, 5–9 mm long, 3–5 mm broad, glabrous or hispidulous at top, apex acute, nearly entire; sinus between calyx lobes narrow and acute. Corolla tubular-campanulate, shallowly lobed, 1.5–2.8 cm long,1–1.5 cm wide, greenish yellow or pale blue, nearly glabrous. Stamens glabrous; filaments ca. 3 mm long, slightly dilated at base; anthers ca. 3 mm long. Capsule 1–1.5 cm long, 0.6–1.2 cm in diameter. Seeds ellipsoid, ca. 1.5 mm long, 1 mm in diameter. Figure IV 1:25A.

Phenology: flowering in August and September; fruiting in October.

Habitats: grassy slopes at an altitude of 2600–3500 m.

Distribution: **endemic** to China: Taiwan (Map IV 1:25).

Additional specimens examined, **CHINA, TAIWAN, Jiayi:** Mt Alishan, 17. Jul. 1982, *C. E. Chang 16348* (TI); loc. eodem, on route from Tatachiaanpu to Paiyunshanchuang, 2700–3500 m, 09. Aug. 1989, *W. P. Leu & W. H. Hu 189* (A, HAST); Mt Yushan, from Tataga-ambu to Paiyunsunjuan, Ufonshan, Kai-shen, 2600 m, 04. Oct. 1985, *M. Minaki et al. 1031* (TI); Mt Yushan, 20. Jul. 1935, *K. Segawa* s. n.(TI); Mt Yushan, Tataka, 2600 m, 05. Sept. 1969, *T. Yamazaki & F. Yamazaki 936* (TI). **Nantou**: 2610 m, 27. Oct. 1992, *Kirkham & Flanagan (ETOT) 149* (K); Yushan National Park, 23°28′N, 120°57′E, 2700–2850 m, 06. Sept. 1991, *T. G. Lammers 8439* (A); loc. eodem, 3000–3100 m, 06. Sept. 1991. *T. G. Lammers 8445* (A, K, KUN, TI). **Taizhong**: Jenai Township, 24°09′58″N, 121°16′45″E, 3000 m, 08. Sept. 1998, *S. L. Kelley et al. 151–98* (A).

26. Codonopsis bomiensis D. Y. Hong, Plant Divers. Resour. 36(3): 285. fig. 1. 2014. Type : China, Tibet, Bomi, Guxiang, 3200 m, 30. Jul. 1965, *J. S. Ying & D. Y. Hong 650812* (holotype: PE; isotypes: PE).

Codonopsis rotundifolia Benth. var. *angustifolia* Nannf. ex Fletcher, Bot. Mag. 167: tab. 131. 1950. syn, nov. Type: China, Tibet, Mainling County, Doshong, 2900 m, 28. Jul. 1938, *F. Ludlow, G. Sherriff & G. Taylor 5377a* (holotype: BM).

Roots carrot - shaped, 10–35 cm long, 0.5– 3 cm in diameter. Caudex short, up to 1.5 cm long. Stems several from one caudex, twining, up to 2 m long, or less frequently ascending, glabrous or sparsely villose, branched, long branches 15– 30 cm long, fertile or sterile; short branches found above long branches and mostly fertile. Leaves alternate; petiole 5– 12 mm long; leaf blade ovate or

FIGURE IV 1:24A *Codonopsis hemisphaerica* P. C. Tsoong ex D. Y. Hong. **a**, root, caudex, and base of stems; **b**, a part of twining stem; **c**, a flower with corolla opened, showing calyx, stamens, disc, style, and stigma.

ovate-lanceolate, 2–12 cm long, 1–4.5 cm broad, adaxially sparsely hispidulous, abaxially sparsely or densely hirsute, base cuneate or obtuse, margin serrate, crenulate, rarely lobed or entire, apex acute or short - acuminate. Flowers terminal on main stem or also on branches; pedicel glabrous, up to 10 cm long. Calyx adnate to hypanthium up to middle, calyx lobes and corolla inserted at the same level; calyx tube semiglobose, glabrous; calyx lobes lanceolate

FIGURE IV 1:24B *Codonopsis hemisphaerica* D. C. Tsoong ex D. Y. Hong. From Sichuan: Baoxing, 06. Aug. 2011, by Yan YU.

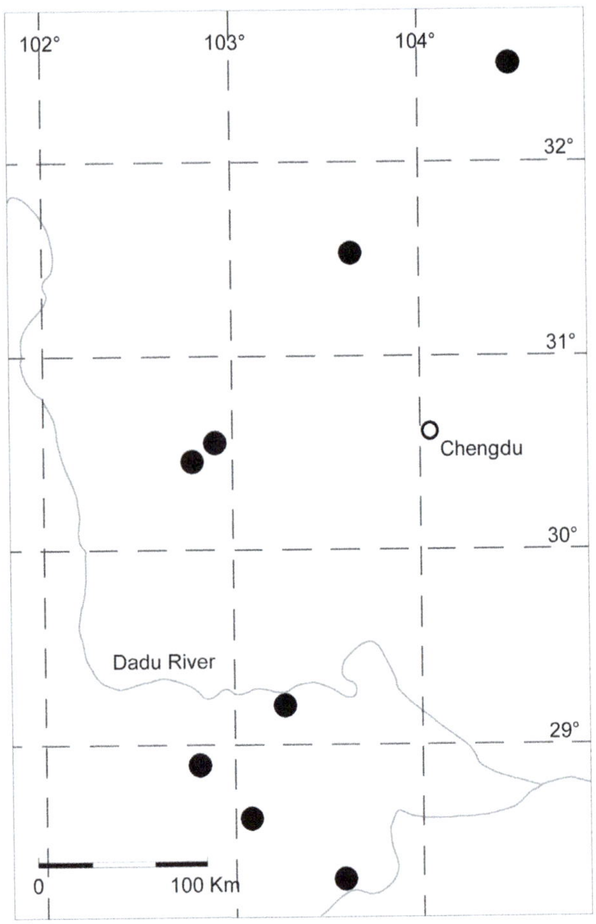

MAP IV 1:24 Distribution of *Codonopsis hemisphaerica* P. C. Tsoong ex D. Y. Hong.

to linear-lanceolate, 1.4–2 cm long, 0.5–1 cm broad, glabrous, apex acute or obtuse, entire or denticulate; sinus between lobes narrow and pointed. Corolla campanulate, pale yellow-green, with purple veins, 2– 3 cm long, 1.5–2 cm wide; corolla lobes deltoid, acute, 5–7 mm long, ciliate. Stamens glabrous; filaments ca. 3 mm long, slightly

FIGURE IV 1:25A *Codonopsis kawakamii* Hayata, redrawn from *Fl. China* 19, 2011. **a**, middle part of a twining stem; **b**, an undihescent capsule; **c**, a stamen; **4**, abaxial surface of a leaf.

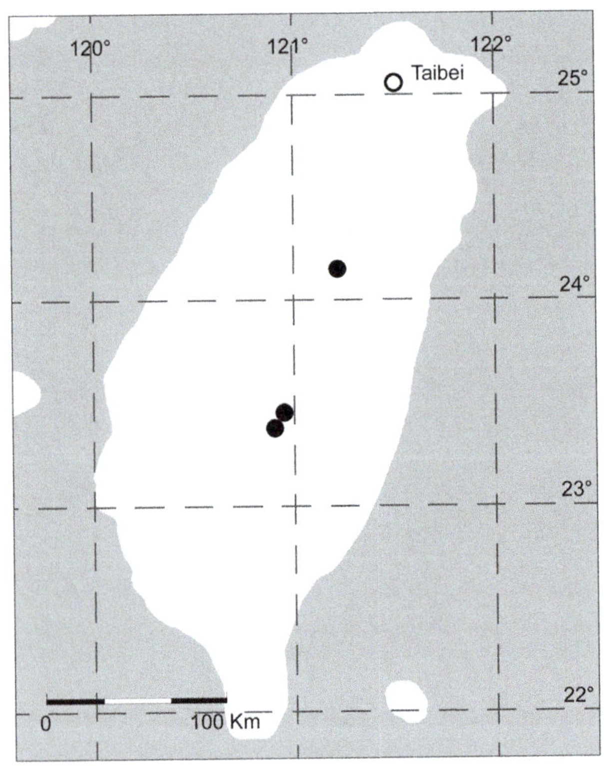

MAP IV 1:25 Distribution of *Codonopsis kawakamii* Hayata.

dilated at base; anthers ca. 3 mm long. Capsule hemispheric in inferior part, conical at superior part, ca. 1.5 cm long, 1.5 cm in diameter. Seeds ellipsoid, brown, ca. 0.8 mm long. Figures IV 1:26A & B.

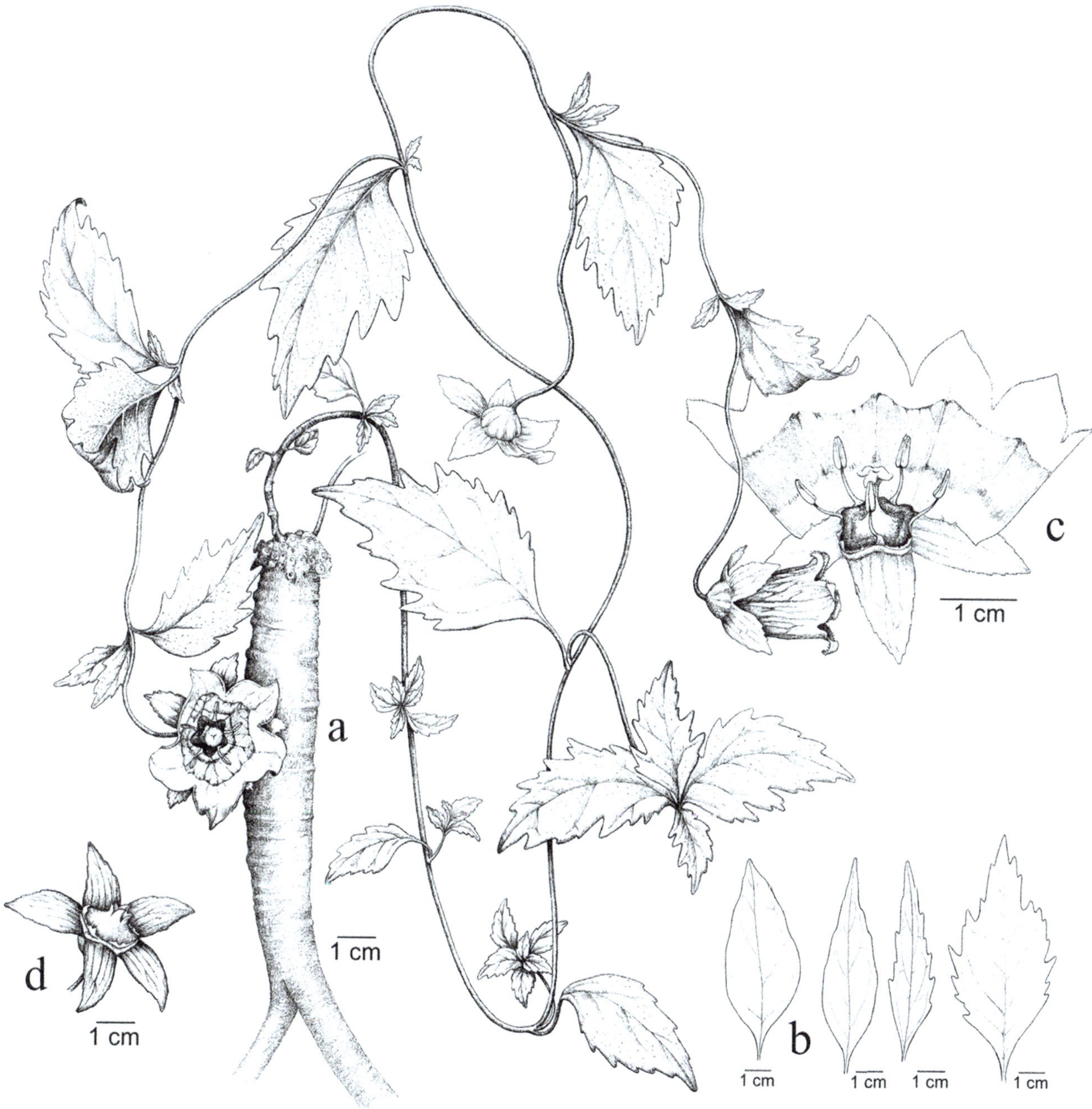

FIGURE IV 1:26A *Codonopsis bomiensis* D. Y. Hong. **a**, habit; **b**, different forms of leaves; **c**, a flower with corolla opened, showing calyx, stamens, disc, style, and stigma; **d**, a 3-loculicidal capsule.

Chromosomes: 2n = 16 (Gadella, 1966, under *C. rotundifolia* var. *angustifolia* Nannf. ex Fletcher).

Phenology: flowering in July and August; fruiting in September and October.

Habitats: forests or scrubs; 2700– 4000 m.

Distribution: **endemic** to China: SE Tibet (Map IV 1:26).

Notes: *Codonopsis bomiensis* is rather similar to *C. henryi* Oliv. and *C. levicalyx* L. T. Shen in leaves which are elliptic and rough-dentate. Partly for this reason, Hong et al., (2011) wrongly identified specimens in Tibet as *C. levicalyx*. Now we know that *C. levicalyx* is conspecific with *C. henryi*, and our new species differs distinctly from *C. henryi* in having the calyx lobes glabrous, not reflexed, and larger, 1.4 –2 cm long, and the corolla campanulate, 1.5–2 cm wide, and with ciliate lobes.

Additional Specimens examined, **CHINA, TIBET, Bomi**: Daxing, 3250 m, 03. Sept. 1982, *B. S. Li & S. Z. Cheng 550* (PE); Qomdo, 3300 m, edges of forests by river, 16. Jul. 1979, *R. F. Huang 778* (HNWP); Ri'along Gou, 2800 m, 13. Jul. 1979, *R. F. Huang 1017* (HNWP); vicinity of Tongmai, 1960 m, among herbs by fields, 21. Jul. 1965,

FIGURE IV 1:26B *Codonopsis bomiensis* D. Y. Hong. **a–f**, from Tibet: Mainling, 04. Sept. 2011, by Xin-Tang MA & De-Yuan HONG. **d**, a flower longitudinally sectioned, showing disc (arrow), stamens, style, and inferior ovary.

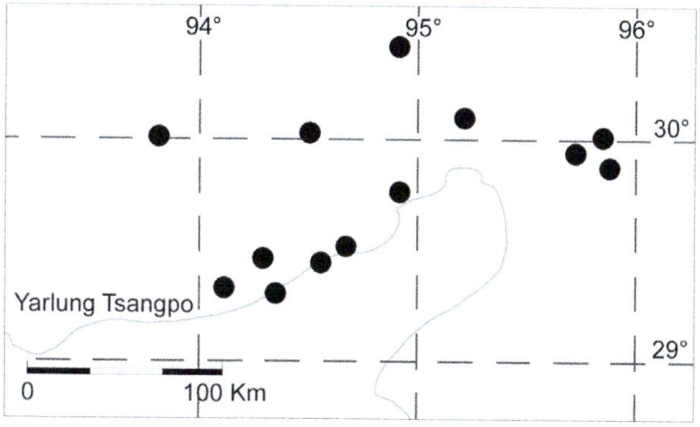

MAP IV 1:26 Distribution of *Codonopsis bomiensis* D. Y. Hong.

P. G. Xiao et al. 2251 (KUN); Yi'ong, 3100–3400 m, 17. Aug. 1983, *B. S. Li et al. 6603* (PE); Yi'ong, Yi'ong Farm, 3500 m, meadows with shrubs, 18. Aug. 1983, *XZ Plant Resources Exped. 3621* (XZ); Yuren, 3100 m, 03. Sept. 1973, *Qinghai-Xizang Exped. 73-1359* (KUN, PE); loc. eodem, 3200 m, 02. Sept. 1973, *Acad. Sin. Qinghai-Gansu Exped. 1050* (PE); Zamu, 2700 m, 13. Jul. 1965, *Y. T. Chang & K. Y. Lang 653* (PE); loc. eodem, 2700 m, herbs by crop fields, 13. Jul. 1965, *P. G. Xiao & G. C. Xia 2183* (KUN); loc. eodem, 2900 m, thickets, 04. Sept. 1983, *XZ Plant Resources Exped. 3972* (XZ); s. loc., 3200 m, 18. Aug. 1960, *G. C. Xia & T. K. Mi 740* (KUN). **Mainling**: Bagar, 3200 m, 25. Aug. 1974, *Qinghai-Xizang Exped. 74-1947* (KUN, PE); Bagar, Nangou, 3070–3300 m, 28. Jul. 1975, *Qinghai-Xizang Suppl. Exped. 750962* (PE); Gongga Harbor, 3200 m, 22. Jul. 1972, *Xizang Chinese Mater. Medic. & Herbs Exped: 4155* (HNWP, PE); Gyala, 29°43′N, 94°56′E, 2835 m, 23. Jul. 1938, *F. Ludlow, G. Sherriff & G. Taylor 5377* (paratype of *C. rotundifolia* var. *angustifolia*: BM); Hongwei Lumbering Yard, 3100 m, 26. Aug. 1974, *Qinghai-Xizang Exped. 74-1954* (KUN, PE); loc. eodem, 26. Aug. 1974, *Qinghai-Xizang Exped. 74-1962* (KUN, PE); loc. eodem, 3100–3200 m, 29. Jul. 1975, *Qinghai-Xizang Suppl. Exped. 751012* (HNWP, KUN, PE); near Jiage, 3100 m, 07. Aug. 1982, *C. C. Ni et al. 2803* (PE, XZ); loc. eodem, 10. Jul. 1972, *Xizang Chinese Mater. Medic. & Herbs Exped. 3635* (HNWP, PE); Lusha, 29°30′ N, 94°30′ E, 2900 m, 01. Oct. 1947, *F. Ludlow et al. 13292* (BM); Lusha, 29°25′N, 94°30′E, 18. Sept. 1938, *F. Ludlow, G. Sherriff & G. Taylor 5377b* (paratype of *C. rotundifolia* var. *angustifolia*: BM); near Mainling Town, 3100 m, 15. Aug. 1975, *Qinghai-Xizang Suppl. Exped.751397* (HNWP, KUN, PE); loc. eodem, 3000 m, 13. Aug. 1982, *C. C. Ni et al. 2924* (PE, XZ); Nonghou La, Babu, 3100 m, 29. Jul. 1974, *Qinghai-Xizang Exped. 74-3713* (KUN, PE); Paiqu, Doxong Shan, 3200 m, forests, 08. Aug. 1973, *HNWP Xizang Exped. 1882* (HNWP); Paiqu, Kaigar-Lamula, 3300–3400 m, 15. Sept. 1989, *R. F. Huang CG-89-517* (HNWP); 3500 m, 29. Jul. 1983, *XZ Plant Resources Exped. 3420* (XZ); near Paka, Kulu Phu Chu, 29°15′ N, 94°25′ E, 29. Jul. 1938, *F. Ludlow et al. 5999* (BM); Temo La, 3050–3960 m, 27. Jul. 1935, *F. Kingdon-Ward 12059* (BM); Tsangbo Valley, 2740–3050 m, 28. Jul. 1924, *F. Kingdon-Ward 6030* (BM, E). **Nyingchi**: Hougouqiao, 3400–3600 m, 08. Aug. 1983, *B. S. Li et al. 06237* (PE); Jomugou, 3500 m, thickets at edges of *Abies* forests, 09. Aug. 1983, *XZ Plant Resources Exped. 3523* (XZ); Lunang, 3050 m, among herbs, 26. Jul. 1965, *P. G. Xiao et al. 2308* (KUN); near Naixi Pharmaceutical Factory, 3200 m, mixed forests, 04. Aug. 1965, *P. G. Xiao & G. C. Xia 2431* (KUN); loc. eodem, 3200 m, 29. Jul. 1965, *P. G. Xiao et al. 2338* (KUN); loc. eodem, 3200 m, among herbs, 28. Jul. 1965, *P. G. Xiao et al. 2311* (KUN); Naixi, 3100 m, 02. Aug. 1965, *Y. T. Chang & K. Y. Lang 1232* (PE); loc. eodem, 31. Jul. 1965, *Y. T. Chang & K. Y. Lang 1187* (KUN, PE); Nambu La, 30°0′ N, 94°30′ E, 3810 m, 01. Sept. 1938, *F. Ludlow et al. 6971* (BM) ; SE of Nyingchi, Hou Shan, 3100 m, 14. Oct. 1960, *G. X. Fu 696* (PE); Shoga Dzong, 30°0′ N, 93°45′ E, 3200 m, *F. Ludlow et al. 14156* (BM).

27. Codonopsis deltoidea Chipp, J. Linn. Soc. Bot. 38:387. 1908; Tsoong, Contr. Inst. Bot. Nat. Acad. Peiping 3(3): 94. fig. 8B. 1935; Anonymous, Icon. Cormophyt. Sin. 4: 379. fig. 6172. 1975; Shen & Hong in Hong, Fl. Reipubl. Popularis Sin. 73(2): 46. 1983. Type: "China: Mt Omi (Sichuan, Emei)", Aug. 1904, *E. H. Wilson 5035* [lectotype here designated: K (K000575333); isolectotypes: BM, K (K000575334 & K000575335), PE]; loc. eodem, 1980 m, Oct. 1903, *E. H. Wilson 3988* (paratypes: K, PE)

Roots carrot-shaped or cylindrical, more or less branched, 15–30 cm long, 1–1.5 cm in diameter. Stems twining, over 1 m long, sparsely villous, gradually glabrescent, branches and branchlets both leafy, sterile or with a terminal flower. Leaves on main stem alternate; petiole 1.5–8 cm long, villous or hirsute; leaf blade deltoid-ovate or broad-ovate, 3–9.5 cm long, 4–10 cm broad, sparsely puberulent or hirsute, adaxially green, abaxially gray-green, sparsely puberulent, hirsute along veins, base shallowly cordate, truncate, rounded, or broad-cuneate, margin dentate, apex acuminate; leaves on branches opposite, but petiole shorter, leaf blade ovate to narrowly ovate. Flowers solitary or sometimes in cymes, terminal on main stem, major branches and branchlets; pedicel less than 2 cm long, villous. Calyx tube adnate to whole hypanthium, semiglobose, glabrous or villous along ribs; calyx lobes ovate, 1.1–1.3 cm long, 0.5–0.7 cm broad, apex acute, margin verrucose-serrate, often white villous-ciliate. Corolla broad-tubular, 2–2.5 cm long, 1–1.5 cm wide, pale yellow-green, with purple veins, rarely white; corolla lobes 1/5–1/4 as long as corolla length, deltoid. Filaments ca. 6 mm long, slightly dilated at base, glabrous; anthers ca. 3 mm long. Capsule semiglobose below, short-conical above, 1.2–1.4 cm long, 0.8–1 cm in diameter, fruiting calyx lobes 1.4–1.9 cm long, 0.8–1.2 cm broad. Seeds ovoid, smooth, yellow-white. Figures IV 1:27A & B.

FIGURE IV 1:27A *Codonopsis deltoidea* Chipp. **a**, root and caudex; **b**, upper part of a twining stem; **c**, a flower with corolla opened, showing calyx, stamens, disc, style, and stigma; **d**, a capsule.

Phenology: flowering and fruiting from July to October.

Habitats: forests, margins of forests, or scrubs; 1000–2700 m.

Distibution: **endemic** to China and confined to W Sichuan (Map IV 1:27).

Additional specimens examined, **CHINA, SICHUAN, Ebian**: Jinhe, sunny slopes, 22. May, 1959, *Sichuan Econ. Plant Surv.* (*Leshan 59*) *1593* (PE). **Emei** (Emei Shan)**:** Baiyanwan, 2050 m, 17. Aug. 1956, *Y. H. Tao 51549* (SZ); Chudian, 2000 m, edges of forests, 26. Jun. 1960, *Sichuan*

FIGURE IV 1:27B *Codonopsis deltoidea* Chipp. **a–c**, from Sichuan: Hongxi, 19. Sept. 2012, by Qiang WANG; **d**, from the same locality, 06. Jul. 2012, by Qiang WANG.

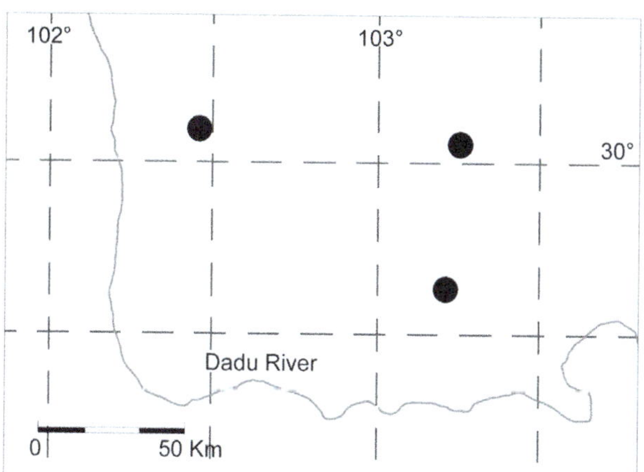

MAP IV 1:27 Distribution of *Codonopsis deltoidea* Chipp.

Medic. Plant Surv. 12409 (PE); Dachengsi, 2400 m, 29. Sept. 1958, *S. Y. Chen et al. 6085* (SM, SZ); Dachengsi, 26. Aug. 1939, *Z. W. Rao 4764* (PE); loc. eodem, 13. Aug. 1935, *T. H. Tu 586* (PE); Huayanding, 2000 m, forests, 01. Aug. 1971, *236 Sichuan Team 0632* (PE); Leidongping, 2400 m, 22. Aug. 1956, *G. H. Yang 51669* (SZ); near Leidongping, 29. Jul. 1952, *H. L. Tsiang & S. S. Chang 31826* (PE, SZ); loc. eodem, 2400 m, 06. Jul. 2012, *Q. Wang W12023* (PE); Leidong-si, 27. Aug. 1957, *S. Y. Chen et al. 4215* (SM, SZ); Luohanpo, 08. Apr. 1952, *W. P. Fang 21144* (SZ); Sandaohe, 30. Sept. 1959, *G. H. Yang 56286* (PE, SZ); Zuantianpo, 1000 m, 19. Jul. 1931, *F. T. Wang 23410* (KUN, PE); loc. eodem, 2100 m, 01. Aug. 1957, *G. H. Yang 56333* (KUN, SZ); loc. eodem, forests, 04. Aug. 1951, *W. P. Fang 20737* (SZ); loc. eodem, 22. Jul. 1952,

H. L. Tsiang & S. S. Chang 31735 (PE, SZ); s. loc., 2000 m, 31. Jul. 1938, *C. H. Wu 100* (PE); s. loc., 2590–2740 m, 1928, 12. Aug. 1928, *W. P. Fang 2855* (PE); s. loc., 04. Aug. 1938, *W. P. Fang 12953* (PE); s. loc., 2400 m, 13. Aug. 1940, *T. C. Lee 3205* (KUN). **Hongya**: s. loc., 2012, *Q. Wang H12086* (PE).

28. Codonopsis rotundifolia Benth. in Royle, Ill. Bot. Himal. 254. t. 62. 1834; Hooker, Bot. Mag. tab. 4942. 1856; Hooker & Thomson, J. Proc. Linn. Soc. Bot. 2: 14. 1857; Clarke in Hooker, Fl. Brit. Ind. 3: 432. 1881; Komarov, Trudy Imp. S. -Peterburgsk. Bot. Sada 29(1): 107. 1908; Chipp, J. Linn. Soc. Bot. 38: 382. 1908; Anthony, Notes Roy. Bot. Gard. Edinburgh 15: 180. 1926; Hong, Novon 20: 423. 2010; Hong, Lammers & Klein in Wu, Raven & Hong, Fl. China 19: 517. 2011. ≡*Wahlenbergia rotundifolia* (Benth.) A. DC. in de Candolle, Prodr. 7: 425. 1839. ≡*Glossocomia rotundifolia* (Benth.) Rupr., Bull. Cl. Phys. - Math. Acad. Imp. Sci. Saint - Pétersbourg 15: 210. 1857. Type: Bentham's tab. cited.

Codonopsis rotundifolia Benth. var. *grandiflora* Hook. f., Bot. Mag. 13: tab. 5018. 1857. Type: Hooker's tab. cited.

Codonopsis lurida Lindl., Edwards's Bot. Reg. 25 (Misc.): 82. 1839. ≡*Glossocomia lurida* (Lindl.) Lindl., Edwards' Bot. Reg. 28: pl. 3. 1842. ≡*Wahlenbergia lurida* (Lindl.) G. Manetti, Cat. Pl. Hort. Medic.: 106. 1842. Type: Lindley's plate cited.

Codonopsis nepalensis Hara, J. Jap. Bot. 53(5):138. fig.2. 1978. syn. nov. Type: C Nepal, Seti Khola, Annapurna Himal., 3200 m, 05. Aug. 1954. *Stainton, Sykes & Williams 6687* (holotype: BM; isotype: E).

Codonopsis longifolia D. Y. Hong, Acta Phytotax. Sin. 18(2): 245. fig. 1.4. 1980. Type: China, S Tibet, Nyalam, near Quxam, 3600–3700 m, 27. Aug. 1972, *Xizang Chinese Mater. Medic. & Herbs Exped. 1563* (holotype: PE).

Roots carrot-shaped, ca. 18 cm long, 1 cm in diameter. Stems over 1 m long, twining, branched, glabrous or sparsely villous. Leaves alternate; petiole 0.5–2.5 cm long, sparsely hirsute or glabrous; leaf blade from deltoid-orbicular to ovate-lanceolate, 3–9 cm long, 2–6 cm broad, both surfaces nearly glabrous or sparsely hirsute along veins, base mostly cordate, less frequently truncate or rounded, margin dentate or entire, apex short-acuminate to obtuse. Flowers nearly leaf-opposed, forming a cyme, or solitary and terminal; pedicel glabrous or villous, 2–20 cm long. Calyx adnate to hypanthium up to middle; calyx tube 10-ribbed, glabrous; calyx lobes broad-elliptic or ovate-lanceolate, 1.5–3 cm long, 0.5–1.5 cm broad, upright, both surfaces glabrous or hispidulous, margin minute-serrate to entire, glabrous or ciliate. Corolla campanulate, 1.8–3 cm long, lobed for 1/3 to 2/5, green, greenish white, pale blue or pinkish white, light purple, purple inside, glabrous. Filaments glabrous, 4 mm long, slightly dilated at base; anthers 3 mm long, glabrous. Capsule globose in inferior part, conical in superior part, 1.5–2.5 cm long, 1.2–2 cm in diameter. Seeds oblong, brown, ca. 1.5 mm long. Figure IV 1:28A & B.

FIGURE IV 1:28A *Codonopsis rotundifolia* Benth. **a**, upper part of a twining stem and a 3-loculicidal capsule; **b**, upper part of a flower with corolla opened, showing stamens, disc, style, and stigma.

FIGURE IV 1:28B *Codonopsis rotundifolia* Benth. From Tibet: Gyirong, 12. Aug. 2010, by Yan YU.

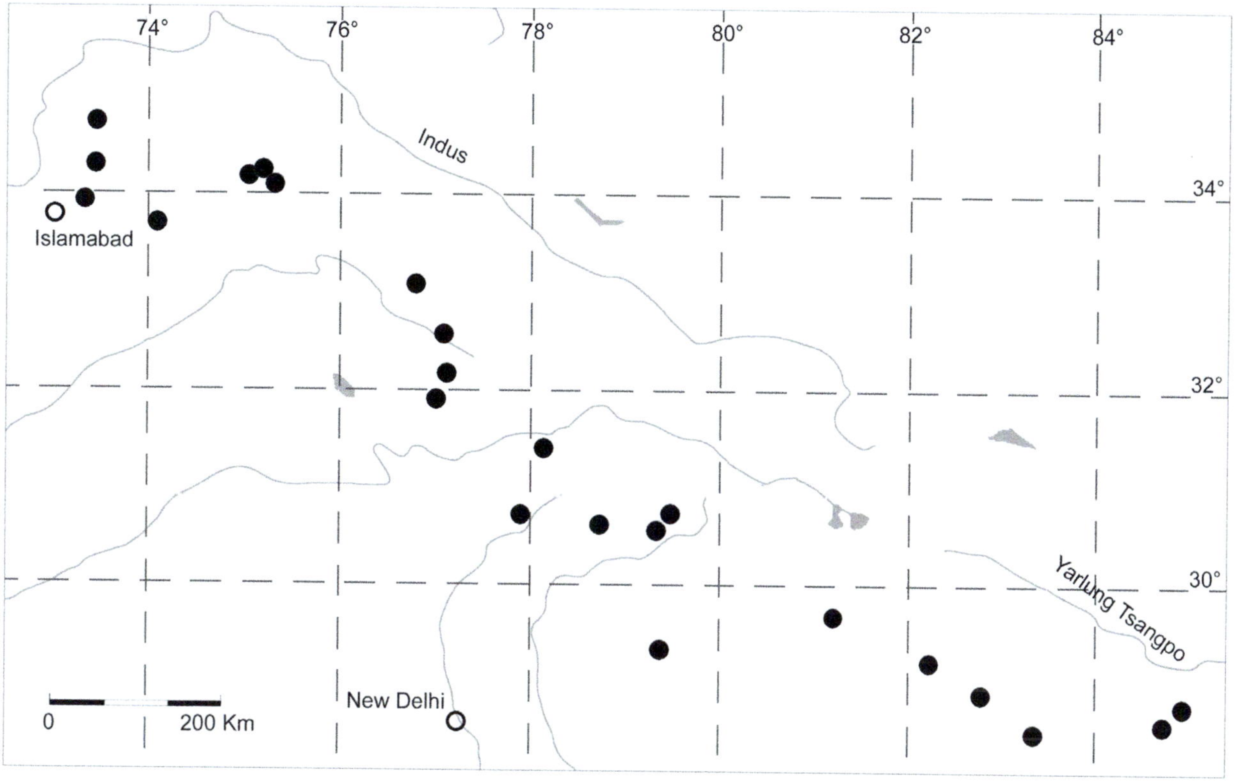

MAP IV 1:28 Distribution of *Codonopsis rotundifolia* Benth.

Chromosomes: 2n=16 (Lammers, 2007a).

Phenology: flowering from July to September; fruiting in August and September.

Habitats: meadows, scrubs, or forests; 1830–3800 m.

Distribution: China (S Tibet), NW India, Nepal and N Pakistan (Map IV 1:28).

Notes: *Codonopsis rotundifolia* Benth. was originally described from W Himalaya (Kashmir and Kumaon) (Clarke, 1881), with stems and leaves usually more or less villous; leaf blade cordate to rounded at the base, and calyx lobes most frequently ciliate. Hara (1978) stated that his plant, *C. nepalensis*, is closely allied to *C. rotundifolia* but differed in having "glabrous stems, petioles and peduncles, acutish entire glabrescent leaves, and glabrous acutish calyx-lobes". For nearly the same reason the present author (Hong, 1980) described *C. longifolia* as new unaware of Hara's publication. Overall examination of a large number of specimens in BM, CAL, E, G, K, KATH, P, PE and W showed that the indumentum in stems, leaves and calyx lobes, and the shape of leaf base varied gradually from the west (N Pakistan) to the east (Nepal and S Tibet). Therefore, I consider it unjustified to divide this group into different species or even subspecies.

Additional specimens examined, **INDIA**, **Himachal Pradesh**: Bashahr State, above Barua, 09. Jul. 1890, *J. H. Lace 377* (CAL, E); Chamba, Tarlok Nath, 2743 m, 10. Aug. 1930, *Koelz 1078* (P); Upper Chenab Valley, 2440 m, 07. Jul. 1879, *R. Ellis 375* (CAL, K); upper Chenab, Dida stream, 2440 m, Aug. 1879, *B. H. Baden-Powell 216* (CAL); upper Chenab, 1880, *R. Ellis 639* (K); Jalauri Pass, Kulu, 1830–2130 m, *Edgeworth 5043* (K); Kulu, 2130 m, 03. Oct. 1887, *G. Watt 9684* (E); Lahul, Gondla, 27. Jul. 1941, *N. L. Bor 9943* (K); Manali, 3050 m, 02. Aug. 1941, *N. L. Bor 15583* (K); Pangi, Gindi, left bank of Chenab River, near Soar, 2740 m, 23. Sept. 1878, *s. coll.256 H* (E); Pangi, Kilar, W of Darwas, 2590 m, 23. Aug. 1897, *J. H. Lace 1643* (E); Pangi, *Stoliczka* s. n.(CAL); Pangi Valley, 1880, *R. Ellis* s. n.(K); loc. eodem, 2750 m, 31. Jul. 1919, *s. coll.* s. n.(K); loc. eodem, 2440 m, 02. Aug. 1899, *Harsukh* s. n.(K); loc. eodem, 2740 m, 10. Aug. 1878, *s. coll 166 E* (E); Sangla, Baspa Valley, Shimla Hill State, 31°25′ N, 78°15′ E, 3660 m, 27. Jul. 1939, *G. Sherriff 7485* (BM, E); Shimla Hills, 06. Aug. 1885, *J. R. Drummond 22628* (E, K); s. loc., 28. Jul. 1888, *J. R. Drummond 22507* (E, K). **Jammu-Kashmir**: Nara Wag, Wangal Valley, 11. Aug. 1940, *P. W. Piufold 248* (BM); Pahalgam, Aug. 1913, *M. A. Evershed* s. n.(BM); loc. eodem, 2290 m, 06. Sept. 1876, *C. B. Clarke 31126* (BM); loc. eodem, 2440 m, 21. Jul. 1876, *C. B. Clarke 29356* (BM); loc. eodem, 2440 m, 05. Sept. 1876, *C. B. Clarke 31113* (K); loc. eodem, Jul. 1925, *R. R. Stewart 7878* (KUH); Poonch, 2130 m, 25. Aug. 1955, *Jan Mohd J11* (BM); Sonamarg, 2590 m, 11. Aug. 1940, *F. Ludlow & G. Sherriff 7903* (BM); Sonamarg, 3200 m, 29. Aug. 1876, *C. B. Clarke 30827* (K); Tangmarg, 1830 m, 16. Aug. 1956, *O. Polunin 56/348* (B, BM, E,); Mt. Trunkhal, summer, 1969,

O. Polunin 9633 (BM); s. loc., Sept. 1876, *C. B. Clarke* s. n.(CAL). **Punjab**: Kunawar, 1886, *J. R. Drummond 22290* (E. K); s. loc., 1885, *J. R. Drummond 22466* (E, K); s. loc. 1922, *J. R. Drummond 25269* (E, K, P). **Uttar Pradesh**: Deoban, 2740 m, Sept. 1898, *J. S. Gamble 27239* (K); Deoban, 2440 m, Jul. 1890, *Keshava Nand & Forest Ranger 268* (E); from Gangharea to Flowers Valley, between Josimath and Badrinath, 3300 m, 15. Aug, 1975, *F. Billiet & J. Leonard 6565* (E, K); Nainital to Srinagar, Baspa Valley, 2740 m, 15. Sept. 1876, *G. E. Benham 31328* (BM); Tehri, Gunges Valley, near Derali, 2440–2740 m, 06. Aug. 1883, *J. F. Duthie 609 (b)* (G); s. loc., *Falconer* s. n.(K).

NEPAL, **Dhawalagiri**: Baglung Distr., Dogadi Khola, 3660 m, 08. Aug. 1954, *Stainton, Sykes & Williams 3797* (BM, CAL, PE); Mustang Distr., Larjung (S of Tukuche), 23. Jul. 1954, *Polunin, Sykes & Williams 1975* (BM, E); Phagure Dhuri: 3810 m, 07. Jul. 1954, *Polunin, Sykes & Williams 3416* (BM, E); Taglung (S of Tukuche), 3660 m, 12. Jul. 1954, *Polunin, Sykes & Williams 1764* (E). **Gandaki**: Gorkha Distr., 28°34′ N, 84°41′ E, 3320 m, 05. Aug. 1994, *M. Suzuki et al. 9485289* (E, KATH). **Karnali**: Bamara, 2750 m, 08. Aug. 1981, *N. P. Manandhar & D. P. Joshi 6879* (KATH); Dolpa Distr., Tarakot, Bheri River, ca. 3350 m, 13. Jul. 1952, *Polunin, Sykes & Williams 2440* (BM, E); Jumla Distr., 2440 m, 01. Jul. 1952, *Polunin, Sykes & Williams 4401* (BM, E, G); Tarakot, 28°50′ N, 82°50′ E, 3000 m, 19. Jul. 1973, S. *Einarsson et al. 2565* (BM); loc. eodem, 3800 m, 19. Aug. 1973, *S. Einarsson et al. 3655* (BM). **Seti**: Bajhang Distr., Thogundanda, 3000 m, 27. Aug. 1972, *M. S. Bista & D. P. Joshi 728* (KATH); loc. eodem, Tapaun-Dhansera, 3740 m, 25, Jul. 1984, *P. R. Shakya et al. 8180* (KATH). **Khaptad National Park**: 3000 m, 03. Aug. 1972, *M. S. Bista & D. P. Joshi 241* (KATH). **Maharigaon**, 3200 m, 15. Jul. 1952, *Polunin, Sykes & Williams 191* (BM).

PAKISTAN, **Baltistan**: Azad, Sonard Neelum Valley, 18. Aug. 1999, *U. H. Zargar786445* (KUH); Baltal, 2740–3050 m, Jul. 1912–1913, *R. R. Stewart* s. n. (KUH); Kishenganga Valley and road to Nanga Parbat via Gangabal Lake, 1950 m, 22. Jul. 1939, *I. D. Stewart 17810* (PE); loc. eodem, Kerau to Reshua, 1680 m, 19. Jul. 1939, *R. R. & L. D. Stewart 17663* (KUH); Surgun, 3200–4000 m, 28. Jul. 1953, *F. Schmid 540* (G). **Hazára**: Bhonja, Kaghan, 18. Jul. 1899, *Inayat* s. n.(CAL); Kagan Valley, Shogran, Jul. 1960, *S. M. A. Kazmi* s. n. (KUH); Murree Hills, Aug. 1907, *R. R. Stewart* s. n.(K); s. loc., *H. A. Deane* s. n.(CAL).

29. Codonopsis reflexa D. Y. Hong, Plant Divers. Resours. 36(3): 289. fig.3. 2014. Type: Nepal, near Dogadi Khola (Dhawalagiri Zone, Baglung Distr.), 3960 m, 11. Aug. 1954, *Stainton, Sykes & Williams 3822* (holotype: PE; isotypes: BM, G).

Plants glabrous throughout except for filaments and style. Roots unknown. Stems twining, over 1 m long, branched at upper part. Leaves all alternate; petiole 1.5–4 cm long; leaf blade ovate-lanceolate, 2.5–7.5 cm long, 1.2–4 cm broad, base cordate, margin serrate or crenate, apex acute or obtuse; those on branches much smaller and rounded at base. Flowers terminal on main stem and major branches. Calyx adnate to whole hypanthium; calyx tube obdeltoid; calyx lobes ovate-lanceolate, 1.4–1.6 cm long, 0.6–0.8 cm broad, broadest below middle, apex acuminate, strongly reflexed. Corolla funnel-shaped, 2.1–4.7 cm long, about 2–4 cm wide, deeply divided to over 3/5 of length, dark blue with a reddish ring near base; corolla lobes ovate-lanceolate, 1.3–2.9 cm long, apex acute. Filaments ca. 5 mm long, dilated at base, dilated part villose-ciliate; anthers 6–7 mm long. Stigma 3 or 2-fid; stigma lobes broad-elliptic. Fruit not seen. Figure IV 1:29A.

Phenology: flowering in August and September.
Habitats: climbing upon shrubs and tall herbs; ca. 3960 m.
Distribution: **endemic** to Central Nepal (Map IV 1:29).

Notes: This newly described species is rather similar to *Codonopsis rotundifolia* in habit with twining stems, long petiole, and large, ovate or ovate-lanceolate leaf blade. However, it differs distinctly from that species in having the corolla deeply divided, to over 3/5 of overall length, the calyx lobes reflexed, and filaments that have the dilated part villose-ciliate.

Additional specimen examined, **NEPAL**, **Dhawalagiri**, Mustang Distr., Kali Kandaki Tal (SW Tukuche), 3500 m, 02. Sept. 1967, *H. J. Lange 61* (B, paratype).

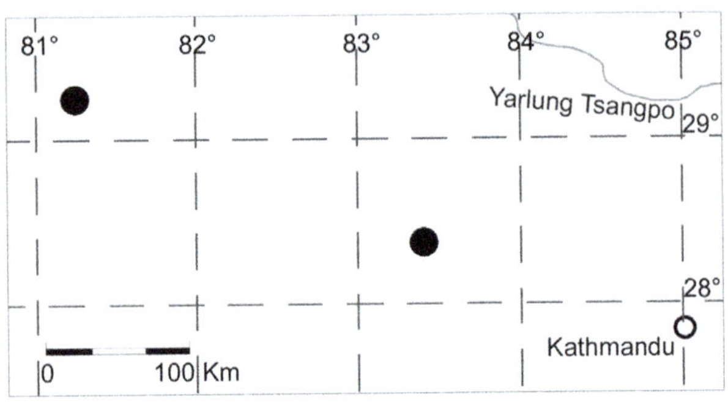

MAP IV 1:29 Distribution of *Codonopsis reflexa* D. Y. Hong.

30. Codonopsis henryi Oliv., Hooker's Icon. Pl. 20: pl. 1967. 1891; Komarov, Trudy Imp. S. -Peterburgsk. Bot. Sada 29(1): 109. 1908; Chipp, J. Linn. Soc. Bot. 38: 386. 1908; Shen & Hong in Hong, Fl. Reipubl. Popularis Sin. 73(2): 47. 1983. Type: "China, Prov. Hupeh (Hubei), Fang Distr. (Fangxian CO)", 1889, *A. Henry 6651* [lectotype here designated: K (K000575336); isolectotypes: BM, CAL, K (K000575337), LE, P, PE]

Codonopsis levicalyx L. T. Shen, Acta Phytotax. Sin. 13(3): 55. pl. 5. fig. 11a & b. 1975; Shen & Hong in Hong, Fl. Reipubl. Popularis Sin. 73(2): 46. 1983; Hong, Novon

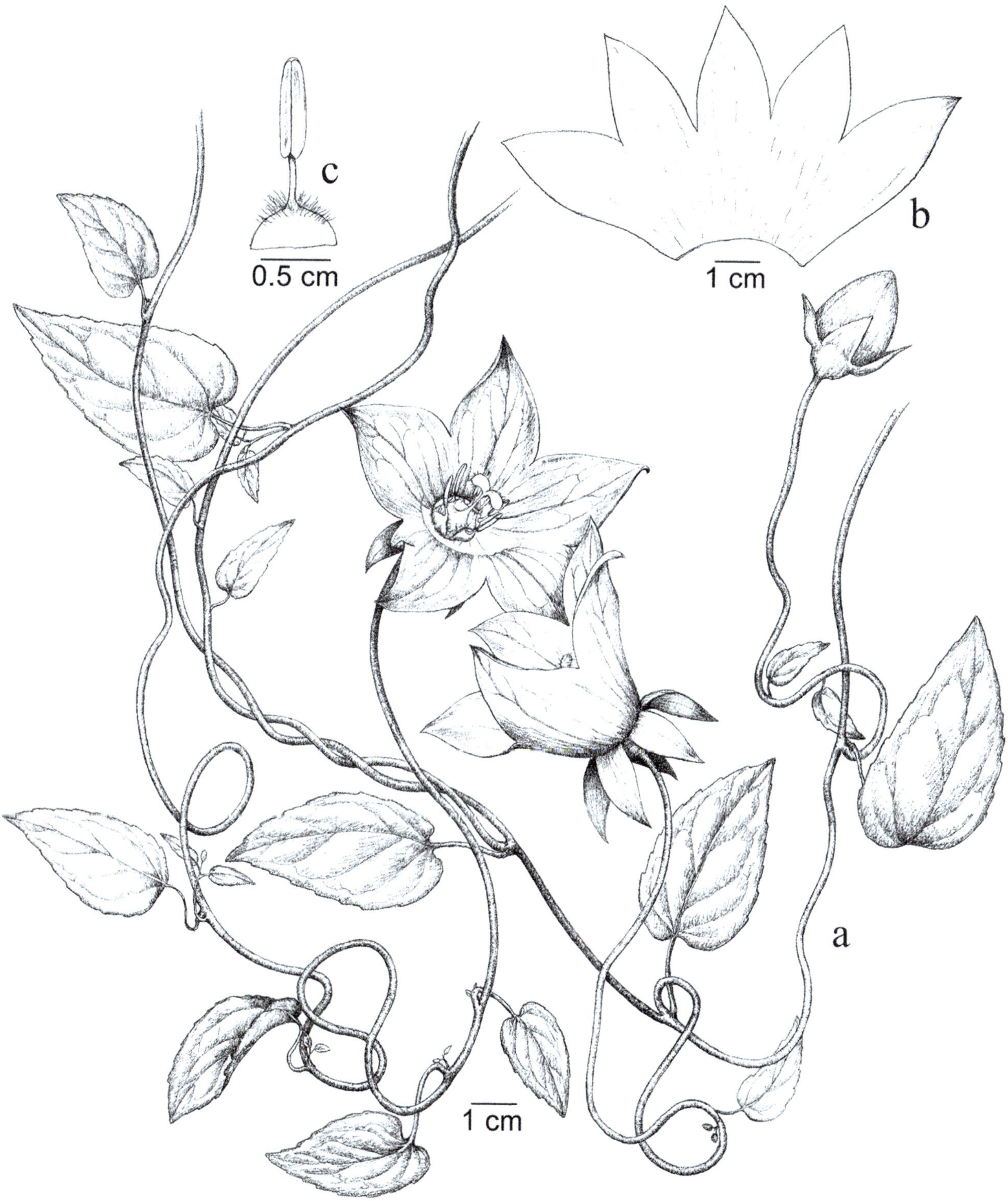

FIGURE IV 1:29A *Codonopsis reflexa* D. Y. Hong. **a**, parts of twining stems; **b**, a corolla opened; **c**, a stamen.

20: 422. 2010; Hong, Lammers & Klein in Wu, Raven & Hong, Fl. China 19: 518. 2011. syn. nov. Type: China, Sichuan, Nanping County, Lingjiang, Xingfu, Baowujia, 2900 m, 13. Oct. 1964, *X. Y. Tang, J. Z. Wu et al.* (N) *64017* (not *6407*) (holotype: PE).

Codonopsis levicalyx L. T. Shen var. *hirsuticalyx* L. T. Shen in Hong, Fl. Reipubl. Popularis Sin. 73(2): 183 (addenda). 1983. syn. nov. TYPE: China, Sichuan, Emei Shan, *M. L. Tu & Y. L. Zeng (Emei) 51-01* (holotype: WCU, n. v.).

Roots carrot-shaped. Stems over 1 m long, twining, branches less than 3 cm long, with 2–4 leaves, sterile or with a terminal flower, sparsely puberulent or glabrescent at internodes. Leaves on main stem alternate, rarely opposite, those on branches nearly opposite; petiole 0.5–2 cm long, puberulent; leaf blade ovate-lanceolate or lanceolate, 3–16 cm long, 1–7 cm broad, adaxially green, puberulent, abaxially gray-green, appressed-puberulent, base decurrent or cuneate, margin conspicuously dentate, apex acuminate. Flowers solitary, terminal on branches; pedicel ca. 1 cm long, puberulent. Calyx tube adnate to hypanthium up to middle, semiglobose, puberulent; calyx lobes reflexed from anthesis, triangular, 6–9 mm long, 3–4 mm broad, apex acute, puberulent and ciliate; sinus between calyx lobes broad and obtuse. Corolla campanulate, or more or less tubular-campanulate, 1.5–3 cm long, 1–1.2 cm wide; corolla lobes triangular, glabrous. Stamens glabrous; filaments ca. 7 mm long, slightly dilated at base; anthers ca. 5 mm long. Capsule ca. 1.5 cm in diameter. Figure IV 1:30A.

Phenology: flowering and fruiting from July to September.

Habitats: grassy slopes or thickets; 2300–3800 m.

Distribution: **endemic** to China: Gansu, W Hubei, and W Sichuan (Map IV 1:30).

Notes: *Codonopsis henryi* Oliv. is a rare species, whose diagnosis has not been well investigated, even in the *Flora of China* (Hong et al., 2011). I reexamined recently the holotypes of *C. henryi* and *C. levicalyx* L. T. Shen and found that the holotype of the latter is actually *Xin-Yue TANG & Zi-Jin WU 64017*, but not *Xin-Yue TANG & Zi-Jie WU 6407*. Therefore, two mistakes were made in the protologue: the number of collection and the name of one of the collectors. In addition, the accurate description of the type *Tang & Wu 64017* (PE) should be: leaves serrate at margin, puberulent on both surfaces; flowers terminal on short branches; calyx lobes reflexed or not, sparsely hispidulous. It should be an element of *C. henryi*.

Additional specimens examined, **CHINA**, **GANSU**, s. loc., 1930, *W. Y. Hsia 8374* (PE). **HUBEI**, s. loc., 1885–1888, *A. Henry 6651A* (E, G). **SICHUAN**, **Danba**, Dongnai Township, 2800 m, *Sichuan Medic. Plant Resources Surv. 20518* (paratype of *C. levicalyx*: WCU). **Lixian**: Luoyonggou, 3800 m, 13. Sept. 1964. *T. L. Dai et al. 64-11* (PE) ; s. loc., 25. Jul. 1958, *C. L. Wu 33187* (PE); s. loc., 20. Sept. 1952, *Z. He & Z. L. Zhou 14139* (SZ). **Luding**: Erlang Shan, 2350 m, 22. Sept. 1953, *W. K. Hu 37900* (paratype of *C. levicalyx* var *hirsuticalyx*: PE). **Wenchuan**: s. loc., Aug. 1961, *X. Y. Tang & S. J. Yue 63-01* (PE).

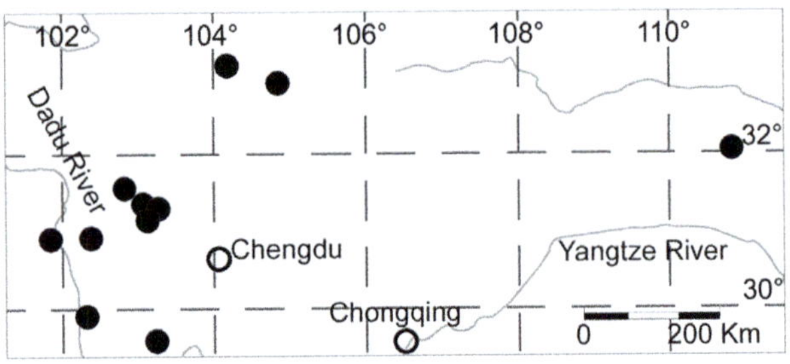

MAP IV 1:30 Distribution of *Codonopsis henryi* Oliv.

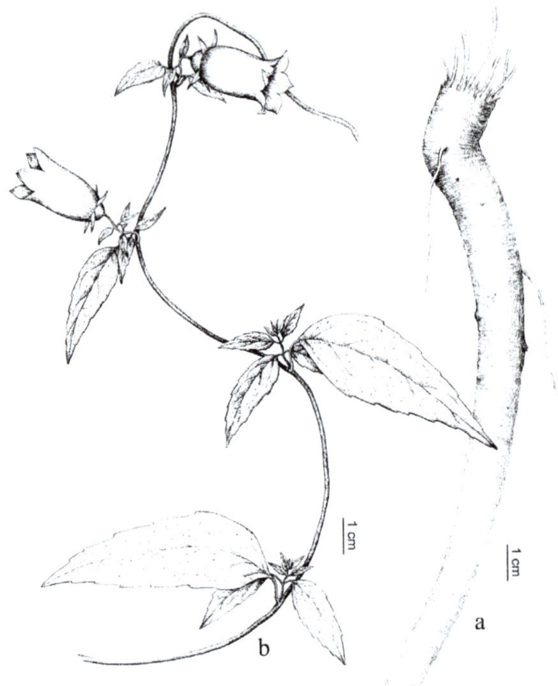

FIGURE IV 1:30A *Codonopsis henryi* Oliv. **a**, root, caudex, and base of stems; **b**, middle part of a twining stem.

31. Codonopsis affinis Hook. f. & Thomson, J. Proc. Linn. Soc. Bot. 2: 12. 1857; Clarke in Hooker Fl. Brit. Ind. 3: 431. 1881; Komarov, Trudy Imp. S. -Peterburgsk. Bot. Sada 29(1): 107. 1908; Chipp, J. Linn. Soc. Bot. 38: 389. 1908; Shen & Hong in Hong, Fl. Reipubl. Popularis Sin. 73(2): 43. pl. 8. fig. 1. 1983; Hong, Lammers & Klein in Wu, Raven & Hong, Fl. China 19: 517. 2011. Type: India (Sikkim), Lachen, 2740 m, 03. Aug. 1849, *J. D. Hooker* s. n.(lectotype here designated: K); loc. eodem, Lachen, 2130–2740 m, 03. Aug. 1849, *J. D. Hooker* s. n.(paratype: K); India (Sikkim), Lachung, 1830 m, 02. Oct. 1849, *J. D. Hooker* s. n.(paratypes: K, PE, W).

Codonopsis affinis Hook. f. & Thomson var. *burmanica* C. B. Clarke in Hooker, Fl. Brit. Ind. 3: 431. 1881. Type: Myanmar, *Griffith 3439* (holotype: K).

Roots much thickened, carrot-shaped, branched. Stems twining, over 2 m long, branched, yellow green or green, nearly glabrous or sparsely white-pubescent. Leaves on main stem and major branches alternate, those on branchlets nearly opposite; petiole 1–6 cm long, pubescent; leaf blade ovate or ovate-oblong, 2.5–16 cm long, 1–9 cm broad, adaxially green, sparsely white-pubescent, abaxially gray-green, densely white-pubescent, base deeply cordate to rounded, margin entire to shallowly crenate, apex short-acuminate. Flowers solitary, terminal, or leaf-opposed; pedicel 1–4 cm long, hispidulous. Calyx tube adnate to nearly whole hypanthium, semiglobose, obscurely 10-ribbed, scaberulose; calyx lobes distant from each other, linear-oblong or linear-lanceolate, 8–17 mm long, 2–3 mm broad, puberulent on both surfaces, apex acute or slightly obtuse, margin waved; sinus between calyx lobes truncate. Corolla broad-campanulate, 1.3–2 cm long, 1–1.6 cm wide, shallowly 5-lobed, yellow-green at lower and middle parts, purple-red at upper part, apically hirsute outside; corolla lobes triangular, often recurved. Filaments slightly dilated at base, glabrous; anthers ellipsoid, ca. 5 mm long, glabrous. Capsule semiglobose or nearly globose below, conical and rostrate above, 1–1.5 cm in diameter, rostrum ca. 4 mm long. Seeds ellipsoid or ovoid, yellow-brown, smooth. Figure IV 1:31A.

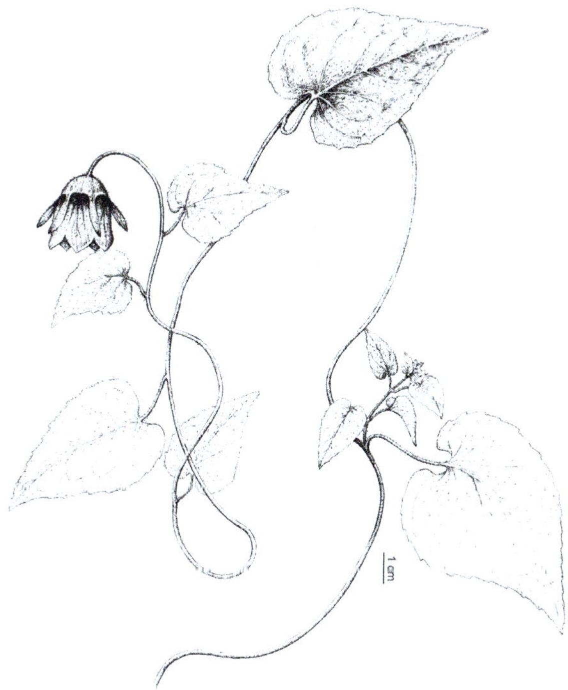

FIGURE IV 1:31A *Codonopsis affinis* Hook. f. & Thomson. **a**, upper part of a twining stem.

Phenology: flowering and fruiting from July to October.

Habitats: forests at an altitude of 2300–3400 m.

Distribution: Bhutan, China (S Tibet), N India, Myanmar, and Nepal (Map IV 1:31).

Additional specimens examined, **BHUTAN**, **Chendebi**: 3050 m, 07. Aug. 1949, *F. Ludlow et al. 17075* (BM, E); loc. eodem, 2440 m, 04. Aug. 1949, *F. Ludlow et al. 19566* (BM). **Gasa** Dzong: 3050 m, 03. Oct. 1949, *F. Ludlow et al. 17402* (BM). **Gichha**: 2440 m, 26. Aug. 1914, *R. E. Cooper 3152 b* (E).

CHINA, **TIBET**, **Mêdog**: 29°22′ N, 95°08′ E, 2000 m, forests, 23. Sept. 2003, *X. F. Gao et al. 7855* (CDBI) ; 29°40′ N, 95°29′ E, edges of forests, 01. Oct. 2003, *X. F. Gao et al. 8071* (CDBI). **Nyalam**: near Quxam, 3200 m, 28. Aug. 1972, *Xizang Chinese Mater. Medic. & Herbs Exped. 1579* (HNWP, PE); near Zham, 3200 m, forests, 15. Aug. 1972, *Xizang Chinese Mater. Medic. & Herbs Exped. 1215* (HNWP, PE); loc. eodem, 2300–2400 m, edges of forests, 16. Aug. 1972, *Xizang Chinese Mater. Medic. & Herbs Exped. 1240* (HNWP). **Zayü**: Luhit F. D., Dreyi to Tezu, 02. Dec. 1957, *R. S. Rao 10907* (CAL); Luhit F. D., Dreyi to Dloeliang, 13. Nov. 1957, *R. S. Rao 10470* (CAL).

INDIA, **Darjiling**: Phalut, 3600 m-Ratho Chu 2100 m-Raman 2400 m, 04. Aug. 1972, *H. Kanai et al. 721440* (BM, TI); Senchal-Tiger Hill-Ghum, 2400 m, 26. Jul. 1969, *H. Hara et al. 69715* (TI). **Sikkim**: Humbo, 3350 m, 24. Oct. 1875, *C. B. Clarke 25551 (K) & 25605* (K); Jumbok, 3050 m, 10. Oct. 1870, *C. B. Clarke 12897A* (CAL, K); Kulhart, 2130 m, 11. Oct. 1870, *C. B. Clarke 12897* (K) & *13003* (K); Lachen, 2440 m, Sept. -Oct. 1909, *Ribu & Rhomoo 3066* (CAL); Lachung, 2740 m, 06. Sept. 1911,

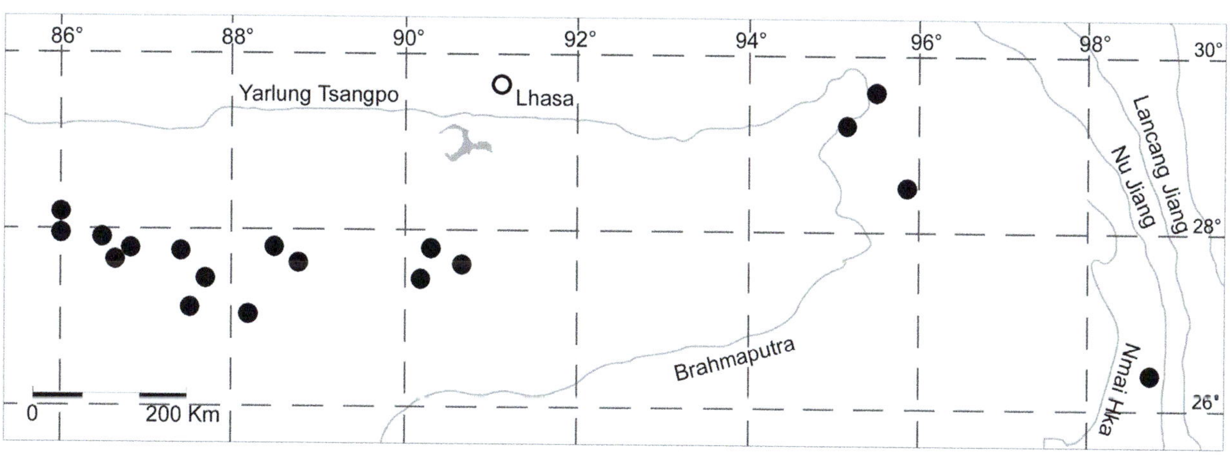

MAP IV 1:31 Distribution of *Codonopsis affinis* Hook. f. & Thomson.

Ribu & Rhomoo 5533 (E); Lachung Valley, 2130 m, 14. Sept. 1892, *G. A. Gammie 1188* (CAL); Rathangoong, Oct. 1908, *Ribu 118* (CAL); Thumbo, 3350 m, 24. Oct. 1875, *C. B. Clarke 25560* (CAL); s. loc., 3350 m, 24. Oct. 1875, *C. B. Clarke 25560A* (BM); s. loc., 1830–3050 m, *J. D. Hooker* s. n.(P, PE).

NEPAL, **Janakpur**: Dolakha Distr., Khare Khola, Patale Pokhar, (4000 m)-a pass (4200 m)- Phedi Kharka (2100 m), 13. Sept. 1983, *H. Ohba et al. 8332037* (KATH). **Kosi**: Dhankuta Distr., 2100 m, 24. Oct. 1963, *H. Hara et al. 6300417* (TI); Sankhuwasabha Distr.: lower Kashwa Khola, above Hedangha, ca. 2300 m, 10. Sept. 1981, *C. Grey-Wilson et al. 4398* (K); Sankhuwasabha Distr., Pamya, 2743 m, Jul. 1966, *Banerjee & P. R. Sakya 5609* (KATH); Sankhuwasabha Distr., Jaljale Himal., Singoa Kharka (4140 m) - Pahakhola (2400 m), 11. Aug. 1991, *H. Ohba & al. 9153427* (TI); Terhathum Distr., Chauki-Gupha Pokhari, 27°15′22″ N, 87°29′ 20″E, 2800 m, 04. Aug. 1999, *Y. Omori et al. 9940005* (E); Terhathum Distr., near Chauki, 27°13′N, 87°28′ E, 2700 m, 29. Aug. 1989, *KEKE 64* (E, K); Terhathum Distr., Tinjure Danda, 27°10′N, 87°29′ E, 2740 m, 07. Sept. 1967, *Williams & Stainton 8407* (BM, K). **Mechi**: Taplejung Distr., Chauke (2700 m) –a pass (3850 m)- Tute (2300 m), 01. Sept. 1977, *H. Ohashi et al. 774168* (TI) & *775515* (TI). **Sagarmatha**: between Namochee and Rukma, ca. 2740 m, 10. May. 1965, *Banerji* s. n.(PE); Solukhumbu Distr., 27°34′N, 86°29′ E, 2550–3400 m, 19. 08. 1985, *H. Ohba et al. 8571813* (E); Solukhumbu Distr., Namche Bazar to Toktok, Monjo, 27°45′56″ N, 86°43′ 23″E, 2780 m, 28. Sept. 2005, *DNEP 3 BX 288* (E); loc. eodem, 3100 m, 14. Oct. 1954, *A. Zimmerman 1674* (G); loc. eodem, Phakding to Luka, 2630 m, 30. Sept. 2005, *DNEP 3 BX 291* (E); loc. eodem, 27°35′N, 86°36–41′ E, 2760 m, 01. Aug. 1997, *M. Wakabayashi et al. 9730063* (E); loc. eodem, 2820 m, 31. Jul. 1997, *M.Wakabayashi et al. 9720068* (E).

32. Codonopsis viridis Wall. in Roxburgh, Fl. Ind. 2 : 103. 1824; de Candolle, Monogr. Camp. 120. 1830; Hooker & Thomson, J. Proc. Linn. Soc. Bot. 2: 12. 1857; Clarke in Hooker, Fl. Brit. Ind. 3: 431. 1881; Komarov, Trudy Imp. S.-Peterburgsk. Bot. Sada 29(1): 106. 1908; Chipp, J. Linn. Soc. Bot. 38: 386. 1908; Hong, Lammers & Klein in Wu, Raven & Hong, Fl. China 19: 517. 2011. ≡*Campanula viridis* (Wall.) Spreng., Cura Post. 78. 1825. ≡*Glossocomia viridis* (Wall.) Rupr., Bull. Cl. Phys. - Math. Acad. Imp. Sci. Saint - Petersbourg 15: 210. 1857. ≡*Wahlenbergia viridis* (Wall.) A. DC. in de Candolle, Prodr. 7: 424. 1839. Type: Nepal, *Wallich 1298* [lectotype here designated: K (K000781220); isolectotypes: E, G, K (K000781221)].

Codonopsis viridis Wall. var. *hirsuta* Chipp, J. Linn. Soc. Bot. 38: 386. 1908. Type: India, Khasia: 1520–1830 m, 11. Aug. 1850, *J. D. Hooker & Thomson* s. n.(fl.) (lectotype here designated: K; isolectotypes: BM, G , W); loc. eodem, s. d., *J. D. Hooker & Thomson* s. n.(fr.) (paratypes: BM, G, K, W); Kalapani, 1370 m, 29. Oct. 1871, *C. B. Clarke 16304* (paratype: BM); Bork Valley, 1530 m, 05. Oct. 1886, *C. B. Clarke 44896A* (paratypes: BM, CAL); India Orient., *Lobb,* s. n.(paratype: W)

Codonopsis griffithii C. B. Clarke in Hooker, Fl. Brit. Ind. 3: 431. 1881. *Codonopsis* sp. Griffith, Notul. Pl. Asiat. 4: 281. 1854 et Icon. Pl. Asiat. t. 482. 1854. Type: Griffith's tab. cited.

Roots unknown. Stems twining, over 1 m long, glabrous, branched. Leaves alternate; petiole 0.5–2 cm long, puberulent; leaf blade ovate to lanceolate, 2–11 cm long, 1–6 cm broad, adaxially hispidulous, abaxially densely hispidulous or hirsute, base cordate to rounded, margin minute-serrate or nearly entire, apex acute to acuminate. Flowers in cymes, extra-axillary, almost leaf-opposed; pedicel 2.5–7 cm long, glabrous at lower and middle parts, pubescent at top. Calyx tube adnate to whole hypanthium, semiglobose, 10-ribbed, glabrous or sparsely pubescent; calyx lobes linear to linear-lanceolate, 1–2 cm long, 2–3 mm broad, reflexed, both surfaces hispidulous entirely or only towards apex, with one or two verrucose denticles on each side; sinus between calyx lobes broad and truncate. Corolla campanulate, 2–3.2 cm long, 2–3 cm wide, lobed to 1/3 of overall length, pale greenish white, flushed and spotted crimson, purple inside, or tube white, lobes deep purple, pubescent, less frequently glabrous outside; corolla lobes deltoid, sometimes recurved. Stamens glabrous; filaments 6 mm long, slightly dilated at base; anthers as long as filaments. Capsule globose in inferior part, conical in superior part, 1–2 cm long, 1.2–1.6 cm in diameter. Seeds oblong, brownish-yellow, ca.0.7 mm long. Figure IV 1:32A.

Phenology: flowering and fruiting from September to November.

Habitats: scrubs on open slopes, or edges of forests; 1220–3000 m. (Nepal, Goketey, 1930, *L. Dhevoj 503* (in BM and E) records 3960–4270 m).

Distibution: Bhutan, China (S Tibet), N India and Nepal (Map IV 1:32).

Notes: I have visited the herbaria BM, CAL, E, G, K, P, W and so forth, but I failed to find *Griffith*'s collection of *C. griffithii* C. B. Clarke. Thus I designate here Griffith's tab. (cited) as the lectotype. Chipp (1908) treated *C. griffithii* C. B. Clarke as a synonym of *C. viridis* var. *hirsuta* Chipp when he published his new variety. He stated that his variety "does not appear to me to be specifically different from *C. viridis* Wall., and Clarke himself stated that it may be only a geographical variety."

Additional specimens examined, **BHUTAN**, **Tobrang**: Trashiyangsi Valley, 2290 m, 01. Sept. 1934, *F. Ludlow & G. Sherriff 917* (BM, E). **Tongsa**: 1830 m, 11. Sept. 1914, *R. E. Cooper 2073* (BM, E); loc. eodem, 2130 m, 30. Aug.

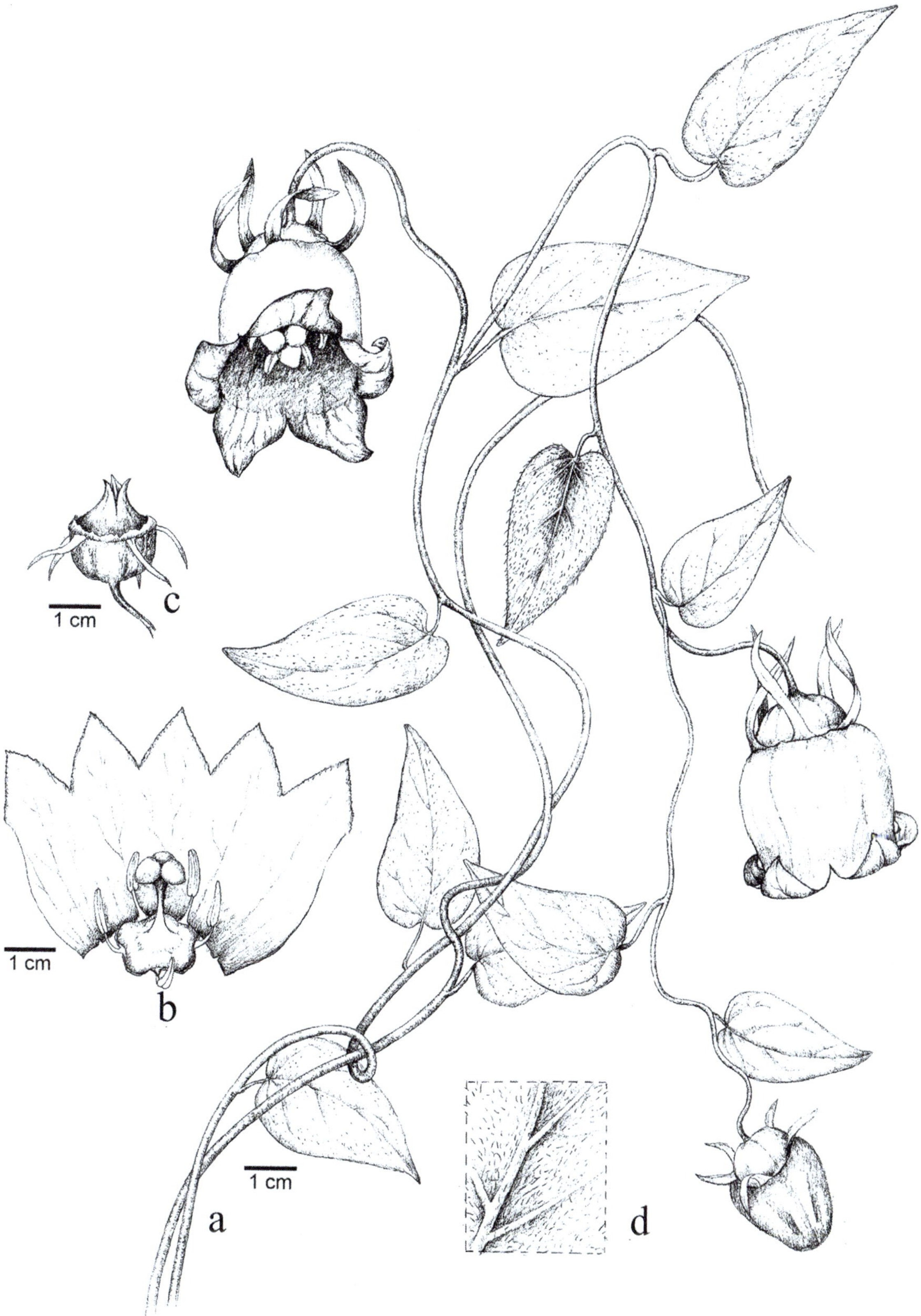

FIGURE IV 1:32A *Codonopsis viridis* Wall. **a**, upper parts of stems; **b**, upper part of a flower with corolla opened, showing stamens, disc, style, and stigma; **c**, a 3-loculicidal capsule; **d**, a part of leaf blade, showing indumentum on abaxial surface.

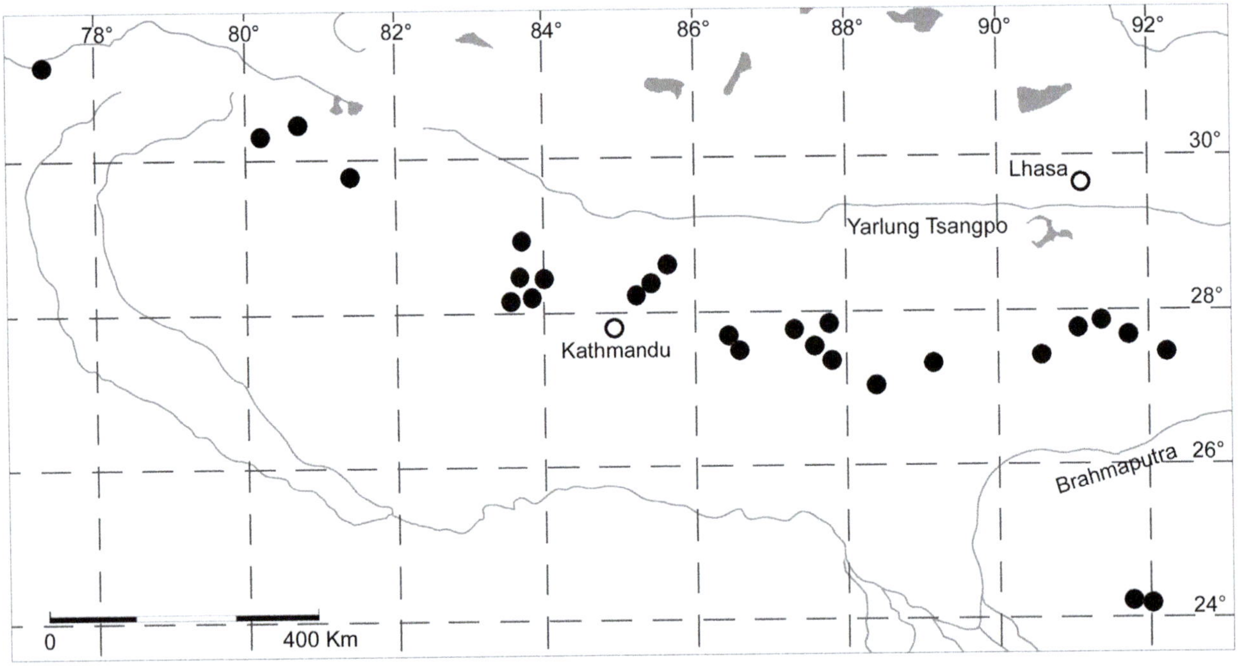

MAP IV 1:32 Distribution of *Codonopsis viridis* Wall.

1949, *F. Ludlow et al.19661* (BM). **Wamlong**: Menthong La, 03. Oct. 1974, *R. Nawang & S. Tshering* s. n.(E). **upper Mo Chu Distr.**, between Tamji and Goen Gaza, 27°52′ N, 89°45′ E, 13. Sept. 1984, *I. W. J. Sinclair & D. G. Long 4962* (E).

CHINA, **TIBET**, **Cona**: Dirang Dzong, 1520–1830 m, 07. Sept. 1938, *F. Kingdon-Ward 14218* (BM); between Gyipu and Kapteng, Nyam Jang Chu, 27°41–42′ N, 91°33–34′ E, 2130 m, 04. Nov. 1938, *F. Ludlow et al. 7063* (BM); Muktur, Tawang Chu, S Tisel, 27°32′33″ N, 91°55′ 56″E, 2440 m, 12. Nov. 1936, *F. Ludlow & G. Sherriff 2834* (BM).

INDIA, **Darjiling**: Victoria Fall, 1980 m, Oct. 1874, *J. S. Gamble 2335 A* (K). **Himachal Pradesh**: Rampur, N Garhwal, 14. Oct. 1938, *Kirah Rain 8857* (E). **Meghalaya**: Boga Pani, 10. Sept. 1885, *C. B. Clarke 40366* (BM, CAL); Khasia Hills, 1220 m, Sept. 1890, *H. Collett 112* (CAL); Mont Khasia, s. loc., 1520–1830 m, 1861, *J. D. Hooker & Thomson* s. n.(BM, CAL, G, K, P, PE, W); Naga Pauce, 1520–1830 m, 04. Aug. 1850, *J. D. Hooker & Thomson* s. n.(BM, K, PE); Tsilong, 1530 m, 12. Oct. 1867, *C. B. Clarke 6300* (K); s. loc., 1530 m, 16. Sept. 1886, *C. B. Clarke 45404* (G). **Uttar Pradesh**: near Khathee, 2130 m, *Madden's 374* (E, K); Kumaon, Kuthi, 2130 m, *R. Strachey & J. E. Winterbottom No3* (BM, K, P, PE); Kumaon, s. d., *Strachey* s. n.(P, PE); Ralam Valley, 16. Aug. 1900, *Inayat 24631* (CAL).

NEPAL, **Bagmati**: Dhading Distr., S of Ganesh Himal, Tiru Danda, 28°5′ N, 85°8′ E, 1830 m, 13. Sept. 1965, *J. D. A. Stainton 5099* (BM); between Khangjung and Syarpagaon, north side of Langtang River, 2200 m, 18. Sept. 1966, *D. H. Nicolson 2409* (BM); Rasuwa Distr., 1900 m, 28°4′6″ N, 85°13′ 46″E, 04. Oct. 2001, *D. G. Long et al. 113* (E); Rasuwa Distr., Grang, 28°03′ N, 85°13′ E, 1800 m, 04. Sept. 1971, *J. F. Dobremez 925* (BM); Rasuwa Distr., Langtang Khola, 2440–2900 m, 23–31. Aug. 1949, *O. Polunin 1812* (BM) ; Thade, 2000 m, 30. Aug. 1977, *N. P. Manandhar & Party 387* (KATH). **Dhawalagiri**: Mustang Distr., 28°36′06″–41′15″ N, 83°35′ 46″–39′09″E, 2180 m, 10. Sept. 1995, *M. Mikage et al. 9552280* (E); Mustang Distr., S of Tukucha, Taglung, Kali Gandaki, 2740 m, 19. Sept. 1954, *Stainton, Sykes & Williams 7929* (BM); Mustang Distr., Lete-Larjung Cobang, ca. 28°30′ N, 83°40′ E, 20. Sept. 1985. *D. P. Joshi et al. 131/85* (KATH); Myagdi Distr., Muna, 1830 m, 06. Sept. 1954, *Stainton, Sykes & Williams 4224* (BM, CAL, PE); Parbat Distr., E of Gorepani ca. 1800 m, 10. Sept. 1973, *C. Grey-Wilson & Phillips 879* (BM, K). **Gandaki**: Annapuna Himal., Seti Khola, 1980 m, 05. Aug. 1954, *Stainton, Syke & Williams 6716* (E); Kaski Distr., Lumle, ca. 28°15′ N, 83°55′ E, 1539 m, 14. Sept. 1985, *D. P. Joshi et al. 11/85* (KATH); Kaski Distr., Mardi Khola: 1680 m, 16. Sept. 1954, *Stainton, Sykes & Williams 8450* (BM, G). **Janakpur**: Dolakha Distr., Yarsa-Jiri, 1980 m, 21. Sept. 1964, *Banerjee, Shrestha & Upadhyay 2879* (KATH); Ramechhap Distr., 1800–2700 m, 11. Sept. 1985, *H. Ohba et al. 853195* (E, TI). **Kathmandu**: Phulchoke, 01. Nov. 1965, *A. D. Schilling 691* (K, TI); loc. eodem, 2400 m, 19. Sept. 1970, *H. Kanai & Ch. Chuma 670071* (TI). **Kosi**: Sankhuwasabha Distr.: above Milke Bhanjang, ca. 3000 m, 01. Sept. 1981, *C. Grey-Wilson 4556* (K); Terhathum Distr., between Chauki and Terhathum, 2200 m, 06. Oct. 1981, *C. Grey-Wilson 4689* (BM, K). **Mechi**: Diorali, Distr. 1900 m, 02. Sept. 1970, *H. Kanai & T. B. Shrestha 672722* (TI); Panchthar Distr., Yektin-Birwa, 28. Nov. 1963, *H. Kanai et al. 6300427* (TI); Yektin-Akasay-Batasay, 30. Nov. 1963, *H.*

Hara et al. 6300428 (TI); Taplejung Distr., Tapletok, 27°27′ N, 87°30′ E, 2300 m, 06. Oct. 1971, *J. F. Dobremez 1388* (BM); Taplejung Distr., Yamphudin, between Kabeli Khola and Omje Khola, 27°28′ N, 87°55′ E, 2400 m, 23. Sept. 1989, *KEKE 899* (E, K); Taplejung Distr., Baroya Khimty-Thakma Khola, 2500 m, 11. Nov. 1963, *H. Hara et al. 6300421* (TI); loc. eodem, 12. Nov. 1963, *H. Hara et al. 6300422* (TI); loc. eodem, 16. Nov. 1963, *H. Hara et al. 6300423* (BM, TI); Thakma Khola-Diorali Bhanjyang-Bhandukay-Yamphodin 17. Nov. 1963, H. *Kanai et al. 6300424* (BM, TI). **Sagarmatha**: Solukhumbu Distr., 2350–2800 m, 04. Sept. 1985, *H. Ohba et al. 8581347* (BM, E, TI). **Seti**: Bajhang Distr., 1660–2230 m, 17. Aug. 1991, *M. Suzuki et al. 60714* (E).

33. Codonopsis farreri J. Anthony, Notes Roy. Bot. Gard. Edinburgh 15: 181. pl. 216. 1926; Shen & Hong in Hong, Fl. Reipubl. Popularis Sin. 73(2): 45. 1983. Type: Myanmar, western flank of the Chimili, N'Maikha-Salween Divide, 26°23′ N, 98°48′ E, 3350–3650 m, Aug. 1924, *G. Forrest 24885* (lectotype here designated: E; isolectotypes: K, P, W); loc. eodem, 26°24′ N, 98°48′ E, 3650 m, Jun. 1925, *G. Forrest 26973* (paratypes: E, K, P, W); Myanmar, Chimili Valley, the uppermost alpine woodlands, 3350 m, 31. Jul. 1919, *Farrer 1144* (paratype: E).

Codonopsis farreri J. Anthony var. *grandiflora* S. H. Huang, Acta Bot. Yunnan. 6: 394. 1984. Type: China, Yunnan, Fugong, 3100–3600 m, 02. Aug. 1979, *Q. Ling 791720* (holotype: KUN).

Codonopsis gombalana C. Y. Wu in Wu & Li, Yunnan Trop, Subtrop. Fl. Res. Rep. 1: 81. pl. 17. 1965; Shen & Hong in Hong, Fl. Reipubl. Popularis Sin. 73(2): 53.1983; Hong, Lammers & Klein in Wu, Raven & Hong, Fl. China 19: 520. 2011, syn. nov. Type: China, Yunnan, Gongshan, Cikai, Heipu Shan, bamboo thickets, 09. Oct. 1940, *K. M. Feng 8295* (holotype: KUN; isotypes: PE).

Stems from 0.5 to 5 m long, twining, rarely ascending or scandent, branched. Leaves on main stem and major branches alternate or nearly opposite, those on branchlets opposite or pseudo-verticillate; petiole 1–1.5 cm long; leaf blade ovate, lanceolate or elliptic, 0.5–4 cm long, 0.5–1.5 cm broad, both surfaces glabrous, base cuneate or rounded, margin entire, apex acute or slightly obtuse. Flowers solitary, terminal; pedicel 5–6 cm long. Calyx tube adnate to whole hypanthium, glabrous; calyx lobes linear-lanceolate, 2–2.6 cm long, 3–5 mm broad, apex acute or slightly obtuse, margin nearly entire. Corolla campanulate, 3.3–5.5 cm long, ca. 3 cm wide, pale green or greenish yellow, with brown-red or purple reticulate veins and with a dark purple patch at base of each of five main veins; corolla tube 2–4 cm long; corolla lobes broad-triangular, obtuse, 1.2–1.6 cm long, 1.2–2 cm broad. Stamens glabrous; filaments ca. 6 mm long, slightly dilated at base; anthers 5–7 mm long. Fruit and seeds not seen. Figures IV 1:33A & B.

Phenology: flowering from June to September.

Habitats: mostly in *Rhododendron* and bamboo thickets, less frequently in forests or meadows, often on granites, less often on limestones; 3000–3700 m.

Distribution: China (NW Yunnan) and NE Myanmar (Map IV 1:33).

Additional specimens examined, **CHINA, YUNNAN, Fugong**: Lishadi Township, 3560 m, bamboo thickets, 12. Aug. 2005, *Gaoligong Shan Biodiversity Survey 27015* (A, CAS); loc. eodem, 3040 m, deciduous broadleaf forests, 08. Aug. 2005, *Gaoligong Shan Biodeversity Survey 26747* (A, CAS); loc. eodem, 3660 m, thickets, 05. Aug. 2005, *Gaoligong Shan Biodiversity Survey 26353* (A, CAS); Lumadeng Township, 3650 m, meadows, 17. Aug. 2005, *Gaoligong Shan Biodiversity Survey 28643* (A, CAS). **Gongshan**: Cikai, Heipu Pass, 3490 m, Bamboo & *Rhododendron* thickets, 12. Aug. 2006, *Gaoligong Shan Biodiversity Survey 32059* (A, CAS); loc. eodem, near Danhatu, 3280 m, thickets, 02. Sept. 2006, *Gaoligong Shan Biodeversity Survey* 34521 (CAS); loc. eodem, Yipsaka Lake, 3560 m, Bamboo & *Rhododendron* thickets, 12. Aug. 2006, *Gaoligong Shan Biodiversity Survey 32082* (CAS); Gongshan, Salween-Kiukiang Divide, Lunguailaka, 3600 m, 14. Sept. 1938, *T. T. Yü 20265* (paratypes of *C. gombalana*: E, KUN, PE).

34. Codonopsis benthamii Hook. f. & Thomson, J. Proc. Linn. Soc. Bot. 2: 14. 1857; Clarke in Hooker, Fl. Brit. Ind. 3: 432. 1881; Komarov, Trudy Imp. S.-Peterburgsk. Bot. Sada 29(1): 113. 1908; Chipp, J. Linn. Soc. Bot. 38: 382. 1908; Hong, Novon 20: 420. 2010; Hong, Lammers & Klein in Wu, Raven & Hong, Fl. China 19: 518. 2011. Type: India, Sikkim, Lachen, 2740 m, 31. Jul. 1849, *J. D. Hooker* s. n.[lectotype designated by Hong 2010: K (K000814213)]; Sikkim, Lachen, 2750–3050 m, 03. Aug. 1849, *J. D. Hooker* s. n.[paratype: K (K000814212)]; Sikkim, 2750–3350 m, s. d., *J. D. Hooker* s. n.[paratypes: K (K000814210 & K000814211)].

Codonopsis macrocalyx Diels, Notes Roy. Bot. Gard. Edinburgh 5: 170. 1912. Type: China, Yunnan, "on the eastern flank of the Lichiang Range" (Lijiang), 27°20′N, 3050–3350 m, Sept. 1906, *G. Forrest 3008* (lectotype designated by Hong, 2010: E; isolectotype: K, PE).

Codonopsis macrocalyx Diels var. *parviloba* J. Anthony, Notes Roy. Bot. Gard. Edinburgh *15:* 183. 1926. Type: China, Yunnan, "on the eastern flank of the Tali range" (Dali), 25°40′N, 2740–3050 m, Jul. 1906. *G. Forrest 3854* (lectotype designated by Hong 2010: K; isolectotype: E); China, Yunnan, "Shweli-Salween divide" (Lushui), 25°48′N, 98°48′E, 3050–3350 m, Jul. 1924, *G. Forrest 24723* (paratypes: E, K, W); "Upper Burma" [N Myanmar], 2440 m, 06. Aug. 1914, *F. Kingdon-Ward 1655* (paratype: E).

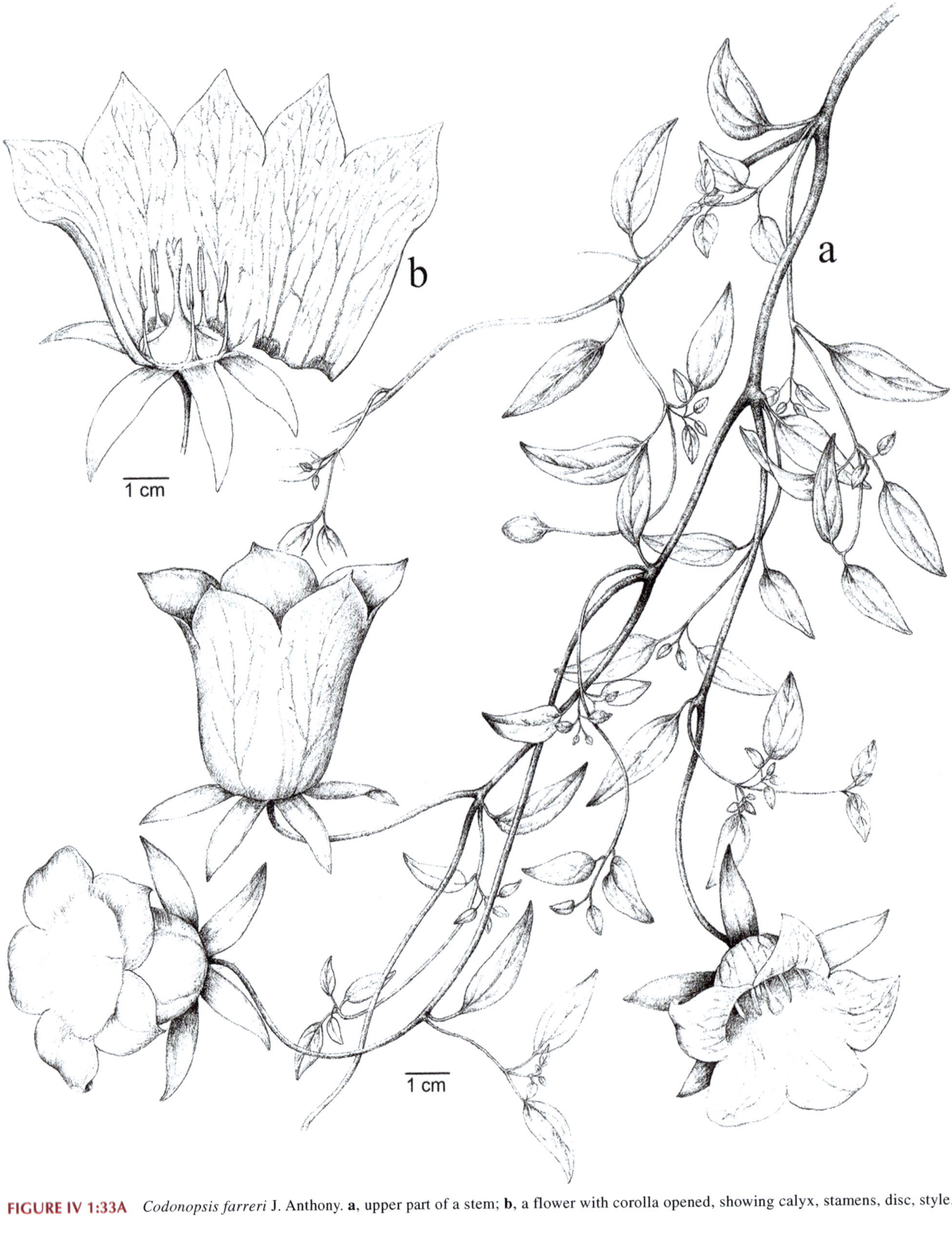

FIGURE IV 1:33A *Codonopsis farreri* J. Anthony. **a**, upper part of a stem; **b**, a flower with corolla opened, showing calyx, stamens, disc, style.

FIGURE IV 1:33B *Codonopsis farreri* J. Anthony. **a–c**, from Yunnan: Gongshan, 30. Jul. 2013, by Qiang WANG.

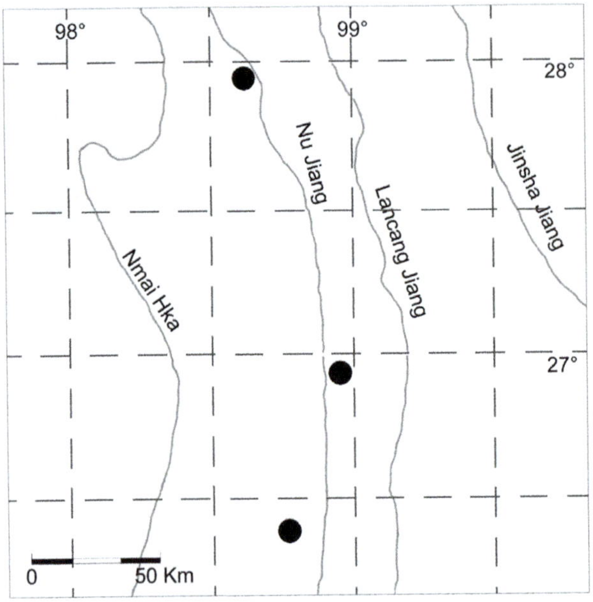

MAP IV 1:33 Distribution of *Codonopsis farreri* J. Anthony.

Codonopsis macrocalyx Diels var. *coerulescens* Hand.-Mazz., Akad. Wiss. Wien Math.-Naturwiss. Kl. Anz. 61 (20): 169. 1924. Type: China, Yunnan, Mekong-Salween Divide ("Schöndsu-la"), 28°4′N, 4025 m, 02. Aug. 1916, *Handel-Mazzetti 9618* (lectotype here designated: WU; isolectotype: E).

Codonopsis pianmaensis S. H. Huang, Acta Bot. Yunnan. 6: 393. fig. 1. 1984; Hong, Lammers & Klein in Wu, Raven & Hong, Fl. China 19: 519. 2010, syn. nov. Type: China, W Yunnan,Lushui County, Pianma, 3500 m, 17. Aug. 1964. *S. K. Wu 8428* (holotype: KUN).

Roots carrot-shaped or cylindrical, 20–30 cm long, 1.5–2.5 cm in diameter. Stems 0.3–2 m long, ascending, scandent, sometimes twining, branched, branches and branchlets leafy, sterile or fertile, sparsely puberulent. Leaves alternate or those on branches nearly opposite; petiole 1–6 cm long, sparsely villous; leaf blade broad-ovate, triangular-ovate, ovate or ovate-lanceolate, 3–10.5 cm long, 1.5–7 cm broad, adaxially green, sparsely puberulent, abaxially gray-white, sparsely or densely puberulent, base shallowly cordate to cuneate, margin irregularly pinnatifid to shallowly lobed, or dentate, crenate, rarely nearly entire, apex acute or obtuse. Flowers solitary, terminal, long-pedicellate; pedicel villous. Calyx tube adnate to whole hypanthium, 10-ribbed, glabrous or sparsely villous; calyx lobes ovate, 1–1.3 cm long, 0.4–0.5 cm broad at anthesis, glabrous, rarely sparsely hispidulous, margin crenate, ciliate, apex acute or obtuse; sinus between calyx lobes narrow and pointed. Corolla tubular, slightly dilated above, 1.8–4 cm long, 1.5–2 cm wide, shallowly lobed for 1/6–1/7 of overall length, yellow, green, yellow-green, or rarely purple, pale brown-red at base, glabrous; corolla lobes triangular, apex acute.

FIGURE IV 1:34A *Codonopsis benthamii* Hook. f. & Thomson. **a**, upper part of a stem; **b**, upper part of a flower with corolla opened, showing stamens, style, and stigma; **c**, different forms of leaves.

Filaments 6–10 mm long, entirely villous, dilated at base; anthers 5–6 mm long, connectives enveloping anthers, villous at back. Capsule 1.2–1.6 cm in diameter; inferior part hemispheric, fruiting calyx lobes up to 2.5 cm long, 1.5 cm broad. Seeds ovoid, brown-yellow. Figures IV 1:34A & B.

Phenology: flowering and fruiting from July to October.

Habitats: grassy slopes, forest margins, scrubs, on granites or marbles; 2100–4300 m.

Distribution: Bhutan, China (SW Sichuan, SE Tibet and NW Yunnan), N India, N Myanmar, and E Nepal (Map IV 1:34).

Additional specimens examined: **BHUTAN, Omta Tso**, 3810 m, 17. Oct. 1949, *F. Ludlow et al. 19839* (BM). **Rinchen Chu**, 2900 m, 06. Aug. 1937, *F. Ludlow & G. Sherriff 3527* (BM, E); loc. eodem, 3200 m, 09. Aug. 1947, *F. Ludlow et al. 17089* (BM). **Rip La**, 2900 m, 16. Jul. 1949, *F. Ludlow et al. 20886* (BM). **Sengor**: between Sengor and Sheridrang, 27°21′N, 91°01′E, 2980 m, 06. Jul. 1979, *A. J. C. Grierson & D. G. Long 2543* (E). **Thimphu**, 27°41′N, 89°38′E, 3210 m, 05. Oct. 1984, *I. W. J. Sinclair & D. G. Long 5566* (E, K). **Upper Kulong Chu** Distr. 2700 m, 04. Aug. 2000, *S. A. Bowes Lyon 13179* (E). **Upper Mo Chu** Distr., 15. Aug. 2003, *S. A. Bowes Lyon 15127* (E); loc. eodem, 3050 m, *F. Ludlow et al. 17379* (BM).

FIGURE IV 1:34B *Codonopsis benthamii* Hook. f. & Thomson. **a**, from Tibet: Zayü, 18. Aug. 2011, by You-Sheng CHEN; **b**, from Yunnan: Gongshan, 30. Jul. 2013, by Qiang WANG; **c**, from Yunnan: Yangbi, 15. Jul. 2009, by Bin LIU.

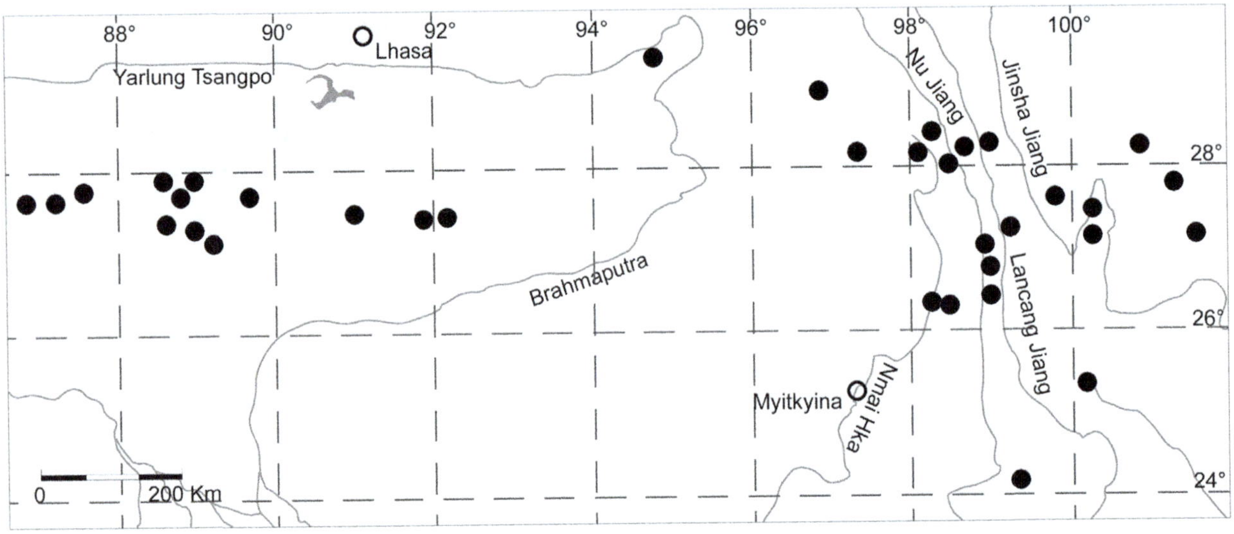

MAP IV 1:34 Distribution of *Codonopsis benthamii* Hook. f. & Thomson.

CHINA, SICHUAN, Muli: 10 km S of Muli, 3050 m, 21. Aug. 1924, *F. Kingdon-Ward 4853* (E); Wo-chio, 2800 m, margins of woods, 05. Aug.1937, *T. T. Yü. 7597* (KUN, PE); Yazui Forest Farm, 3400 m, margins of *Abies* forests, 19. Aug. 1983, *Qinghai-Xizang Exped. 13071*(KUN, PE). **Yanbian**: Zemulong, 27°17′N, 101°32′E, 3700–3800 m, 13. Sept. 2002, *X. F. Gao et al. 3443* (CDBI, PE). **TIBET, Cona**: Dirang Dzong, 2130–2440 m, 10. Sept. 1938, *F. Kingdon-Ward 14228* (BM, PE). **Markam**: Cawarong (Tsa-wa-rung), Chi-na-tung, 2800 m, thickets, Aug. 1935, *C. W. Wang 65204* (PE); E of Salween River, N of Alulaka, Muli-la, 2740 m, 06. Jul. 1932, *J. F. Rock 22359* (E, PE). **Mêdog**: Lage, Duoxiang La, 3000 m, forests, 24. Oct. 1992, *H. Sun et al. 0788* (KUN). **Yadong**: Asang Monitory station, 27°21′N, 88°58′E, 3000 m, forest margins, 08. Jul. 2013, *PE-Tibet Exped. 2945* (PE); Chumbi Valley, 22. Jul. 1877. *G. King 4270* (CAL); Chumbi & Phali, Sim-Doong-Tang, 11, Sept. 1882, *Dungboo* s. n.(K). **Zayü**: Arza of Shangzayü to Rawu, *Abies* forests, 3900 m, 13. Aug. 1973, *Qinghai-Xizang Exped. 73–1069* (KUN, PE); Ridong Distr., edges of *Abies* forests, 08. Sept. 1982, *Qinghai–Xizang Exped. 10111* (KUN, PE); Walung, Seing-hku Valley, 28°10′N, 97°20′E, 3960 m, 04. Oct. 1926, *F. Kingdon-Ward 7531* (K). s. loc., alpine meadows, 3350–3660 m, Sept. 1913, *F. Kingdon-Ward 1116* (E). **YUNNAN**, s. loc., *McLaren's 317* (E, K); W Yunnan, *G. Forrest 29709* (PE). **Bijiang**: Biluo Shan 4000 m, grassy slopes, 27. Aug. 1934, *H. T. Tsai 58236* (KUN, PE, SZ); loc. eodem, 27 Aug. 1934, *H. T. Tsai 58256* (PE, SZ); Biluo Shan, 3750 m, crevices by stream, 01. Aug. 1964, *S. K. Wu 8701* (KUN); Luoxue Shan, 4300 m, grassy slopes, 12. Sept. 1964, *S. K. Wu 8784* (KUN). **Dali**: 25°20′N, 3050 m, Aug. 1912, *G. Forrest 8954* (E, K). **Dêqên**: Doker-la, 28°15′N, 3800–4150 m, 17. Sept. 1915, *Handel-Mazzetti 8121* (W, WU); loc.eodem, 3200 m, 03–5. Aug. 1935, *C. W. Wang 65118* (KUN, PE); loc. eodem, 3600–4000 m, grassy slopes, 01. Aug. 1940, *K. M. Feng 6012* (KUN, PE); loc. eodem, 01. Aug. 1940, *K. M. Feng 5994* (KUN, PE); Junzila, 3600 m, grassy slopes, 06. Oct. 1938, *T. T. Yü 23257* (KUN, PE); Mekong-Salween Divide, 28°20′N, Aug. 1917, *G. Forrest 14680* (E); Mekong-Salween Divide, Sila, 4000 m, grassy slopes, 19. Aug. 1937, *T. T. Yü 22408* (KUN, PE); Mekong-Salween Divide, Sewalongba, 3400 m, grassy slopes, 26. Aug. 1938, *T. T. Yü 22506* (KUN, PE); loc. eodem, 4000 m, 03. Sept. 1938, *T. T. Yü 22673* (E, KUN, PE); Mekong-Salween Divide, E of Yuragan, Nake-la, 3810 m, alpine medows, Jun–Jul. 1932, *J. F. Rock 23127* (E, PE); loc. eodem, Jun–Jul. 1932, *J. F. Rock 23443* (PE); by Mekong, 3700–3800 m, grassy slopes, 20. Aug. 1940, *K. M. Feng 6803* (KUN, PE); Pilo Shan, Che-tse-lo, 4000 m, alpine medows, 15. Sept. 1934, *H. T. Tsai 58634* (KUN, PE); loc. eodem, 27. Aug. 1934, *H. T. Tsai 58256* (KUN, PE); loc. eodem, summit, 27. Aug. 1934, *H. T. Tsai 58236* (PE); Slaza, 3800 m, grassy slopes, 11. Oct. 1938, *T. T. Yü 22796* (KUN, PE). **Fugong**: Lumadeng Township, 2650 m, meadows, 17. Aug. 2005, *Gaoligong Shan Biodiversity Survey 28650* (CAS); loc. eodem, 3650 m, meadows, 17. Aug. 2005, *Gaoligong Shan Biodiversity Survey 28673* (A, CAS); loc. eodem, 3730 m, 07. Aug. *Gaoligong Shan Biodiversity Survey 26602* (A, CAS). **Gongshan**: Binzhongluo, 2880 m, meadows, 01. Sept. 2006, *Gaoligong Shan Biodiversity Survey 32891* (A, CAS); loc. eodem, 3750 m, meadows, 19. Aug. 2006. *Gaoligong Shan Biodiversity Survey 31224* (CAS); loc. eodem, 3470 m, wetlands, 18. Aug. 2006, *Gaoligong Shan Biodiversity Survey 31187* (CAS); loc. eodem, 3470 m, coniferous forests, 17. Aug. 2006, *Gaoligong Shan Biodiversity Survey 31149* (A, CAS); loc. eodem, 3470 m, 18. Aug. 2006, *Gaoligong Shan Biodiversity Survey 31195* (CAS); loc. eodem, 3800 m, meadows, 29. Aug. 2006. *Gaoligong Shan Biodiversity Survey 31615* (A, CAS); loc. eodem, 3780 m, meadows, 24. Aug. 2006,

Gaoligong Shan Biodiversity Survey 31381 (A, CAS); loc. eodem, 3560 m, meadows, 29. Aug. 2006. *Gaoligong Shan Biodiversity Survey 32812* (A, CAS); Champu-tung, Lung-pan-la, 3000 m, thickets, Oct. 1935, *C. W. Wang 67453* (KUN, PE); Champu-tung, 2700–3400 m, thickets by stream, 09. Sept. 1940, *K. M. Feng 7596* (KUN, PE); Cikai, 3220 m, meadows, 24. Aug. 2006. *Gaoligong Shan Biodiversity Survey 34115* (A, CAS); loc. eodem, 3010 m, 12. Aug. 2006, *Gaoligong Shan Biodiversity Survey 33148* (A, CAS); loc. eodem, 3030 m, 21. Aug. 2006, *Gaoligong Shan Biodiversity Survey 33866* (CAS); loc. eodem, meadows, 13. Aug. 2006, *Gaoligong Shan Biodiversity Survey 32096* (CAS); loc. eodem, 3250 m, thickets, 24. Aug. 2006, *Gaoligong Shan Biodiversity Survey 34077* (CAS); loc. eodem, 3250 m, thickets, 24. Aug. 2006, *Gaoligong Shan Biodiversity Survey 34030* (CAS); loc. eodem, 3030 m, 21. Aug. 2006. *Gaoligong Shan Biodiversity Survey 33845* (CAS); Cikai, Heipu Shan, 12. Oct. 1940, *K. M. Feng 8404* (KUN, PE); Upper Kiujiang Valley, Lungtsahmuru, 3750 m, 09. Aug. 1938, *T. T. Yü 19818* (KUN, PE); Salween-Kiukiang Divide, Tsukuai, 3600 m, grassy slopes, 16. Oct. 1938, *T. T. Yü 20708* (E, KUN, PE); Salween-Kiukiang Divide, Tsuhwang, 3100 m, thickets, 17. Sept. 1938, *T. T. Yü 20348* (KUN, PE). **Lijiang** (Yulong): 3100 m, Aug. 1914, *C. Schneider 2302* (K, PE); 27°30′N, 3350 m, Jul. 1914, *G. Forrest 12797* (E, PE). **Shangri-la** (Zhongdian): Chiren, 3200 m, woods, 24. Jul. 1937, *T. T. Yü 12402* (KUN, PE); Wengshui, 28°12′N, 99°58′E, 4000 m, grassy slopes by forest, 03. Sept. 2010, *Kham Exped. 10–3273* (PE); 3500–3600 m, 12. Aug. 1914, *Handel-Mazzetti 4780* (W, WU); Xiaozhongdian, 3800 m, 29. Jul. 1939, *K. M. Feng 1886* (KUN, PE); loc. eodem, 3300 m, open *Quercus* thickets, 30. Jul. 1939, *K. M. Feng 1916* (KUN, PE). **Weixi**: Yezhi, 3600 m, grassy slopes, Aug. 1935, *C. W. Wang 68519* (PE); loc. eodem, Aug. 1935, *C. W. Wang 68421* (PE); Kangpu, 3500 m, Jul. 1935, *C. W. Wang 64533* (PE). **Yongde**: Daxue Shan, 3200 m, grassy slopes 12. Aug. 2002, *E. D. Liu 5808* (KUN).

INDIA, Sikkim: Chola Range, 2440 m, 23. Sept. 1892, *G. A. Gammie 1281* (CAL) ; Chola Range-Nathu La, 4270 m, 29. Sept. 1892, *G. A. Gammie 1322* (CAL) ; Gangtok, 09. Aug. 1877, *G. King 4439* (CAL) ; loc. eodem, 08. Aug. 1877, *G. King's Collector 4440* (CAL); Lachen, 2440 m, Sept-Oct. 1909, *Ribu & Rhomoo 3016* (CAL); loc eodem, 2680 m, 08. Aug. 1913, *R. E. Cooper 394* (E); Lachung, Jul. 1879, *G. King's Collector* s. n.(CAL); Laghep, 3350 m, 15. Aug. 1913, *R. E. Cooper 509* (E); Selu Valley, 3050 m, 03. Sept. 1892, *G. A. Gammie* s. n.(CAL); Yakla, 3050 m, 20. Oct. 1869, *C. B. Clarke 10224* (CAL, K); loc. eodem, 2750 m, 15. Oct. 1869, *C. B. Clarke 10001* (K); loc. eodem, 3050 m, 15. Oct. 1869, *C. B. Clarke 9879* (CAL) & *C. B. Clarke 9879A* (K); Zemu Valley, 2440 m, 07. Jul. 1909, *W. W. Smith & G. H. Cave 986* (B, CAL, E, K, P); s. loc., 2740–3350 m, 1861, *J. D. Hooker* s. n.(BM, CAL, G, K, W).

MYANMAR, Chaungmaw Hka Valley, 2440–2740 m, *F. Kingdon-Ward 3284* (E); Chimili Valley, 3050–3350 m, 17. Aug. 1919, *RA 1219* (E); Di Chu Valley, 3350 m, 26. Jul. 1926, *F. Kingdon-Ward 7193* (K); N' Maikha-Salween Divide, 26°20′N, Jun. 1931, *G. Forrest 29709* (E).

NEPAL, Kosi: Sankhuwasabha Distr. 27°35-40′N, 87°10-15′E, 2540-3500 m, 15. Jul. 1988, *M. Suzuki et al. 8820505* (E, KUN); loc. eodem, 2600 m, 15. Jul. 1988, *M. Suzuki et al. 8850372* (BM); loc. eodem, 2750 m, 15. Jul. 1988, *M. Suzuki et al. 8860189* (BM); loc. eodem, 3350 m, *Rhododendron* forests, 20. Aug. 1981, *C. Grey-Wilson 4102* (K); loc. eodem, Bakang, above Sekidim, ca. 2800 m, 17. Aug. 1981, *C. Grey-Wilson et al. 4070* (BM, K); Sankhuwasabha, between Tashigaon and Kauma, 27°37′N, 87°14′E, 2590 m, 25. Sept. 1991, *D. G. Long et al. 238* (E); loc. eodem, near Kauma, 27°39′N, 87°31′E, 3670 m, 26. Sept. 1991, *D. G. Long et al. 296* (E). **Mechi**: Taplejung Distr., Topke Gola – Shewaden, 2700 m, 21. Aug. 1977, *H. Ohashi et al. 772671* (TI). **Sagarmatha**: Kasua Khola, 3050 m, 05. Aug. 1971, *L. W. Beer et al. 9524* (BM); Solukhumbu Distr., 27°34′N, 86°48′E, 2880 m, 04. Aug. 1997, *M. Wakabayashi et al. 9730115* (E); loc. eodem, 27°35′N, 86°48′E, 3050 m, 04. Aug. 1997, *M. Wakabayashi et al. 9710121* (E); loc. eodem, 27°38′N, 86°48′E, 3300 m, Aug. 1997, *M. Wakabayashi et al. 9715247* (E).

35. Codonopsis tubulosa Kom., Trudy Imp. S.-Peterburgsk. Bot. Sada 29(1): 112. tab. 2. fig. 3. 1908; Chipp, J. Linn. Soc. Bot. 38: 387. 1908; Anthony, Notes Roy. Bot. Gard. Edinburgh 15: 182. 1926; Tsoong, Contr. Inst. Bot. Nat. Acad. Peiping 3(3): 103. fig. 13. 1935; Anonymous, Icon. Cormophyt. Sin. 4: 378. fig. 6170. 1975; Shen & Hong in Hong, Fl. Reipubl. Popularis Sin. 73(2): 48. 1983. Type: China, Yunnan, Mengtze (Mengzi), Gran Mts, 2130 m, *A. Henry 10167* (lectotype here designated: K; isolectotypes: A, E, PE).

Codonopsis accrescenticalyx H. Lév., Cat. Pl. Yunnan 24. 1915. Type: China, Yunnan, "pâturages des montagnes à Tcha-ho", 2600 m, Jul. 1912, *E. E. Maire* s. n.(holotype: E).

Codonopsis pilosa Chipp, J. Linn. Soc. Bot. 38 : 388. 1908. Type : China, Yunnan, Mengtsz (Mengzi), 1830–2130 m, 28. Aug. 1895, *W. Hancock 379* (holotype: K).

Codonopsis bifoliata L. T. Shen in sched. Type: Sichuan, Puge, 2500 m, shady slopes, 07. Aug. 1960, *Sichuan Medic. Plant Resources Surv. 25204* (PE).

Roots carrot-shaped, simple or branched below middle, 10–20 cm long, 0.5–2 cm in diameter. Stems usually 1–3 m long, twining, rarely ascending, occasionally erect, branched; branches and branchlets both leafy, sterile or with a terminal flower, nearly glabrous or sparsely villous. Leaves opposite, tending to alternate distally; petiole 1–17 mm long, puberulent; leaf blade ovate to lanceolate, 3–12 cm long, 1–6 cm broad, adaxially green, hispidulous, abaxially gray-green, sparsely to usually densely white-puberulent, base cuneate or obtuse, margin crenate or nearly entire, occasionally deeply incised, apex acute or obtuse. Flowers terminal; pedicel 1–6 cm long, villous. Calyx tube adnate to whole hypanthium, semiglobose, nearly glabrous to densely villous; calyx lobes ovate-lanceolate, ovate to broadly ovate-deltoid, 1–1.8 cm long, 0.5–1.4 cm broad, adaxially glabrous, abaxially sparsely villous, margin crenate and verrucose-toothed, ciliate-villous, apex obtuse. Corolla tubular, slightly dilated above, shallowly lobed, 2–3.7 cm long, 1–1.6 cm wide, greenish yellow, base often purple, glabrous; corolla lobes ca. 1/6 as long as corolla total length, deltoid, acute at apex, very minute-ciliate. Filaments ca. 10 mm long, dilated at base, ciliate above; anthers 3–5 mm long, connective densely villous. Capsule globose below, conical above, 1.5–2.2 cm long, 1–1.3 cm in diameter. Seeds ovoid, brown-yellow.

Distribution: This species is widely distributed from the east in NE Guangxi of China at 111°E, to the west at 93°E, across 18 degrees of latitude (Map IV 1:35).

Recently Dash and Mao (2011) described *Codonopsis vadsea* as new from Yarlung Zangbo-Brahmaputra region, but I recognize it here at subspecific rank based on their description and figure.

1a. Corolla 2–3.7 cm long, corolla lobes glabrous; petiole 1–11 mm long………………35a. subsp. *tubulosa*

1b. Corolla 1.8–2.2 cm long, corolla lobes sparsely minute-ciliate; petiole 10–17 mm long………………5b. subsp. *vadsea*

35a. Codonopsis tubulosa subsp. tubulosa

Stems usually 1–3 m long, twining, rarely ascending. Petiole 1–11 mm long. Corolla 2–3.7 cm long, glabrous. Figures IV 1:35A & B.

Phenology: flowering from July to September; fruiting in September and October.

Habitats: shrubby slopes, forests, or meadows; 1400–3600 m.

Distribution: China (NE Guangxi, W Guizhou, SW Sichuan, and Yunnan) and N Myanmar.

Additional specimens examined, **CHINA**, **GUANGXI**, **Qianzhou:** Yongnan, 1400 m, grassy slopes, 30. Aug. 1964, *J. H. Huang 195* (GXMI). **GUIZHOU**, **Jinkou**, Fangjin Shan, 2250 m, 07. Aug. 1985, *W. X. Zhang et al. 077* (PE).

FIGURE IV 1:35A *Codonopsis tubulosa* Kom. subsp. *tubulosa*. **a**, a section of stem, showing opposite leaves and indumentum.

Nayong: Juren Distr., 26. Jul. 1959, *Bijie Exped. 482* (PE); loc. eodem, 1900 m, 03. Aug. 1959, *Bijie Exped. 707* (PE). **Panxian**: Bada Shan Pasture, 2100 m, canyon, 23. Aug. 1959, *Anshun Exped. 979* (PE); loc. eodem, 2300 m, canyon thickets, 19. Aug. 1959, *Anshun Exped. 907* (PE). **SICHUAN**, **Baoxing**, Ganyang, 4000 m, 12. Aug. 1936, *K. L. Chu 3558* (PE, SZ). **Butuo**: Jeluo Township, 3000 m, 14. Aug. 1960, *s. coll. 25196* (SM); Tuojo Distr., 2600 m, 24. Aug. 1959, *Sichuan Econ. Plant Surv. (Liang) 5926* (CDBI, KUN, PE). **Dechang**: 15 km from Dashan Township to the city, 2300 m, thickets, 01. Aug. 2010, *E. D. Liu et al. 2558* (KUN). **Ganluo**: Azhai, 2600 m, forests, 14. Jul. 2008, *X. J. He & C. S. Cao 184624* (SZ); Haitang, 21. Jul. 1959, *Sichuan Econ. Plant Surv. (Liang) 4292* (PE); loc. eodem, grassy, slopes, 19. Jul. 1959, *Sichuan Econ. Plant Surv. (Liang) 4052* (CDBI); Huaguo Shan, 2700 m, forests, 16. Jul. 2008, *X. J. He & C. S. Cao 184891* (SZ); Tianba, 2600 m, forests, 19, Jul. 2008, *X. J. He & C. S. Cao 193187* (SZ). **Hongxi**: Waxi, 2650 m, 06. Aug. 1959, *Sichuan Econ. Plant Surv.(Liang) 1644* (CDBI, KUN, PE). **Huidong**: Lunan Township, 15. Jun. 1959, *S. K. Wu 1038* (KUN); loc. eodem, 22. Jun. 1959, *S. K. Wu 1843* (KUN); Sunjiaqing, sunny slopes, 16. Jun. 1959, *S. K. Wu 680* (KUN). **Huili**: Mt Longdschu-schan, 17. Sept. 1914, *Handel-Mazzetti 5206* (W, WU); Maomao Township, 2400 m, 01. Jul. 1960, *X. Z. Wang 60–1037* (SZ). **Jinyang**: Boluoliangzi, forests, 2900 m, 19. May. 1959, *Sichuan Econ.Plant Surv.(Liang) 3141* (CDBI, PE, SZ); loc. eodem, 2800 m, *Sichuan Econ.Plant Surv.(Liang) 3143* (CDBI, PE); Gaofeng, 2700 m, sunny slopes, 28. Aug. 1976, *Sichuan Plant Surv. 14649* (CDBI, PE); Nanwa Distr., 2500 m, 16. Aug. 1978, *s. coll. 603* (SM). **Leibo**: Dahegou, 1250 m, edges of forests, 24. Jul. 1983, *C. S. Cao & J. B. Shi 119347* (CDBI, SZ); Dongzigou, 2340 m, 26. Jul. 1983, *M. Y. He & C. M. Tan 121542* (CDBI, SZ); Erba, 1250 m, forests, 25. Sept. 1983, *C. S.*

FIGURE IV 1:35B *Codonopsis tubulosa* Kom. subsp. *tubulosa*. **a–d**, from Sichuan: Muli, 21. Aug. 2012, by Zhuo ZHOU.

Cao & C. M. Tan 118094 (CDBI, SZ); near Gudui, 2400 m, 26. Jun. 1959, *Sichuan Econ. Plant Surv.* (*Liang*) *1006* (CDBI, PE); Sanba, 1250 m, forests, 15. Aug. 1983, *C. S. Cao & J. B. Shi 119484* (SZ). **Meigu**: s. loc., 18. Aug. 1959, *Sichuan Econ.Plant Surv.(Liang) 1652* (CDBI, KUN, PE). **Mianning**: Tuowu, 2800 m, 08. Aug. 1978, *Mianning Exped. 468* (SM); YeleTownship, Zier Shan, 2600 m, 25. Jul. 1959, *Sichuan Econ.Plant Surv.* (*Xi*) *2146* (KUN, PE). **Miyi**: Puwei Township, 2500 m, 17. Aug. 1958, *Z. He 11305* (SZ). **Muli**: 27°30′N, 101°E, 3350 m, Aug. 1922, *G. Forrest 22075* (E, K); Mt. Gibboh, 3050 m, Aug. 1928, *J. F. Rock 16935* (E); 2740 m, 21. Aug. 1905, *F. Kingdon-Ward 4563* (E, K); 917 Forest Farm, 3200 m, thickets, 13. Oct. 1993, *Q. E. Yang 93016* (PE); Baidiao Shan, 3100 m, 24. May. 1964, *J. D. Li (Muli) 64-2* (PE). **Puge**: Li'an, 2550 m, 07. Aug. 1979, *Puge Exped. 498* (SM); Tuomugou, Huangzhu Shan, 2500 m, forests, 25. Aug. 1959, *Sichuan Econ. Plant Surv.* (*Liang*) *5487* (CDBI, PE). **Tianquan**: Erlang

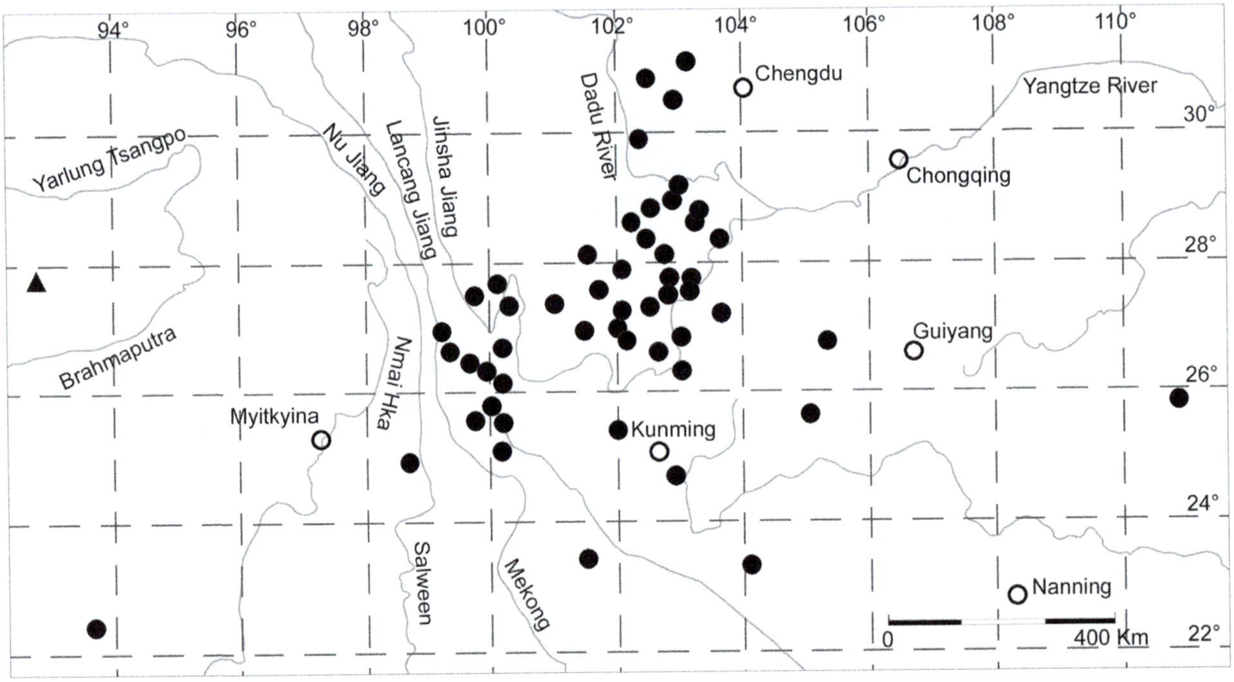

MAP IV 1:35 Distribution of *Codonopsis tubulosa* Kom. Dots: subsp. *tubulosa*; triangles: subsp. *vadsea* (S. S. Dash & A. A. Mao) D. Y. Hong.

Shan, *H. L. Tsiang 37516* (PE, SZ). **Wenchuan**: Wolong, Dengshengou, 2700 m, 09. Aug. 2007, *X. J. He et al. 150018* (SZ) & *150076* (SZ). **Xichang**: 2900 m, upon bushes, 10. Aug. 1932, *T. T. Yü 1293* (PE). **Xide**: Mianshan, 2400 m, forests, 28. 07. 2008, *X. J. He & C. S. Cao 195663* (SZ); Muku, 2700 m, 31. Jul. 2008, *X. J. He & C. S. Cao 197004* (SZ); Nibo, 2400 m, 01. Aug. 2008, *X. J. He & C. S. Cao 196778* (SZ). **Yanbian**: Dapingzi, 2800 m, 24. Aug. 1978, *Yanbian Exped. 338* (SM). **Yanyuan**: Pingchuan to Yanyuan, near the pass, 3168 m, open *Quercus* shrubs, 04. Aug. 2010, *E. D. Liu et al. 2583* (KUN); Yanjing Township, Mt Yuanbao, 2600–3000 m, 11. Jun. 1964, *T. L. Dai et al. 64-13* (PE); s. loc., 2400 m, thickets and grasses, 14. Sept. 1964, *F. G. Zhu & J. Hong 001* (PE). **Yuexi**: Bao'an, 2700 m, forests, 13. Jul. 1959, *Sichuan Econ.Plant Surv. (Liang) 3866* (CDBI, KUN, PE); Xinmin, 2500 m, forests, 25. Jul. 2008, *X. J. He & C. S. Cao 185510* (SZ). **Zhaojue**: Jiefang Gou, 2810 m, 03. Sept. 1979, *Zhaojue Surv. 956* (SM). **YUNNAN**, s. loc., 1933–1934, *Y. Tsiang 11427* (K); s. loc.,1920, *J. Cavalerie 7421* (E, K) & *7434* (E). **Chengjiang**: Lianghuang Shan, 3500 m, 23. Aug. 1939, *H. Wang 41670* (PE). **Dali**: Cangshan, Zhonghesi to Xiangshiya, 05. Oct. 1946, *T. N. Liou 21228* (PE); loc. eodem, 05. Oct. 1946, *T. N. Liou 21151* (PE) & *21229* (PE); Cangshan Temple to Zhonghesi, 10. Oct. 1946, *T. N. Liou 21467* (PE); Chienchuan-Mekong Divide, 26°34′N, 99°40′E, 3350–3660 m, Aug. 1922, *G. Forrest 21991* (E, K); E side of Diancang Shan, 25°42′N, 100°8′E, 12. Jul. 1984, *Sino-Amer. Bot. Exped. 999* (B, E, PE); Dianchang Shan, Zhonghefeng, 2050–3100 m, 03. Aug. 1990, *Murata et al. 21* (A); 25°40′ N, 3050–3400 m, Aug. 1906, *G. Forrest 3853* (E, K, PE); 2400 m, pine forests, 28. Jul. 1933, *H. T. Tsai 53899* (PE); Huadianba, ca. 3000 m, 28. Aug. 1984, *K. Iwatsuki et al. 1441* (KUN); loc. eodem, 2900 m, thickets, 10. Jul. 1984, *s. coll. 225* (KUN); Mingkwang Valley, 25°15′N, 1830–2130 m, Jun. 1912, *G. Forrest 8356* (E, K); Zhonghesi to Beigou, 05. Oct. 1946, *T. N. Liou 21307* (PE); Zhongheshi to Beiyakou, 08. Oct. 1946, *T. N. Liou 21405* (PE); Zhonghesi to Liangshanheyiding, 09. Oct. 1946, *T. N. Liou 21428* (PE) & *21436* (PE); Zhonghesi, Banshan, Aug. 1941, *H. C. Wang 1176* (PE); Zhonghe-si to Zhonghefeng, 3400 m, subalpine bushes with grasses, 25. Jul. 1960, *Yunnan NW Exped. 10513* (KUN, PE); 2800–3000 m, 10. Aug. 1913, *Limpricht 985* (W); s. loc., *McLaren 133* (E, K) & *985* (W); s. loc., 2740–3050 m, Aug. 1910, *G. Forrest 7098* (E, K). **Dongchuan**: Fazhe Forestry Centre, 3140 m, 06. Aug. 2009, *E. D. Liu et al. 2208* (KUN); loc. eodem, 2400 m, roadsides, 31. Jul. 2008, *H. Peng et al. 8986* (KUN) & *9061* (KUN); Shekuai Township, 3100 m, thickets, 26. Jul. 2008, *H. Peng et al. 8035* (KUN); loc. eodem, 2890 m, thickets, 28. Jul. 2008, *H. Peng et al. 8546* (KUN). **Eryuan**: Fengyu to Liantie Pass, 2946 m, forests, grassy slopes, 18. Aug. 2010, *E. D. Liu 2750* (KUN). **Heqing**: Songqui, Ma'er Shan, 3500 m, mixed forests, 29. Aug. 1936, *R. C. Ching 24086* (KUN); loc. eodem, 3350 m, pastures, 23 Jul. 1963, *Jinshajiang Exped. 63–6034* (KUN). **Lanping**: 3000 m, thickets, 17. Aug. 1933, *H. T. Tsai 53774* (KUN, PE,SZ). **Lijiang** (Yulong): Lameiyong, Sept. 1986, *Z. T. Wang 869-377* (PE); Laschiba, mixed forests, 03. Nov. 1939, *R. C. Ching 21881* (KUN, PE); NW of Lijiang Snow Range, Jul. 1923, *J. F.*

Rock 9798 (E); Ludian, Lidiping, Machang, 3000 m, 22, Jul. 1956, *P. I. Mao 166* (KUN, PE, SZ); loc. eodem, 3100 m, 26. Sept. 1956, *P. I. Mao 319* (KUN, PE, SZ); Yangtze bend, 27°45′N, Aug. 1913, *G. Forrest 10791* (E, K); Yung-peh-Yungning, 27°20′N, 100°48′E, 3350 m, Aug. 1922, *G. Forrest 22086* (E, K, W). **Mengzi**: Mts in north, 2130 m, s. d. *A. Henry 10167A* (K). **Qiaojia**: Dacu to Ganghaizi, 2700 m, 16. Aug. 2003, *H. Wang et al. 03–0794* (KUN); Sanchahe, 1800 m, 14. Aug. 2003, *H. Wang et al. 03–0681* (KUN). **Tengchong** (Tengyueh): s. loc., *E. B. Howell 23* (E, K). **Weixi**: s. loc., s. d. *G. Forrest 30505* (PE). **Wenshan**: Laojun Shan, 2935 m, thickets, 21. Jul. 1993, *Y. M. Shui 2851* (PE); loc. eodem, 2930 m, 21. Jul. 1993, *Y. M. Shui 2850* (PE). **Wuding**: Kuanpo, 2600 m, evergreen forests, 14. Aug. 2002, *H. Li et al. 0212* (KUN); Madeping, by fields on slopes, 14. Nov. 1958, *S. Q. Huang 292* (KUN). **Yangbi**: Yangbi River basin, 3050–3660 m, Aug. 1922, *J. F. Rock 6210* (E); Zhongshan, sunny shrubs, 24. Aug. 1984, *K. Iwatsuki et al. 874* (KUN); Zhongshan, 2600 m, grasslands, 25. Aug. 1984, *K Iwatsuki 1137* (KUN). **Zhaotong**: Qiaomaidi, 2300 m, by river, 10. Aug. 1974, s. coll. s. n.(KUN). **Cong-tchouau**: s. loc., 2690 m, Oct. 1913, *E. E. Maire 556* (E); s. loc., 1907, *F. Ducloux 235* (E); 2700 m, Sept. 1913, *E. E. Maire 569* (E) & *869* (E); NE Yunnan, San-Kia, s. loc., 2700 m, *E. E. Maire* s. n.(W). **MYANMAR**, Chin Hills, 1956, *F. Kingdon-Ward 22813* (K).

35b. Codonopsis tubulosa subsp. **vadsea** (S. S. Dash & A. A. Mao) D. Y. Hong, comb. & stat. nov. Basionym: *Codonopsis vadsea* S. S. Dash & A. A. Mao, Nelumbo 53: 17. fig. 1. 2011. Type: Yarlung Zangbo-Brahmaputra region, Kurung Kumey District, Mt Vadse, 3800 m, 04. Aug. 2010, *A. A. Mao & S. S. Dash 31846* (holotype: CAL, n. v.; isotype: CAL, n. v.).

Stems 0.5–1 m long, ascending or erect. Petiole 1–1.7 cm long. Corolla 1.8–2.2 cm long, corolla lobes very minutely ciliate.

Phenology: flowering from July to September.
Habitats: subalpine to alpine grassy slopes; 3700–4000 m.
Distribution: confined to Kurung Kumey in Yarlung Zangbo-Brahmaputra region.

Notes: According to Dash and Mao (2011), the subspecies has erect stems, but according to our field observation, in *Codonopsis* the stems are usually slender, and thus they can not be erect when they reach 1 m long. It is more likely that they are ascending or scandent.

36. Codonopsis microtubulosa Z. T. Wang & G. J. Xu, Acta Phytotax. Sin. 31(2): 184. fig. 1. 1993; Hong, Lammers & Klein in Wu, Raven & Hong, Fl. China 19: 516. 2011, pro syn. Type: China, Sichuan, Nanping (Jiuzhaigou), 2500 m, 23. Jul. 1986, *Z. T. Wang & B. Y. Yü 867206* (holotype: CPU; isotype: PE).

FIGURE IV 1:36A *Codonopsis microtubulosa* Z. T. Wang & G. J. Xu, showing root, caudex, a stem, and basal part of the other two stems.

Herbs perennial, pilose throughout. Roots carrot-shaped, 8–21 cm long, 0.5–1 cm in diameter, simple or sometimes branched below middle. Caudex often short, 2–4 mm long, 6–10 mm in diameter. Stems 1 to several from one caudex, scandent or twining, 45–90 cm long, branched from base, less often simple. Leaves alternate, but those at lower and middle parts of stem usually opposite; petiole 2–12 mm long; leaf blade cordate-ovate, 1.1–3.6 cm long, 0.6–3.3 cm broad, base shallowly to deeply cordate, apex obtuse, margin crenate or dentate. Flowers solitary, terminal on both main stem and branches; pedicel 1.1–2.9 cm long. Calyx tube adnate to whole hypanthium, semi-globose, 3.5–4 mm long, 5.2–6 mm in diameter, nearly glabrous; calyx lobes oblong-lanceolate, 6–8 mm long, 2–4 mm broad, both surfaces glabeous, apex obtuse, margin entire, villous-ciliate. Corolla tubular, 1–1.5 cm long, 0.5–1 cm wide, glabrous, lobed to 1/5; corolla lobes 2.5–4.2 mm long, 1.5–2.5 mm broad, apex acute. Fruit unmature. Figure IV 1:36A.

Notes: We reduced this species as a synonym of *C. pilosula* in *Fl. China* (Hong et al., 2011). Recently we examined two collections in P, *Farges 1366* and *Farges* s. n. The former has four individuals and the latter has 30 individuals. Thus, one of the collections came from a rather large population and therefore the material is adequate for judging the status of *C. microtubulosa*. It is a very distinct species, differing from all other species in having long pilose hairs, and from *C. benthamii* and *C. tubulosa*, which are also of tubular corolla and twining stems, in having much smaller corolla only 1–1.5 cm long, 0.5–1 cm wide (vs. 1.8–4 cm long, 1–1.8 cm wide).

Phenology: flowering in July and August.
Habitats: ca. 2500 m.
Distribution: **endemic** to China: N Chongqing and N Sichuan (Map IV 1:36).

Additional specimens examined, **CHINA**, **CHONGQING**, **Chengkou**; s. loc. 03. Aug. 1896, *R. P. Farges 1366* (P); s. loc. s. d. *R. P. Farges* s. n. (P).

37. Codonopsis subglobosa W. W. Smith, Notes Roy. Bot. Gard. Edinburgh 8: 108. 1913; Nannfeldt, Acta Horti Gothob. 5: 28. 1930 & Notes Roy. Bot. Gard. Edinburgh 16: 159. 1931; Tsoong, Contr. Inst. Bot. Nat. Acad. Peiping 3(3): 100. 1935; Anonymous, Icon. Cormophyt. Sin. 4: 381. fig. 6176. 1975; Shen & Hong in Hong, Fl. Reipubl. Popularis Sin. 73(2): 42. 1983; Hong, Lammers & Klein in Wu, Raven & Hong, Fl. China 19: 516. 2011. Type: China, Yunnan, "the eastern flank of the Lichiang Range" (Lijiang), 27°20′N, 2740 m, Sept. 1910, *G. Forrest 6684* (lectotype here designated: E; isolectotypes: BM, K).

Roots carrot-shaped or cylindrical, 30–50 cm long, 1.5–8 cm in diameter. Caudex short or rarely up to 40 cm long. Stems twining, often over 2 m long, branched, yellow-green or green, sparsely white-hirsute. Leaves on main stem and branches alternate, those on branchlets nearly opposite; petiole 0.5–2 cm long, sparsely white-hirsute; leaf blade broad-ovate, ovate to narrowly ovate, 0.5–3 cm long, 0.5–2.5 cm broad, adaxially green, hispidulous, abaxially gray-green, hirsute, base shallowly cordate to rounded, margin slightly waved or shallowly crenate, apex obtuse or acute. Flowers solitary, terminal on branchlets, or leaf-opposed; pedicel hirsute. Calyx tube adnate to nearly whole hypanthium, semiglobose, with 10 conspicuous radiate ribs, ribs sparsely white-hirsute, rarely glabrous; calyx lobes distant to each other, nearly orbicular or rhomboid-ovate, 9–13 mm long, 6–8 mm broad, abaxially white-hirsute, rarely glabrous, veins conspicuous, margin serrate and ciliate, often revolute. Corolla globose-campanulate, 2–2.5 cm long, ca 2.5 cm wide, pale yellow-green, apex dark red-purple; corolla lobes broad-triangular. Filaments slightly dilated at base. Stigma 3-fid; each stigma segment 2-lobed. Capsule semiglobose below, conical or rostrate above. Seeds ellipsoid or ovoid, yellow-brown. Figures IV 1:37A & B.

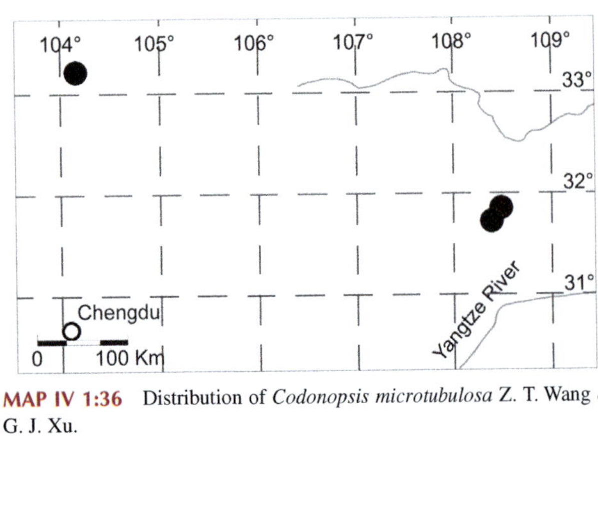

MAP IV 1:36 Distribution of *Codonopsis microtubulosa* Z. T. Wang & G. J. Xu.

Phenology: flowering from July to September; fruiting from September to November.
Habitats: stony mountain meadows or scrubs; 2400–4100 m.
Distribution: **endemic** to China: W Sichuan, E Tibet, and NW Yunnan (Map IV 1:37).

Additional specimens examined, **CHINA**, **SICHUAN**, **Baiyu**: Samar-Garyi, 30°43′ N, 99°05′ E, 3110 m, grassy slopes by forest, 24. Jul. 2010, *Kham Exped. 10–0552* (PE). **Batang**: Dong Distr., Yarigong, 3200 m, *Quercus* thickets, 14. Jul. 1964, *J. D. Li 64-37* (PE); Rongguo Shan, 3800 m, grassy slopes, 20. Jul. 1964, *J. D. Li et al. 64-46* (PE); Yarigong, 29°25′ N, 99°22′ E, 3072 m, waste fields, 21. Jul. 2010, *Kham Exped. 10–0404* (PE); Zangqênrong, 2800–3000 m, sunny thickets, 19. Jul. 1964, *J. D. Li et al. 64-45* (PE); Zongza, 3000 m, 16. Jul. 1964, *J. D. Li 64-44* (PE, SM). **Danba**: Dasang, 2600 m, 29. Jul. 1960, *F. D. Fu & M. L. Chen 0426* (CDBI); Taiping Township, Heifending, 3000 m, 13. Jul. 1964, *T. L. Dai et al. (Danba) 64-8* (PE, SM). **Daocheng**: Daocheng to Yading, 3400 m, by road, 03. Aug. 2012, *F. S. Yang & Y. L. Liu 075* (PE); Gongga-ling, near temple, 3100 m, open thickets, 23. Aug.

FIGURE IV 1:37A *Codonopsis subglobosa* W. W. Smith, redrawn from *Icon. Cormophyt. Sin.* 4. 1975. A section of a twining stem.

1937, *T. T. Yü 12887* (KUN, PE); Julong Township, 3400 m, thickets or edges of forests, 28. Jul. 1971, *Z. G. Liu 0431* (CDBI); Julong, 3900 m, thickets, 20. Jul. 1973, *Sichuan Veget. Surv. 2178* (CDBI, KUN, PE); Shengmu Township, Gêwa Village, 3500 m, 02. Aug. 1964, *J. D. Li & Z. M. Xu 64-53* (PE, SM). **Dawu**: Baimei, 07. Aug. 1986, *Z. T. Wang 869-355* (PE); Qingdaqiao, 3240 m, rocky grasses, 17. Jul. 1964, *T. L. Dai (Qianning) 64-5* (PE, SM); Wari Distr., Muru Township, 3500 m, thickets, 25. Aug. 1974, *Sichuan Plant Surv. 5901* (CDBI, PE); s. loc., 3700 m, grasslands, 25. Jul. 1964, *T. L. Dai (Qianning) 64-24* (PE). **Dêrong**: Deep Gorge of Ding Qu, 28°46′54″ N, 99°18′27″ E, 2600–2900 m, 21. Jul. 2004, *D. E. Boufford et al. 31033* (A). **Jiulong**: Tanggo, 3240 m, edges of forests, 13. Jul. 1980, *Sichuan Plant Surv. 28280* (CDBI). **Kangding**: s. loc., 1893, *J. A. Sourie 129* (G, K, P); s. loc., 1893, *A. Soulie 125* (P); 2660–4110 m, *A. E. Pratt 531* (K, P). **Litang**: Litang Qu Valley, 29°42′ N, 100°33′ E, 4022 m, thickets, 06. Sept. 2010, *Kunming-Edinburgh Exped. to Sichuan 183* (E). **Muli**: Bo'ao, Baidiao, 2700 m, by river, 22. Sept. 1959, *S. K. Wu 3182* (KUN); Bo'ao, Liziping, 2500 m, 17. Sept. 1980, *C. S. Cao et al. 8770*

FIGURE IV 1:37B *Codonopsis subglobosa* W. W. Smith. **a**, from Sichuan: Daocheng, 03. Aug. 2012, by Fu-Sheng YANG; **b**, s. loc., 19. Sept. 2008, by Bo XU; **c–f**, from Sichuan: Muli, 21. Sept. 2011, by De-Yuan HONG. **d**, showing 3 stigma segments; **e**, showing 4 stigma segments, each of which is 2-lobed; **c**, showing elongated caudex and base of stems.

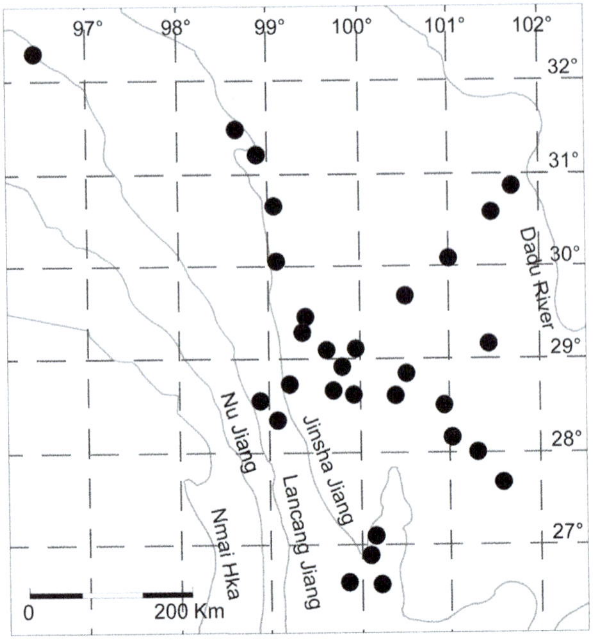

MAP IV 1:37 Distribution of *Codonopsis subglobosa* W. W. Smith.

(CDBI, SZ); Dingdonggou to Bowa Forest Farm, 2550 m, thickets, 14. Aug. 1978, *C. S. Cao et al. 7521* (CDBI, SZ); Ker Township, 28°11′27″N, 101°06′49″E, 2900 m, thickets, 22. Sept. 2011, *D. Y. Hong et al. H11035* (PE); Liziping Township, Baicaoping, 2750 m, thickets by road, 21. Sept. 2011, *D. Y. Hong et al. H11029* (PE); Shuiluo Township, 2350–2850 m, thickets, 24. Oct. 1959, *S. K. Wu 3482* (KUN); Wachang, 2850 m, thicket, 27. Oct. 1982, *Sichuan Plant Surv. 526* (CDBI); Wa'ertsa, 3100 m, scrubs, 03. Nov. 1937, *T. T. Yü 14794* (E, PE); Yalientsa, 3000 m, thickets, 02. Sept. 1937, *T. T. Yü 14151* (KUN, PE); s. loc., 2740 m, 21. Aug. 1905, *F. Kingdon-Ward 4559* (E). **Xiangcheng**: Qianjin Township, 3800 m, thickets, 17. Jul. 1972, *Z. G. Liu 1484* (CDBI); Rawu, Rachong, forests, 16. Aug. 1981, *Qinghai-Xizang Exped. 4128* (CDBI); Reda, 3450–3650 m, 29°6′11″ N, 99°37′ 55″E, 16. Jul. 1998, *D. E. Boufford et al. 28793* (E); near Shagong, 3113 m, thickets by road, 12. Aug. 2010, *E. D. Liu 2689* (KUN); near the city, 3060 m, 25. Jul. 1964, *J. D. Li & Z. M. Xu 50* (SM); near No.4 Distr. Government, 3400 m, thickets, 11. Aug. 1981, *Qinghai-Xizang Exped. 4641* (PE); 2500–3200 m, Jul. 1964, Ezu & *T. F. Lan 006* (PE). **Yajiang**: s. loc., 2800 m, 10. Aug. 1964, *J. D. Li & Z. M. Xu 64-60* (SM); s. loc., 04. Jul. 1930, *Huangkai et al. 676* (PE). **Yanyuan**: Bailing Shan 06. Sept. 1986, *Z. T. Wang 869-383* (PE); Yanyuan to Kwapi, 27°45′N, 2900–3200 m, 03. Oct. 1914, *Handel-Mazzetti 5456* (E, K, W, WU). **TIBET, Jomda**: Gamtog to Gyorxung, 07. Sept. 1951, *S. X. Jia 284* (PE). **YUNNAN, Dêqên**: Baima Shan, 3600 m, 08. Aug. 1976, *Wu 4364* (KUN); vicinity of Dêqên town, 3300 m, valleys, 22. Sept. 1986, *H. Sun & Z. G. Qian 710* (KUN); 3660–3960 m, 01. Aug. 1913, *F. Kingdon-Ward 930* (E). **Heqing**: Heqing Valley, 2440 m, *G. Forrest 83* (E). **Jianchuan**: Jianchuan to Yangcheng, 2300 m, 01. Oct. 1962, *C. C. Lu* s. n. (KUN). **Lijiang** (Yulong): Baisa, roadsides, open thickets, 10. Sept. 1939, *R. C. Ching 30602* (KUN, PE); Baisa, thickets by fields, 25. Sept. 1959, *K. M. Feng 22742* (HNWP, KUN, PE); between Dali and Lijiang, 2800–2900 m, 23. Sept. 1916, *Handel-Mazzetti 10058* (G. W, WU); Laschiba, open thickets, 29. Aug. 1939, *R. C. Ching 21374* (KUN, PE); between Lijiang and Dali, Aug. 1922, *J. F. Rock 6289* (E, PE); Tai-mgo-Koo, mixed forests, 21. Sept. 1939, *R. C. Ching 21695 A* (KUN, PE); Xuesongcun, 3000 m, thickets, 22. Sept. 1937, *T. T. Yü 15601* (KUN, PE); Yuhu Township, 2500 m, 04. Sept. 1955, *K. M. Feng 21345* (KUN, PE, SZ); Yulong Snow Range, Sept. 1987, *B & L 12405* (E); Yulong Snow Range, open pastures, 18. Sept. 1939, *R. C. Ching 30611* (KUN); s. loc., 2800 m, 03. Aug. 1914, *C. Schneider 2114* (G, K). **Shangri-la** (Zhongdian): Dongwang, 2800 m, 21. Aug. 1988, *J. S. Yang 88-118* (KUN, PE).

38. Codonopsis bragaensis Grey-Wilson, Plantsman, 12 (2): 99. fig. in p. 72. 1990. Type: Nepal, Gandaki Zone, Manang Distr., Marsyandi Valley, slopes of Annapurna 3 to the SW of Braga, c. 3800 m, July 1989, *C. Grey-Wilson et al.* s. n.(holotype: K).

Basal parts not seen. Stems procumbent, twining above, about 1 m long, branched with leafy branches from lower half, unbranched or with few-leaved branches from upper half, densely pubescent. Leaves alternate, but those near base of stems opposite: petiole less than 2 mm long; leaf blade ovate to ovate-lanceolate, 1–2.9 cm long, 0.9–2.4 cm broad, pubescent on both surfaces, base rounded to slightly cordate, margin entire, apex obtuse to subobtuse. Flowers solitary, terminal on main stem, occasionally a lateral flower found in axil of uppermost leaf. Calyx tube adnate to whole hypanthium, pubescent; calyx lobes lanceolate-triangular, 0.9–1.2 cm long, margin entire but rather undulate. Corolla campanulate, 1.6–2.4 cm long, white flushed with mauve, with blue veins, covered with white setae; corolla lobes somewhat spreading, broad-triangular, ca. 6 mm long. Stamens glabrous; filaments and anthers each ca. 5 mm long. Fruit not seen. Figure IV 1:38.

Phenology: flowering in July.

Habitats: low scrubs of *Berberis*, *Rosa sericea* and *Potentilla fruticosa*; ca. 4000 m.

Distibution: **endemic** to Central Nepal (Map IV 1:38).

Notes: *C. bragaensis* was locally abundant in the type locality, but it has not been found elsewhere so far. The species is distinct from all the others in having an entirely pubescent calyx, lanceolate-triangular calyx lobes, and a corolla which is densely covered with white setae and has a characteristic coloration.

FIGURE IV 1:38 *Codonopsis bragaensis* Grey-Wilson, showing a section of twining stem and a terminal flower. *Redrawn from Grey-Wilson, 1990.*

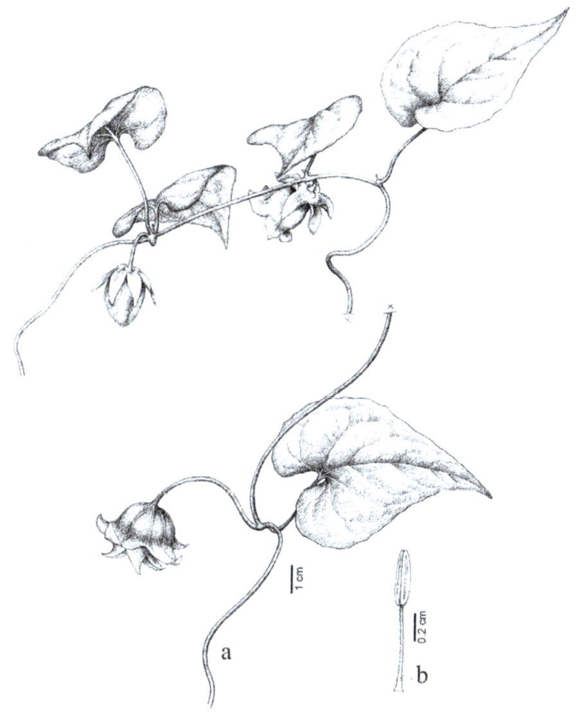

FIGURE IV 1:39A *Codonopsis cordifolioidea* P. C. Tsoong. **a**, two sections of a twining stem, showing a flower, a flower bud, and a 3-loculicidal capsule; **b**, a stamen.

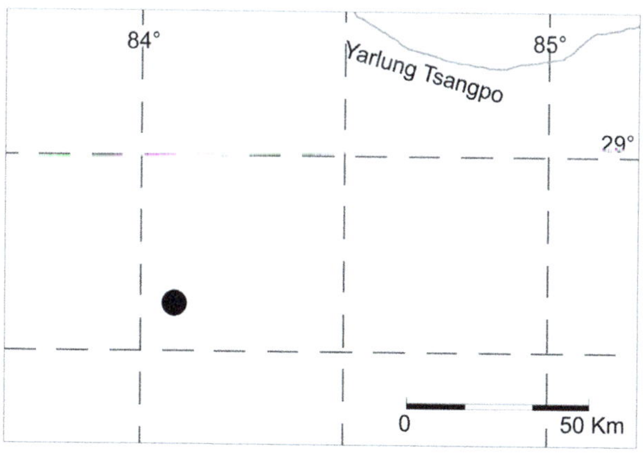

MAP IV 1:38 Distribution of *Codonopsis bragaensis* Grey-Wilson.

39. Codonopsis cordifolioidea P. C. Tsoong, Contr. Inst. Bot. Nat. Acad. Peiping 3(3): 93. fig. 7 & pl. 12. 1935; Shen & Hong in Hong, Fl. Reipubl. Popularis Sin. 73(2): 38. 1983; Hong, Lammers & Klein in Wu, Raven & Hong, Fl. China 19: 515. 2011. Type: China, Yunnan, Shangpa County (Fugong), woods, 08. Oct. 1933. *H. T. Tsai 54814* [holotype: PE(00032074); isotypes: PE (00032071 & 00032075)].

Basal parts not seen. Plants glabrous, or nearly glabrous, or only leaves sometimes very sparsely hispidulous. Stems up to 5 m long, twining, with a few, very short, branches. Leaves on main stem widely spaced, alternate, ca. 10 cm distant between two leaves; petiole filiform, 3–9 cm long; leaf blade broad-ovate, 4–10 cm long, 3–7 cm broad, veins conspicuous, lateral veins from base, adaxially green, abaxially gray-green, base nearly cordate, with a square sinus, margin nearly entire, apex short-acuminate or acute; leaves on branches usually only 2 and opposite, petiole less than 1 cm long, similar to those on main stem in shape, but sinus not square. Flowers solitary, nearly opposite to leaves; pedicel 3–6 cm long. Calyx tube adnate to hypanthium up to middle, semiglobose; calyx lobes deltoid-lanceolate, 7–10 mm long, 5–6 mm broad, entire, apex acuminate; sinus between calyx lobes narrow. Corolla campanulate, 1.7–1.8 cm long, 1–1.2 cm wide, 5-lobed to middle, purple, dark blue or white but purple-red inside throat and limb; corolla lobes lanceolate-deltoid. Filaments ca. 5 mm long, dilated at base. Capsule semiglobose in lower half, rostrate above, 1.3–1.8 cm in diameter. Seeds ellipsoid, brown. Figure IV 1:39A.

Phenology: flowering and fruiting from July to October.
Habitats: forests at an altitude of 1400–2200 m.
Distribution: **endemic** to China: NW Yunnan (Map IV 1:39).

Additional specimens examined, **CHINA, YUNNAN, Fugong**: Lishadi Township, 1830 m, evergreen broad-leaved forests, 11. Aug. 2005, *Gaoligong Shan Biodiversity Survey 26972* (CAS); s. loc., 20. Sept. 1933, *H. T. Tsai 56554* (paratypes: PE). **Gongshan**: Drungjiang, 1750 m, secondary forests, 24. Aug. 1982, *Qinghai-Xizang Exped.*

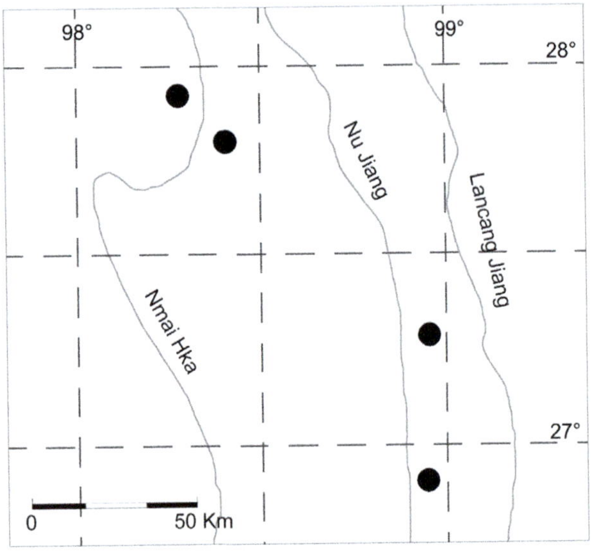

MAP IV 1:39 Distribution of *Codonopsis cordifolioidea* P. C. Tsoong.

9562 (KUN, PE); W bank of Drungjiang, 1400 m, forests, 08. Dec. 1990, *Drungjiang Exped. 925* (KUN); Drungjiang, Meiliwang, 1800 m, evergreen forests, 10. Jan. 1991, *Drungjiang Exped. 1936* (KUN); Drungjiang Township, 1600 m, 30. Dec. 1990, *Drungjiang Invest. Team 1446* (CAS); loc. eodem, 1400 m, 08. Dec. 1990, *Drungjiang Invest. Team 925* (CAS); loc. eodem, 1460 m, 30. Oct. 2004, *Gaoligong Shan Biodiversity Survey 21121* (CAS) ; Gameilai River, 1400 m, thickets, 30. Dec. 1990, *Drungjiang Exped. 1446* (KUN); W slope of Gaoligong Shan, grasses at edges of evergreen broad-leaved forests, 2200 m, 28. Jul. 1982, *Qinghai-Xizang Exped. 8817* (KUN, PE).

40. Codonopsis micrantha Chipp, J. Linn. Soc. Bot. 38: 382. 1908; Anthony, Notes Roy. Bot. Gard. Edinburgh 15: 180. 1926; Tsoong, Contr. Inst. Bot. Nat. Acad. Peiping 3(3): 97. fig. 10. 1935; Wu in Wu & Li, Yunnan Trop. Subtrop. Fl. Res. Rep. 1: 81. 1965; Anonymous, Icon. Cormophyt. Sin. 4: 377. fig. 6168. 1975; Shen & Hong in Hong, Fl. Reipubl. Popularis Sin. 73(2): 42. 1983; Hong, Lammers & Klein in Wu, Raven & Hong, Fl. China 19: 516. 2011. Type: China, Yunnan, Yunnanfu (Kunming), *P. Ducloux 513* (holotype: K).

Campanumoea violifolia H. Lév., Cat. Pl. Yunnan 24. 1915. Type: China, Yunnan, "haies - plaine de Tche-Hay", 2900 m, Aug. 1911, *E. E. Maire* s. n.(holotype: E ; isotype : E)

Melothria violifolia H. Lév., Cat. Pl. Yunnan, 65. 1916. Type: China, Yunnan,"haies de la plaine á Ta-Kiao", 2900 m, Aug. 1912. *E. E. Maire* s. n.(holotype: E).

Roots long-cylindrical, 20–30 cm long, 0.5–1 cm in diameter. Stems over 1 m long, twining, branched, yellow-green or green, sparsely villous or nearly glabrous. Leaves alternate, opposite on branches; petiole 2–5 cm long; leaf blade cordate, 2–6 cm long, 2.4–4 cm broad, adaxially green, glabrous or sparsely white-hirsute, abaxially gray-green, white-hirsute or nearly glabrous, base deeply cordate, basal lobes slightly incurved, sinus broad and obtuse to nearly square, margin crenate or serrate, apex obtuse or acute. Flowers solitary, terminal or leaf-opposed; pedicel 1–3 cm long, glabrous. Calyx tube adnate to hypanthium up to near top, semiglobose; calyx lobes triangular-lanceolate, 1–1.4 cm long, 0.4–0.6 cm broad, apex acute, glabrous or sparsely ciliate; sinus between calyx lobes narrow and pointed. Corolla campanulate, 7–9 mm long, 0.7–0.9 cm wide, divided to middle, white; corolla lobes triangular, glabrous or ciliate. Stamens glabrous; filaments ca. 3 mm long, slightly dilated at base; anthers ca. 2 mm long. Ovary almost entirely inferior to corolla. Capsule semiglobose below, conical and rostrate above, 8–10 mm in diameter, superior part very short, 2.5–3.5 mm long. Seeds ovoid, slightly compressed, short-tailed, ca. 1 mm long, 0.5 mm wide, brown-yellow. Figures IV 1:40A & B.

FIGURE IV 1:40A *Codonopsis micrantha* Chipp, redrawn with permission from *Icon. Cormophyt. Sin.* 4. 1975. **a**, lower-middle part of a stem; **b**, a section of a twining stem.

Phenology: flowering and fruiting from July to October.

Habitats: mountain scrubs or open forests; 1300–2900 m.

Distribution: **endemic** to China: SW Guizhou, SW Sichuan, and Yunnan (Map IV 1:40).

FIGURE IV 1:40B *Codonopsis micrantha* Chipp. From Sichuan: Panzhihua, 13. Sept. 2010, by Yun-Qing HAO.

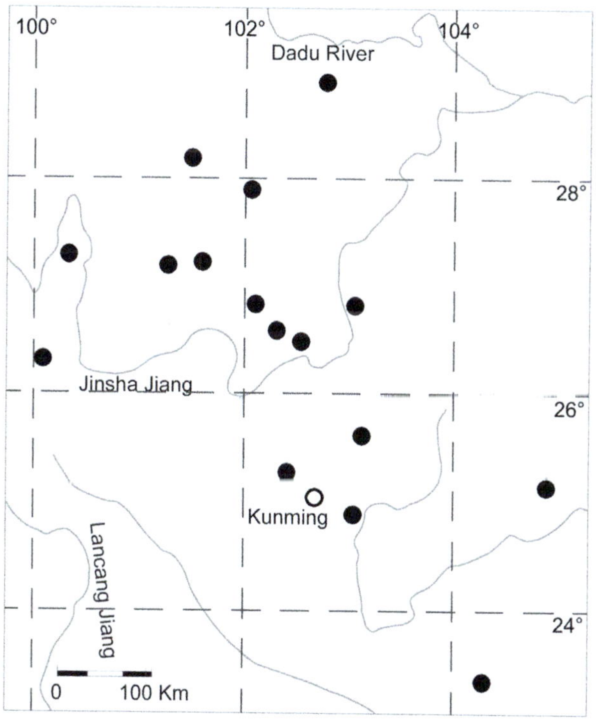

MAP IV 1:40 Distribution of *Codonopsis micrantha* Chipp.

Additional specimens examined, **CHINA**, **GUIZHOU**, **Xingyi**: Baling, 1300 m, by cultivated fields, 09. Aug. 1960, *Z. S. Zhang & Y. T. Chang 7320* (PE). **SICHUAN**, **Ganluo**: s. loc., 1800 m, 10. Sept. 1979, *s. coll. 867* (SM). **Huidong**: Matou Shan, 2200 m, 16. Aug. 1961, *T. L. Dai & S. P. Tang 11843* (SM); Xincun, 2350 m, 14. Sept. 1961, *s. coll. 11936* (SM). **Huili**: s. loc., 1950 m, 20. Sept. 1914, *Handel-Mazzetti 5231* (E, K, W, WU). **Miyi**: Nanba, 2300 m, meadows by stream, 02. May. 1959, *S. K. Wu (Sichuan Econ.Plant Surv.) 454* (KUN). **Muli**: Bowa, Baidiao, 2300 m, roadsides, 15. Sept. 1959, *S. K. Wu (Sichuan Econ. Plant Surv.) 2700* (KUN). **Xichang**: Luoji Shan, 21. Aug. 1978, *N. C. Guo 445* (SM). **Yanbian**: Hongbao Township, 27°07′ N, 101°34′ E, 2400 m, thickets, 17. Sept. 2002, *X. F. Gao et al.3615* (CDBI); Qinghe Township, 1300 m, thickets, 25. Sept. 2002, *X. F. Gao et al. 3939* (CDBI, PE); Yankou Township, 27°06′ N, 101°18′ E, 2220 m, 19. Sept. 2002, *X. F. Gao et al. 3630* (CDBI). **YUNNAN**, **Eryuan**: Niujie, Huoyan Shan, forests, 24. Sept. 1975, *B. Y. Qiu 61065* (KUN). **Fuming**: Yongding, Laoqing Shan, 2200 m, mixed forests, 19. Oct. 1964, *B. Y. Qiu 596099* (KUN, PE). **Kunming**: Heliu Wan, in shrubs, 18. Aug. 1950, *P. I. Mao 57* (PE). **Lijiang** (Yulong): Yulong Snow Range, Baishuihe, 2600 m, mixed forests, 18. Aug. 1942, *K. M. Feng 9072* (KUN, PE). **Qiaojia** (Ca-Kiao): 2900 m, 1913, *E. E. Maire 754* (E). **Songming**: Guodong, 25. Aug. 1957, *B. Y. Qiu 54996* (KUN, PE). **Wenshan**: Laojun Shan, 2300 m, thickets 04. Aug. 1993, *Y. M. Shui 3264* (PE).

Section 3. Lanceolatae D. Y. Hong, sect. nov. Type: *Codonopsis lanceolata* (Sieb. & Zucc.) Trautv. (*Campanumoea lanceolata* Sieb. & Zucc.)

Roots carrot-shaped, ellipsoid, or globose. Stems always twining. Leaves at top of branches usually 3 or 4 in a pseudo-verticillate fascicle. Seeds winged. Plants not fetid.

Two species in E and NE Asia: E China, Japan, Korea Peninsula, and the Far East of Russia. They grow in meadows, shrubs and forests at lowlands below 2400 m.

41. Codonopsis lanceolata (Sieb. & Zucc.) Trautv., Trudy Imp. S.-Peterbursk. Bot. Sada 6: 46. 1879; Hemsley, J. Linn. Soc. Bot. 26: 5. 1889; Komarov, Trudy Imp. S.-Peterburgsk. Bot. Sada 29(1): 109. 1908; Chipp, J. Linn. Soc. Bot. 38: 379. 1908; Anthony, Notes Roy. Bot. Gard. Edinburgh 15: 182. 1926; Tsoong, Contr. Inst. Bot. Nat. Acad. Peiping 3(3): 95. 1935; Anonymous, Icon. Cormophyt. Sin. 4: 383. fig. 6180. 1975; Shen & Hong in Hong, Fl. Reipubl. Popularis Sin. 73(2): 37. 1983; Hong, Lammers & Klein in Wu, Raven & Hong, Fl. China 19: 515. 2011. ≡*Campanumoea lanceolata* Sieb. & Zucc., Fl. Jap. 1: 174. tab. 91. 1835. ≡*Glossocomia lanceolata* (Sieb. & Zucc.) Maxim., Bull. Acad. Imp. Sci. Saint-Pétersbourg 31: 62. 1886. Type: Siebold & Zuccarini's tab. cited (lectotype here designated).

Codonopsis lanceolata (Sieb. & Zucc.) Trautv. var. *emaculata* Honda, Bot. Mag. Tokyo 50: 436. 1936. *Codonopsis lanceolata* (Sieb. & Zucc.) Trautv. f. *emaculata* (Honda) H. Hara, Enum. Sperm. Jap. 2: 98. 1952. Type: not designated.

Codonopsis lanceolata (Sieb. & Zucc.) Trautv. var. *ohmurae* T. Koyama, J. Jap. Bot. 32: 61. 1957. Type: Japan, Honshu, Totomi Prov., Shibukawa Spa, serpentinous rocks, oak forests, 02. Sept. 1956, T. *Ohmura* 16743 (holotype: TI).

Glossocomia hortensis Rupr., Bull. Cl. Phys.-Math. Acad. Imp. Sci. Saint-Pétersbourg 15: 209. 1857. Type: not designated.

Campanumoea japonica Sieb. ex Morr., Belg. Hort. 337. 1863. Type: not designated.
Codonopsis bodinieri H. Lév., Fl. Kouy-Tchéou. 57. 1914 & Cat. Pl. Yunnan, 24. 1915. Type: China, Guizhou, Guiyang (Kouy-Yang, mont. du Collége, gorges de Yang-Pa), 20. Jul. 1898, *Bodinier* 2437 (lectotype here designated : E) ; Guizhou, Tsin Gai, Gai Kao-Po, Jul. 1903. *Bodinier* 1180 (paratype: E).
Codonopsis yesoensis Nakai, Rep. Veg. Daisetsu Mts.: 56. 71. 1930. Type: not designated.

Plants glabrous throughout or occasionally sparsely villous on stems and leaves. Roots usually carrot-shaped, 10–20 cm long, 1–6 cm in diameter. Caudex nearly cylindrical. Stems twining, over 1 m long, yellow-green tinged purplish, often branched. Leaves on main stem mostly alternate, those at end of branches usually in a group of 3 or 4, nearly opposite and pseudo-verticillate, lanceolate or narrowly rhomboid-ovate; petiole 1–5 mm long; leaf blade rhomboid-ovate, narrowly ovate or elliptic, 3–10 cm long, 1.5–4.5 cm broad, adaxially green, abaxially gray-green, veins obvious, base attenuate, margin usually entire or sparsely crenate, apex acute or obtuse. Flowers solitary or paired on top of branchlets; pedicel 1–9 cm long. Calyx tube adnate to whole hypanthium, semiglobose; calyx lobes ovate-deltoid, 1.3–3 cm long, 0.5–1.2 cm broad, acute, entire; sinus between calyx lobes acute or gradually becoming broader after anthesis. Corolla campanulate, 2–4 cm long, 2–3.5 cm wide, shallowly lobed; corolla lobes deltoid, recurved, 0.5–1 cm long, yellow-green or milk-white, with purple spots. Stamens glabrous; filaments subulate, 4–6 mm long, slightly dilated at base. Capsule semiglobose below, rostrate at top, 2–2.5 cm in diameter. Seeds ovoid, winged, brown. Figures IV 1:41A & B.

Chromosomes: 2n=16 (Ge & Wang, 1990; Nishikawa, 1985; Probatova et al., 2011; Yoo & Lee, 1989).

Phenology: flowering and fruiting from July to September.

Habitats: bushes, edges of forests, or open woods; below 2400 m.

Distribution: East Asia: China, Japan, Korea Peninsula and Russia (the Far East) (Map IV 1:41).

Additional specimens examined, **CHINA**, **ANHUI**, **Dongzhi**: Xiangling, 380 m, 23. Sept. 1997, *C. M. Tan 971202* (PE). **Huangshan:** Chüishihlin to Tangkou, 06. Aug. 1935, *T. N. Liou & P. C. Tsoong 2136* (PE); Maopong, forests, 25. Oct. 1951, *Huadong Working Station 6141* (PE); Songgúen, 700 m, 14. Aug. 1973, *G. S. Zhou et al. 512* (PE); near Tangkou, 10. Aug. 1935, *T. N. Liou & P. C. Tsoong 2342* (PE); near Yixiantian, 31. Oct. 1951, *Huadong Working Station 6441* (PE); s. loc., 27. Jul. 1930, *S. S. Chien 1160* (W). **Jiuhua Shan**: Dongya-si to Huixiangge, 16. Oct. 1951, *Huadong Working Station 5894* (PE); s. loc., 540 m, 02. Aug. 1934, *C. S. Fan & Y. Y. Li 195* (E, K). **BEIJING**, s. loc., 1877, *Bredshneider 477* (K). **Changping**: Heishanzhai, 13. Aug. 1970, *Bei Yao 67* (PE); near Mingling, 230 m, 10. Aug. 1931, *K. M. Liou 672* (PE). **Fangshan**: Dizang'en, 27. Aug. 1956, *PE Herbarium 3414* (PE); Shangfang Shan, 400 m, damp & shaded places, 05. Sept. 1932, *W. Y. Hsia 3255* (PE); loc. eodem, 09. Sept. 32, *W. Y. Hsia 3336* (PE); loc. eodem, 03. Oct. 1934, *T. N. Liou 8292* (PE); Yuegezhuang, 14. Jun. 1971, *Beijing Chinese Mater. Medic. Exped. (Fang Shan) 228* (PE). **Fengtai** : Xiaoloumeng, Dagou, 12. Aug. 1953, *C. G. Yang 1423* (PE). **Haidian**: Jinshan-si, *C. F. Li & W. K. Hsia Hopei Exped. 1930* (PE). **Mentougou**: forest farm, 17. Aug. 1998, *Beijing Normal Univ. 972086* (PE). **Miyun**: Potou Forest Farm, ca. 1000 m, thickets, 03. Oct. 2011, *D. Y. Hong et al. H11045* (PE); Baichengzi, 28. Aug. 1972, *s. coll. Mi 826* (PE). **FUJIAN**, **Ninghua**: Zhongsha, 11. Sept. 1958, *Fudan Univ. W Fujian Exped. 92642* (KUN). **GUANGDONG**, **Hongkong**: Ma On Shan, 10. Aug. 2003, *S. Y. Hu & P. P. H. But 23977* (K); s. loc., 20. Aug. 2000, *S. Y. Hu & S. Liu. 133* (K, KUN); s. loc., 28. Aug. 1993, *P. Chow 138* (KUN). **Huiyang**: Lianhua Shan, *W. T. Tsang 25568* (E). **Lechang**: Heo-tze-ling, 17. Oct. 1928, *Y. Tsiang 1380* (E); Pingshi, open woods, 19. Aug. 1942, *S. H. Chun 2177* (KUN). **Luofu Shan**: 9–27. Aug. 1917, *E. D. Merrill. 10753* (K). **Renhua**: Changjiang Township, 1670 m, grasslands, 26. Aug. 1958, *L. Deng 7218* (KUN); **Ruyuan**: s. loc., forests, 05. Oct. 1933, *S. P. Ko 53319* (KUN). **GUANGXI**, **Lingchuan**: Xinzhai Township, 1100 m, roadsides, 11. Nov. 1963, *X. Y. Chen 53790* (KUN). **HEBEI**, **Fuping**: Wayaozi Village, 25. Aug. 1934, *K. M. Liou 3384* (PE). **Pingshan**: Muchang Managerment Area, 900 m, *s. coll.* s.n. (PE). **Xiaowutai Shan**: Yangjiaping, ca. 1100 m, 18. Aug. 1921. *H. Smith 461* (E, LD);Yangjiaping, Donggou, 25. Sept. 1951, *C. G. Yang 494* (PE); Yangjiaping, Wayao, 02. Aug. 1951, *C. G. Yang 173* (PE); Yangjiaping, Sept. 1935, *Y. Liu 11432* (PE); No. 6 Distr., Lijiapu Village, 10. Aug. 1955, *C. G. Yang 1674* (PE). **Xinglong**: Wenzhanggou, shaded slopes, 27. Jul. 1956, *PE Herbarium 2199* (PE). **Xuanhua:** s. loc., 21. Jul. 1930, *H. Serre* s. n.(W). **Zhuolu**: s. loc., 25. Sept. 1951, *C. G. Yang 494* (KUN). **HEILONGJIANG**, **Acheng**: Yuquan, 28 May 1954, *G. Z. Wang 1890* (IFP). **Dailing**: Dongshan, *Quercus-Lespedeza* thickets, 04. Aug. 1959, *W. Wang et al. 789* (PE). **Mudanjiang**: 1959, *Mudanjiang Exped. 905* (PE). **Ning'an**: Jingpohu, 24. Jul. 1980, *J. Z. Wang & X. H. Yu 3262* (IFP); loc. eodem, 310 m, 12. Aug. 1931, *H. W. Kung (K) 2065* (PE). **Shangzhi**: Weihe Distr., 28. Aug. 1953, *G. Z. Wang 1659* (IFP). **HENAN**, **Baofeng**: s. loc., 1959, *General Exped. 18034* (KUN, PE). **Huixian**: Baiyun-si, thickets, 06. Jul. 1959, *Xinxiang Normal College 4071* (PE). **Linxian**: 1500 m, 08. Aug. 1959, *Xinxiang Normal College 4309* (PE). **Luanchuan**: 1959, *General Exped. 20459 (PE) & 22603* (PE). **Lushi**: Laojun Shan, 680 m, 05. Aug. 1935, *K. M. Liou 4963* (PE). **Shangcheng**: Yuzidian, Jingangtai, 700

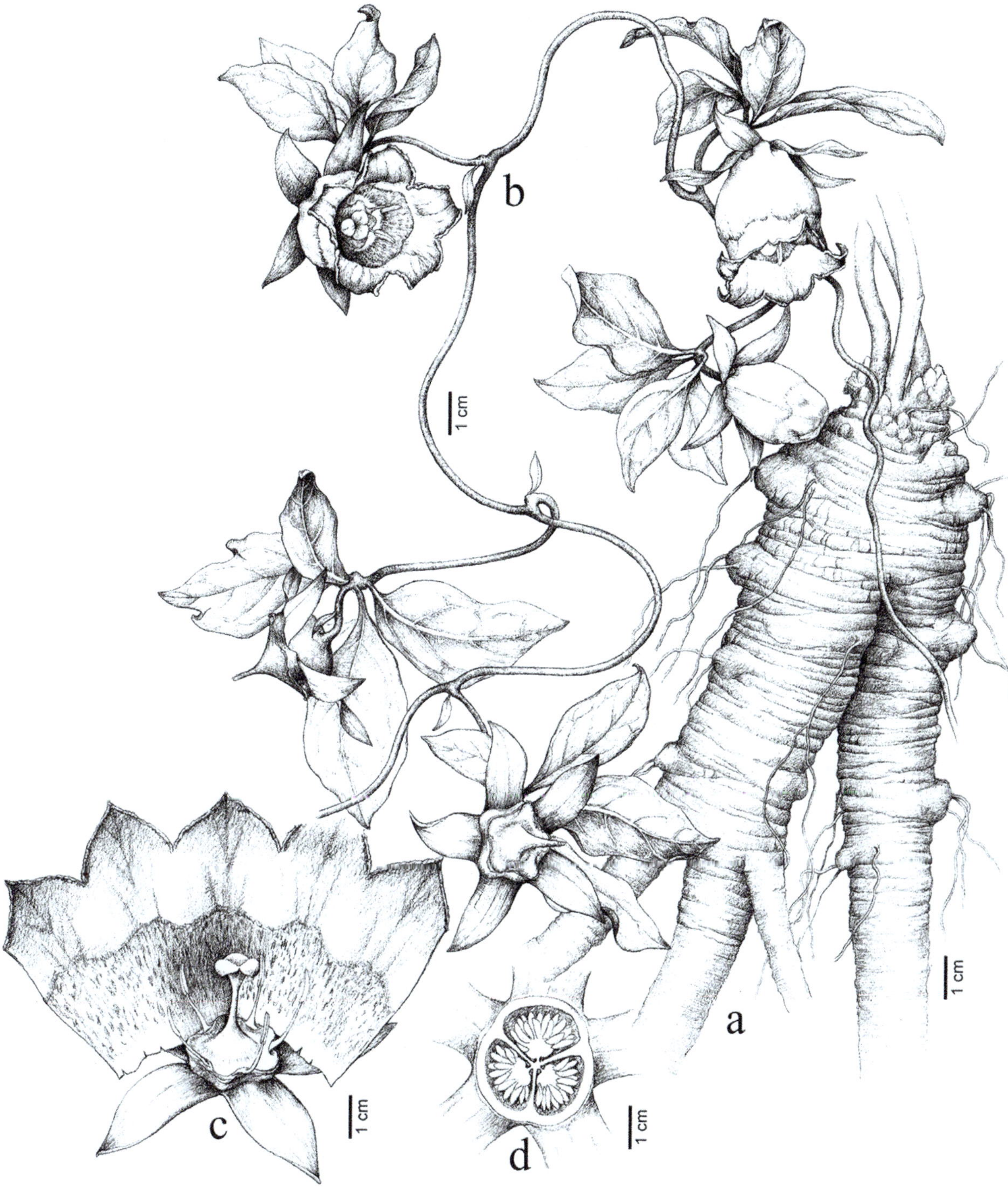

FIGURE IV 1:41A *Codonopsis lanceolata* (Sieb. & Zucc.) Trautv. **a**, root, caudex, and base of stems; **b**, a part of twining stem; **c**, a flower with corolla opened, showing calyx, stamens, disc, style, and stigma; **d**, a cross section of ovary, showing 3 locules and numerous ovules.

m, 1959, *s. coll. 71* (PE). **Songxian**: Longchiman, 900 m, 27. Aug. 1983, *Plant Resources Exped. L0327* (PE); loc. eodem, 05. Aug. 1983, *Plant Resources Exped. L0040* (PE); Longwangmiaogou, 1100 m, 24. Aug. 1959, *s. coll. 35061* (PE). **Xinxian**: Yuchongshangou, 20. Jul. 1989, *Xinxiang Normal College 8227* (PE). **Xinyang**: s. loc., s. d., *H. B. Zhou* s. n.(PE) ; s. loc., s. d. *H. B. Zhou 0515* (PE). **Xixia**: Laojian-ling, 18. Aug. 1960, *K. C. Kuan & T. L. Dai 1709* (PE). **HUBEI**: W Hubei, Aug. 1907, *E. H. Wilson 2290* (E, K, W); loc. eodem, Aug. 1907, *E. H. Wilson 2569* (E); s. loc., *A. Henry 6527* (E, G, K). **Yichang**: Dalao-ling, 250 m, secondary broad-leaved forests, 05, Sept. 2007, *C. M. Zhao*

FIGURE IV 1:41B *Codonopsis lanceolata* (Sieb. & Zucc.) Trautv. **a** & **e**, from Beijing: Miyun, 12. Jun. 2011 (e) and 14. Aug. 2011 (a), by Lian-Zhong FU; **b**, from Jilin: Changbai Shan, 05. Sept. 2009, by Yao ZHOU; **c** & **d**, from Beijing: Miyun, 03. Oct. 2011, by De-Yuan HONG; **f**, from Jilin: Tonghua, 29. Apr. 2009, by Yao ZHOU; **c**, showing disc (arrow) and ovary which is inferior to corolla, but half-inferior to calyx.

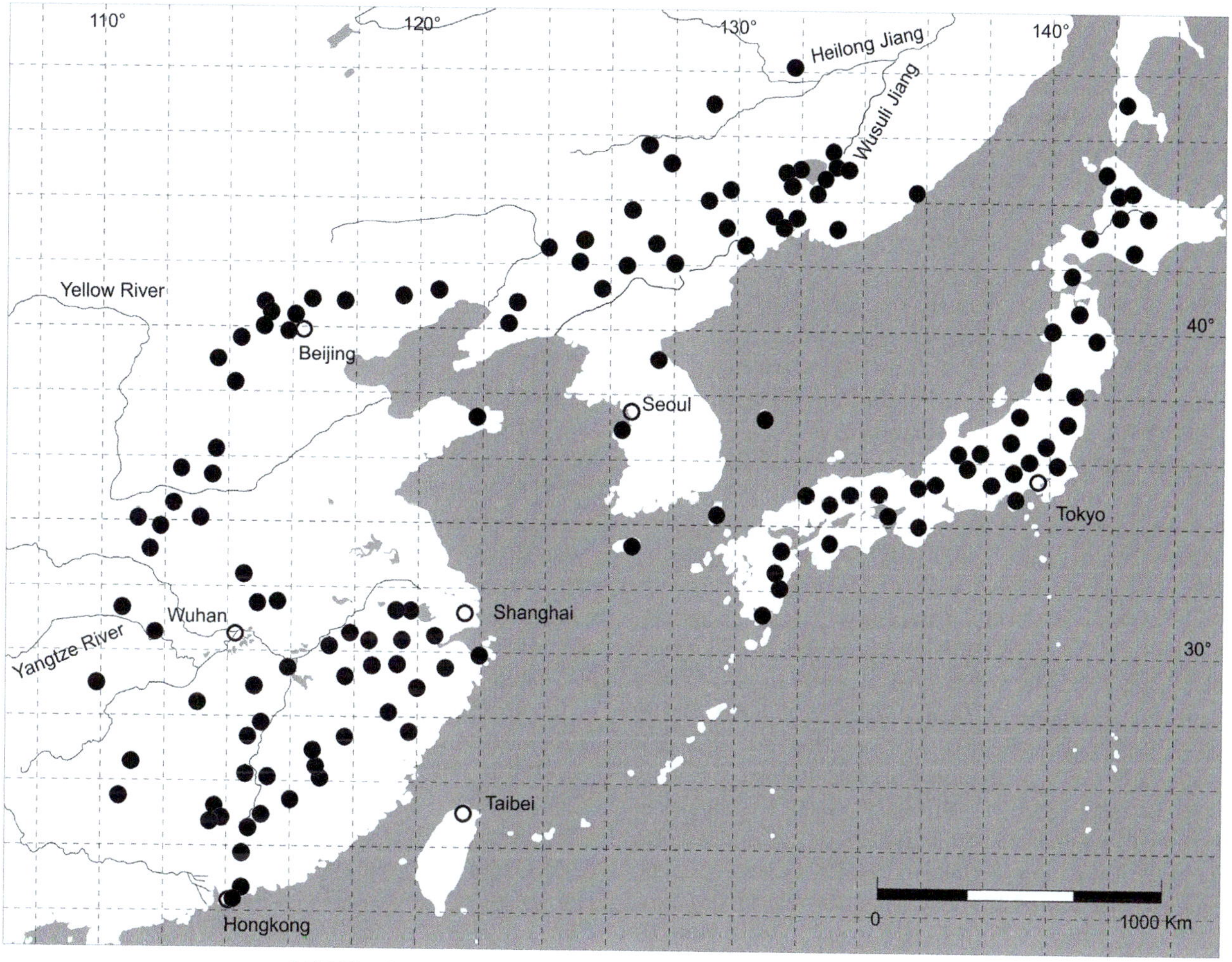

MAP IV 1:41 Distribution of *Codonopsis lanceolata* (Sieb. & Zucc.) Trautv.

Ex 0482 (PE); s. loc., 1889, *A. Henry 2342* (K). **Xingshan**: *E. H. Wilson 2564* (K, W). **HUNAN**, **Changning**: 560 m, 26. Jul. 1935, *C. S. Fan & Y. Y. Li 357* (G, W). **Changsha**: Yolu Shan, 150 m, 09. Sept. 1917, *Handel-Mazzetti 11345* (K, W). **Xinning**: Ziyun Shan, 1300 m, 05. Sept. 1984, *Z. Y. Li et al. 158* (E, PE); loc. eodem, 1150 m, by water, 15. Oct. 1962, *L. H. Liu 15172* (KUN); s. loc., 650 m, 21. Sept. 1935, *C. S. Fan & Y. Y. Li 525* (G, W). **Yongshun**: Xiaoxi, Yuyang Shan, forests, 29. Aug. 1991, *X. L. Yu 91893* (KUN). **JIANGSU**: **Liyang**: Shenxijie, 200 m, forests, 11. Aug. 1956, *F. X. Liu et al. 2608* (KUN, PE). **Yixing**: Longchi Shan, 150 m, evergreen forests, 04. Aug. 1960, *W. Z. Fang et al. 83* (KUN, PE, SZ). **JIANGXI**, **Anfu**: Wugong Shan, 800 m, forests 18. Aug. 1959, *S. S. Lai et al. 1950* (KUN). **Fenyi**: Dagang Shan, 600 m, 22. Aug. 1985, *K. Yao 9168* (K). **Guangchang**: Tangfang, thickets, by stream, 09. Oct. 1962, *J. Z. Song 2378* (KUN); s. loc., edges of forests, 27. Sept. 1958, *Q. M. Hu 5278* (KUN, PE). **Longnan**: 16–30. Sept. 1934, *S. K. Lau 4458* (G). **Lushan**: Guling, 1000 m, 01. Oct. 1992, *M. X. Nie 92376* (K, LD); Guling-Gebirge, 1100 m, Sept. 1908, *A. K. Schindler 391* (K); Taiyicun, roadsides in forests, 16. Oct. 2002, *Lushan Bot. Gard. Exped. Y 0050* (PE); s. loc., 31. Aug. 1922, *A. N. Steward 2685* (E). **Ninggang**, Maoping, 700 m, open woods, 09. Oct. 1965, *S. S. Lai et al. 5270* (KUN). **Qiannan**: 1–29. Aug. 1934, *S. S. Lau 4176* (G). **Ruijin**: Rentian, grasses, 16. Aug. 1958, *Q. M. Hu 4300* (PE); Tuoxia Shan, grasslands with shrubs, 07. Aug. 1958, *S. S. Lai 4042* (KUN). **Shicheng**: Gaoping Township, ravines, 19. Sept. 1958, *S. S. Lai 5029* (KUN). **Suichuan**: s. loc., 500 m, thickets by stream, 06. Oct. 1963, *S. J. Yue et al. 4633* (KUN). **Wuyuan**: s. loc., 17. Aug. 1925, *R. C. Ching & E. D. Merrill 8892* (K). **Xingguo:** Liangcun, thickets, 21. May. 1960, Inst. Bot. 339 (PE). **Xiushui**: Huangsangang, forests, 31. Aug. 1963, *S. S. Lai 3370* (KUN). **Yanshan**: Shilong, 27°51′ N, 117°45′E, 1300 m, 31. Aug. 1964, *Z. P. Jian et al. 401289* (PE); Wuyi Shan, 27°49′ N, 117°43′E, 1100 m, 26. Aug. 1964, *Z. P. Jian et al. 400968* (PE); loc. eodem, grasses, 15. Aug. 1958, *S. S. Lai 4070* (KUN, PE). **JILIN**, Mukden to Kirin, between Hui Fa River and Kirin, May–Aug. 1886, *H. E. M. James* s. n.(K). **Antu**: Erdaobaihe, Baoma River, 800 m, 12. Aug. 1963, *s. coll. 1092* (PE); 25 km E of Sandaobaihe, 880 m, 16. Aug.

1959, *Yanbian No. 2 Team 73* (PE). **Changbai Shan**: Naitou Shan, 1070 m, forests, 01. Sept. 1959, *Yanbian No.1 Team 537* (PE). **Emu**: 350 m, 21. Sept. 1931, *H. W. Kung (K) 2346* (PE). **Hunchun**: Xitumengzi thickets at edges of *Quercus* forests, 20. Jul. 1959, *P. Y. Fu et al. 831* (PE). **Hunjiang**: Linjiang, Mao'er Shan, 730 m, 12. Aug. 1952, *T. N. Liou et al. 1984* (IFP, PE). **Jilin City**: Longtan Shan, edges of forests, 27. Aug. 1950, *Y. L. Chang 929* (IFP, PE). **Tonghua**: near Tonghua City, 16. Aug., *J. Sato 9041* (PE); 40 km SW of Tonghua, thickets, 15. Aug. 1956, *Sino-Germ. Exped. 429* (PE). **Wangqing**: Chunyang to Laomiao, edges of forests, 08. Aug. 1959, *P. Y. Fu 985* (PE). **Wusong**: Xiaopucun, thickets, 14. Sept. 1959, *Tonghua Team 279* (PE). **LIAONING**, **Anshan**: Mts Qianshan, *R. E. Faber 1611* (K); loc. eodem, 28. Sept. 1950, *Q. T. Li et al. 2586* (PE); loc. eodem, 200 m, 30. Jul. 1930, *H. W. Kung 830* (PE). **Beizhen** : Lüshan, 30, Sept. 1953, *C. Wang 2825* (IFP); Beizhen, 26. Jun. 1961, *J. Wei 100* (IFP). **Benxi**: Caohekou, 05. Aug. 1910, *Y. Yabe* s. n.(PE) ; Caohezhang, 13. Aug. 1965, *C. Q. Lin 948* (PE). **Chaoyang** : Heshangmazi, 26. May 1961, *P. Y. Fu 202* (IFP). **Huanren**: Huanggou, *Quercus* thickets, 14. Aug. 1959, *C. S. Wang et al. 3894* (PE); Huanren, Laotudingzi, 23. Aug. 1964, *S. C. Cui & Y. C. Zhu 231* (IFP); Huanren, Taiyuzi, 24. Aug. 1964, *S. C. Cui & Y. C. Zhu 241* (IFP). **Kaiyuan**: Kaiyuan, Bakesu, 04. Aug. 1974, P. Y. Fu & Y. L. Chang 3058 (IFP); Majiazhai, Shuangtai-ling, thickets, 19. May. 1983, *G. X. Fu & Y. Ma 012* (PE). **Lingyuan**: Fogedong, shaded slopes, 10. Sept. 1959, *General Exped 188* (PE); Lingyuan, Hekan, 12. Aug. 1983, *C. F. Fang 2946* (IFP). **Qingyuan**: Dabiangou, 18. Jun. 1959, *W. Wang et al. 608* (IFP). **Xifeng**: Helong, 01. Sept. 1959, *X. P. Yu 171* (IFP); Xifeng, Yingchang, 06. Jun. 1961, *Y. C. Deng 134* (IFP). **Xiuyan**: Tangchi Township, forests, 09. Sept. 1959, *W. Wang et al. 1429* (IFP, PE). **SHANDONG**, **Kunyu Shan**: 09. May. 1959, *T. Y. Zhou 731* (KUN). **SHANXI**, **Wutai**: Chenyongshe, 2400 m, 02. Aug. 1959, *K. C. Kuan & Y. L. Chen 02570* (PE); Jinggangku, 26. Jul. 1959, *K. C. Kuan & Y. L. Chen 1670* (PE). **Yangcheng**: Sang-Linshu, 800 m, 17. Sept. 1983, *T. W. Liu & Z. F. Zen 1317* (PE). **ZHEJIANG**, **Chun'an**: Wangfu, rocks in forests, 03. Sept. 1958, *s. coll. 30293* (PE). **Deqing**: Mogang Shan, Aug. 1929, *T. N. Liou 1524* (PE). **Hangzhou**: Mt Filaifeng, 04. Sept. 1958, *s. coll. 0862* (PE); loc. eodem, forests, 02. Sept. 1957, *S. Y. Chang 1373* (PE). **Kaihua**: Suzhuang, 260 m, grasses, 22. May 1959, *Zhejiang Plant Resources Surv. 26062* (PE); Waixishang, 07. Aug. 1958, s. coll. 29964 (PE). **Lin'an**: Changhua, Longtang Shan, 22. Aug. 1957, *X. Y. He 26330* (PE); loc. eodem, 11. Aug. 1957, *X. Y. He 25272* (PE). **Lishui**: Yunfeng, 15. Aug. 1959, *S. Y. Chang 6293* (PE). **Longquan**: 850 m, 19. Aug. 1958, *R. H. Shan 5762* (PE). **Ningbo**: 1877, *W. Hancock 81* (K). **Qingyuan**: Zhuping, 975 m, 24. Aug. 1924, *R. C. Ching 2353* (PE). **Taisun**: Wuyan-ling, forests, 05. Aug. 1958, *S. Y. Chang 3622* (PE). **Tianmu Shan**: Dashuwang, 21. Sept. 1958, *s. coll. 30955* (PE); Laodian to Dixang'en, 20. Aug. 1952, *P. C. Keng & T. Hong 267* (KUN); Wuliting, 03. Sept. 1984, *G. X. Fu & R. S. Ni 132* (PE); loc. eodem, *K. K. Tsoong 680* (PE) ; Xianding, thickets at edges of forests, 22. Aug. 1957, *X. Y. He 24652* (PE); 400 m, glades, 23. Aug. 1959, *Zhejiang Plant Resources Surv. 28930* (PE); s. loc., 26. Jul. 1930, *T. N. Liou 7977* (PE); s.loc., 28. Sept. 1947, *Y. W. Law 1030* (K); s. loc., 28. Oct. 1951, *s. coll. 10185* (PE); s. loc., 13. Sept. 1979, *s. coll. 182-4 (93)*. **Tiantai Shan**: Dongchakou, 302 m, thickets, *X. Y. He 27947* (PE); Ximaopong, grasses in forests, 11. Sept. 1957, *X. Y. He 27975* (PE); s. loc., 1958, *G. R. Chen 2333* (KUN, PE).

JAPAN, **Aomori**: Aomori-shi, Mts Hakkoda, 850 m, 20 Jul. 1978, *J. Murata & H. Ohba 5678* (TI). **Fukushima**: Mt Iide, 1600–2100 m, 02. Aug. 1979, *T. Yamazaki et al. 128* (TI); Minami-aizu-gun, 1600–1926 m, 27. Aug. 1983, *T. Yahara et al. 2723* (PE). **Gifu**: Mino, 27. Aug. 1922, *Kishiota 3199* (PE); loc.eodem, 22. Aug. 1927, *K. Shiota 3200* (PE); loc. eodem, 28. Aug. 1935, *K. Shiota 9088* (PE). **Gunma**: Tano-gun, Manba-machi, 800 m, 02. Sept. 1975, *J. Murata et al. 823* (PE). **Guwassan**: 28. Sept. 1897, *Faurie 694* (WU). **Hakodate**: Hakodate, 1861, *Maximowicz* s. n.(K); Jesso Isl. near Hakodate, 1861, *Albrecht* s. n.(K); loc. eodem, 20. Aug. 1897, *Faurie 693* (WU). **Hiroshima**: s. loc., 250 m, *H. Li et al.* s. n.(KUN). **Hokkaido**: Asahigawa, 17. Aug. 1923, *Y. Narita* s. n.(B); Mt Daisetsu, 1100 m, *Picea jezoensis* forests, 02. Sept. 1974, *T. Yamazaki 743* (TI); S Hokkaido, 1884, *W. P. Brooks* s. n.(PE); Kamikawa-gun, Higashikawa-cho, 20. Sept. 1997, *Kirkham et al.'s Exped. to Hokkaido (EHOK) 86* (K); Mt Moiwa, 03. Sept. 1912, *S. Tamaki* s. n.(TI); Sapporo, Mt Moiwayama, 25. Aug. 1933, *H. Hara 101* (TI); loc. eodem, 25. Aug. 1931, *H. Hara 4428* (TI); loc. eodem, Aug. 1929, *P. H. Dorsett & W. J. Morse 1177* (B, PE); Soya-shicho, Toyotomi-cho, 31. Aug. 1977, *D. E. Boufford & E. W. Wood 19829* (PE); Teshio, 16. Aug. 1934, *H. Hara 4335* (TI); E of Uryu River, 44°12′24″N, 142°8′34″E, 135 m, 23. Sept. 1997, *Kirkham et al.'s Exped. to Hokkaido (EHOK) 52* (K). **Jakao** Mt, 28. Aug. 1903, *H. Takeda 301* (K). **Hondo**: Sinano, Mt. Sirouma, 21. Aug. 1936. *K. Hisauti* s. n.(LD). **Hyogo**: Yumemaemachi, Abiko, Mt Seppikosan, 250 m, 30. Sept. 1978, *J. Murata 6450* (PE, TI); s. loc., 1868, *Wawra* s. n.(W). **Kanagawa**: Hakone, Mt Kimtoki, 700–1000 m, 30. Sept. 1961, *M. Furuse 39335* (K); Mts Tanzawa, 10. Sept. 1978, *K. Midorikawa 202* (TI); loc. eodem, ca. 700 m, 28. Sept. 1980, *T. Kubo 218* (PE). **Musashi**: Omiya, 20. Aug. 1909, *s. coll.* s. n.(LD). **Nagano**: Kamiminochi-gun, Nishiyama, 1180–1200 m, 24, Aug. 1976, *H. Ohashi & Y. Fukuda* s. n.(TI); Shinano, ca. 1150 m, 25. Sept. 1958, *M. Furuse* s. n.(PE). **Nagasaki**: Is. Tsushima, Shimoagata-gun, 20 m, 25. Oct. 1973, *Y. Tateishi 173* (TI). **Nanokawa**: Tosa, 03. Sept. 1889, *K. Watanabe* s. n.(PE). **Nara**: Yoshino-gun, Mt Daifugen-dake in Mt Ohmine, 1250 m, sunny shrubby slopes, 12. Aug. 1977, *H. Koyama & M. Hotta 5424* (E, PE).

Niigata: Kitauonuma-gun, 750 m, 22. Aug. 1964, *H. Kanai et al. 8912* (TI). **Ohita**: Teradoko, 900 m, 05. Oct. 1997, *H. Koyama 10202* (PE). **Okayama**: Sugoson Tsukubo-gun, woods, 27. Oct. 1935, *K. Uno* s. n.(PE). **Saitama**: Agano, shady places, 17. Sept. 1967, *F. Yamazaki 10678* (PE). **Sendai**: s. loc., 26. Sept. 1923, *F. Ilhiba* s. n.(G). **Shiga**: Ika-gun, 07. Sept. 1980, *H. Ohba & S. Akiyama* s. n.(TI) ; Sakada-gun, between Kasiwabara (300 m) – Mt Ryozen (1084 m), 24. Sept. 1984, *H. T. Im 2739* (PE); Takashima-gun, 100–200 m, 14. Sept. 1982, *H. Ohashi et al. 8676* (TI). **Shimane**: Taishacho, 13. Oct. 1958, *H. Hara* s. n.(TI). **Shizuoka**: near Hiratsuka Gotenba City, Mt Fuji, ca. 1000 m, 26. Aug. 1978, *M. Tsuji et al. 685* (PE); Tagata-gun, Amagi Pass, ca. 700 m, 30. Sept. 1983, *K. Inone & H. T. Im 2234* (TI). **Tochigi**: Hondo, Shimotsuke, ca. 1400 m, 05. Sept. 1964, *M. Furuse 42421* (K); Nikko-shi, ca. 900 m, 21. Aug. 1979, *K. Inoue 1324* (KUN, TI). **Tokyo**: Shiroyama, Hachioz-shi, 250 m, edges of forests, 03. Oct. 1965, *F. Yamazaki 9882* (PE); Mt Takanosu, Nishitama-gun, 1750 m, sunny grasslands 31. Aug. 1987, *H. Funakoshi 523* (TI). **Yamanashi**: Minamikoma-gun, 400–700 m, 22. Oct. 1978, *Y. Tateishi et al. 4560* (TI); Minamikoma-gun, W of Mt Tenshi, 750–1000 m, 22. Oct. 1978, *Y. Tateishi et al. 4589* (TI); Minamitsurus-gun, Mt Mitsutoge, 1400 m, 18. Oct. 1986, *H. T. Im 7761* (TI). **Yezo Island**: s. d. *Maries* s. n.(PE); 07. Sept. 1903, *S. Arimoto* s. n.(PE).

KOREA PENINSULA, **Chiisan**, Aug. 1904, *R. K. Smith 1143* (TI). **Hallaisan**, 1900 m, 13. Aug. 1908, *Taquet 1491* (E, K); loc. eodem, 09. Aug. 1910, *Taquet 4399* (E). **Kangwong** Prov., Kumgang, 30. Jul. 1928, *K. Kondo* s. n.(TI). **Taebaengi** Isl., 36°56′N, 126°10′E, 23. Aug. 1984, *B. R. Yinger et al. 2581* (K). **Ullung Isl.** (S Korea), 310 m, 19. Oct. 1989, *Kirkham, Flanagan & Boyle (KFBX) 157* (K). **Yengsil**, 1000 m, 17. Aug. 1908, *Taquet 1492* (K).

RUSSIA, Amur, Kap-Sau Distr., 28. Jul. 1897, *V. L. Komarov 1493* (K); Amur, 23. Aug. 1895, *V. L. Komarov* s. n.(W). **Chabarovskiy**, left bank of Amur, 09. Aug. 1910, *I. V. Kuznetsov 642* (LE). **Sakhalin**, Tsunijin, Aug. 1923, *Y. Narita* s. n. (TI). **Vladivostoksk**, Muravyov Amurskiy Peninsula, 43°15′N, 131°02′E, 23. Aug. 1929, *V. Transhel 910* (LE); coast of Amurskiy Bay, 43°23′N, 132°E, 20 Jul. 1909, *E. S. Kuznetsova* s. n. (LE); Shmakovsk, 45°N, 133°23′E, 19. Aug. 1927, *L. Nekrasova 303* (LE); Popova Island, 42°58′N, 131°43′E, 18. Aug. 1928, *M. F. Grishko 559* (LE); Suchan River basin, 43°N, 133°15′E, 15. Jul. 1913, *A. Bulavina 648* (LE); Lower reaches of Tyutiche River, 44°N, 136°E, 20. Jul. 1909, *s. coll. 347* (LE); Hankaisk, 44°40′N, 131°50′E, 14. Aug. 1928, *D. P. Vorobyov* s. n. (LE); coast of Manchuria, 44–45°N, 1859, *C. Wilford* s. n.(W).

42. Codonopsis ussuriensis (Rupr. & Maxim.) Hemsl., J. Linn. Soc., Bot. 26: 6. 1889; Komarov, Trudy Imp. S.-Peterburgsk. Bot. Sada 29(1): 111. 1908; Chipp, J. Linn. Soc. Bot. 38: 380. 1908; Anthony, Notes Roy. Bot. Gard. Edinburgh 15: 182. 1926; Shen & Hong in Hong, Fl. Reipubl. Popularis Sin. 73(2): 38. pl. 7. fig. 1–2.1983; Hong, Lammers & Klein in Wu, Raven & Hong, Fl. China 19: 515. 2011. ≡*Glossocomia ussuriensis* Rupr. & Maxim., Bull. Cl. Phys.-Math. Acad. Imp. Sci. S. -Pétersbourg 15: 209. 1857. ≡*Glossocomia lanceolata* (Sieb. & Zucc.) Maxim. var. *ussuriensis* (Rupr. & Maxim.) Regel, Index Hort. Sem. Petrop. 1886: 92. 1886. ≡*Codonopsis lanceolata* (Sieb. & Zucc.) Trautv. var. *ussuriensis* (Rupr. & Maxim.) Trautv., Trudy Imp. S.-Peterburgsk. Bot. Sada 6: 47. 1879. Type: "Manshuria", 28. Jul. 1857, *Maximowicz* s. n. (lectotype: LE); loc. eodem, 31. Jul. 1855 & 11. Aug. 1855, *Maximowicz* s. n. (paratypes: LE); loc. eodem, 18. Aug. 1856, *Maximowicz* s. n. (paratype: LE).

Codonopsis minima Nakai, Veg. Quelp. 85. 1914 & Bot. Mag. (Tokyo) 29: 7. 1915. ≡*Codonopsis ussuriensis* (Rupr. & Maxim.) Hemsl. var. *minima* (Nakai) T. Lee, Seoul. Nat. Univ., Coll. Agric. Rese 11(2–1): 3. 1986. Type: Korea, Yengsil, 1000 m, forests, 17. Aug. 1908, *Taquet 1490* (lectotype here designated: TI; isolectotypes: E, K); Korea, Yengsil. 1000 m, forests, 9. Jul. 1910, *Taquet 4398* (paratype: E).

Codonopsis ussuriensis (Rupr. & Maxim,) Hemsl. f. *viridiflora* J. Ohara, J. Phytogeogr. Taxon. 33: 72. 1985. Type: not designated.

Glossocomia lanceolata (Sieb. & Zucc.) Maxim. var. *obtusa* Regel, Bull. Cl. Phys.-Math. Acad. Imp. Sci. Saint.-Pétersbourg 15: 223. 1857. Type: not designated.

Plants glabrous throughout, or stems and leaves sparsely villous. Roots nearly globose or ellipsoid, 2–4 cm long, 1–3 cm in diameter, gray-yellow. Stems twining, green, white or dark purple. Leaves on main stem alternate, lanceolate or rhomboid-ovate, smaller, those at top of branches usually in a group of 3–5, pseudo-verticillate; petiole less than 3 mm long; leaf blade lanceolate or oblong, 3–5 cm long, 1.5–2.5 cm broad, adaxially green, glabrous, abaxially gray-green, glabrous or sparsely villous, base attenuate, margin entire, apex acute or obtuse. Flowers solitary, terminal on slender branches; pedicel 2–5 cm long, bracts single, small, lanceolate or narrowly rhomboid-ovate. Calyx tube adnate to whole hypanthium, semiglobose; calyx lobes narrowly lanceolate or ovate-deltoid, 1–2 cm long, 0.6–0.8 cm broad, margin entire, apex acute. Corolla somewhat urceolate, 2–3 cm long, 1.5–2.5 cm wide, shallowly lobed; corolla lobes deltoid, dark purple, insides with remarkable black-purple stripes or black spots. Stamens glabrous; filaments 3–5 mm long, slightly dilated at base; anthers 2–4 mm long. Capsule semiglobose in inferior part, rostrate in superior part. Seeds ovoid, narrow-winged, dark brown, lucid. Figures IV 1:42A & B.

Chromosome: 2n = 16 Jee et al., 1989, under *C. minima*; Yoo & Lee, 1989, under *C. minima*).

Phenology: flowering and fruiting in July and August.

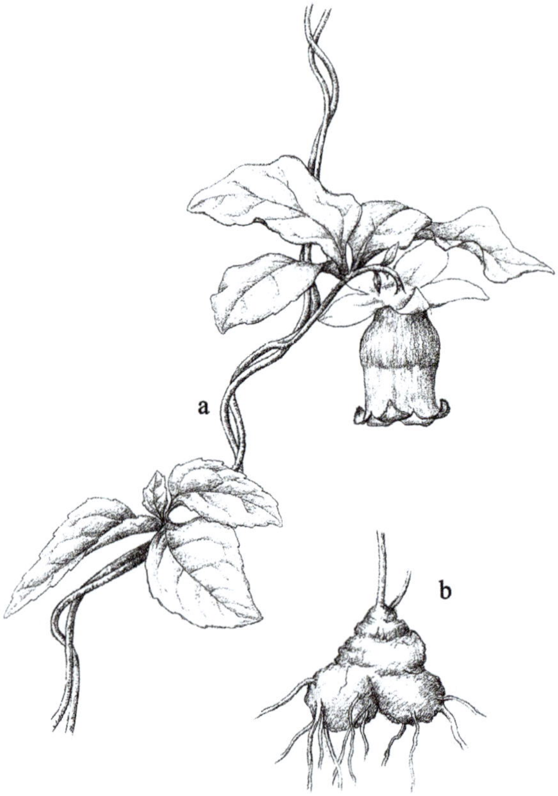

FIGURE IV 1:42A *Codonopsis ussuriensis* (Rupr. & Maxim.) Hemsl. **a**, two sections of twining stems; **b**, a root and basal part of two stems.

Habitats: ravines, moist meadows, mostly in sandy soils; below 1000 m.

Distribution: East Asia: NE China, Japan, Korea Peninsula, and Russia (the Far East) (Map IV 1:42).

Notes: We can see from the citation of literature that some authors treated *Codonopsis ussuriensis* as a variety in *C. lanceolata*. The two species are distinct from each other in roots (see Figures IV 1: 41A & B and IV 1: 42A & B) and in seeds (see Figures II 2: 4, K–P). However, these two characters are often unavailable, and thus it is not easy to identify them correctly. Based on examination of a large number of specimens, I carried out a statistical analysis in the Herbarium of Komarov's Inst. Bot. (LE). I used 14 available specimens with carrot-shaped roots (*C. lanceolata*) and 16 available specimens with globose roots (*C. ussuriensis*). We found that the two species are distinct in length since their variation in corolla length is statistically discontinuous, though continuous in extremes, and their differences in length and width of leaf blade are nearly distinct. The conclusion should be that the two species are not only distinct but can also be distinguished by length of corolla and size of leaf blade.

Additional specimens examined, **CHINA**, **HEILONGJIANG**, ad fluv. Amur, 1854, *L. Schrenk* s. n.(W); loc. eodem, 1855, *R. Maack 388* (PE); Amur medius, 1891, *Korshinsky* s. n.(G); Amur, *Maximoviez* s. n.(K). **Hailun**: No 17 Distr., hilly grassy slopes, Jul. 1953, *NE China Exped. 376* (PE). **Hulin**: Jul. 1955, *Wastes Exped.* s. n (IFP). **Loubei**: Xingdong to Damaha River, meadows in forests, fl. orange-red, 16. Aug. 1955, *C. S. Wang 40* (IFP). **Mishan**: by Muling He, Aug. 1955, *R. Y. Li 103* (IFP). **Mudanjiang**: s. loc., 1959, s. coll. *0573* (PE). **Nenjiang**: to Heihe (187 km), 350 m, 07. Aug. 1956, *G. X. Cui & C. Q. Lin 253* (PE). **Ning'an**: Dongjingcheng, edges of mixed forests, 380 m, fl. pale purple, 18. Jul. 1975, *T. T. Yü & L. T. Lu 75298* (PE). **Yichun**: meadows, fl. pale purple, 25. Jul. 1956, *Sino-German Exped. 7728* (PE); loc. eodem, meadows, 25. Jul. 1956, fl. pale purple-red, *T. N. Liou 7709* (IFP); loc. eodem, meadows, 25. Jul. 1956, *T. N. Liou et al. 7729* (IFP, SZ). **Yilan**: Tuchengzi, meadows, 03. Aug. 1959, *Y. L. Chang 1903* (IFP, SZ). **JILIN**, **Changbai Shan**: Baihe, fl. purple, 25. Jul. 1963, *W. Wang et al. 2196* (PE); Erdaobaihe, 675 m, 28. Jul. 1986, *D. Y. Hong, W. H. Sauer et al. 32993* (PE). **Dunhua**: 270 m, Aug. 1951, *T. N. Liou 3516* (PE). **Hunchun**: Chunhua, edges of *Quercus* forests or meadows with shrubs, 19 Jul. 1959, *P. Y. Fu et al. 828* (IFP, PE); loc. eodem, meadows, fl. yellow with purple spots, 11. Aug. 1958, *C. S. Wang* 2071 (IFP). **Tonghua**: s. loc., *J. Sato 8983* (PE) 40 km SW of Tonghua, mixed forests, fl. purple, 15. Aug. 1956, *Sino-German Exped. 476* (IFP, PE) & *429* (IFP). **Wangqing**: Chunyang to Laomiao, fl. purplish green, 08. Aug. 1958, *P. Y. Fu. 985* (IFP). Oniodo Valley, 03. Aug. 1896, *V. L. Komarov 1492* (K). **LIAONING**, **Benxi**: Caohekou, forests, fl. with purple-red spots, 24. Aug. 1956, *C. S. Wang 307* (IFP). **Fengcheng**: Jiguan Shan, forests, 16. May. 1958, *D. C. Zhao et al. 541* (IFP). **Zhuanghe**: Xianrendong, Aug. 1981, *Y. F. Zhao 158* (IFP).

JAPAN, Hokkaido, Honsu, Shikoku, and Kyushu (from Iwatsuki et al., Fl. Jap. IIIa. 1993) . **Aichi**, Tokyo Univ. Forest, Seto-Shi, 16. Oct. 1984, *H. Funakoshi 646* (TI). **Hakodate**: Yeso, 14. Aug. 1878, *R. Yatabe* s. n.(G). **Hokkaido**: Hidaka, 11. Aug. 1934, *H. Hara 102* (TI); Koitoi, Shiranuka-cho, Shiranuka-gun, 3 m alt., 13. Aug. 1981, *K. Takita 772* (A); Kushiro, Aug. 1925, *S. Saito S264* (TI) & *S. Saito T318* (TI); Kushiro-shi, Otanoshige-shoro, near sea, 26. Aug. 1980, *F. Yamazaki 2600* (KUN, PE, SZ); Nokkeushi, 13. Aug. 1937, *K. Uno 21570* (PE); s. loc., 13. Aug. 1937, *K. Uno 2157* (G). **Ibaraki**, Suikaido-shi, along Kokaigawa, 24. Jul. 1995, *T. Yamazaki* s. n.(TI). **Musashi**: Shibuya, 17. Aug. 1909, *H. Sakurai* s. n.(E). **Nagano**: Mt Kirigamire, 31. Jul. 1951, *S. Suauki 1345* (PE). **Tokyo**: s. loc., Aug. 1910, *s. coll.* s. n.(E).

KOREA PENINSULA, s. loc., Aug. 1901, *U. Faurie 348* (G). **Cheju-do**, 700 m, 13. Aug. 1913, *T. Ishidoya 46* (TI). **Hamgyongnam-do**, 09. Jun. 1909, *T. Nakai* s. n. (TI). **Kangkoi**: 03. Aug. 1911, *R. G. Mills* s. n. (PE). **Kangwon-do**, Ch'angdo, 09. Aug. 1902, *J. Uchiyama* s. n. (TI). **Kankyo** Prov.: Sempo, 566–866 m, 03. Aug. 1917, *E. H. Wilson 1848* (K). **P'yonganbuk-do,** *s. coll.* s. n. (TI). **Seoul**, Han-Tschu-Wan, 04. May 1894, *M. A. Sontag* s. n. (LE).

FIGURE IV 1:42B *Codonopsis ussuriensis* (Rupr. & Maxim.) Hemsl. **a–c**, from Jilin: Changbai Shan, 30. Aug. 2008 (a). 23 Aug. 2009 (b) and 02. Sept. 2008 (c), by Qing-Wen SUN.

RUSSIA, **Amursk**: s. loc., *Maximowicz* s. n.(W); s. loc., 03. Jul. 1895, *V. L. Komarov* s. n.(W); Amursk, near Radde Village, 48°59′N, 130°60′E, 13. Aug. 1926, *E. Selivanova* s. n. (LE); Bureya Valley, 49°30′N, 129°40′E, 21. Jul. 1914, *A. N. Krishtofovich 934* (LE); loc. eodem, 49°30′N, 129°40′E, 21. Jul. 1914, *A. N. Krishtofovich 936* (LE). **Khabavosk**, Nekrasovsk, 47°93′N, 135°54′E, 13. Aug. 1927, *O. Malyutina-Kiselyova* s. n. (LE); near Sheremetevka Village, 47°21′N, 134°16′E, 21. Aug. 1928, *V. I. Skobunov 603* (LE). **Kunashir** Island, near Alyohino Village, 44°05′N, 145°44′E, 16. Aug. 1965, *E. Egorova 3148* (LE); loc. eodem, 13. Aug. 1960, *A. M. Chernyaeva* s. n. (LE). **Primorsk**, Hanka Lake, 45°N, 133°E, Aug. 1867, *N. M. Przhevalsky* s. n. (LE); Mangugai Valley, 43°15′N, 131°23′E, 18. Jul. 1913, *N. Dyukina 766* (LE). **Vladivostok**: near Popov Isl., 18. Aug. 1988, *S. Kharkevich & T. Buch 738* (K); Shmakovsk, Stepanovka Village, 44°95′N, 133°48′E, 28. Jul. 1927, *E. Vasilkova 200* (LE).

Section 4. Leptocodon (Hook. f. & Thomson) D. Y. Hong, stat. nov. ≡*Codonopsis* subg. *Leptocodon* Hook. f. &

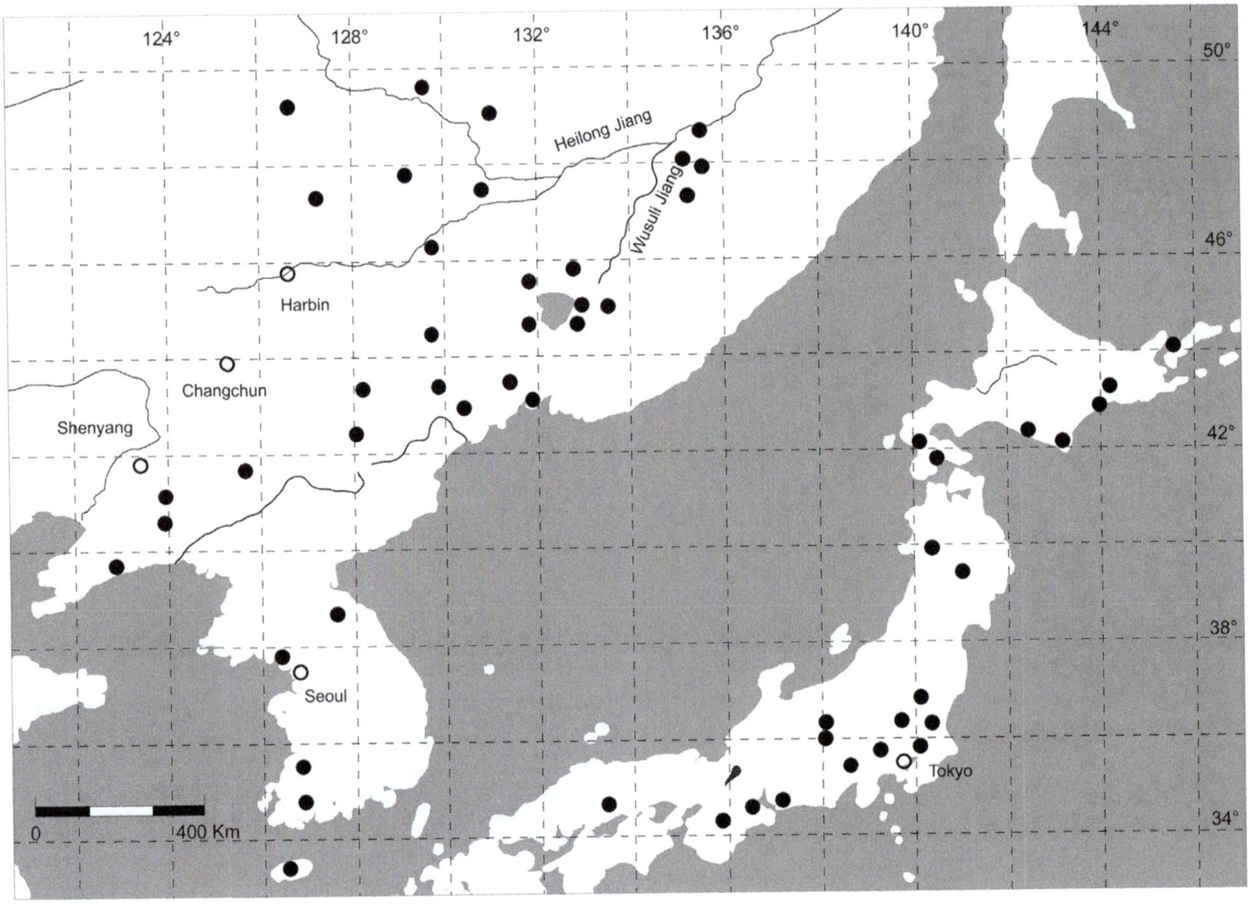

MAP IV 1:42 Distribution of *Codonopsis ussuriensis* (Rupr. & Maxim.) Hemsl.

Thomson, Ill. Himal. Pl. t. 16, A. 1855. ≡*Leptocodon* (Hook. f. & Thomson) Lem., Ill. Hort. 3. Misc. 49. 1856; Hook. f. & Thomson, J. Proc. Linn. Soc. Bot. 2: 17. 1857; Clarke in Hooker, Fl. Brit. Ind. 3: 430. 1881; Tsoong, Contr. Inst. Bot. Nat. Acad. Peiping 3(3): 104. 1935; Hong, Fl. Reipubl. Popularis Sin. 73(2):74. 1983. Type: *Codonopsis gracilis* Hook. f. & Thomson.

Herbaceous vines, strongly fetid. Leaves alternate, those on branches often opposite or nearly opposite. Flowers solitary, extra-axillary, less frequently axillary or leaf-opposed. Calyx tube entirely adnate to hypanthium, broad-obconical. Corolla long-tubular. Stamens alternate with 5 free, lamellar glands. Ovary semi-inferior, with superior part narrowly conical, over 3 times longer than inferior part, 3-locular; stigma 3-fid, segments clavate. Capsule apically 3-loculicidal. Seeds many.

Two species, found in the Himalaya and Hengduan Mts: Bhutan, SW China, N India, N Myanmar, and E Nepal. They grow in thickets and forests at middle altitudes from 1100 to 3350 m.

43. Codonopsis gracilis Hook. f. & Thomson, Ill. Himal. Pl. t. 16, A. 1855; Grey-Wilson, Plantsman 12(2): 81. 1990. ≡*Campanumoea gracilis* (Hook. f. & Thomson) G. Nichols., Ill. Dict. Gard. 1: 259. 1884. ≡*Leptocodon gracilis* (Hook. f. & Thomson) Lem., Ill. Hort. 3: Misc. 49. 1856; Clarke in Hooker, Fl. Brit. Ind. 3: 430. 1881; Tsoong, Contr, Inst. Bot. Nat. Acad. Peiping, 3(3): 104. fig. 14. 1935; Hong, Fl. Reipubl. Popularis Sin. 73(2): 74. 1983. Type: India, Sikkim, Lachung, 1830–2130 m, 15. Aug. 1849, *J. D. Hooker* s. n. [lectotype here designated: K (K000781242)]; Sikkim, Lachen, 2130–2750 m, 03. Oct. 1849, *J. D. Hooker* s. n. [paratype: K (K000781243)]; Sikkim, Lachen, 1830–2440 m, May. 1849, *J. D. Hooker* s. n. [paratypes: BM, K (K000781244)].

Plants glabrous throughout, rarely sparsely villous on young leaves and pedicel. Roots unknown. Stems slender, over 1 m long, branched. Leaves alternate, rarely those on branchlets opposite, long-petiolate; leaf blade ovate-orbicular, 0.9–2.5 cm long, 0.9–2.5 cm broad, membranous, adaxially green, abaxially gray, base truncate or obtuse, margin sinuate-crenate, apex obtuse. Flowers solitary, pendent; pedicel slender, 1–5 cm long, straight or curved. Calyx glabrous; calyx tube 2–3 mm long; calyx lobes reflexed, narrowly oblong, 5–8 mm long, 1–2 mm wide, base attenuated, stalk-like, margin with 1 or 2 pairs of callose serratures. Corolla tubular, 2.8–3.5 cm long,

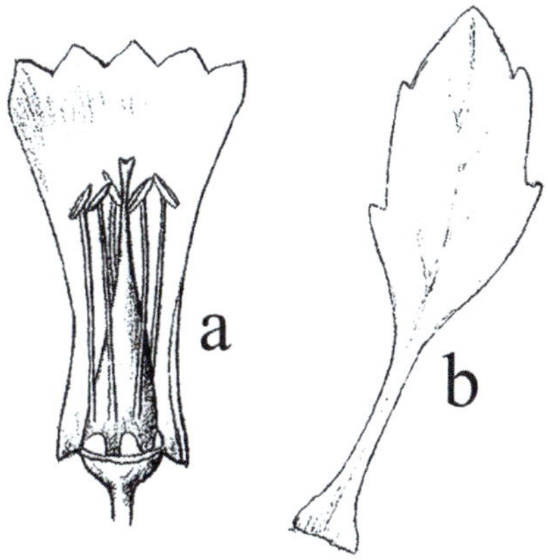

FIGURE IV 1:43A *Codonopsis gracilis* Hook. f. & Thomsom. **a**, a flower with calyx removed and corolla opened, showing stamens, half-superior ovary, style, stigma, and 5 glands which are alternate with stamens; **b**, a calyx lobe distinctly different from that in *C. hongii* Lammers (see Figure IV 1: 44A).

FIGURE IV 1:43B *Codonopsis gracilis* Hook. f. & Thomson. From Tibet: Dinggyê, 31. Jul. 2013, by Lei XIE.

shallowly lobed, blue, violet or purple; corolla tube ca. 4 mm wide; limb 1–1.5 cm wide; corolla lobes triangular, 4–5 mm long. Filaments ca. 2 cm long. Style ca. 2 cm long. Capsule semiglobose or obconic, inferior part 4–5 mm long; superior part conical or ovoid, 8–10 mm long. Seeds red-brown, linear-oblong, smooth, 1.5–1.8 mm long. Figures IV 1:43A & B.

Phenology: flowering and fruiting from August to October.

Habitats: forests or thickets; 1100–2600 m,

Distribution: SW China (SW Sichuan, S Tibet, and W Yunnan), N India (Sikkim), W Myanmar, and E Nepal (Map IV 1:43).

Additional specimens examined, **CHINA SICHUAN, Huili**: s. loc., 2400 m, 02. Oct. 1958, *Z. He et al. 11532* (SZ); s. loc, 2050 m, 22. Oct. 1914, *Handel-Mazzeti 5638* (E, WU). **Miyi**: Pingshan, Longganzi, 2300 m, 19. Sept. 1978, *s. coll. 393* (SM). **Muli**: Chabulang, Waerzhai, 2700 m, thickets, 07. Sept. 1983, *Qinghai-Xizang Exped. 13642* (PE); Milikigon, 2600 m, thickets, 04. Sept. 1937, *T. T. Yü 14165* (A, KUN, PE); Taoba Township, 2200 m, 30. Aug. 1978, *C. S. Cao et al. 8195* (SZ); Wanglang to Chabulang, 2600 m, thickets, 16. Sept. 1983, *Qinghai-Xizang Exped. 14108* (PE); 29°50′N, 101°E, 3050–3350 m, thickets, Aug. 1922, *G. Forrest 22158* (E); s. loc., 2740 m, 21. Aug. 1909, *F. Kingdon-Ward 4755* (E). **Yanbian**: s. loc., 2500 m, bushes, 07. Oct. 1932, *T. T. Yü 1659* (A, PE). **TIBET, Dinggyê**, Chengtang Township, Gamagou, 27°51′N, 87°24′E, 2426 m, forests, 13. Jul. 2013, *PE-Tibet Exped. 3217* (PE). **YUNNAN, Binchuan**: Jizhu Shan, 2000 m, pine forests, Aug. 1963, *Jishajiang Exped 63–6660* (KUN); Pianjing to Shazijie, 17. Oct. 1946, *T. N. Liou 21716* (PE). **Dali**: Dali to Tengyueh, 25°30′N, 2440–2740 m, Sept. 1903, *G. Forrest 987* (E); W of Dali, Mekong watershed, on the way to Youngchang and Tengyueh, 1230–2440 m. Sept–Oct. 1922, *J. F. Rock 6835* (E); 26°40′N, 2740 m, Sept. 1913, *G. Forrest 11028* (E). **Gongshan**: Drungjiang, Meiliwang, 1800 m, evergreen forests, 07. Jan. 1991, *Drungjiang Exped. 1642* (KUN). **Heqing**: Lianping, 2600 m, thickets, 19. Aug. 1940, *R. C. Ching 23755* (KUN, PE). **Lijiang** (Yulong): Jiazi Township, Baishuihe, 1900 m, thickets by river, 16. Sept. 1955, *K. M. Feng 21540* (KUN, PE); W of Lijiang, near Ngaza, Aug. 1923, *J. F. Rock 10579* (A). **Yongde**: Weibo, 2200 m, roadsides, 01. Sept. 2002, *E. D. Liu 5996* (KUN); Wumulong Township, Yingchangjie, 2400 m, grassy slopes, 01. Sept. 2003, *E. D. Liu & D. C. Yang 247* (KUN). **Yongning**: mountain summit, s. d., *McLaren 210* (E).

INDIA, Sikkim: Choongthung, Sept. 1903, *Prain* s. n.(E); North Distr. 1930 m, 1966, *Edinburgh Exped. to N Sikkim (1966) 249* (E).

MYANMAR, West Central Burma, Mt Victoria, 2500 m, 26. Oct. 1956, *F. Kingdon-Ward 22750* (BM).

NEPAL, Kosi: Arun Valley, Hatiar, 1830 m, 28. Aug. 1956, *J. D. A. Stainton 1437* (BM); Dhankuta Distr., 1100 m, 24. Oct. 1963, *H. Hara et al. 6300441* (BM, K); Sankuwasabha Distr., 2020 m, 18. Aug. 1998, *S. Noshino et al. 9830059* (E) ; Terhathum, Chittre, 27°06′N, 87°25′E, 2200 m, 16. Aug. 1972, *J. F. Dobremez (DBR NEP) 1474* (BM). **Mechi**: E Nepal, Tute (2300 m)–Sindua (2100 m), 02. Sept. 1977, *H. Ohashi et al. 775528* (BM, KUN, SZ); Taplejung Distr., Ghunsa Khola, 2650 m, 06. Sept. 1990, *KEKE 273* (E, K).

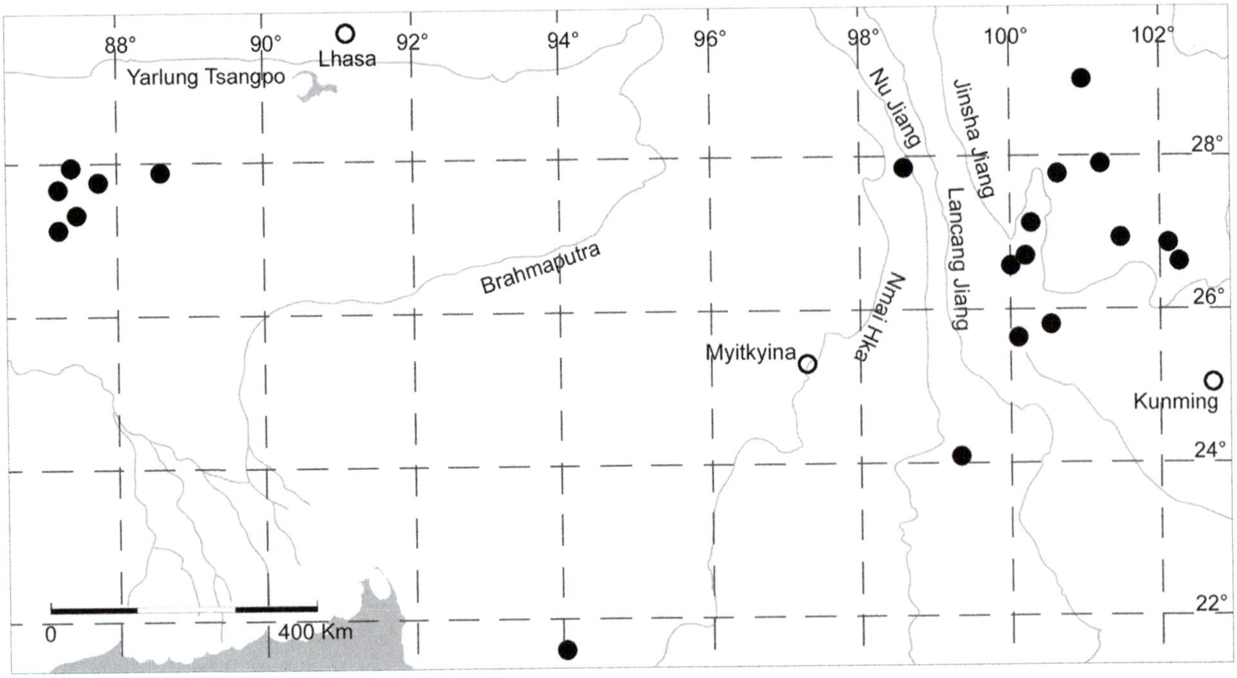

MAP IV 1:43 Distribution of *Codonopsis gracilis* Hook. f. & Thomson.

44. Codonopsis hongii Lammers, Novon 11: 67. 2001. ≡*Leptocodon hirsutus* D. Y. Hong, Acta Phytotax. Sin. 18: 246. fig. 2: 1–2, 1980; Hong, Fl. Reipubl. Popularis Sin. 73(2): 76. pl. 12: 1–2. 1983. ≡*Codonopsis hirsuta* (D. Y. Hong) K. E. Morris & Lammers, Novon 9: 387. 1999, not (Hand.-Mazz.) D. Y. Hong & L. M. Ma (1992). Type: China, Tibet, Zayü, Shangzayü, 2500 m, 09. Aug. 1973, *Qinghai-Xizang Exped. 73–1014* (holotype: PE).

Roots slender, not thickened. Stems up to 5 m long, hirsute on young parts. Leaves on main stem alternate, those on branchlets almost opposite, all leaves long-petiolate; leaf blade broad-ovate or nearly orbicular, 1.5–3 cm long, 1.5–2.5 cm broad, abaxially villous, rarely extremely sparsely villous, base truncate or rounded, margin gross-dentate, apex rounded. Flowers solitary, nearly leaf-opposed; pedicel slender, over 3 cm long, glabrous. Calyx tube hemispheric, 10-ribbed; calyx lobes ovate, overlapping each other at anthesis, separated at fruiting, upright, 5–7 mm long, 3–4 mm broad, hirsute abaxially and on margin, base cordate or nearly truncate, not stalk-like, margin sinuate-crenate. Inferior part of capsule semi-globose, 8–10 mm in diameter; superior part long-conical, ca. 10 mm long. Seeds ellipsoid, ca. 1 mm long. Figures IV 1:44A & B.

Phenology: flowering from July to September; fruiting from September to November.

Habitats: mixed forests, thickets, or grasslands, on granites; 1200–3350 m.

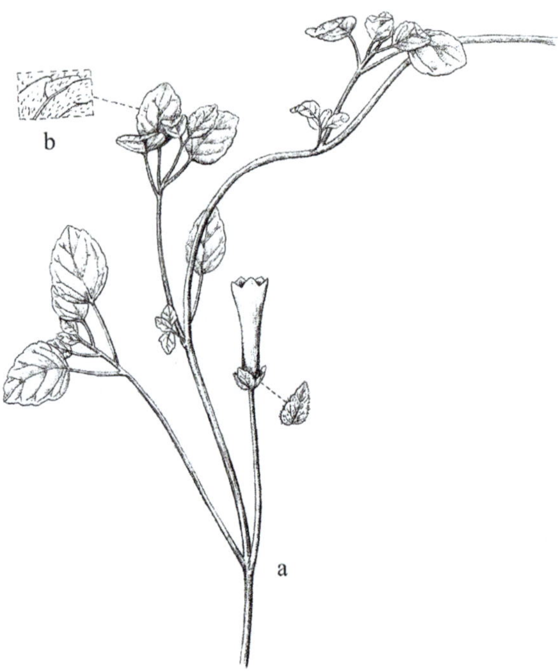

FIGURE IV 1:44A *Codonopsis hongii* Lammers. **a**, a section of twining stem with a flower, showing a calyx lobe; **b**, abaxial surface of leaf blade, showing indumentum. *From Acta Phytotax. Sin. 18: 1980.*

Distribution: Bhutan, China (S Tibet and NW Yunnan), N India (Sikkim), and N Myanmar (Map IV 1:44).

Additional specimens examined, **BHUTAN, Chendebi**: Deuchung, Khoma Chu, 2290 m, 28. Sept. 1949, *F. Ludlow et al. 21479* (BM, E); Kancham Runakla, 1860 m, 21. Aug.

FIGURE IV 1:44B *Codonopsis hongii* Lammers. **a–d**, from Tibet: Cona, 30. Aug. 2011, by Xin-Tang MA & De-Yuan HONG.

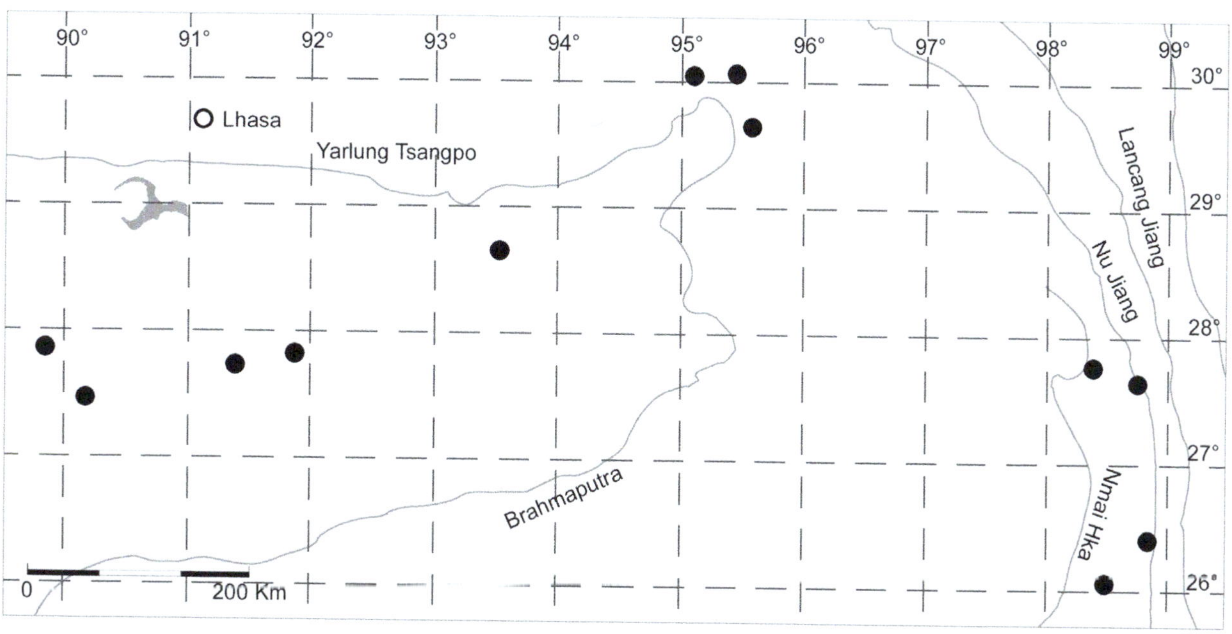

MAP IV 1:44 Distribution of *Codonopsis hongii* Lammers.

1914, *R. E. Cooper 3264* (BM, E); 2440 m, 29. Aug. 1949, *F. Ludlow et al. 19658* (BM) . **Tobrang**: between Trashiling and Chendebi, 2290 m, 06. Aug. 1949, *F. Ludlow et al. 17059* (BM); Tashiling, Jangsa, 2590 m, 10. Sept. 1914, *R. E. Cooper 2059* (E) ; Trashiyangsi Valley, 2440 m, 29. Aug. 1934, *F. Ludlow & G. Sherriff 905* (BM, E); 2130 m, 25. Sept. 1949, *F. Ludlow et al. 21222* (BM, E). **Upper Mo Chu** Distr.: Goen Gaza - Gaza Zam, 27°54′N, 89°45′E, 2450 m, 13. Sept. 1984, *I. W. J. Sinclair & D. G. Long 4980* (E, K).

CHINA, TIBET, **Bomi**: Qomdo, 30°04′N, 95°34′E, 2800 m, 01. Sept. 2011, *D. Y. Hong et al. H11019* (PE); Tangmai, 2150 m, thickets, 15. Sept. 1980, *C. C. Ni et al. 1649* (PE); loc. eodem, edges of thickets, 11. Sept. 1976, *Wu 5761* (KUN); loc. eodem, 1960 m, grasses by road, 27. Oct. 1975, *Sichuan Inst. Chinese Mater. Medic. Exped. 322* (SM); loc. eodem, 2000 m, *FLPH Tibet Exped. 12–2002* (PE); loc. eodem, 1965, *Y. T. Chang & K. Y. Lang 889* (paratypes : PE); Yi'ong, 2400 m, 11. Jul. 1965, *J. S. Ying & D. Y. Hong 65–1272* (paratype: PE). **Cona**: Mama-Tea Farm, 2600–2700 m, roadsides, 26. Aug. 1975, *Qinghai-Xizang Suppl. Exped. 75–1663* (KUN, PE); loc. eodem, 27°51′N, 91°47′E, 2532 m, 30. Aug. 2011, *D. Y. Hong et al. H11009* (PE). **Lhunzê**: Zara-Maja, 2780 m, 07. aug. 2013, *Y. S. Chen et al. 13–0678* (PE). **Mêdog**: 52K to 80K, 3000m, 19. Sept. 2012, *FLPH Tibet Exped. 12–1898* (PE). **YUNNAN**, **Gongshan**: Drungjiang Township, 1660 m, evergreen broad-leaved forests, 31. Oct. 2004, *Gaoligong Shan Biodiversity Survey 21262* (CAS, E); loc. eodem, 31. Oct. 2004, *Gaoligong Shan Biodiversity Survey 21299* (CAS, E); loc. eodem, 1800 m, 07. Jan. 1991, *Drungjiang Invest. Team 1642* (CAS); Lazao, Salween Valley, grassy slopes, 29. Oct. 1940, *K. M. Feng 8673* (PE).

INDIA: **Sikkim**, Karponang, 2940 m, 20. Sept. 1916, *G. H. Cave* s. n. (E).

MYANMAR: Western flank of the Chimi-li, N' Maikha-Salween Divide, 26°23′N, 98°48′E, 3050–3350 m, Aug. 1924, *G. Forrest 24881* (E, K); Hpimaw Hill, 1980 m, 05. Sept. 1919, *RA 1289* (E); above Kang Fang, 1220–1530 m, 20. Aug. 1914, *F. Kingdon-Ward 1887* (E); Kang Fang (Ngawchang Valley), 1530 m, 19. Sept. 1924, *F. Kingdon-Ward 3683* (E).

Section 5. Campanumoea (Bl.) D. Y. Hong, stat. nov. ≡*Campanumoea* Bl., Bijdr. Fl. Ned. Ind. 13: 726. 1826; Moeliono & Tuyn in van Steenis, Fl. Males. Ser. 1. 6: 118. 1960. pro syn. Type: *Codonopsis javanica* (Bl.) Hook. f. (*Campanumoea javanica* Bl.)

Roots carrot-shaped. Stems twining. Leaves alternate or opposite, leaf blade deeply cordate. Calyx epigynous or hypogynous; corolla always epigynous. Fruit a berry. Plants not fetid.

Two species, widely distributed in E and SE Asia from Bhutan to Indonesia, Philippines and Japan. They grow in thickets, forests, and forest margins (thickets or herbs) at altitudes from 250 to 3200 m.

45. Codonopsis javanica (Bl.) Hook. f., Ill. Himal. Pl. 116. tab. 16B. 1855; Moeliono & Tuyn in van Steenis, Fl. Males. Ser. 1. 6: 119. 1960; Lammers in Peng, Phytogeogr. Bot. Inventory Taiwan, Inst. Bot. Acad. Sin. Monogr. ser. 12: 46. 1992. ≡*Campanumoea javanica* Bl., Bijdr. 727. 1826; de Candolle, Monogr. Camp. 119. 1830; Clarke in Hooker, Fl. Brit. Ind. 3: 435, 1881; Tsoong, Contr. Inst. Bot. Nat. Acad. Peiping, 3(3): 110. 1935; Hong, Fl. Reipubl. Popularis Sin. 73(2): 70. 1983; Hong, Lammers & Klein in Wu, Raven & Hong, Fl. China 19: 526. 2011. ≡*Campanula javanica* (Bl.) D. Dietr., Syn. Pl. 1: 758. 1839. Type: Java Occid., Mons Megamondong, *Blume* s. n. (holotype: P).

Codonopsis cordata Hassk., Nat. Tijd. N. I. 10: 9. 1856. ≡*Campanumoea cordata* (Hassk.) Miq., Sum. 599. 1862. Type: Sumatra, *J. E. Teysmann* s. n.(holotype: K).

Codonopsis cordifolia Kom., Trudy Imp. S.- Peterburgsk. Bot. Sada, 29(1): 108. 1908. Type: China, Yunnan, "meridiem versus a 'Red River' prope Manmei", 1830 m, *A. Henry 9634 A* (not preserved ?).

Campanumoea labordei H. Lév., Bull. Soc. Agr. Sci. Arts Sarthe 39:323. 1904 & Fedde Rep. Spec. Nov. 6: 376. 1909. Type: China, "Kweichow" (Guizhou), "environs de Tsin-gay à Tchao-se", 07. Sept. 1899, *Laborde & Bodinier 2899* (holotype : E).

Roots thickened, carrot-shaped. Stems twining, multi-branched, glabrous, up to 3 m long. Leaves opposite, rarely alternate, long-petiolate; leaf blade cordate or cordate-ovate, 2.6–8 cm long, 2–7.5 cm broad, glabrous or sometimes abaxially sparsely villous, margin dentate or crenate, or rarely entire. Flowers solitary, axillary, glabrous throughout. Calyx free from hypanthium, cleft nearly to base; calyx lobes lanceolate or ovate, rarely narrowly triangular, 8–20 mm long. Corolla epigynous, campanulate, cleft to middle, white or yellow-green, purple or reddish inside. Stamens 5. Stigma 4- or 5-fid; ovary and berry 5-locular. Berry violet or greenish white suffused with red, globose, 0.7–2.6 cm in diameter. Seeds irregular in shape, surface reticulate.

Distibution: Bhutan, China, NE India, Indonesia, Japan, Laos, Myanmar, Thailand, and Vietnam (Map IV 1:45).

1a. Corolla 1.5–3.2 cm long; berry purple, 1.5–2.6 cm in diameter .... 45a. subsp. *javanica*

1b. Corolla 0.6–1.7 cm long; berry greenish white suffused with red, 0.7–1.5 cm in diameter .... 45b. subsp. *japonica*

45a. Codonopsis javanica subsp. **javanica**

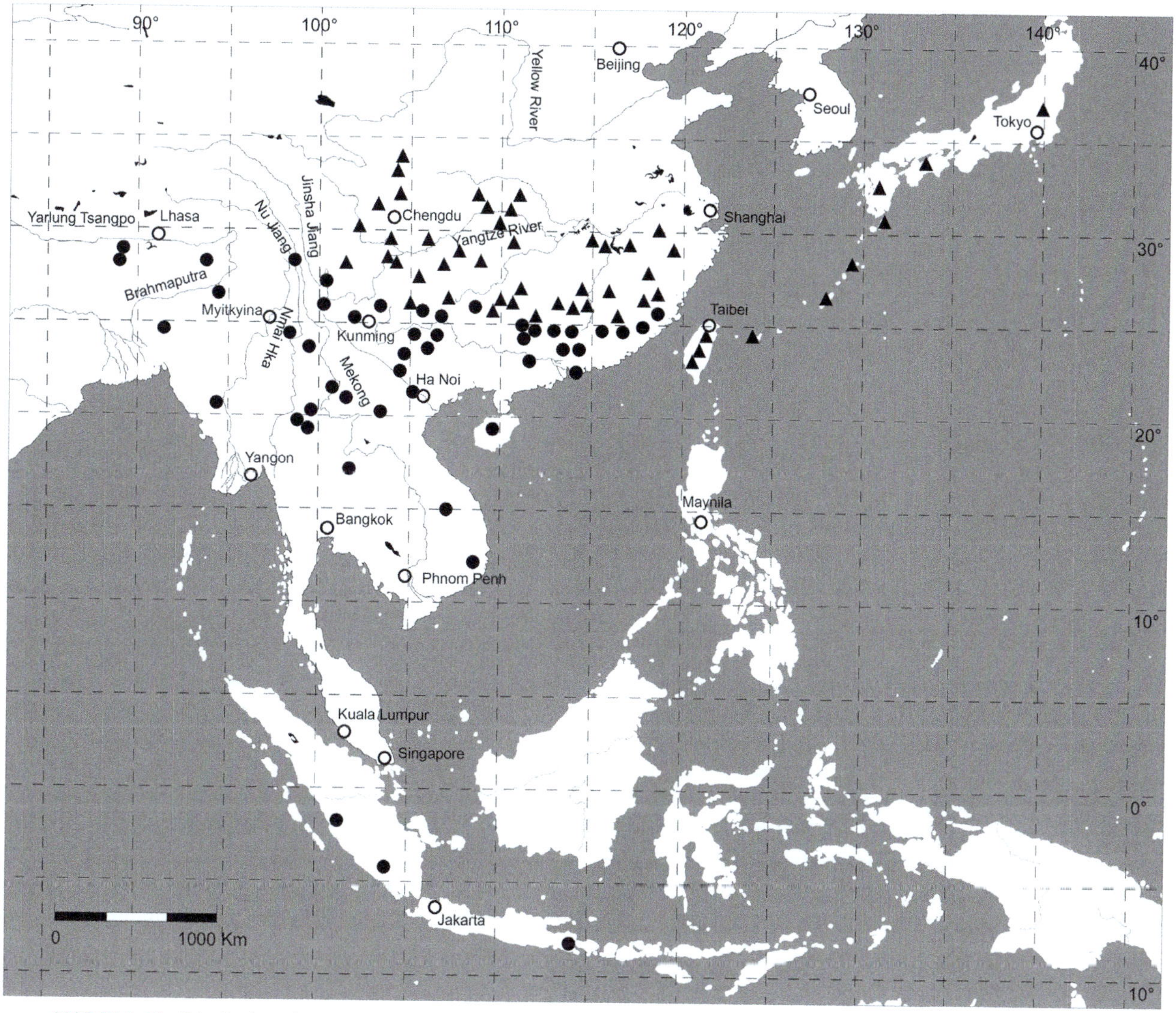

MAP IV 1:45 Distribution of *Codonopsis javanica* (Bl.) Hook.f. Dots: subsp. *javanica*; triangles: subsp. *japonica* (Makino) Lammers.

Leaf blade 3.5–8 cm long, 2.2–7.5 cm wide. Calyx lobes lanceolate, 1–2 cm long. Corolla 1.5–3.2 cm long. Berry purple, 1.5–2.6 cm in diameter. Figures IV 1:45A & B.

Phenology: flowering and fruiting from May to November. **Habitats**: forests, thickets, or forest margins ; 270–3200 m. **Distribution**: E and SE Asia: Bhutan, S China, N India, Indonesia, Laos, Myanmar, Thailand, and Vietnam.

Additional specimens examined, **BHUTAN**, **Chungkar**: 2130 m, 24. Nov. 1938, *F. Ludlow et al. 6782* (BM). **Shali**: Trashiyangsi Valley, 1980 m, 11. Oct. 1934, *F. Ludlow & G. Sherriff 1053* (BM, E). **Punakha**: Gichha, 2440 m, 08. Aug. 1914, *R. E. Cooper 2705* (E).

CHINA, FUJIAN, Dehua: Chenban, Oct. –Nov. 1974, *s. coll. 0274* (PE). **Nanjing**: Hexi, Caijia Shan, 510 m, 10. Sept. 1974, *s. coll. 236–6 01360* (PE). **GUANGDONG, Conghua**: s. loc., 600 m, open woods, 03. 12. 1958, *L. Deng 8824* (PE, SZ). **Dabu**: Yanshang Township, 850 m, dense forests, 23. Jun. 1957, *L. Deng 5383* (PE). **Deqing**: Mobin, 12. Aug. 1958, *Y. G. Liu 1996* (PE). **Heping**: Daba, 270 m, forests, 13. Jan. 1998, *M. C. Tan 971890* (PE). **Hong Kong**: Castle Peak, 09. Dec. 2000, *S. Y. Hu & Paul But 23527* (A, K). **Huaji**: Tongshan, *W. T. Tsang 22919* (PE). **Luofu Shan**: Huanglongguan, thickets, 22. Aug. 1930, *N. Q. Chen 41659* (SZ). **Maixian**: Yinna Shan, thickets, *W. T. Tsang 21500* (PE). **Qujiang**: Baiyun nunnery, scandent on herbs, 11 Oct. 1930, *S. P. Ko 50827* (KUN, PE, SZ). **Wengyuan**: Qingyun Shan, s. d., *S. K. Lau 2233* (PE); Zhangyuan Township, 07. Nov. 1935, *S. K. Lau 25143* (PE, SZ). **Xinfeng**: Huangpo Township, 650 m, thickets, 11. Oct. 1958, *L. Deng 8029* (KUN, PE, SZ). **Yangshan**: Xilu Township, 1300 m, 21. Nov. 1958, *P. C. Tam 60302* (PE). **Yingde**: Wentang Shan, 08. Dec. 1957, *C. Huang 163796* (KUN, PE). **Yunfu**: Yunwu Shan, 23. Oct. 1934, *C. Wang 32675* (PE, SZ). **GUANGXI, Bose**: Bulin, Weikuan Shan, 480 m, s. d., *Baise Exped. 1512* (PE); Dawang Shan, 800 m, 11. Sept.

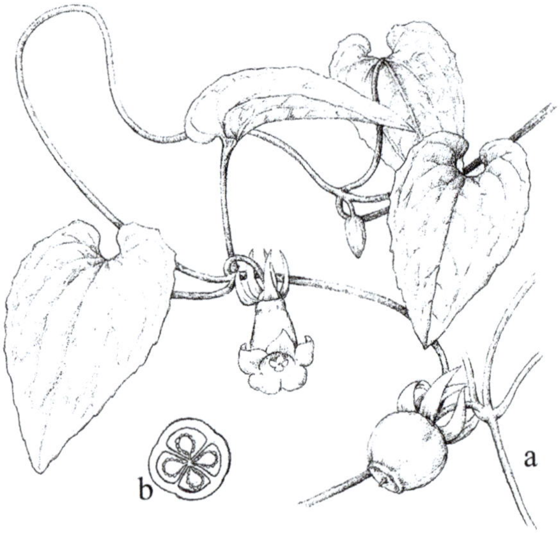

FIGURE IV 1:45A *Codonopsis javanica* (Bl.) Hook.f., subsp. *javanica*, redrawn from *Fl. China* 19. 2011. **a**, a section of twining stem with a flower and a berry, which shows that ovary is superior to calyx; **b**, a cross section of ovary. *From Flora of China Vol. 19, 2011.*

1953, *Guangxi Exped. 3598* (PE). **Lingyun**: Laoshanjiao, 29. Jul. 1937, *S. K. Lau 28799* (PE, SZ). **Hexian**: Lisong, grasses by rood, 10. Oct. 1959, *Z. T. Li 604126* (PE). **Napo**: Baihe Township, evergreen forests on rocky slopes, 22. Oct. 1997, *H. Akiyama et al. 1308* (KUN); Defu, ca. 1300–1400 m, 24. Oct. 1997, *H. Akiyama et al. 1433* (KUN); Pingmeng, 600–1100 m, 19. Oct. 1997, *H. Akiyama et al. 137* (KUN). **Tian'e**: Lupaizhen, 11 Sept. 1989, *Beijing Team 897026* (PE). **Yangshuo**: Jibo Distr., Dashuitian, thickets in ravines, 08. Nov. 1963, *Z. Y. Chen 53335* (KUN). **Zhaoping**: Ling Wun, dense woods, 29. Jul. 1937, *S. K. Lau 28799* (PE). Me-Kon, She-feng Dar Shan, S. Nanning, 792 m, thickets, 06. Nov. 1928, *R. C. Ching 8469* (PE). On Tak, 980 m, woods, 20. Sept. 1935, *S. P. Ko 55775* (PE). Tsin Kung Shan, N. H. Yen, 1070 m, thickets, 20. Aug. 1928, *R. C. Ching 7028* (PE). Wujiang, forests, 25. Oct. 1959, *Z. T. Li 604215* (KUN); Yaoshan, 19. Oct. 1936, *C. Wang 40281* (PE). **GUIZHOU**, **Anlong**: Huali, 1180 m, thickets, 19. May. 1960, *Guizhou Exped. 3624* (PE); Xianheping, 1300–1500 m, thickets, 12. Oct. 1998, *Sino-Japanese Exped. 100487* (KUN); loc. eodem, 1300–1500 m, evergreen forests, *Sino-Japanese Exped. 100713* (KUN); s. loc., 500 m, open scrubs, 20. Oct. 1930, *Y. Tsiang 7442* (PE). **Ceheng**: Qiaoma Forest Farm, 450–900 m, thickets, 30. Oct. 1991, *Yunnan-Guizhou-Guangxi Floristic Surv. 50914* (KUN); Rongju Township, 1500 m, thickets, 04. Sept. 1958, *Z. Y. Cao 533* (PE). **Rongjiang**: Jihua, 880 m, roadsides, 19. Aug. 1985. *Zhang 51789* (KUN). **Wangmo**: Leyuan, 680 m, thickets, 25. Aug. 2003, *G. F. Wang 1–0754* (PE). **Xingren**: Diandou, 1300 m, grasslands, 26. Aug. 1960. *Guizhou Exped. 8629* (PE); Haizi, 1400–1450 m, thickets, limestones, 10, Oct. 1998, *S. K. Wu et al. 100469* (KUN) & *100471* (KUN); Longchang, 1400 m, 18. Aug. 1960, *Guizhou Exped. 7862* (PE); Mabaoshu, 1200–1500 m, thickets, limestones, 09. Oct. 1998, *S. K. Wu et al. 100463* (KUN) & *100464* (KUN); Xiashan Township, 1340 m, forest edges, 14. Aug. 1986, *C. Z. Dang & P. Dang 586* (PE). **Xingyi**: Pogang, 1240–1450 m, thickets, limestones, 05. Oc. 1998, *S. K. Wu et al. 100402* (KUN), *100407* (KUN), *100411* (KUN) & *100816* (KUN). **Zhenfeng**: s. loc., 1600 m, dense forests, 26. Aug. 1960 , *Guizhou Exped. 8522* (PE). **HAINAN**, **Danxian**: s. loc., *W. T. Tsang 899* (PE). **Fan Yah**, 910 m, 1932–1933, *N. K. Chun & C. L. Tso 44146* (PE). **Hongmao Shan**, S of Fan Ta, forests, *Tsang & Fung 624* (PE). **Loktung**, dense woods, 02. Jul. 1936, *S. K. Lau 27387A* (PE). **Wuzhi Shan**: Fandongcun, ca. 910 m, 24. Oct. 1932, *N. Q. Chen 44146* (PE). **TIBET**, **Ukhrul**: 1520–1830 m, 03. Aug. 1948, *F. Kingdon-Ward 17895* (BM, PE). **YUNNAN**, **Baoshan**: summit of Gaoligong Shan, 2050 m, 16. Nov. 2000, *H. Li et al. 13114* (CAS); Lisushan to Huangcaoping, May 1941, *H. C. Wang 792* (PE); Mangkuan Township, 1350 m, 18. Oct. 1998, *H. Li et al. 10616* (CAS); loc. eodem, 1350 m, 18. Oct. 1998, *H. Li et al. 10550* (CAS). **Cangyuan**: Balao, Jinhe, 1200 m, 07. Sept. 1980, *S. T. Li 80–1048* (KUN); Mengyong, 1850 m, secondary forests, 04. Sept. 1980, *S. T. Li 80–979* (KUN). **Dali**: Tengyueh (Tengchong) -Dali road, 25°25′N, 2130–2440 m, Sept. 1903, *G. Forrest 1109* (E); s. loc., 05. Apr. 1946, *T. N. Liou 14230* (KUN). **Eshan**: s. loc., 1840 m, thickets, 14. Oct. 1988, *Eshan Exped. 831* (KUN); No 4 Distr., Haibin, 1600–1700 m, ravines, 02. Oct. 1958, *S. K. Wu 180* (KUN). **Fumin**: Fumin to Qinglong Gorge, 1770 m, secondary forests, 07. Dec. 2006, *Y. Du et al. SCSB-A-000360* (KUN); by Qingshuihe, 2100 m, thickets, 25. Oct. 1964, *B. Y. Qiu 596407* (KUN). **Fugong**: Lumadeng Township, 1780 m, limestones, 11. Aug. 2005, *Gaoligong Shan Biodiversity Survey 26999* (A, CAS); Pihe to Gongshan, 29. Sept. 1980, *J. S. Yang* s. n.(KUN); s. loc., 2200 m, ravines, 14. Oct. 1933, *H. T. Tsai 54429* (KUN, PE, SZ); s. loc., 1800 m, by stream, 09. Oct. 1933, *H. T. Tsai 54837* (KUN, PE, SZ); s. loc., 2800 m, forests, 11. Oct. 1934, *H. T. Tsai 58722* (KUN, PE, SZ). **Gengma**: Fubei Township, 1300 m, secondary forests, 04. Dec. 1958, *T. P. Zhu 0468* (KUN). **Gongshan**: Cikai, 1500 m, subtropical evergreen broad-leaved forests, 09. Oct. 2002, *Gaoligong Shan Biodiversity Survey 15876* (CAS). **Guangnan**: Zhetu to Xiwei, 1480–1520 m, herbs by road, 05. Nov. 1965, *Wenshan Exped. (65) 233* (KUN, PE). **Hekou**: Yaoshan, 22. Nov. 2003, *X. Yang 03010* (KUN); loc. eodem, 1100 m, 29. Oct. 1992, *Y. M. Shui 962* (PE). **Jiangchuan**: Dongbin Township, 1700 m, roadsides, 31. Oct. 1989, *Yuxi Exped. 1323* (KUN). **Jianshui**: s. loc., 21. Mar. 1941, *T. N. Liou 18370* (PE). **Jingdong**: Huiyao, 2000 m, river banks, 11. Nov. 1939, *M. K. Li 2283* (KUN); Jiding Shan to Modaohe, 1920 m, edges of forests, 06. Nov. 1963, *Z. H. Yang 101453* (KUN); Sen-cha-tsun, 1600 m, thickets, 09. Oct. 1939, *M. K. Li 0378* (KUN); Wenpu Township, 1800 m, roadsides, 04. Oct. 1956, *B. Y. Qiu 52420* (KUN, PE); loc. eodem, 1700 m, 30. Sept. 1956, *B. Y.*

FIGURE IV 1:45B *Codonopsis javanica* (Bl.) Hook. f. subsp. *javanica*. **a** & **d**, from Guangxi: Daxing, 16. Oct. 2008, by Xin-Tang MA; **b**, from Guangxi: Chongzuo, 15. Oct. 2008, by Yu-Fen DU; **c**, from the same locality as b, 15. Oct. 2008, by Xin-Tang MA.

Qiu 52660 (KUN, PE); loc. eodem, 1800 m, thickets by road, 18. Aug. 1959, *S. G. Xu 5147* (KUN); Xinmin, 1800 m, grasslands with shrubs, 14. Nov. 1993, *H. Peng 2001* (KUN). **Jinghong**: Bajiaoqing, 840 m, thickets, Oct. 1936, *C. W. Wang 79732* (KUN, PE); Damenglong, 1800 m, grassy slopes, Sept. 1936, *C. W. Wang 78419* (KUN, PE); Kunge, 1260 m, thickets, Oct. 1936, *C. W. Wang 79266* (KUN, PE); Menglong to Jiluo Township, 1033 m, edges of Monsoon forests, 20. Nov. 2006, *T. Zhang et al. SCSB-B-000261* (KUN); Mengsong, 1200 m, open woods, 09. Nov. 1956, *H. T. Tsai 80044* (PE); Xiaomengyang, 1100 m, woods, Sept. 1936, *C. W. Wang 75848* (KUN, PE). **Jinping**: Fen-shui-ling, 1800–2000 m, roadsides, 11. Oct. 1996, *S. K. Wu et al. 3173* (KUN); loc. eodem, 2100 m, secondary forests, 11. Oct. 1996, *S. K. Wu et al. 4059* (KUN); Leidashu, 2000 m, 07. Dec. 1958, *H. W. Li 370* (KUN); Wutai Shan, 1400 m, roadsides, 07. Oct. 1996, *S. K. Wu et al. 3124* (KUN). **Kunming**: Heilongtan, 26. Oct. 1938, *K. M. Feng 18* (KUN); loc. eodem, 1945, *K.*

M. Feng 10821 (KUN); West Hill, 3200 m, 25. Oct. 1957, *B. Y. Qiu 55343* (KUN); West Hill, Sujiacun, 1941, *T. N. Liou 19326* (PE); West Hill, Kao-chiao, 1900 m, 1946, *T. N. Liou 14822* (PE); West Hill, Huatingsi, 24. Nov. 1945, *T. N. Liou 14467* (PE); loc. eodem, 22. Sept. 1946, *T. N. Liou 20885* (PE); loc. eodem, 05. Oct. 1986, *Z. T. Wang 8610–411* (PE); West Hill, Taihuasi, 2100 m, 23. Aug. 1946 *T. N. Liou 16700* (PE); loc. eodem, 14. Nov. 1945, *T. N. Liou 14326* (KUN, PE); loc. eodem, 25. Aug. 1945, *T. N. Liou 14230* (PE); loc. eodem, edges of forests by road, 16. Sept. 1985, *Y. L. Chen & C. Grey-Wilson 85–34* (PE) ; s. loc., Oct. 1940, *T. N. Liou 19581* (PE). **Lingcang**: Mayidui, 1300 m, forests by stream, 23. Aug. 1957, *J. S. Xin 281* (KUN); Mengwang, 1950 m, thickets, 03. Sept. 1957, *J. S. Xin 666* (KUN, PE); Mengwang to Botuo, 1900 m, thickets, 03. Sept. 1957, *J. S. Xin 672* (KUN); Poshan, 2400 m, thickets, 08. Oct. 1938, *T. T. Yü 17955* (KUN, PE). **Longchuan**: s. loc., 1350 m, thickets on edges of forests, 27. Aug. 1976, *S. J. Pei 14131* (KUN). **Longling**: Longjiang Township, 2011 m, evergreen broad-leaved forests, 21. Aug. 2003, *Gaoligong Shan Biodiversity Survey 17303* (CAS); Zhen'an, 1830–2000 m, oak forests, 05. Dec. 1958, *C. Chen 760* (KUN). **Lüchun**: Erpu to Banpo, 1300 m, forest edges by road, 22. Oct. 2000, *Y. M. Shui 14034* (KUN); Erpu to Xiaoheijiang, 850 m, secondary forests, 18. Oct. 2000, *Y. M. Shui 13243* (KUN); Lüchun to Baqiashan, 1600 m, 17. Sept. 1973, *D. D. Tao 0049* (PE); Mayu Hydropower Station, 840–1300 m, evergreen forests, 30. Oct. 1995, *S. K. Wu et al. 2099* (KUN); loc. eodem, rain forests. 31. Oct. 1995, *S. K. Wu et al. 1002* (KUN) & *2133* (KUN). **Lushui**: Yaojiaping to Jiangqiao, 2000 m, grassy slopes with shrubs, 21. Aug. 1964, *S. K. Wu 8506* (KUN). **Luxi** (Mangshi): s. loc., 1280 m, forests, 04. Sept. 1958, *s. coll. 580185* (KUN). **Malipo**: Pan-chia-chü, 1600–1700 m, open thickets, 31. Oct. 1947, *K. M. Feng 12640* (KUN, PE). **Menghai**: Manzhen, 1100–1200 m, open woods, 17. Nov. 1955, *Kunming Working Station 7328* (PE); near the city, forest edges, 13. Sept. 1973, *S. Y. Bao 0021* (KUN); s. loc., thickets, 14. Sept. 1959, *H. T. Tsai 59–10405* (KUN); s. loc., thickets, 13. Sept. 1958, *K. L. Le 540* (KUN). **Mengla**: Houlin, 700 m, waste slopes, 19. Oct. 1959, *H. T. Tsai 59–11022* (KUN); Shangmeng, 800–900 m, forests, s. d., *Z. H. Yang 12003* (KUN); s. loc., 1150 m, 03. Sept. 1991, *Z. H. Tsi 91–432* (A). **Mengzi**: Bei-chu-hoa, 1800 m, wood margin, 19. Sept. 1939, *C. W. Wang 83253* (KUN, PE); Moo-bei, 1860 m, 11. Oct. 1939, *C. W. Wang 83392* (KUN, PE); S of Red River, 1830 m, *A. Henry 9634 A* (PE); s. loc., 1520 m, *A. Henry 9634* (PE); s. loc., 1730 m, wood margin, 06. Dec. 1932, *H. T. Tsai 52365* (KUN, PE). **Pingbian**: Mau-mau-tou, 1380 m, 17. Sept. 1939, *C. W. Wang 81814* (KUN, PE) ; Shiban Township, 1400 m, thickets, 14. Oct. 1954, *K. M. Feng 5013* (KUN, PE); Yaoshan, Baiyan Township, 1200 m, grassy slopes, 05. Jul. 1959, *P. I. Mao 2372* (KUN, PE); loc. eodem, 1300 m, grassy slopes, 16. Sept. 1954, *K. M. Feng 4495* (KUN, PE, SZ); s. loc., 1200 m, thickets, 17. Aug. 1934, *H. T. Tsai 61530* (PE, SZ) ; s. loc., 1400 m, 10. Sept. 1934, *H. T. Tsai 61797* (PE, SZ). **Qiubei** : s. loc., woods, 22. Dec. 1932, *H. T. Tsai 51428* (KUN). **Ruili**: Mengyou , ca. 900 m, thickets, 10. Dec. 1958, *C. Chen 521* (KUN). **Shiping**: Daleng Shan, 2400 m, mountain summit, 01. Dec. 1958, *S. K. Wu 946* (KUN). **Shuangbai**: Haizidi, oak forests, 30. Sept. 1958, *S. C. Huang 0124* (KUN). **Simao**: Maliping, 1000 m, open woods, 03. Oct. 1955, *P. I. Mao 6416* (KUN, PE, SZ); Zhengdong, 800 m, 20. Sept. 1955, *P. I. Mao 6139* (SZ). **Songming**: Aziying, Guodong, 2100–2200 m, forests, 10. Oct. 1953, *Acad. Sin. Kunming Station 50323* (KUN); loc. eodem, 2300 m, 08. Nov. 1988, *D. Z. Li 88201* (KUN). **Tengchong**: Guangming, 2000 m, broad-leaved forests, 01. Aug. 1980, *S. T. Li 80–378* (KUN); Jietou Township, N of Datang, 1950 m, secondary evergreen broad-leaved forests, 27. Oct. 1998, *H. Li et al. 11037* (CAS); loc. eodem, 1560 m, 26. Oct. 1998, *H. Li et al. 10996* (CAS); Mangkuan, Township, 1525 m, forests, 10. Sept. 2003, *Gaoligong Shan Biodiversity Survey 19074* (CAS); loc. eodem, 1590 m, 09. Sept. 2003, *Gaoligong Shan Biodiversity Survey 18948* (CAS); Qushi, 1930–2120 m, pine forests, 12. Sept. 1960, *W. Q. Yin 60–1034* (KUN); Qushi, Baijiahe, 1740 m, *Castanea* forests, 21. Sept. 1960, *W. Q. Yin 60–1251* (KUN); Qushi, Laodameng, 1470 m, secondary forests, 06. Dec. 2007, *Y. Z. Zhou 2–141* (KUN); Qishi Township, 1510 m, 29. Oct. 1998, *H. Li et al. 11227* (CAS); Sanyuan Township, 1560 m, remnant secondary evergreen broad-leaved forests, 03. Nov. 1998, *H. Li et al. 11431* (CAS, E); Wuhe Township, 1900 m, 26. Aug. 2003, *Gaoligong Shan Biodiversity Survey 17960* (CAS); s. loc., 1924–1925, *G. Forrest 25119* (PE). **Wenshan**: Byng-bah, 1750 m, mixed forests, 10. Aug. 1947, *K. M. Feng 11064* (KUN, PE); Laojun Shan 1800 m, thickets, 25. Aug. 1961, *S. K. Wu 61–3792* (KUN); loc. eodem, 2000 m, secondary forests, 22. Sept. 1992, *Y. M. Shui & X. H. Li 168* (KUN); loc. eodem, 2100 m, edges of forests, 20. Aug. 1985, *Z. H. Tsi et al. 578* (PE). **Xichou**: Fadou, 1385 m, thickets, 03. Nov. 2007, *E. D. Liu & L. Chen 1955* (KUN); loc. eodem, 1300 m, bushes, hillsides, 08. Nov. 1939, *C. W. Wang 85489* (KUN, PE); Shiang-pyg-shan, 1500 m, among herbs, 03. Sept. 1947, *K. M. Feng 11593* (KUN, PE). **Xishuangbanna**: Menglun, 1100 m, forests, 17. Sept. 1959, *S. J. Pei 59–10334* (KUN). **Yangbi**: Pingpo, 06. Nov. 1946, *T. N. Liou 22521* (PE); Xiapingpo to Goutouqing, 06. Nov. 1946, *T. N. Liou 22533* (PE). **Yanshan**: Bar-garh, 1100 m, thickets, 13. Nov. 1939, *C. W. Wang 84857* (PE); Lung-tang-tzai-Shan, 1350 m, thickets, 25. Sept. 1939, *C. W. Wang 83746* (KUN, PE); Pie-shih-eih, 1200 m, thickets or grasses, 27. Oct. 1939, *C. W. Wang 84787* (PE); Shih-moh-la, 1200 m, grassy slopes, 14. Oct. 1939, *C. W. Wang 84377* (KUN, PE); To-kou-tzai, 1200 m, thickets, 02. Oct. 1939, *C. W. Wang 84134* (KUN). **Yongde**: Wumulong Township, 2210 m, cliffs, 17. Aug. 2003, *S. S. Zi 287* (KUN); Wumulong, Yinchangjie, 2400 m, thickets, 19. Aug. 2003, *E. D. Liu & D. C. Yang 216* (KUN). **Yuanjiang**: Nanxi, 1380 m, forests, 06. Nov. 1964, *Y. H. Li 5899* (KUN); Sanquan, 1700 m, 12. Nov.

1958, *S. K. Wu 770* (KUN). **Yunlong**: Caoj zhen, thickets by woods, 2300 m, 06. Nov. 1993, *Q. E. Yang 9305* (PE). **Yingjiang**: Fulian Township, 1450 m, thickets, 03. Dec. 1958, *S. K. Wen 580827* (KUN); s. loc., 1400 m, Aut. 1952, *R. C. Ching 50693* (PE). **Weixi**: by Lancangjiang, 03. Nov. 1940, *K. M. Feng 8923* (KUN, PE); W bank of Lancang River, 2200 m, 04, Nov. 1963, *C. C. Lu 63238* (KUN). **Yunlong**: Caojian, 2406 m, thickets, 23. Sept. 2006, *A. H. Li et al. SCSB-A-000172* (KUN). **Zhenkang**: Mali-ling, 1650 m, 25. Aug. 1938, *T. T. Yü 17474* (PE).

INDIA, **Darjiling**: Rishap, 1070 m, 18. Aug. 1870, *C. B. Clarke 12388 A* (K); Runguet, 1220 m, Aug. 1881, *J. S. Gamble 9766* (K). **Meghalaya**: Jowai, 1530 m, Aug. 1892, *G. King's Collector* s. n.(E); Jowai, 1070 m, 30. Aug. 1886, *C. B. Clarke 44761A* (BM); Mamloo, 910 m, 02. Oct. 1867, *C. B. Clarke 5490* (BM); Naga Pauce, 1220–1830 m, 01. Aug. 1850, *J. D. Hooker & T. Thomson* s. n.(K); above Naga Pauce, 1220–1830 m, 19. Oct. 1850, *J. D. Hooker & T. Thomson* s. n.(K); s. loc., 1220–1830 m, s. d. *J. D. Hooker & T. Thomson* s. n.(P, PE); s. loc., 1220–1830 m, 17. Oct. 1850, *J. D. Hooker & T. Thomson* s. n.(K); s. loc., 1220–1830 m, 24. Sept. 1850, *J. D. Hooker & T. Thomson* s. n.(K). **Nagaland**: Lushai Hills, 1520 m, Oct. 1927, *N. E. Pany 315A* (K); loc. eodem, 610 m, 23. Oct. 1927, *N. E. Pany 315A* (K); loc. eodem, 910–1220 m, Sept. 1928, *W. J. L. Wenger 225* (K); Naga Hills, Dzulake, 1530 m, 31. Aug. 1935, *N. L. Bor 6220* (K). **Sikkim**: Lechong, 910 m, 04. Sept. 1874, *C. B. Clarke 684* (K).

INDONESIA, Java: SE Java, 1880, *H. O. Forbes 788* (BM, PE) & *1035* (BM, PE); Kawa Manukwads Jarut, 08. Feb. 1915, *H. N. Ridley* s. n.(K); near Kawah Tjibuni, foot path, 13. May, 1940, *C. Holstvoogd 388* (PE); Lemarang, 10. Jun. 1896, *Koorders 28042* (K); Madioen, Ngebel, ca. 900 m, 27. May, 1896, *Koorders 23946* (K); Pasoeroean, Pad Smeroehoeve near Sendoera, 1000–2000 m, 12. Jun. 1935, *C. G. G. J. van Steenis 7309* (PE); s. loc., *H. O. Forbes 988* (K); s. loc., 1846, *Lobb 105* (BM, K). **Sumatra**: Bengkulu, Bukit Kaba, 1600 m, 30. Aug. 1973, *J. Dransfield 3659* (K); s. loc., 1880, *H. O. Forbes 2223* (BM).

LAOS, **Attopeu** Prov., Boloven Plateau, 1100 m, 05. Oct. 1928, *M. Poilane 15789* (P); loc. eodem, 07. Oct. 1928, *M. Poilane 15891* (P). **Huaphanh**: 20°4′37″N, 103°27′56″E, 1580 m, 21. Oct. 2002, *K. Homsombath & M. F. Newman 1423* (E, P).

MYANMAR, **Hpimaw**: Ngawchang Valley 1530 m, 19. Sept. 1923, *F. Kingdon-Ward 3667* (E); Hpemaw Hill, 2620 m, 13. Sept. 1910, *R. F. 1309* (E); Ngawchang Valley, 1220 m, 23. Aug. 1919, *R. F. 1273* (E); Nyitadi, 2130 m, 23. Aug. 1920, *R. F. 1838* (E). **Maymyo Plateau**: s. loc., 1070 m, 31. Aug. 1908, *J. H. Lace 4265* (E, K); loc. eodem, 1070 m, Sept. 1911, *J. H. Lace* s. n.(E). **Mindat**: 1220–1520 m, open thickets, 26. Jul. 1956, *F. Kingdon-Ward 22527* (BM, E, PE); loc. eodem, 1220 m, 21. Aug. 1956, *F. Kingdon-Ward 22881* (BM); loc. eodem, 1220–1530 m, 08. Aug. 1956, *F. Kingdon-Ward 22551* (BM). **N'Maikha-Salween Divide**: Nwar-Hka Valley, 27° N, 1220–1830 m, 04. Sept. 1914, *F. Kingdon-Ward 1927* (E); s. loc., 1924–1925, *G. Forrest 24892* (E, K, PE). **Satawgaw**: s. loc., 1924–1925, *G. Forrest 24969* (E, K, PE) & *25394* (E, K, PE). **Shan Jolin**: Pwehla, 1220 m, Nov. 1888. *H. Collett 969* (K). **Shan States**: Keng Tung, 910 m, Sept. 1909, *R. W. MacGregor 794* (E). **Thandaung**: s. loc., Oct. 1934, *F. G. Dickason 5162* (PE); N Burma, N Triangle (Hkinlum), 1220 m, 07. Oct. 1953, *F. Kingdon-Ward 21413* (BM, E). Upper Burma, 1939, *R. Kaulback 412* (BM, PE). Chaungmaw Hka, 910–1220 m, 09. Aug. 1911, *F. Kingdon-Ward 3491* (E).

THAILAND, **Angka**: ca. 1610 m, 26. Sept. 1927, *H. B. G. Garrett 467* (K). **Chiangmai**: Chiengdao, 13. Oct. 1926, *A. P. G. Kerr 317* (BM); Inthanon, ca. 1800 m, 17. Dec. 1890, Dy Phon & Smitinand 71/90 (P); Nang Ka, 05. Nov. 1930, *A. P. G. Kerr 3368* (BM, K); Soutep, 910–1220 m, 04. Sept. 1910, *A. P. G. Kerr 1371* (BM, K) ; S side of Suthep, 18°50′N, 98°54′E, ca. 1210 m, 02. Dec. 1959, *L. B. Abbe et al. 9294* (K, PE); loc. eodem, obandened field, ca. 1450 m. 11. Sept. 1958, *Th. Sorensen et al. 4909* (PE); s. loc., 1200–1500 m, 01. Nov. 1967, *B. L. Burtt 5601* (E). **Chiangrai**: Prov. 20°22′N, 99°50′E, 1300 m, evergreen forests, 30. Aug. 1983, *E. F. Anderson 5209* (PE). **Phitsanulok**: Phu Miang, 1200–1650 m, thickets, 04. Oct. 1967, *T. Shimizu et al. 11667* (E, K, PE); Phu Miang, Rhu Rom Rot, 1200–1600 m, 03. Oct. 1967, *T. Shimizu et al. 11576* (E, K, PE). **Unerimai**: Mê Tûn, ca. 1200 m, 03. Jul. 1922, *A. P. G. Kerr 6214* (BM, K).

VIETNAM, **Haut Donnai** Prov., Massif du Bi Doup, 2000 m, 16, Oct. 1940, *M. Poilane 50996* (P). **Kontum** Prov., Dak Gley Distr., 1100–1200 m, 14, Nov. 1995, *L. Averyanov et al. VH 1607* (P); NW slopes of Mts Ngoc Linh, 1600 m, 02, Mar. 1995, *L. Averyanov et al. VH 345* (P). **Lam Dong** Prov., Lac Duong Distr., 12°07′N, 108°36′E, 04, Nov. 1997, *L. Averyanov & N. Q. Binh VH3865* (P). **Sa Pa**, 3–4 km from Sa Pa, 1450 m, grassy slopes, 14. Dec. 1964, *Sino-Vietnamese Exped. 480* (KUN).

45b. Codonopsis javanica subsp. **japonica** (Makino) Lammers, Bot. Bull. Acad. Sin. (Taipei) 33(3): 285. 1992. ≡*Campanumoea javanica* Bl. var. *japonica* Makino, Bot. Mag. (Tokyo) 22: 155. 1908. ≡*Campanumoea javanica* subsp. *japonica* (Makino) D. Y. Hong, Fl. Reipubl. Popularis Sin. 73(2): 71. 1983. ≡*Campanumoea japonica* Maxim., Bull. Acad. Imp. Sci. Saint-Pétersbourg 12: 67. 1867, non Sieb. ex Morr., Belgique Hort. 13: 337. 1863. ≡*Campanumoea maximowiczii* Honda, Bot. Mag. (Tokyo) 50: 389. 1936. Type: Japan, Nagasaki, 1863, *Maximowicz* s. n.(holotype: K).

Leaf blade 2.6–8 cm long, 2.2–6.5 cm broad. Calyx lobes ovate or lanceolate, or rarely narrowly triangular, 0.8–1.2(–1.7) cm long. Corolla 0.6–1.3 (–1.7) cm long. Berry greenish white suffused with red, 0.7–1.5 cm in diameter. Figure IV 1:45C.

FIGURE IV 1:45C *Codonopsis javanica* (Bl.) Hook. f. subsp. *japonica* (Makino) Lammers. **a**, **d–f**, from Sichuan: Emei, 19. Sept. 2011, by De-Yuan HONG; **b**, from Guangxi: Xing'an, Maoér Shan, 07. Oct. 2010, by You-Sheng CHEN; **c**, from Sichuan: Emei, 19. Sept. 2011, by Qiang WANG & De-Yuan HONG.

Phenology: flowering in August and September; fruiting from September to November.

Habitats: thickets or forest margins; 200–2000 m.

Distribution: widely distributed in E Asia: China (N Guangdong, N Guangxi, NE Yunnan, Taiwan, and northwards to S Anhui, S Zhejiang, Hubei and Sichuan) and Japan (northwards to Honshu).

Additional specimens examined, **CHINA**, **ANHUI**, s. loc., *M. B. Deng 4848* (PE). **CHONGQING, Beibei**: Jingyun Shan, 01. Nov. 1940, *C. Pei 7508* (PE). **Chengkou**: Baichi Shan, 1200 m, edges of forests, 17. Sept. 1958, *T. L. Dai 102759* (PE, SZ); loc. eodem, 635 m, 25. Sept. 1958, *T. L. Dai 103078* (PE, SZ); loc. eodem 600 m, 28. Sept. 1958, *T. L. Dai 103195* (PE, SZ); Taoyuan Township, 765 m, 16.

Oct. 1958, *T. L. Dai 103768* (PE,SZ). **Fengjie**: Jiezhuyuan, 900 m, 23. Sept. 1964, *H. F. Zhou & H. Y. Su 110509* (PE, SZ); Shuangdian Township, 13. Aug. 1958, *M. Y. Fang 23987* (PE, SZ); Xinhe, 1000 m, 08. Oct. 1964, *H. F. Zhou & H. Y. Su 111368* (SZ); Xinhe, 1100 m, 21. Oct. 1964, *H. F. Zhou & H. Y. Su 111549* (PE, SZ); Zhuyuan, 800 m, forests by road, 10. Oct. 1964, *H. F. Zhou & H. Y. Su 110757* (SZ); loc. eodem, 850 m, 07. Oct. 1964, *H. F. Zhou & H. Y. Su 110752* (SZ). **Fuling**: Zili, 260 m, thickets, 30. Sept. 2007, *Z. Y. Liu 180942* (PE). **Nanchuan**: Bailuping, 800 m, forests, 21. Oct. 1957, *G. F. Li 64854* (PE, SZ); Jinfo Shan, 1380 m, 02. Sept. 1957, *G. F. Li 63941* (KUN, PE, SZ); loc. eodem, s. d., *G. F. Li 64699* (KUN, SZ); loc. eodem, 950 m, 17. Nov. 1957, *G. F. Li 65058* (KUN, PE, SZ); loc. eodem, 1945, *F. T. Wang 10985* (PE); Xiaohe, Dongjiagou, 900 m, 19. Oct. 1957, *J. H. Xiong & Z. L. Zhou 94063* (PE, SZ). **Wulong**: Baima Shan, 1700 m, thickets, 28. Sept. 2007, *Z. Y. Liu 180505* (PE); Huangying, 260 m, thickets by forests, 30. Oct. 2009, *Z. Y. Liu 184429* (PE); Tongzi, 11. Sept. 1978, *W. H. Wang 3590* (SZ). **Wushan**: Baolong, Daxihe, 272 m, 10. Sept. 2009, *Sanxia Exped. 3725* (PE). **FUJIAN, Nanping**: Mandang Shan, 300 m, 06. Oct. 2000, *G. S. He 10173* (PE); s. loc. 300 m, 26. Aug. 1958, *Fudan Univ. Exped. 70059* (KUN). **Shanghang**: Qiu Shan, 800 m, edges of forests, 09. Sept. 1987, *L. K. Ling 3591* (PE). **Shaowu**: s. loc., *H. C. Zhou 5048* (SZ) & *6238* (SZ). **Wuyi Shan**: Yanghuangkeng, 700 m, dense forests, 08. Sept. 1979, *Wuyi Exped. 01394* (PE). **Yong'an**: s. loc., 07. Dec. 1932, *D. S. Wang 1309* (PE). **Zhangping**: Guilin, thickets, 25. Nov. 1932, *R. Ling* s. n.(PE). **GANSU, Wenxian**: Bikou, 800 m, by stream, 08. Sept. 1959, *Z. Y. Zhang 14699* (PE); Fanba, Gaojiagou, 1000 m, under rocks, 01. Sept. 1976, *J. Q. Fu 2800* (PE). **GUANGDONG, Liannan:** Jinkeng Township, 330 m, open woods, 08. Sept. 1958. *P. C. Tam 59247* (KUN, PE). **Lianxian**: Yangshan, 10. Sept. – 06. Dec. 1930, *S. P. Ko 50829* (PE); Yao'an, Tianguang Shan, 760 m, open woods, 29. Oct. 1958, *P. C. Tam 59973* (KUN, PE). **Lecang**: Wushan, 400–650 m, bushes path side, *Y. Tateishi et al. 1537* (PE). **Renhua**: s. loc., 700–900 m, open woods, 30. Aug. 1958, *L. Deng 7345* (PE). **Ruyuan**: Dalingjiao, 23. Nov. 1934, *S. P. Ko 54833* (PE, SZ); Xiyunsi, Dashiyan, 25. Nov. 1933, *S. P. Ko 53795* (SZ). **Shixing**: Luoba Township, open woods 05. Aug. 1958, *L. Deng 7024* (PE). **GUANGXI, Longsheng**: Daluo Township, thickets, 20. Nov. 1957, *Z. T. Li & Y. C. Chen 600063* (KUN, PE); Pingshui Township, 950 m, open woods, 16. Nov. 1955, *Guangfu Forest Farm Exped. 1120* (PE). **Rongshui**: Sanfang, Jiuwan Shan, open woods, 03. Aug. 1958, *S. H. Chun 16048* (KUN, PE). **Xing'an**: Mao'er Shan, 850 m, roadsides, 14. Nov. 1956, *S. L. She 900211* (PE). **GUIZHOU, Bijie**: s. loc., woods, 21. Sept. 1930, *Y. Tsiang 9015* (PE). **Chishui**: 07. Nov. 1938, *P. C. Tsoong* s. n.(PE). **Dafang**: Jiulong Shan, 2000 m, 18. Aug. 1959, *Bijie Exped. 1023* (PE). **Dejiang**: Qinglongzhen, 850 m, thickets, 24. Sept. 2004, *M. X. An 3854* (PE); Shaba Forest Farm, 1000 m, 16. Aug. 1959, *N. Guizhou Exped. 2682* (PE). **Guiyang**: Shibanshao, Babian Shan, 1200 m, 24. Oct. 1956, *Sichuan-Guizhou Exped. 2049* (PE, SZ); s. loc., 1300 m, thickets, 28. Oct. 1956, *Sichuan-Guizhou Exped. 2284* (PE); s. loc., 07. Sept, 1899, E. Bodinier 2699 (P). **Huaxi**: Siyazhai, 1300 m, 20. Aug. 1958, *Z. Y. Cao 377* (PE). **Jiangkou**: Taiping Township, 500 m, grassy slopes with shrubs, 01. Jul. 1988, *Wulingshan Exped. 1400* (KUN). **Pu'an**: Pubei Forest Farm, 1000 m, 08. Sept. 1959, *Anshun Exped. 1420* (PE). **Sandu**: Gundou to Bajie, 303–500 m, roadsides, 27. Oct. 1980, *Y. K. Li & D. F. Huang 8533* (PE). **Shibing**: Yuntai Shan, 900 m, edges of secondary forests, 31. Aug. 1988, *Wulingshan Exped. 3553* (KUN). **Songtao**: Lengjiaba, NE side of Fangjing Shan, 1000 m, forests, 5–9. Oct. 1986, *Sino-Amer. Guizhou Bot. Exped. 2157* (PE). **Tongren**: Mayan Township, 700 m, thickets, limestones, 08. Jul. 1988, *Wulingshan Exped. 1501* (KUN). **Tongzi**: s. loc., 450 m, dense woods, 27. May 1930, *Y. Tsiang 5151* (PE). **Weng'an**: Huangping, 700 m, roadsides, 03. Sept. 1959, *Libo Exped. 2036* (KUN, PE); Zhucang, 800 m, thickets, 18. Sept. 1959, *Libo Exped. 2396* (KUN); loc. eodem, 900 m, 11. Sept. 1959, *Libo Exped. 2080* (PE). **Yanhe**: Malinhe, 720 m, forests, 30. Sept. 2003, *J. W. Xiao 5911* (PE). **Yinjiang**: Fanjing Shan, 400–500 m, light woods, 22. Dec. 1930, *Y. Tsiang 7793* (PE); Huguoshi, W side of Fanjing Shan, 850–1300 m, forests, 23. Sept. 1986. *Sino- Amer. Guizhou Exped. 1485* (PE). **Zunyi**: Jinding, 1100 m, forests, 04. Sept. 1956, *Sichuan-Guizhou Exped. 1399* (PE, SZ). **HUBEI**, s. loc., 1885–1888, *A. Henry 4437* (PE). **Badong**: Luoping, 1000 m, 03. Oct. 1952, *Q. L. Chen et al. 1837* (E, PE). **Enshi**: Xintang, Hongnan Township, 310 m, 25. Nov. 1958, *H. J. Li 8906* (PE); s. loc. 31. Oct. 1934. *H. C. Chow 1829* (PE). **Hefeng**: Jima, 03. May. 1959, *F. S. Peng 251* (PE); Longjingba, 350 m, among herbs, 17. Sept. 1958, *H. J. Li 5953* (PE); s. loc., 250 m, by rocks, 20, Oct. 1955, *H. J. Li 8126* (PE, SZ). **Jianshi**: s. loc., 04. Sept. 1934, *H. C. Chow 1517* (PE). **Lichuan**: Maoba, 30. Oct. 1951, *L. Y. Dai & Z. H. Qian* (E) *1691* (PE); vicinity of Zhuangjiawan, 1500 m, 07. Oct. 1980, *1980 Sino-Amer. Exped. 2009* (PE). **Shennongjia**: Laojun shan, 1000–1250 m, 31. Aug. 1980, *1980 Sino-Amer. Exped. 455* (PE). **Tongshan**: Taiping Shan, 480 m, forests, 03. Oct. 1997, *C. M. Tan 971367* (PE); loc. eodem, 450 m, forests, 04. Oct. 1997, *C. M. Tan 971407* (PE). **Xianfeng**: Qingshuitang, 1100 m, among herbs, 30. Sept. 1958, *H. J. Li 9436* (PE). **HUNAN, Baojing**: Huangliandong, forests, 23. Sept. 1958, *L. H. Liu 9939* (KUN, PE). **Dong'an**: Shunhuang Shan, thickets, 04. Oct. 1948, *Y. Liu & Father 00893* (PE); loc. eodem, 700 m, 21. Oct. *J. K. Liu 815* (PE). **Fenghuang**: Liangdouyang Township, 650 m, 17. Sept. 1988, *Wulingshan Exped. 1169* (PE). **Sangzhi**: Shayuan to Nanmuping, 420–520 m, 10. Jul. 1988, *Beijing Exped. to Hunan 3687* (PE); Tianping Shan, forests, 27. Aug. 1988, *Beijing Exped. to Hunan 4097* (PE); Yangliuxi, 220 m, thickets, 12. Sept. 1984,

Hunan-Guizhou Exped. 1494 (KUN); s. loc., 1200 m, grasslands, 30. Jun. 1984, *Hunan-Guizhou Exped. 0987* (KUN). **Shimen**: Shanyaopuhe, 450 m, thickets, 11 Jun. 1989, *Hupingshan Exped. 0131* (PE). **Wugang**: Mt Yunshan, 700 m, thickets, 03. Oct. 1963, *L. H. Liu & G. Z. He 16121* (KUN, PE). **Xinning**: Malingtong, 600 m, 13. Oct. 1935, *C. S. Fan & Y. Y. Li 606* (PE); Ziyun Shan, 1000 m, roadsides, 29. Oct. 1962, *L. H. Liu 15389* (KUN, PE, SZ). **Yizhang**: Mt Mangshan, 1700 m, 30. Sept. 1956, *L. H. Liu* s. n.(KUN); loc. eodem, 600 m, 10. Oct. 2005, *B. Z. Xiao 4832* (PE); loc. eodem, 800 m, 30. Sept. 2005, *B. Z. Xiao 4775* (PE); loc. eodem, 600 m, roadsides by stream, 11. Nov. 1961, *L. H. Liu 10282* (PE). **Yongshun**: Xiaoxi, 400 m, thickets, 24. Oct. 1982, *H. Li et al. 2033* (KUN); loc. eodem, 400 m, forests, 03. Aug. 1983, *Hunan-Guizhou Exped. 3197* (KUN); s. loc. evergreen broad-leaved forests by stream, 29. May 1988, *Beijing Exped. to Hunan 0489* (PE). **Zhijiang**: Dashuao Township, 600 m, 13. Oct. 1988, *Wulingshan Exped. 2026* (PE). **JIANGXI, Chongyi**: Taoliao, 890 m, by road in forests, 02. Nov. 2002, *Lushan Bot. Gard. Exped. C138* (PE). **Dexing**: Damao Shan, thickets, 30. Sept. 1958, *s. coll.5234* (PE). **Guangcang**: Tangfang, thickets, 26. Sept. 1958, *C. M. Hu 5147* (KUN, PE). **Jing'an**: Sanzhuacang Township, 350 m, roadsides in forests, 11. Nov. 2004, *Lushan Bot. Gard. Exped. DJA 2004 0503* (PE). **Longnan**: Chenglong, *X. M. Mo 21031* (PE). **Quannan**: Qutou, *X. M. Mo 21360* (PE). **Shicheng**: Jinhua Shan, thickets, 20. Sept. 1958, *C. M. Hu 5038* (KUN, PE); Jinkou Township, Qiling, edges of forests, 21. Aug. 1958, *C. M. Hu 4492* (KUN, PE). **Suichuan**: Qiling, 600 m, forests, 04. Nov. 1965, *S. S. Lai et al. 5466* (KUN). **Tonggu**: Daduan Township, 650 m, forest trail, 27. Oct. 2004, *Lushan Bot. Gard. Exped. DTG 2004 00287* (PE); Hongsu, 317 m, 28. Oct. 2004, *Lushan Bot. Gard. Exped. WNB 2004 00278* (PE). **Wuning**: Luoxi, 620 m, forests, 24. Aug. 1998, *J. H. Zhang 98763* (KUN). **SICHUAN, Anxian**: Taipingchang, 25. Aug. 1958, *W. K. Hu 13301* (SZ). **Dujiangyan**: Lingyan Shan, 750 m, 25. Oct. 1956, *X. Li 47003* (PE); Qingcheng Shan, 30. Aug. 1958, *C. L. Wu 33801* (PE); loc. eodem, 900 m, 31. Sept. 1956, *X. Li 47179* (PE); loc. eodem, 800 m, 04. Nov.1956, *X. Li 47255* (PE); loc. eodem, 1300 m, 06. Nov. 1938, *T. P. Wang 10015* (PE). **Emei:** Bailongdong, 19. Nov. 1956, *Y. B. Yang 53249* (SZ); Baoguo-si, 700 m, 14. Nov. 1958, *Y. L. Qiao 835* (PE, SZ); Dàesi, 800 m, 29. Oct. 1952, *W. K. Hu 4* (SZ); Erping, 1000m, 06. Nov. 1956, *G. F. Li 53171* (SZ); Guanxinpo, 1300 m, 18. Oct. 1957, *G. H. Yang 57568* (SZ); 6. Hongchunping, 14. Sept. 1952, *J. H. Xiong 32542* (SZ); Hongchunping, 1150 m, edges of forests, 27. Aug. 1964, *K. C. Kuan et al. 1826* (PE); below Huangwan, 480 m, 27. Oct. 1956, *G. F. Li 53015* (SZ); Huayanding, 13. Dec. 1940, *W. P. Fang 15720* (SZ); Longmendong, 550 m, 08. Oct. 1941, *C. L. Zhou 4738* (SZ); Wannian-si, 01. Nov. 1952, *H. L. Tsiang & S. S. Chang 33416* (SZ); loc. eodem, 23. Oct. 1952, *H. L. Tsiang & S. S. Chang 33322* (SZ); loc. eodem, 1956, *G. F. Li 52296* (SZ); Wannian-si, 1050 m, 19. Sept. 2011, *D. Y. Hong et al. H11022* (PE); loc. eodem, 01. Nov. 1952, *J. H. Xiong et al. 33416* (PE); loc. eodem, 05. Nov. 1952, *J. H. Xiong et al. 33506* (PE, SZ); loc. eodem, 23. Oct. 1952, *J. H. Xiong et al. 33322* (PE) ; loc. eodem, Oct. 1936, *W. C. Cheng 6729* (PE) ; loc. eodem, 04. Oct. 1958, *S. Y. Chen et al. 6190* (SZ); Xiadaping, 1200 m, 14. Oct. 1956, *G. F. Li 52690* (SZ); Xixiangchi, 20. Oct. 1938, *T. N. Liou 10353* (PE) ; Zhongfengsi, 900 m, 03. Sept. 1957, *G. H. Yang 75047* (KUN, PE, SZ); loc. eodem, 04. Oct. 1938, *Z. W. Yao 3181* (PE, SZ); Zhuping, 800 m, 12. Sept. 1957, *G. H. Yang 57260* (PE, SZ); 1200 m, 20. Aug. 1940, *T. C. Lee 3365* (KUN); 1800m, 10. Oct. 1940. *T. C. Lee 3888* (KUN); 1100 m, 21. Aug. 1938, *C. Y. Chiao & C. S. Fan 340* (PE); s. loc., 1930, *Z. W. Rao 4282* (SZ); s. loc., 1955, *G. F. Li 53105* (SZ); s. loc., 27. Nov. 1946, *W. K. Hu 9076* (SZ). **Leibo**: Mahuyan, dense forests, 22. May 1959, *Sichuan Econ. Plant Surv. (Liang) 6269* (PE); Xining, 1200 m, 14. Nov. 1983, *C. S. Cao 122644* (CDBI, SZ). **Leshan**: Sanxi, 18. May. 1959, *Sichuan Econ. Pl. Surv. (Le) 8025* (KUN). **Lixian**: Laisugou, 3100 m, *D. P. He 46460* (SZ). **Lushan**: s. loc., 18. Oct. 1936, *K. L. Chu 4038* (SZ). **Mabian**: s. loc., 1934, *T. T. Yü 4180* (PE). **Miyi:** s. loc., 17. Aug. 1958, *Z. He et al. 11285* (SZ). **Pengshui**: Dojiagui, by stream, *s. d., s. coll. 4126* (PE). **Pingshan**: Longhua, 650 m, 05. Jun. 1959, *s. coll., 886* (SZ). **Pingwu**, Yaodu, 650 m, 08. Nov. 1937, *K. T. Fu 2359* (PE). **Tiangquan**: Jiaojiping, 1175 m, 31. May. 1959, *Sichuan Econ. Plant Surv. (Ya) 521* (PE); Zhongzhen, 02. Nov. 1951, *W. K. Hu & Z. He 11799* (PE). **Xingwen**: Xianfeng Township, 750 m, 22. May 1959, *Sichuan Econ. Plant Surv. (Yi) 1254* (KUN). **Yingjin**: Datian Township, 29. Nov. 1951, *W. K. Hu & Z. He 12113* (PE). **TAIWAN** (from *Fl. Taiwan*, 1998): **Ilan**, **Taoyuan**, **Xinzhu**, **Miaoli**, **Taizhong**, **Nantou**, **Jiayi**, **Gaoxiong** and **Taidong**. **YUNNAN**, **Suijiang**: Ershisigang, 1500 m, forests, 08. Oct. 1973, *K. M. Feng et al. 73–231* (KUN). **ZHEJIANG**, **Longquan**: Dasai Shan, 07. Oct. 1961, *s. coll. 8924* (PE).

JAPAN, C to W Honshu (westwards from Ibaraki), Shikoku, and Kyushu (from Iwatsuki et al., *Fl. Jap.* IIIa. 1993); **Tokyo Pref.**: Hachiooji, Mt Takawo, 22. Oct. 1991, *M. Furuse 55830* (K); loc. eodem, 350–450 m, 03. Dec. 1991, *M. Furuse 55831* (K); Iwaya-san, NE side, 21. Jun. 1938, *F. C. Greatrex* s. n.(K); between Iwaya-San and Nerreshitoge, 25. Aug. 1938, *F. C. Greatrex* s. n. (K); Kurogi, 23. Nov. 1938, *F. C. Greatrex* s. n.(K).

46. Codonopsis inflata Hook. f. & Thomson, Ill. Himal. Pl. t. 16, C. 1855. ≡*Campanumoea inflata* (Hook. f. & Thomson) C. B. Clarke in Hooker, Fl. Brit. Ind. 3: 436. 1881; Hong in Wu, Fl. Xizang. 4: 583, 1985. Type: India, Sikkim, 1520–2130 m, *J. D. Hooker* s. n. [lectotype here designated: K (K000814187); isolectotypes: CAL, P, PE]; loc. eodem, 1830 m, *J. D. Hooker* s. n. [paratype: K (K000814188)]; loc. eodem, Lachung, *J. D. Hooker* s. n. [paratype: K (K000814190)].

FIGURE IV 1:46A *Codonopsis inflata* Hook. f. & Thomson. A section of twining stem with a berry, which shows that the ovary is inferior also to calyx. *From Flora of China Vol. 19, 2011.*

Roots slightly thickened, branched. Stems twining, almost glabrous. Leaves alternate; petiole 3–6 cm long; leaf blade ovate to ovate-lanceolate, 4–10 cm long, 2–5 cm broad, adaxially glabrous or sparsely puberulent, abaxially glaucous, puberulent, base deeply cordate, margin entire or sinuate, apex acuminate. Flowers solitary, almost leaf-opposed, or in a terminal and few-flowered cyme; pedicel 2–3 cm long. Calyx adnate to hypanthium up to top; calyx tube obdeltoid-conical, puberulent; calyx lobes and corolla inserted at same level, ovate-triangular, ca. 1 cm long, margin often revolute. Corolla urceolate 2.5–3 cm long, cleft to 1/3, pale yellow or pale green, often purple at margin and along veins. Stamens 5; filaments glabrous. Style glabrous; stigma 3-lobed; ovary 3-locular. Berry yellow-green, globose, 1.2–1.5 cm in diameter. Seeds ovoid-globose, surface reticulate. Figures IV 1:46A & B.

Phenology: flowering and fruiting from August to November.

Habitats: evergreen broad-leaved forests or herb communities at forest margins; 900–2750 m.

Distibution: Bhutan, China (SE Tibet), N India (Sikkim & Darjiling) and E Nepal, (Map IV 1:46).

Additional specimens examined, **BHUTAN, Dumsong**: 1830 m, 17. Nov. 1875, *C. B. Clarke 26420C* (K).

CHINA, TIBET, Mêdog: Baibung, 1500 m, thickets, 08. Nov. 1992, *H. Sun et al.1117* (KUN); Buqiong Hill, 1000 m, shrubs, 15. Nov. 1992, *H. Sun et al.1367* (KUN); Dagmo, 1900 m, 06. Mar. 1993, *H. Sun et al. 4267* (KUN); Dagmo to Gutang, 1700 m, evergreen broad-leaved forests, 02. Sept. 1974, *Qinghai-Xizang Exped. 74–4988A* (KUN, PE); De'ergong, 1600 m, herbs at edges of broad-leaved forests, 20. Aug. 1974, *Qinghai Xizang Exped. 74–4445* (KUN, PE); Gelin, 2000 m, 26. Dec. 1992, *H. Sun et al. 2668* (KUN); loc. eodem, 1800 m, evergreen forests, 12. Sept. 1974, *s. coll. 2499* (PE); Hanmi, 1100 m, grasslands with shrubs, 22. Oct. 1992. *H. Sun et al. 0489* (KUN); loc. eodem, 2100 m, forests, 23. Oct. 1992, *H. Sun et al. 0538* (KUN); Mêdog Distr., evergreen forests, 05. Aug. 1974, *Qinghai-Xizang Exped. 1588* (KUN, PE); near the town, 29°19′N, 95°19′E, 1256 m, 02. Aug. 2012, *D. Y. Hong et al. H12001* (PE) ; Xianaba to Dagmo, 1700–1800 m, evergreen forests, 26. Oct. 1982, *B. S. Li & S. Z. Cheng 01358* (PE); Xiran, 900 m, fields, 23. Nov. 1992, *H. Sun et al. 1517* (KUN); Xiran, the back hill, 1750 m, forests, 07. Dec. 1992, *H. Sun et al. 1971* (KUN); loc. eodem, 2000 m, bushes, 07. Dec. 1992, *H. Sun et al. 2007* (KUN); Zangbo Gorge (Gampe La), 1520–1830 m, 12. Sept. 1924, *F. Kingdon-Ward 6371* (K). **Zayü**: Luhit Valley, 28°0′N, 96°15′E, 1220 m, 12. Jul. 1922, *F. Kingdon-Ward 7695* (K); Therolian, Luhit Valley, 1220 m, 03. Nov. 1950, *F. Kingdon-Ward 20283* (BM).

INDIA, Darjiling: Batasi (2100 m) – Mani Bhanjyang (2000 m), 09. Aug. 1972, *H. Kanai et al. 723171* (TI); Dhobÿhera, 1520 m, Aug. 1881, *J. S. Gamble 9740* (K); Jospokri, 2440 m, 15. Dec. 1912, *G. H. Cave* s. n.(E); Kurseong, 1370 m, Apr. 1845, *J. S. Gamble 538A* (K); Labdah, 1530 m, 09. Nov. 1918, *G. H. Cave* s. n.(E); Palmajua (2200 m) – Batasi (2100 m), 08. Aug. 1972, *H. Kanai et al. 723172* (TI); Rungbee, 24. Sept. 1875, *C. B. Clarke 24788B* (K); Rungbee, 1830 m, 24. Sept. 1875, *C. B. Clarke 26760B* (BM); loc. eodem, 1370 m, Oct. 1872, *J. S. Gemble 2336A* (K); Tonglu, 2740 m, 10. Dec. 1912, *G. H. Cave* s. n.(E); s. loc., 07. Aug. 1923, *J. M. Cowan* s. n.(E); s. loc., 2430 m, Sept. 1881, *J. S. Gamble 9854* (K); s. loc., *Griffith 3437* (K). **Sikkim:** Chateng, 2440 m, 06. Nov. 1917, *G. H. Cave* s. n.(E); Gangtok, Penlong La, 2100 m, 13. Sept. 1964, *H. Hara & K. C. Pradhan* s. n.(TI); s. loc., 1680 m, 26. Sept. 1884, *C. B. Clarke 35908* (BM); s. loc., 1520–2130 m, *J. D. Hooker* s. n (P, PE).

NEPAL, Mechi: Mai Pokhari, 04. Dec. 1963, *H. Hara et al. 6300416* (BM, TI); Mai Pokhari, 27°0′N, 87°57′E, 2100 m, 28. Sept. 1971, *J. F. Dobremez (DBR NEP) 1212* (BM).

HIMALACODON D. Y. Hong & Q. Wang

Hong & Wang in Wang et al., J. Syst. Evol. 52(5): 548. 2014. *Codonopsis* subg. *Obconicicapsula* D. Y. Hong, Acta Phytotax. Sin. 18(2): 245. 1980. Type: *Himalacodon dicentrifolius* (C. B. Clarke) D. Y. Hong & Q. Wang (= *Wahlenbergia dicentrifolia* C. B. Clarke).

Roots carrot-shaped, branched. Stems erect, many branched, each branch bearing a terminal flower. Calyx tube adnate to hypanthium up to top and hypanthium adnate to ovary nearly up to top, ovary almost entirely inferior, obconical, calyx lobes and corolla inserted at same level. Corolla campanulate, shallowly 5-lobed. Disc or glands lacking. Stamens entirely glabrous; filaments slightly dilated at

FIGURE IV 1:46B *Codonopsis inflata* Hook. f. & Thomsom. All from Tibet: Mêdog, 02. Aug. 2012, **a**, **c**, **d**, **f**, **h** & **i**, by De-Yuan HONG; **b**, **e**, **g**, by Xin-Tang MA; **f**, a longitudinal section of flower, showing disc and ovary inferior to both calyx and corolla; **f** & **g**, showing disc.

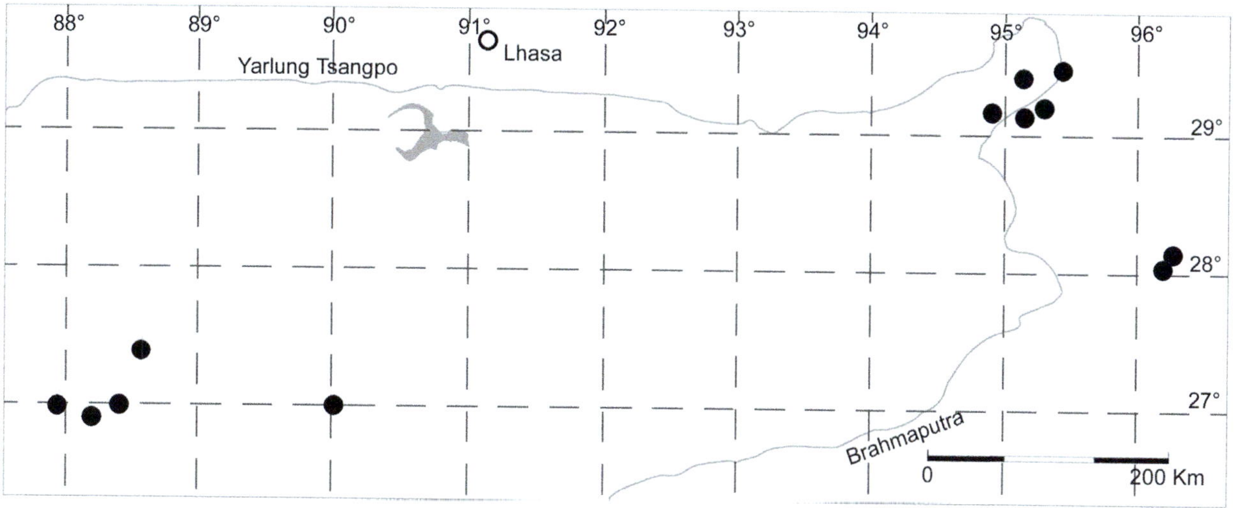

MAP IV 1:46 Distribution of *Codonopsis inflata* Hook. f. & Thomson.

base. Ovary 3-locular, ovules many; style clavate, densely glandular-hairy; stigma 3-fid, stigma lobes nearly triangular, broadest at base. Fruit a loculicidal capsule, obconical. Seeds many, narrowly ovoid, compressed. Plants not fetid. Pollen grains 5-colpate; colpi short with colpus length / polar axis length ≤ 0.4; colpus membrane granulose; spinules on sexine small, ≤ 1 μm long. Chromosomes 2n=14.

The plant in *Himalacodon* was first placed in *Wahlenbergia* with a question mark by Clarke (1881), but it has been treated as a member of *Codonopsis* by Smith (1912). Hong (1980) recognised it as a subgenus of *Codonopsis* based on its obconical hypanthium and fruit, and compressed seeds. Eddie et al.'s (2003) molecular phylogeny based on ITS sequences shows that *Codonopsis dicentrifolia* (C. B. Clarke) W. W. Smith is more closely related to *Cyananthus* than to *Codonopsis*. Our recent molecular phylogenetic analyses indicate that this species forms a major clade with *Codonopsis* subg. *Pseudocodonopsis* Kom. and *Codonopsis purpurea* Wall., separated from another major clade consisting of the core *Codonopsis*. The observation on pollen morphology shows that the former clade has short-colpate pollen, whereas the latter has long-colpate pollen (Hong & Pan, 2012). Counting chromosomes tells us that the species under study possesses 2n=14 (vs. 2n=16 in *Codonopsis*). Furthermore, *Codonopsis* is of disc or 5 free glands alternate with stamens, while *Codonopsis dicentrifolia* has no disc, nor glands.

The genus *Pseudocodon* D. Y. Hong & H. Sun is characteristic of rotate corolla, mostly globose roots, mostly twining stems, and long spinules on exine of pollen grains. Therefore, *Himalacodon* is distinct also from *Pseudocodon*.

Our genus is the most closely related to *Pankycodon* D. Y. Hong & X. T. Ma. They are paraphyletic in our molecular trees (Wang et al., 2013 & 2014). However, they can not be clumped into a single genus not only phylogenetically, but also morphologically. In *Pankycodon* leaves on main stem are opposite, seeds lunate and strongly compressed with a wide wing and hypanthium is higher than ovary by 1–2 mm, whereas in *Himalacodon* leaves are alternate, seeds narrowly ovoid, slightly compressed, wingless, and hypanthium is as high as ovary. Wang et al. (2014a) provides more information about the genus and its allies.

A monospecific genus endemic to the eastern Himalaya, confined to China (S Tibet), India (Sikkim) and Nepal (Map IV 2:1).

1. **Himalacodon dicentrifolius** (C. B. Clarke) D. Y. Hong & Q. Wang, in Wang et al., J. Syst. Evol. 52 (5): 549. 2014. ≡*Wahlenbergia ? dicentrifolia* C. B. Clarke in Hooker, Fl. Brit. Ind. 3: 430. 1881. ≡*Campanopsis dicentrifolia* (C. B. Clarke) Kuntze, Revis. Gen. Pl. 2: 379. 1891. ≡*Codonopsis dicentrifolia* (C. B. Clarke) W. W. Smith, Rec. Bot. Surv. India 4: 388. 1913; Shen & Hong in Hong, Fl. Reipubl. Popularis Sin. 73(2): 65. pl. 8. fig. 3. 1983; Hong in Wu, Fl. Xizang. 4:582. 1985; Hong, Lammers & Klein in Wu, Raven & Hong, Fl. China 19: 524. 2011. Type: India, Sikkim, Phulloot, 3350 m, 27. Oct. 1875, *C. B. Clarke 25722 A* (holotype: K).

Plants totally glabrous. Roots carrot-shaped, more or less branched, up to 40 cm long, 3 cm in diameter. Stems erect, 30–60 cm tall; branched at upper part. Leaves mostly alternate, those on branches nearly opposite; petiole 2–6 mm long; leaf blade ovate or ovate-oblong, 1.1–4.5 cm long, 0.7–2.2 cm broad, abaxially purple or red, base cuneate or rounded, margin entire or sinuous, sometimes slightly revolute, apex acute or obtuse. Flowers solitary, terminal on main stem and branches; pedicel 0.5–3 cm long. Calyx tube adnate to hypanthium nearly to top, obconical, 5–7 mm long, 7–10 mm wide; calyx lobes nearly linear, 7–10 mm long, ca. 1 mm broad, margin entire, apex acuminate; sinus between calyx lobes broad and obtuse. Corolla campanulate, 2.5–3.5 cm long, 2.5–3.5 cm wide, 5-lobed to ca. 1/3, blue, blue-purple or pale blue; corolla lobes broad-ovate,

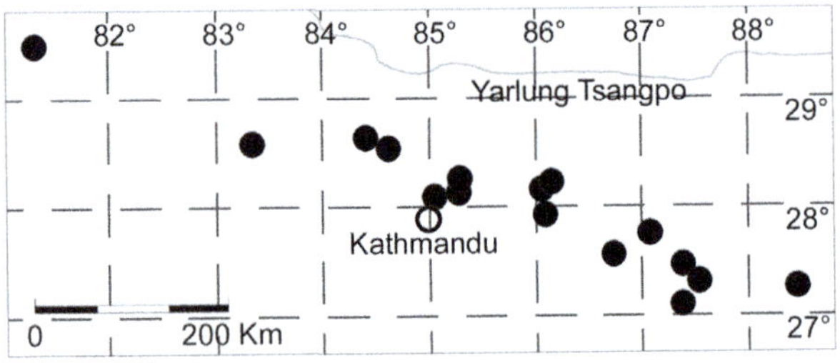

MAP IV 2:1 Distribution of *Himalacodon dicentrifolius* (C. B. Clarke) D. Y. Hong & Q. Wang [*Codonopsis dicentrifolia* (C. B. Clarke) W. W. Smith].

0.8–1.3 cm long, 1–1.5 cm broad, apex acute. Filaments slightly dilated below, slender above, 3–4 mm long; anthers ca. 7–8 mm long. Capsule shortly conical in superior part, obconical or obdeltoid-conical in inferior part, obscurely 10-ribbed, 1.2–1.5 cm long, 1–1.2 cm in diameter. Seeds ovoid, compressed, wingless. Figures IV 2:1A & B.

Chromosomes: 2n=14 (Wang et al., 2014a).

Phenology: flowering and fruiting from August to October.

Habitats: growing on rocks by broad-leaved forests or in thickets at altitudes from 2700 to 4100 (–5200) m.

Distribution: E Himalaya: China (S Tibet), N India (Sikkim), and Nepal (Map IV 2:1).

Notes: When Clarke (1881) described this species as new, he placed a question marker before '*dicentrifolia*', which clearly shows that he had reservation in treating the species as a member of *Wahlenbergia*.

Additional specimens examined: **CHINA**, **TIBET**, **Gyirong**: Gyirong Town, near Todang Fall, 3100–3300 m, *Quercus* forests, 11. Sept. 1990, *B. S. Li et al. 13685* (PE); Sale Township, Labi-Sale, 3100 m, on rocks, 01. Sept. 1990, *B. S. Li et al. 13357* (PE). **Nyalam**: near Quxam, 3300 m, 29. Aug. 1972, *Xizang Chinese Mater. Medic. & Herbs Exped. 1598* (HNWP, PE); loc. eodem, 3400 m, 01. Sept. 1981, *C. C. Ni et al. 1906* (PE, XZ), loc. eodem, 3200 m, on cliffs, 27. Aug. 2011, *D. Y. Hong et al. H11004* (PE). **Tingri:** Rongxar, 2700 m, deciduous broad-leaved forests, 02. Aug. 1959, *s. coll. 749* (PE).

INDIA, **Sikkim**: Gantong, 3650–3960 m, 30. Aug. 1913, *R. E. Cooper 748* (E); loc. eodem, 3960 m, 01. Oct. 1913, *R. E. Cooper 987* (E); loc. eodem, 3350 m, 18. Aug. 1910, *W. W. Smith 4381* (E); Saudekbu, Oct. 1908, *Ribu 371* (E).

NEPAL (W): **Seti**: Bajura Distr./Bajhang Distr., 29°33′N, 81°23′E, 2190–3350 m, 16. Aug. 1991, *M. Suzuki et al. 9170642* (E). Barpak, 4570–5200 m, on rocks, 1928, *L. Dheuoj 52* (BM); Pongsing, 1830–4220 m, 1929, *L. Dhevoj 93* (BM). **Tsumje**: 3500 m, 24. Jul. 1953, *S. Nakao s. n.* (TI). **NEPAL** (C): **Bagmati**: S of Gurjakhani, 3650 m, 16. Aug. 1954, *Stainton, Sykes & Williams 3864* (BM, E, G, P, PE); Rasuwa Distr., 28°15′N, 85°07′E, 3500 m, 07. Aug. 1994, *F. Miyamoto et al. 9420197* (KUN); Rasuwa Distr., Chilime Kharka-Chilimagean, ca. 3350 m, 26–28. Jul. 1949, *O. Polunin 1454* (BM, E); Rasuwa Distr., Satsae Khola, 28°00′N, 85°10′E, 3000 m, 13. Aug. 1970, *J. F. Dobremez 610* (BM); Sindhupalchok Distr.,Tembathang, 4110 m, 04. Sept. 1927, *Clive Wigram 106* (E). **Dhawalagiri**: Myagdi - Baglung Distr., Jalja La, 3250 m, 16. Sept. 1996, *M. Mikage et al. 9684221* (BM). **Gandaki**: Manang Distr., around Bimtang, 28°38′04″N, 84°28′20″E, 3500 m, 11. Aug. 1994, *M. Suzuki et al. 9470377* (E, KATH); loc. eodem, 28°34′N, 84°40′E, c. 3500 m, 05. Aug. 1994, *M. Suzuki et al. 9485294* (E). **Janakpur**: Dolakha Distr., S of Rupte, 28°04′N, 86°06′E, 3200 m, 25. Aug. 1974, *J. H. de Haas 2556* (BM). **NEPAL (E)**: **Kosi**: Sankhuwasabha Distr., Milke Danda, Gidde, 3450 m, 20. Jul. 1991, *H. Ohba et al. 9120118* (BM, TI); loc. eodem, ca. 3350 m, 05. Oct. 1971, *L. W. Beer et al. 10083* (BM); loc. eodem, ca. 3350 m, 04. Oct. 1971, *L. W. Beer et al. 10074* (BM); 27°44′N, 87°12′E, 3510 m, 29. Sept. 1991, *D. G. Long et al. 397* (E); loc. eodem, Tinjure Danda, ridge between Door Pani and Chauki, 27°10′N, 87°26′E, 2960 m, 29. Aug. 1989, *KEKE 62* (E, K); loc. eodem, N slopes of Jaljala Himal., 27°33′N, 87°26′E, 3430 m, 19. Oct. 1991, *D. G. Long et al. 899* (E); loc. eodem, Milke Danda ridge, 07. Oct. 1980, *Fedais* s. n. (K); Terhathum Distr., Arun Valley, Maghang Khola, E of Num, 87–88°E, 3350 m, *J. D. A. Stainton 1725* (BM, E); Basantpur (Terhathum), 1670 m, shrub on rock crevices, 13. Aug. 1976, *N. P. Manandhar 317* (KATH); Bheding, 3650–3960 m, 1930, *L. Dhevoj 0335* (BM, E); Barse, SE of Barse, 3350 m, 18. Oct. 1954, *Stainton, Sykes & Williams 9084* (BM); Bimtakothi, 3810 m, 26. Aug. 1950, *D. G. Lowndes 1455* (BM, E, G, P); near Chauki, 27°15′N, 87°29′E, c. 2800 m, 04. Aug. 1999, *Y. Omori et al. 9940003* (E); loc. eodem, Chauki, 27°11′N, 87°29′E, 3000 m, 17. Aug. 1972, *J. F. Dobremez 1546* (BM, G). **Mechi**: Taplejung Distr., Hile Kharka (3530 m) – Mane Bhanjyang (3490 m), 27°24′N, 87°27′E, 07. Aug. 1999, *Y. Omori et al. 9950022* (E). **Sagarmatha**: Solukhumbu Distr., Tashing Dingma (3300 m) – Thuli Kharka (4100 m), 27°37′N, 86°47′E, 27. Aug. 1997, *M. Wakabayashi et al. 9715260* (E); loc. eodem, 27°36′N, 86°44′E, 29. Aug. 1997, *M. Wakabayashi et al. 9720315* (E); loc. eodem, Chhatarwa (4010 m) – a pass (3840 m) – Kurke (3525 m), 27°39′N, 86°47′E, 29. Aug. 1995, *F. Miyamoto et al. 9580477* (E).

PANKYCODON D. Y. Hong & X. T. Ma

Hong & Ma in Wang et al., J. Syst. Evol. 52(5): 549. 2014. Type: *Pankycodon purpureus* (Wall.) D. Y. Hong & X. T. Ma (*Codonopsis purpurea* Wall.)

Roots attenuate. Stems pendulous from rocks or trees. Leaves opposite. Flowers solitary, terminal on main stem and branches. Hypanthium apex free from ovary by 2 mm. Calyx tube adnate to whole hypanthium, i. e. up to the base of corolla. Corolla campanulate, (4- or) 5-lobed to near middle. Stamens (4 or) 5, glabrous, slightly dilated at base. Ovary nearly totally inferior to both calyx and

FIGURE IV 2:1A *Himalacodon dicentrifolius* (C. B. Clarke) D. Y. Hong & Q. Wang. **a**, root and basal part of two stems; **b**, upper part of a stem; **c**, a flower with corolla opened, showing calyx, stamens, style, and stigma; **d**, a cross section of ovary, showing 3 locules.

FIGURE IV 2:1B *Himalacodon dicentrifolius* (C. B. Clarke) D. Y. Hong & Q. Wang. All from Tibet: Nyalam. **a** & **c–e**, 27. Aug. 2011, by Xin-Tang MA & De-Yuan HONG; **b**, 26. Aug. 2011, by De-Yuan HONG.

corolla, 3-locular; ovules many; style clavate, densely glandular-hairy above; stigma 3-fid, lobes deltoid-lanceolate at early anthesis, later lanceolate. Fruit a capsule, 3-loculicidal. Seeds many, lunate, strongly compressed, widely winged. Plants not fetid. Pollen 5-colpate; colpi short with colpus length / polar axis length ≤ 0.4; colpus membrane smooth; spinules small, ≤ 1 μm long. Chromosomes 2n=14.

Wang et al. (2014a) provides the reason to separate *Codonopsis purpurea* Wall. as a new genus from the rest of *Codonopsis*, and explains its relationships with *Pseudocodon* and *Himalacodon*. The reader may see a brief statement of this issue under *Himalacodon*.

Monospecific genus, distributed in Bhutan, China (S Tibet and W Yunnan), N India, and Nepal (Map IV 3:1).

1. **Pankycodon purpureus** (Wall.) D. Y. Hong & X. T. Ma in Wang et al., J. Syst. Evol. 52(5): 549. 2014. ≡*Codonopsis purpurea* Wall. in Roxburgh, Fl. Ind. 2: 105. 1824; Hooker & Thomson, J. Proc. Linn. Soc. Bot. 2: 12. 1857; Clarke in Hooker, Fl. Brit. Ind. 3: 431. 1881; Komarov, Trudy Imp. S. - Petersburgsk. Bot. Sada 29(1): 113. 1908; Chipp, J. Linn. Soc. Bot. 38: 390. 1908; Shen & Hong in Hong, Fl. Reipubl. Popularis Sin. 73(2): 52. pl. 7. fig. 3. 1983; Haridasan & Mukherjee in Hajra & Sanjappa, Fasc. Fl. Ind. 22: 74. 1996; Hong, Lammers & Klein in Wu, Raven & Hong, Fl. China 19: 520. 2011. ≡*Campanula purpurea* (Wall.) Spreng., Cura Post. 78. 1825. ≡*Wahlenbergia purpurea* (Wall.) A. DC. in de Candolle, Prodr. 7: 425. 1839. Type: Nepal, Gossain Than, *Wallich 1299* [lectotype designated by Hong in Wang et al., 2014: K; isolectotypes: BM, E, G, LD, P, W].

Plants glabrous throughout. Roots attenuate, yellow-white, smooth. Stems several from one caudex, few-branched, 30–80 cm long. Leaves opposite or sometimes alternate towards apex; petiole 1–10 mm long; leaf blade ovate or lanceolate, 2–12 cm long, 1.5–5 cm broad, base rounded or cuneate, margin nearly entire or sinuous, apex acute or obtuse. Flowers terminal or leaf-opposed at top. Calyx tube adnate to whole hypanthium, semiglobose-obconical; calyx lobes (4 or) 5, triangular-lanceolate, 1–2 cm long, 0.6–1 cm broad, margin entire, apex acute. Corolla campanulate or funnelform-campanulate, 2–3.5 cm long, 3–4 cm wide, dark red-purple, (4- or) 5-lobed to middle; corolla lobes triangular. Stamens glabrous; filaments ca. 6 mm long, slightly dilated; anthers ca. 5 mm long. Capsule obovoid-conical below, conical above, 2–3 cm long, 1.5–2 cm in diameter; valves 5–8

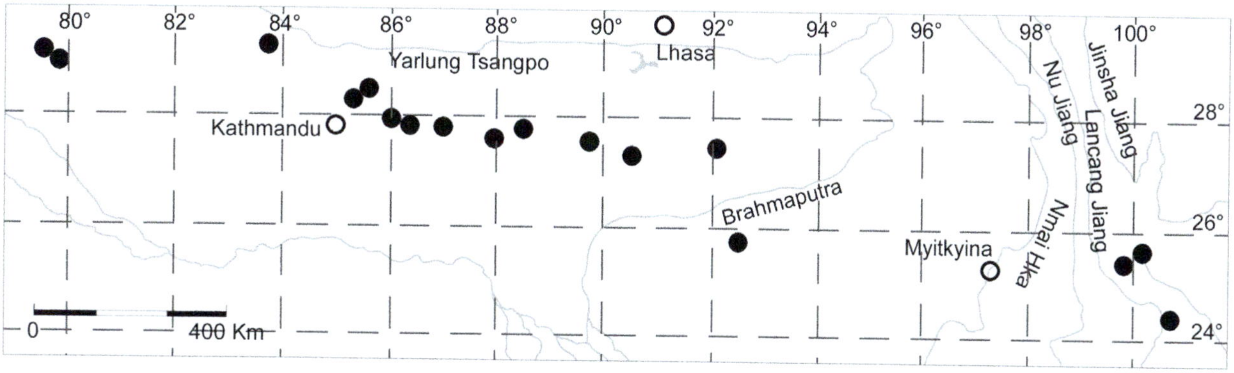

MAP IV 3:1 Distribution of *Pankycodon purpureus* (Wall.) D. Y. Hong & X. T. Ma (*Codonopsis purpurea* Wall.).

mm long. Seeds lunate, strongly compressed, wide-winged, yellow, ca. 2 mm long, 1 mm wide. Figures IV 3:1A & B.

Chromosomes: 2n=14 (Wang et al., 2014a).

Phenology: flowering and fruiting from September to October.

Habitats: grassy slopes, scrubs, often pendent from rocks, sometimes epiphytic on trunks in forests at an altitude of 2000–3300 m.

Distribution: Bhutan, China (S Tibet, W Yunnan), N India, and Nepal (Map IV 3:1).

Additional specimens examined, **BHUTAN**, **Dotena**, Thsiubu Chu, 27°40'N, 89°42'E, 17. Oct. 1949, *F. Ludlow et al. 17535* (BM). **Tongsa**, 1980 m, 11. Sept. 1914, *R. E. Cooper 2071* (E); loc. eodem, 1830 m, 11. Sept. 1914, *R. E. Cooper 2074* (E); loc. eodem, 1830 m, 02. Aug. 1937, *F. Ludlow & G. Sherriff 3497* (BM); loc. eodem, 27°28'N, 90°30'E, 19. Jul. 1937, *F. Ludlow & G. Sherriff 3463* (BM); Tongsa, Mangde Chu, 1980 m, 03. Aug. 1949, *F. Ludlow et al. 19558* (BM, E); loc eodem, 27°30'N, 90°30'E, 2130 m, 15. Oct. 1949, *F. Ludlow* et al. *19826* (BM); Tongsa, near Tongsa Bridge, 27°30'N, 90°30'E, 2130 m, 05. Aug. 1949, *F. Ludlow et al. 17046* (BM, E).

CHINA, TIBET, Cona: Senge Dzong, 2130-2440 m, 28. Aug. 1938, *F. Kingdon-Ward 14180* (BM). **Nyalam:** between Zham and Friendship Bridge, 2000 m, 19. Aug. 1972, *Xizang Chinese Mater. Medic. & Herbs Exped. 1308* (HNWP, PE); loc. eodem, 2066 m, on rocks, fl. purple-red, 27. Aug. 2011, *D. Y. Hong et al. H11003* (PE). **YUNNAN**, s. loc., s. d.,*G. Forrest 30948* (E, PE). **Dali**: 25°40'N, 2740–3050 m, Sept. 1910, *G. Forrest 6818* (BM, E). **Jingdong**: Jinding Shan, 2200 m, 29. Aug. 1959, *S. G. Xu 5285* (KUN, PE); loc eodem, 1920 m, 06. Nov. 1963, *Z. H. Yang et al. 101452* (KUN). **Yangbi**: Meixi, Malutang, 2380 m, 09. Aug. 1963, *Zhongdian Exped. 63-4236* (KUN). SW China, s. loc., 1933-4, *Y. Tsiang 11528* (PE).

INDIA, **Meghalaya**: Lailankote, 1520 m, 26. Sept. 1886, *C. B. Clarke 44864D* (BM); E Lailankote, 1520 m, 25. Sept. 1886, *C. B. Clarke 44864* (CAL); loc. eodem, 1520 m, Oct. 1890, *C. B. Clarke* s. n. (BM, G); near Lailangkote, 1530 m, 1890, *H. Collett* s. n. (K); 1220 m, Aug. 1890, *H. Collett 30* (CAL); loc. eodem, 1520 m, 26. Sept. 1886, *C. B. Clarke 44864A* (K); loc. eodem, 1520 m, 26. Sept. 1886. *C. B. Clarke 44864C* (K); Nongkream, 1220 m, Sept. 1890, *H. Collett 90* (CAL). **Sikkim**: Cheungtong, 1830 m, 10. Aug. 1913, *R. E. Cooper 444* (E). **Uttar Pradesh**: Kumaon, Lohashal, 1830 m, s. d. *R. Strachey & J. E. Winterbottom 1* (K); Nainital Distr., *M. K. Wall 1901* (BM).

NEPAL, s. loc., *s. coll. 224* (CAL). **Bagmati**: between Dhunche and Singdum Gompa, 1900m, 22. Aug. 1972, *H. Kanai et al. 723166* (BM, KUN, TI); Gram to Dunche, ca. 2000 m, 21. Aug. 1972, *H. Kanai et al. 721740* (TI) & *723167* (TI); Grang, E side of Trisuli River, 1800 m, 15. Sept. 1966, *D. H. Nicolson 2357* (BM, KATH); Langdeng, 3050 m, 01. Sept. 1931, *K. H. Sharma E126* (BM); Langtang Khola, 2440–2900 m, 23–31. Aug. 1949, *O. Polunin 1805* (BM, E, G, P); Rasuwa Distr., 2600, 28. Aug. 2001, *F. Shrestha & N. Joshi 152/2001* (KATH); Rasuwa Distr., near Bokejhund, 1950 m, moist and rocky places, 01. Oct. 1977, *N. P. Manandhar & Party 389* (KATH); Rasuwa Distr., Dhunche, Trisoli Valley, 28°8'N, 85°20'E, 1680 m, 02. Oct. 1965, *J. D. A. Stainton 5189* (BM, KATH). **Gandaki**: Kaski Distr. 29°15'N, 83°50'E, 2050 m, 20. Aug. 1988, *M. Suzuki et al. 8860475* (BM); Kaski Distr., 1900 m, 20. Aug. 1988, *M. Suzuki et al. 10831* (KATH); Marsiandi Valley, Dharamsala, 2440 m, 30. Aug. 1950, *D. G. Lowndes 1481* (BM, CAL, E). **Janakpur**: 27°51'N, 86°13'E, 1200 m, 14. Oct. 1954, *A. Zimmerma 1253* (G); Dolakha Distr., 1120–2030 m, 15. Sept. 2007, *S. Noshiro et al 20710072* (E); Dolakha Distr., Kalingchok, Thala to Tale Bisauna, ca. 2500 m, 10. Sept. 1970, *H. Kanai et al. 672793* (KATH, TI). **Kathmandu**: Phulchoke, SW of Kathmandu, 2440 m, on damp mossy rocks in dense oak forests, 27. Aug. 1965, *A. D. Schilling 601* (K); Phulchoki, 2400 m, 09. Aug. 1969, *H. Kanai et al. 693388* (TI); loc. eodem, 2400 m, 04. Sept. 1966, *H. J. Lange 138* (B); Thankot, 12 km W of Kathmandu, 1830 m, 1–3. Aug. 1956, *K. de B. Codrington 1* (BM). **Mechi**: Taplejung Distr., Shewaden-Papung, ca. 2200 m, pine forests, 22. Aug. 1977, *H. Ohashi et al. 773971* (TI); Taplejung Distr., Tamur Valley, Mewa Khola, 2440 m, 09. Aug. 1956, *J. D. A. Stainton 1305* (BM, E). **Sagarmatha**: Solukhumbu Distr. 1950–3100 m, 03. Aug. 1997, *M. Wakabayashi et al. 9730094* (E). **Precise geographic position unclear**: Bhuji Khola, 2440

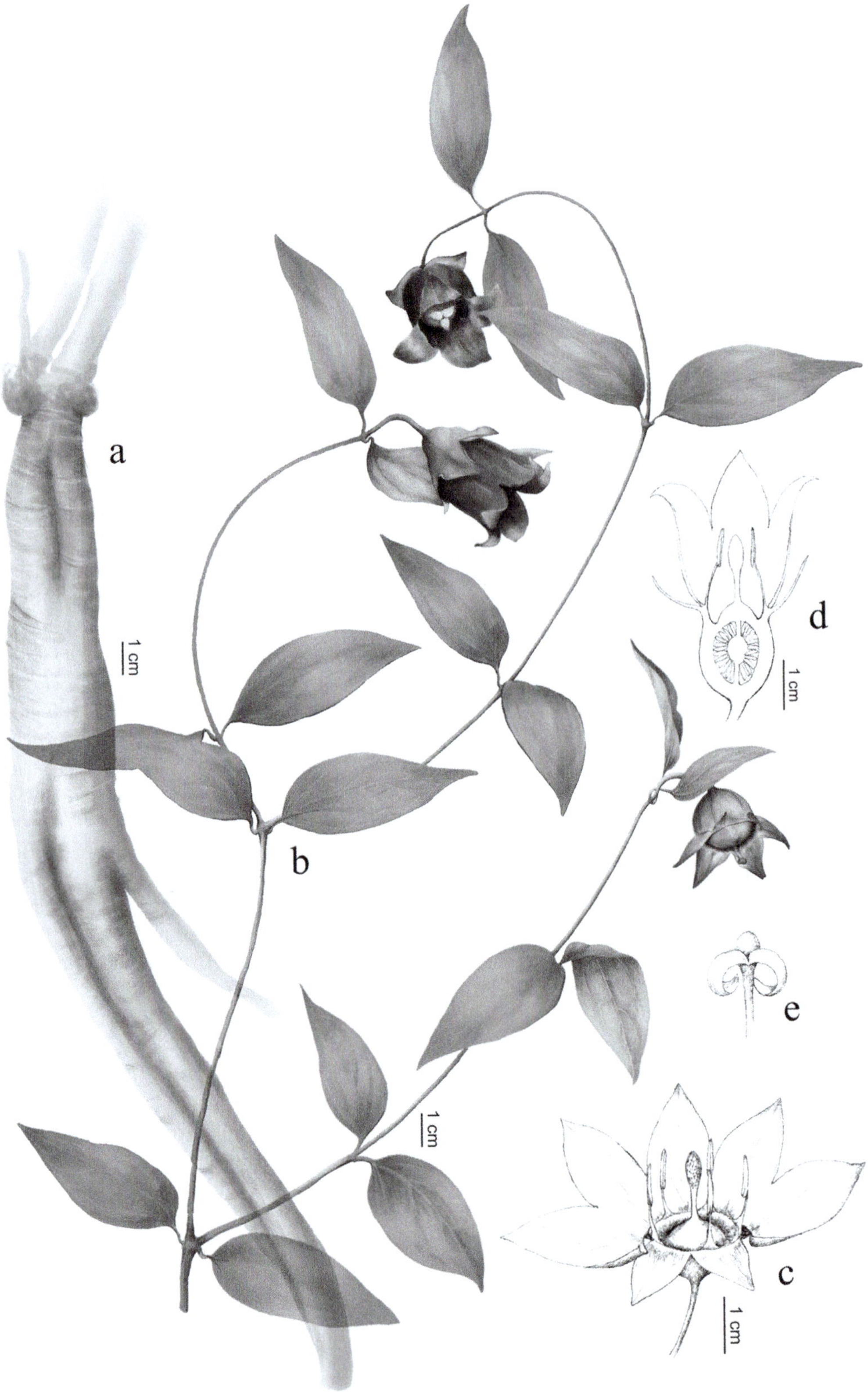

FIGURE IV 3:1A *Pankycodon purpureus* (Wall.) D. Y. Hong & X. T. Ma. **a**, root and basal parts of 3 stems; **b**, upper part of a stem; **c**, a flower with corolla opened, showing calyx, stamens, and style; **d**, a longitudinal section of flower, showing hypanthium higher than ovary; **e**, style and stigma.

FIGURE IV 3:1B *Pankycodon purpureus* (Wall.) D. Y. Hong & X. T. Ma. All from Tibet: Nyalam, 27. Aug. 2011. **a**, **c**–**e** & **g**, by Xin-Tang MA & De-Yuan HONG; **b** & **f**, by De-Yuan HONG.

m, 15. Oct. 1954, *Stainton, Sykes & Williams 9045* (BM, E); Changtang to Redak, 2250 m, 08. Aug. 1977, *H. Ohashi et al. 774870* (TI); Simbhanjyang, 2286 m, rock crevices, 15. Aug. 1961, *S. B. Malla 224* (KATH); loc. eodem, 2490 m, shady rocks, 18. Aug. 1964, *Rajbhandari & Upadhya 2240* (KATH); Terra Bence, 2740–3050 m, 1930, *L. Dhevoj 0427* (BM, E); Ulleri Pass, 2300 m, 28. Aug. 1967, *H. J. Lange 12* (B, K).

PSEUDOCODON D. Y. Hong & H. Sun

Hong & Sun in Wang et al., J. Syst. Evol. 52(5): 546. 2014; Wang & Hong, Phytotaxa 204 (1): 60. 2015. ≡*Codonopsis* subg. *Pseudocodonopsis* Kom., Trudy Imp. S. -Peterburgsk. Bot. Sada 29(1): 102. 1908; Tsoong, Contr. Inst. Bot. Nat. Acad. Peiping 3(3): 90. 1935; Shen & Hong in Hong, Fl. Reipubl. Popularis Sin. 73(2): 66. 1983; Hong, Lammers & Klein in Wu, Raven & Hong, Fl. China 19: 524. 2011. Type: *Pseudocodon convolvulaceus* (Kurz) D. Y. Hong & H. Sun (*Codonopsis convolvulacea* Kurz).

Herbs mostly twining, much less frequently ascending. Roots globose or ellipsoid, but in *Pseudocodon petiolatus* they elongated, attenuate and constricted at middle part. Stems single from caudex. Leaves alternate, rarely opposite. Flowers solitary, less frequently in a cyme of 2–4, terminal. Calyx tube entirely adnate to hypanthium, obovoid or obconical; calyx lobes 5, inserted at same level as corolla, entire. Corolla 5-cleft to near base, and thus rotate. Stamens 5; filaments short, base strongly dilated, dilated part semiorbicular and densely lanate at periphery adaxially. Ovary nearly entirely inferior to both calyx and corolla, 3-locular; stigma 3-fid, stigma lobes ovate. Fruit a 3-loculicidal capsule, base obovoid or obconical, upper part conical. Seeds many, oblong or ellipsoid, wingless, brown-yellow. Plants not fetid. Pollen grains 5-colpate; colpi very short, with colpus length / polar axis length 0.2–0.3; sexine with long spinules (≥2 μm). Chromosomes 2n=14.

The genus *Pseudocodon* differs distinctly from *Codonopsis* in a number of characters. The corolla is cleft to near the base and rotate (vs. shallowly lobed and campanulate or tubular); the disc is absent (vs. always present with various forms); filaments are always abruptly dilated at their base into a semiorbicular and densely lanate lamella, rarely ciliate (vs. slightly dilated and glabrous or villous-ciliate); the roots are globose, less frequently ellipsoid, rarely elongated (vs. carrot-shaped or nearly columnar, rarely long-ovoid); pollen is of short colpi, with colpus length/polar axis length 0.2–0.3 (vs. ≥0.7), smooth colpus membrane (vs. spinulose) and high sexine spinules (≥2 μm vs. ≤ 1 μm). Furthermore, in our molecular phylogenetic tree (Wang et al., 2013 & 2014 a) it forms a monophyletic clade with *Himalacodon* D. Y. Hong & Q Wang (*Codonopsis* subg. *Obconicicapsula* D. Y. Hong) and *Pankycodon* D. Y. Hong & X. T. Ma (*Codonopsis purpurea* Wall.), but separated from the clade consisting of the core *Codonopsis*. Phylogenetic relationships among these two clades and the *Cyananthus* clade are unclear, probably due to the result of a fast radiation. The relationships among *Pseudocodon*, *Himalacodon* D. Y. Hong & Q. Wang, and *Pankycodon* D. Y. Hong & X. T. Ma have been discussed by Wang et al. (2014 a), and briefly introduced under *Himalacodon*.

The taxonomy of *Pseudocodon* is problematic (Wang & Hong, 2015). In the genus only *P. rosulatus* has been stable in taxonomic status and is distinct from all the other species. *P. petiolatus* D. Y. Hong is newly described species. The rest species form the *Codonopsis convolvulacea* Kurz complex and the number of species in the complex has been controversial among authors. Nine taxa, i. e., *C. convolvulacea*, *C. convolvulacea* var. *heterophylla* C. Y. Wu (1965), *C. efilamentosa* W. W. Smith, *C. forrestii* Diels, *C. forrestii* var. *hirsuta* P. C. Tsoong ex L. T. Shen, *C. graminifolia* H. Lév., *C. limprichtii* Lingelsh. & Borza, *C. limprichtii* var. *hirsuta* Hand.-Mazz., and *C. vinciflora* Kom., had been described before "Fl. Reipubl. Popularis Sin." 73(2), where they were grouped by Shen and Hong (in Hong, 1983) into one species, *C. convolvulacea* with seven varieties, while Hong & Ma (1992) recognized four species, *C. convolvulacea* (with three subspecies, subsp. *convolvulacea*, subsp. *forrestii* and subsp. *vinciflora*), *C. efilamentosa*, *C. graminifolia* and *C. hirsuta*. Wang & Xu (1993) described a new species, *Codonopsis retroserrata*, which was said closely related to *C. forrestii*. Grey-Wilson (1995) published his taxonomic revision of this complex, recognizing six species. He separated *C. efilamentosa*, *C. forrestii*, and *C. vinciflora* Kom. as independent species, raised *C. limprichtii* var. *pinifolia* to specific status, *C. pinifolia* (Hand. - Mazz.) Grey-Wilson, and described *C. nepalensis* Grey-Wilson as new (= *C. grey-wilsonii* J. M. H. Shaw). On the basis of Grey-Wilson's paper, Lammers and Klein (2010) made a further taxonomic revision of *Codonopsis* subg. *Pseudocodonopsis* Kom. For the *C. convolvulacea* complex, they recognized *C. limprichtii*, *C. hirsuta* and *C. mairei* H. Lév. as separate species, and reduced *C. pinifolia* (Hand.-Mazz.) Grey-Wilson as a synonym of *C. limprichtii*. In addition, they described *C. macrophylla* with unknown origin as new. The controversy in taxonomy of the *Codonopsis convolvulacea* complex is clearly shown in "*Flora of China*" (Hong, Lammers & Klein, 2011: Vol.19, p. 525), where only three species, *C. convolvulacea* with four subspecies, *C. graminifolia* and *C. hirsuta*, are recognized. However, the two co-authors, Lammers and Klein, state their viewpoint in the note. They disagree with Hong's viewpoint and argue for their treatment of nine species (Lammers & Klein, 2010). Which treatment is relatively reasonable, i. e., how many species should be recognized, one, three, four, six or nine? I realize that this complex is very variable, particularly in petiole length, leaf shape and size of corolla, and that any taxonomy without field observations and statistical analysis of characters can hardly be justified, as Grey-Wilson (1995) admitted "no treatment of *Pseudocodonopsis* is going to be entirely satisfactory and any treatment will have its critics." The present treatment follows Wang and Hong (2015).

Eight species, extending from SW China westwards to Nepal via N Myanmar and Bhutan (Map IV 4:0).

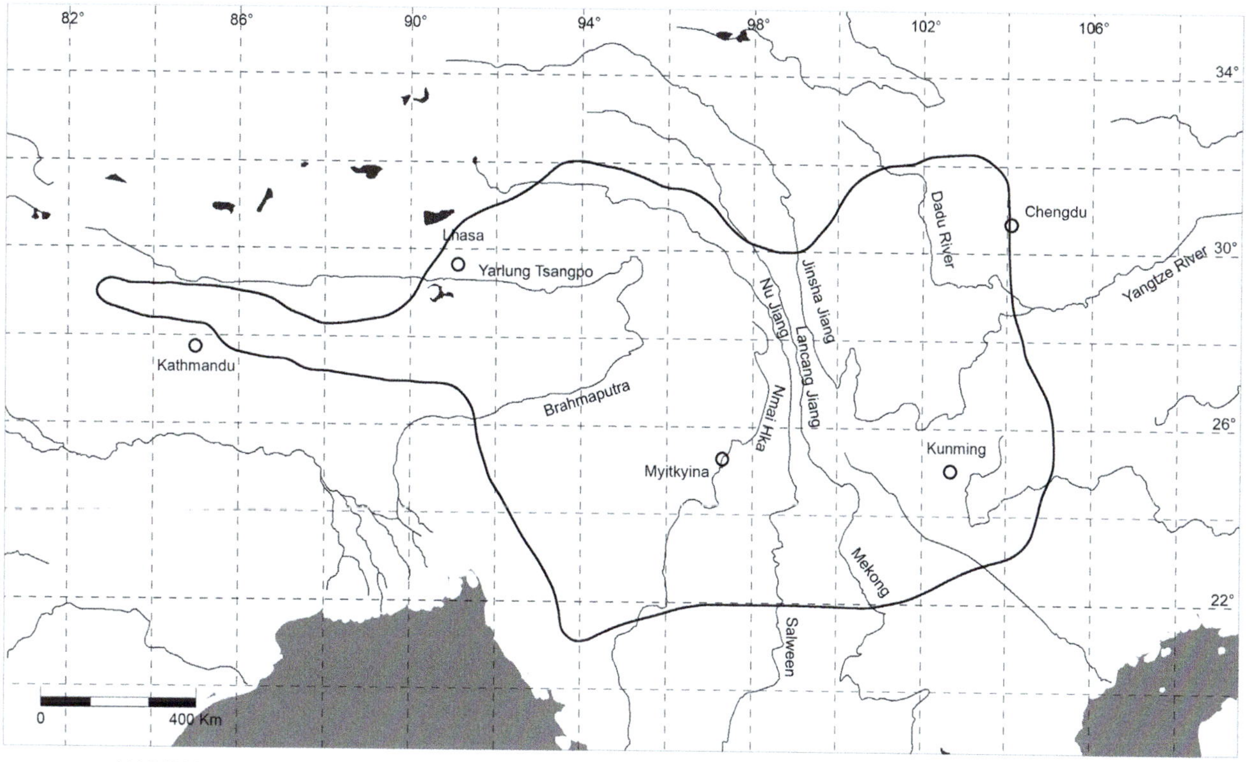

MAP IV 4:0 Distribution of the genus *Pseudocodon* D. Y. Hong & H. Sun (*Codonopsis* subg. *Pseudocodonopsis* Kom.).

1a. Basal leaves rosulate, persistent; stems erect or ascending, rarely procumbent, never twining..8. *P. rosulatus*

1b. Basal leaves absent; stems nearly always twining.

2a. Stems and leaves more or less white-hirsute; leaves at least partially opposite...........................4. *P. hirsutus*

2b. Stems and leaves glabrous; leaves all alternate

3a. Leaf blade filiform or linear, less than 5 (–9) mm wide, mostly aggregated at lower part of stem; stems less than 1 m long............................5. *P. graminifolius*

3b. Leaf blade cordate-ovate to linear-lanceolate, 0.4–8 cm wide, evenly distributed, stems usually over 1 m long.

4a. Calyx lobes strongly reflexed; roots elongate, attenuated, with a constriction at middle part; petiole 1.5–2.5 cm long, nearly as long as leaf blade............................7. *P. petiolatus*

4b. Calyx lobes upright; roots globose or ellipsoid; petiole 0.2–6(–10) cm long, much shorter than leaf blade.

5a. Leaf blade retroserrate, setose on margin............................6. *P. retroserratus*

5b. Leaf blade entire, crenate, dentate or serrate, without setae on margin.

6a. Leaf blade membranous, usually serrate or dentate, very occasionally entire; calyx lobes subulate or subulate-lanceolate, 3–10 mm long, 1.5–4 mm broad, rarely up to 16 mm long, 6 mm broad............................2. *P. vinciflorus*

6b. Leaf blade thin-papery, rarely membranous, entire or dentate; calyx lobes subulate to ovate-lanceolate; 6–18 mm long, 2.2–10 mm broad.

7a. Leaf blade dentate; corolla often with a red or dark red-purple ring at base............................3. *P. grey-wilsonii*

7b. Leaf blade mostly entire; corolla without a colourful ring at base.........1. *P. convolvulaceus*

1. Pseudocodon convolvulaceus (Kurz) D. Y. Hong & H. Sun in Wang et al., J. Syst. Evol. 52(5): 548. 2014; Wang & Hong, Phytotaxa 204 (1): 61. 2015. =*Codonopsis convolvulacea* Kurz, J. Bot. 11: 195. 1873; Komarov, Trudy Imp. S.-Peterburgsk. Bot. Sada 29(1): 102. 1908; Chipp, J. Linn. Soc. Bot. 38: 389. 1908, p. p., excl. synonym; Anthony, Notes Roy. Bot. Gard. Edinburgh 15: 177. 1926, p. p., excl. synonyms & vars. *limprichtii*, *hirsuta*, & *pinifolia*; Tsoong, Contr. Inst. Bot. Nat. Acad. Peiping, 3(3): 90. fig. 6. 1935, p.p., excl. var. *pinifolia*; Shen & Hong in Hong, Fl. Reipubl. Popularis Sin. 73(2): 67. 1983, p.p., quoad type variety & var. *forrestii*, Hong & Ma in Wang & Wu,

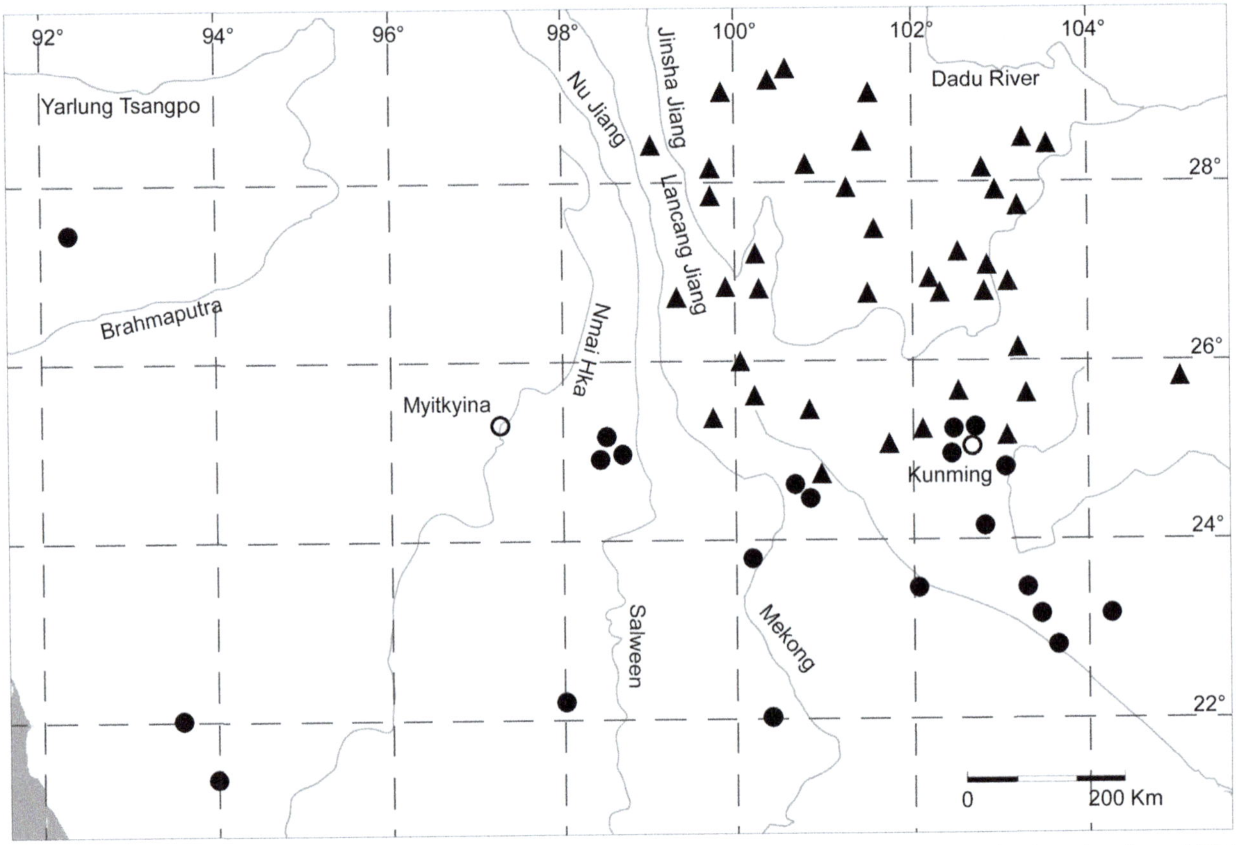

MAP IV 4:1 Distribution of *Pseudocodon convolvulaceus* (Kurz) D. Y. Hong & H. Sun. Dots: subsp. *convolvulaceus*; triangles: subsp. *forrestii* (Diels) D. Y. Hong.

Vasc. Pl. Henduan Mountains 2: 1960. 1994. Type: China, Yunnan, Mengzi, 1520–1830 m, s. d., *A. Henry 9425* (neotype designated by Hong, 2010: PE; isoneotypes: E, K, LE).

Roots ovoid-globose or ovoid, 2.5–5 cm long, 1–1.5 cm in diameter. Stems twining or nearly ascending, usually with a few branches, up to more than 1 m long, glabrous. Leaves alternate or sometimes opposite, evenly distributed along stems; petiole from absent to 6 (–10) cm long; leaf blade linear-lanceolate to cordate-ovate, 2–10 cm long, 0.4–8 cm broad, base cuneate, rounded or slightly cordate, margin entire or serrate, apex obtuse, acute or acuminate. Flowers solitary, terminal on main stem and branches, sometimes in a cyme of 2–4; pedicel 2–12 cm long, glabrous. Calyx tube adnate to whole hypanthium, obconical, 3–7 mm long, 4–10 mm wide, glabrous, 10-ribbed; calyx lobes usually lanceolate, less frequently subulate or long-ovate, 6–18 mm long, 2.2–10 mm broad, glabrous, apex acuminate or acute; sinus between calyx lobes narrow, pointed or slightly obtuse. Corolla pale blue or blue-purple; corolla lobes elliptic, 1–5 cm long, 0.6–2 cm wide, apex acute. Filaments 1–3 mm long; anthers 4–5 mm long. Superior part of capsule short-conical, valves ca. 4 mm long; inferior part obconical, 1–1.6 cm long, ca. 8 mm in diameter. Seeds numerous, oblong, ca. 1.5 mm long, brown-yellow.

Distribution: China (Yunnan, Guizhou and SW Sichuan) and N Myanmar (Map IV 4:1).

Notes: Kurz (1873) did not cite any specimen when he described *Codonopsis convolvulacea* as new, and I failed in finding any specimen which seems to have been examined by Kurz in all the herbaria I visited (see Preface). Thus I designated *A. Henry 9425* from Mengzi, Yunnan Province, as the neotype, which has at least four duplicates.

1a. Leaf blade lanceolate to linear-lanceolate, entire, base rounded to cuneate; petiole shorter than 3 mm; calyx lobes 6–10 mm long, 2.2–4 mm broad; corolla lobes 1–3 cm long..2a. subsp. convolvulaceus

1b. Leaf blade ovate-lanceolate to linear-lanceolate, entire or serrate, base slightly cordate to cuneate; petiole 0.3–6 (–10) cm long; calyx lobes 10–18 mm long, 2.5–10 mm broad; corolla lobes 1.8–5 cm long.................2b. subsp. *forrestii*

1a. Pseudocodon convolvulaceus subsp. **convolvulaceus**

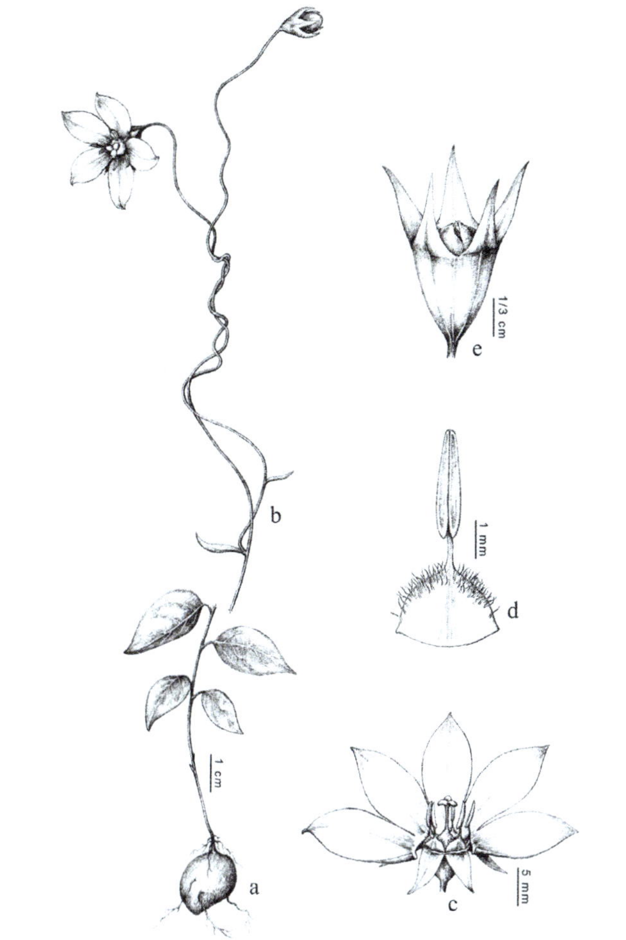

FIGURE IV 4:1A *Pseudocodon convolvulaceus* (Kurz) D. Y. Hong & H. Sun subsp. *convolvulaceus*. **a**, globose root and lower part of stem; **b**, upper part of the stem; **c**, a flower with corolla opened, showing calyx, stamens, style, and stigma; **d**, a stamen; **e**, a 3-loculicidal capsule.

Leaves nearly sessile or with petiole shorter than 3 mm; leaf blade lanceolate to linear-lanceolate, 2–7 cm long, 0.5–1.5 cm broad, papery, entire, base rounded to cuneate. Calyx lobes 6–10 mm long, 2.2–4 mm broad. Corolla lobes 1–3 cm long. Figures IV 4:1A & B.

Phenology: flowering in August and September; fruiting from September to November.

Habitats: grassy slopes, thickets, or forests, on granites; 1100–2500 m.

Distribution: China (S & SW Yunnan and SE Tibet) and Myanmar.

Additional specimens examined, **CHINA, TIBET, Cona:** Tembang, 1830–2130 m, 16. Jul. 1938, *F. Kingdon-Ward 13943* (BM). **YUNNAN, Jingdong**: Anzhao, 1850 m, grassy slopes, 23. Oct. 1993, *H. Peng 1520* (KUN), *2387* (KUN) & *8871* (KUN); Taizhong, Dashui Gou, 1600 m, pine forests, 28. Sep. 1959, *S. G. Xu 5408* (PE). **Kunming**: Bot. Gard. Bot. Inst. Kunming, 1950 m, wild, 28 Aug. 1998, *Y. Luo 0045* (KUN); the hill behind Kunming Inst. Bot., 02. Sept. 1975, *B. Y. Qiu 60084* (CDBI); Heilongtan, 28. Aug. 1991, *S. K. Wu et al. 1434* (KUN); Heilongtan, ravines, 21. Sept. 1955, *B. Y. Qiu 50273* (KUN); Qiongzhu-si, Daguanqing, 2200 m, thickets, 15. Sep. 2010, *D. Y. Hong, H. Sun et al. H10011* (PE); West Hill, 2050 m, Aug. 1989, *T. J. Zhang 0018* (PE); West Hill, 1941, *T. N. Liou 20020* (PE); loc. eodem, 1941, *T. N. Liou 19916* (PE); West Hill, Huating-si, 2000 m, 16. Aug. 1946, *T. N. Liou 16671* (PE); loc. eodem, 2200 m, 23. Aug. 1953, *K. M. Feng 50137* (PE, SZ); West Hill, Kao-Chiao, 1930 m, 08. Aug. 1945, *T. N. Liou 14048* (PE); West Hill, below Taihuasi, 2000 m, 25. Aug. 1945, *T. N. Liou 14208* (PE). **Lincang**: Mengwang Distr., 1600–2100 m, thickets, 02. Sep. 1957, *J. S. Xin 654* (PE). **Longyang Distr.**: Lujiang Township, Bawan Village, 1800 m, coniferous forests, 01. Sept. 2003, *Gaoligong Shan Biodeversity Survey 18391* (CAS); loc. eodem, 1350 m, coniferous forests, 31. Aug. 2003, *Gaoligong Shan Biodiversity Survey 18278* (CAS). **Menghai**: s. loc., 1150 m, 03. Sept. 1991, *Z. H. Tsi 91-410* (A, K, PE). **Mengzi**: Heilongtan, 1300 m, grassy slopes, 09. Sept. 1939, *C. W. Wang 81656* (PE); Tongshan, 1400 m, 26. Sept. 1939, *C. W. Wang 81731* (PE); 1680–1980 m, Sept. 1893, *W. Hancock 74* (K); s. loc., 1520–1830 m, 28. Aug. 1895, *W. Hancock 385* (K); s. loc., 22. Sept. 1895, *W. Hancock* s. n. (K). **Pingbian**: Shuban-po, 1400 m, 24. Sept. 1939, *C. W. Wang 82077* (PE). **Songming**: Aziying, Tubacun, 2000–2100 m, Aug. 1989, *T. J. Zhang 0013* (PE) & *0014* (PE); loc. eodem, 2100 m, Aug. 1989, *T. J. Zhang 0010* (PE). **Tengchong**: Qishi Township, Longkou Village, scrubs, 29. Oct. 1998. *H. Li et al. 11205* (CAS); Sanyun Township, Pingdi Village, 1335 m, forests, 30. Aug. 2003, *Gaoligong Shan Biodiversity Survey 18057* (CAS); Shweli Valley, 25°20′N, 1830 m, Aug. 1913, *G. Forrest 11993* (E); Teng-Yueh, *E. B. Howell 86* (E); Volcanic mouth, 1800 m, 01. Aug. 1980, *S. T. Li 80–342* (PE). **Tonghai**: Xiushan, by village, 15. Aug. 1989, *Yuxi Exped. 0878* (KUN). **Wenshan**: Laojun Shan, grasses, 1850 m, 14. Aug. 1985, *Z. H. Tsi et al. 409* (PE) ; loc. eodem, 2500 m, open woods, 29. Sep. 1992, *Y. M. Shui 454* (KUN). **Yiliang**: Yangzhonghai, 1800 m, 06. Sep. 1977, *B. Y. Qiu 771252* (CDBI, KUN, PE); Yiliang to Qixing Cun, pine forests, 27. Aug. 1975, *B. Y. Qiu 60723* (CDBI, KUN). **Yuanjiang**: Yebai, 1250–1600 m, thickets, 16. Nov. 1965, *W. Q. Ying 1962* (PE). **Zhenkang**: Maliling, 1800 m, grassy slopes, 24. Aug. 1938, *T. T. Yü 17456* (PE).

MYANMAR, **Mindat**: s. loc., 1520 m, 31. Aug. 1956, *F. Kingdon-Ward 226010* (BM). **Shan States**, S Shan States, 1910, *W. G. Robatrow 158* (K); s. loc., Nov. 1903–Jan. 1904, *C. H. Melville 51045* (BM). **Loi Mrue**: s. loc., 1680 m, 1909. *R. W. MacGregor 67* (CAL). **Saga**: s. loc., 1893, *Abdul Khalil* s. n. (CAL). **Hakha**: Chin Hills, Khusa Road 4 km from Hakha, ca. 1830 m, 18. Sept. 1910, *F. E. W. Venning 103A* (K).

FIGURE IV 4:1B *Pseudocodon convolvulaceus* (Kurz) D. Y. Hong & H. Sun subsp. *convolvulaceus*. **a–c**, from Yunnan: West Hills of Kunming, 15. Sept. 2010, by De-Yuan HONG. **a**, the arrow indicates stem and leaves.

1b. Pseudocodon convolvulaceus subsp. **forrestii** (Diels) D. Y. Hong in Wang & Hong, Phytotaxa 204 (1): 61. 2015. ≡*Codonopsis forrestii* Diels, Notes Roy. Bot. Gard. Edinburgh 5: 171. 1912; Anthony, Notes Roy Bot. Gard. Edinburgh 15: 179. 1926; Tsoong, Contr. Inst. Bot. Nat. Acad. Peiping 3(3): 91. 1935; Lammers & Klein, Bot. Stud. (Taipei) 51: 559. 2010. ≡*Codonopsis convolvulacea* subsp. *forrestii* (Diels) D. Y. Hong & L. M. Ma in Li & Zhu, Fl. Sichuan. 10: 546. 1992; Hong, Lammers & Klein in Wu, Raven & Hong, Fl. China 19: 525, 2011. ≡*Codonopsis convolvulacea* var. *forrestii* (Diels) Ballard, Bot. Mag. 162: tab. 9581. 1939. Type: China, Yunnan, on the banks of Yangtze, 1520–2130 m, Sept. 1904, *G. Forrest 48* (lectotype designated by Hong in Wang & Hong, 2015: E; isolectotype: PE).

Codonopsis forrestii Diels var. *hirsuta* P. C. Tsoong & L. T. Shen ex L. T. Shen, Acta Phytotax. Sin. 13(3): 55. 1975. Type: China, Sichuan, Yanyuan, 2400–2500 m, *F. G. Zhu* (*Yanyuan*) *004* (not preserved ?).

Codonopsis efilamentosa W. W. Smith, Notes Roy. Bot. Gard. Edinburgh 8: 107. 1913. ≡*Codonopsis convolvulacea* Kurz var. *efilamentosa* (W. W. Smith) L. T. Shen in Hong, Fl. Reipubl. Popularis Sin. 73(2): 69. 1983. Type: China, Yunnan, "the eastern flank of the Lichiang Range" (Lijiang), 27°30′N, 3050 m, Jul. 1910, *G. Forrest 6258* (holotype:E).

Codonopsis limprichtii Lingelsh. & Borza, Repert. Spec. Nov. Regni Veg. 13: 391. 1914; Lammers & Klein, Bot. Stud. (Taipei) 51: 555, 2010. p. p., excl. syn. ≡*Codonopsis convolvulacea* Kurz var. *limprichtii* (Lingelsh. & Borza) J. Anthony, Notes Roy. Bot. Gard. Edinburgh 15: 178. 1926. Type: China, Yunnan: "Talifu (Dali), an schattigen Plätzen unter Kiefern vor Ai-ning-po, bei Pu-peng", 2500 m, 21. Aug. 1913. *Limpricht 851* (holotype: WRSL, n. v. ; isotype: WU).

Codonopsis mairei H. Lév., Cat. Pl. Yunnan, 24. 1915. Type: China, Yunnan, "paturages des mont. à Ta-hai », 3200 m. Jul. 1911, *E. E. Maire* s, n. (holotype :E).

Leaves all petiolate with petiole 0.3–6 (–10) cm long; leaf blade cordate-ovate, ovate-lanceolate to linear-lanceolate, 2–10 cm long,0.4–8 cm broad, papery, glabrous, entire or serrate, base cordate to cuneate. Calyx lobes 1–1.8 cm long, 2.5–10 mm broad. Corolla lobes 1.8–5 cm long. Figure IV 4:1C.

Chromosomes: 2n=14 (Wang et al., 2014a).

Phenology: flowering from July to September; fruiting from September to November.

Habitats: forests, thickets, grassy slopes, or edges of forests; 1500–4000 m.

Distribution: SW China (W Guizhou, N Yunnan, and SW Sichuan) and N Myanmar.

Notes: 1) I treat here *Codonopsis limprichtii* Lingelsh. & Borza as a synonym of *Pseudocodon convolvulaceus* subsp. *forrestii* (*Codonopsis forrestii* Diels), which has cordate-ovate, ovate-lanceolate or occasionally linear-lanceolate leaf blade. However, Lammers and Klein (2010) treated *C. limprichtii* as an independent species with *C. graminifolia*

FIGURE IV 4:1C *Pseudocodon convolvulaceus* (Kurz) D. Y. Hong & H. Sun subsp. *forrestii* (Diels) D. Y. Hong. **a**, **b** & **e–g**, from Yunnan: NE Kunming, 15. Sept. 2010, by De-Yuan HONG; **c**, from Sichuan: Daocheng; 03. Aug. 2012, Fu-sheng YANG; **d**, from Yunnan: Shangri-la, 16–17. Sept. 2010, by De-Yuan HONG. **d** & **g**, showing variation of leaves within population.

H. Lév. and *C. limprichtii* var. *pinifolia* Hand. -Mazz. (= *C. pinifolia* Grey-Wilson, invalid nom.) (both with linear leaf blade) as its synonyms. Actually, the protologue of *C. limprichtii* Lingelsh. & Borza describes "Folia...... inferiora ovalia, parce obtuse dentate, ad 3 cm longa, 2 cm lata, superiora lanceolata, integerrima, ad 5 cm longa et 1 cm lata". And I have examined the isotype of *C. limprichtii* at WU (*Limpricht 851*), which has leaves ovate to linear-lanceolate, 1.5–3.2 cm long, 0.3–1.3 cm broad. Therefore, *C. limprichtii* is consistent with *C. forrestii*, but differs distinctly from *C. graminifolia* or *C. limprichtii* var. *pinifolia*. According to my field observation *C. graminifolia* H. Lév. (= *Pseudocodon graminifolius*) is a distinct species, with totally glabrous and short stems, and glabrous and linear leaves which are aggregated at the lower part of stem.

Additional specimens examined, **CHINA**, **GUIZHOU**, **Pu'an**: Pubei Forest Farm, 1800 m, 08. Sept. 1959, *Anshun Exped. 1437* (PE). **SICHUAN**, **Butuo**: Shaluo, 2400 m, 09.

Aug. 1979, *s. coll. 711* (SM). **Daocheng**: Daocheng to Yading, 3500 m, herbs with shrubs, 03. Aug. 2012, *F. S. Yang & Y. L. Liu 071* (PE); Riwa, Lamuge, 3900 m, edges of thickets, 30. Jul. 1973, *s. coll. (Dao) 2417* (KUN); Valley E of Daocheng, 29°2′N, 100°32′E, 3800 m, 10. Sept. 2010, *Kunming & Edinburgh Exped. to Sichuan 244* (E). **Huidong**: Baisha Township, 3000 m, by fields, 19. Jun. 1959, *Sichuan Econ. Plant Surv. (Xi) (59) 774* (KUN); Tangtang, 2100 m, 07. Aug. 1978, *Huidong Exped. 543* (SM). **Huili**: Yimen, Chahe Township, grassy slopes, 03. Oct. 1958, *Z. He et al. 11619* (SM, SZ). **Jinyang:** Bingdi, 3000 m, 30. Aug. 1976, *Sichuan Plant Surv. 14899* (CDBI); Duiping, 1700 m, oak forests, 19. Aug. 1976, *Sichuan Plant Surv. 14856* (CDBI). **Jiulong**: Tanggo, 3200 m, thickets, *Y. B. Yang 23013* (CDBI). **Leibo**: Majingzi, 700 m, 18. Sept. 1979, *Leibo-Chinese Mater. Medic. Resources Surv, 1220* (SM). **Meigu**: Bapu, 2100 m, 13. Aug. 1979, *s. coll. 540* (SM). **Miyi**: Wanqiu, 1300 m, 13. Oct. 1963, *Miyi Working Team 194* (CDBI); s. loc., 1200 m, 19. Oct. 1982, *Z. A. Liu 23957* (CDBI). **Muli**: Bo'ao, 2200 m, 27. Aug. 1978, *C. S. Cao et al. 8316* (CDBI, SZ); Dingdonggou to Bo'ao Forest Farm, 2600 m, 14. Aug. 1978, *D. S. Cao et al. 7520* (CDBI, SZ); Muli-Yongning, 3100–3500 m, 29. Jul. 1915, *Handel-Mazzetti 7267* (WU); Sanjia Township, 2310 m, 10. Oct. 1955, *W. X. Liu 51191* (SZ); Sihe Commune, 2800 m, 27. Aug. 1978, *C. S. Cao et al. 8149* (CDBI, SZ); Wanglang to Chabulang, broad-leaved forests, 16. Sept. 1983, *Qinghai-Xizang Exped. 14111* (KUN, PE); Yazui Forest Farm, 3200 m, thickets, 18. Aug. 1983, *Qinghai-Xizang Exped. 12954* (PE); Distr. No. 1, Meteorologic Station, 2700 m, thickets, 17. Sept. 1978, *C. S. Cao & G. Hu 8769* (SZ); 913 Forest Farm, Zhani, 2600 m, thickets, 23. Aug. 1983, *Qinghai-Xizang Exped. 13331* (PE); s. loc., 2740 m, 04. Aug. 1921, *F. Kingdon-Ward 4545* (E); s. loc., 2740 m, 09. Aug. 1921, *F. Kingdon-Ward 4756* (E); s. loc., 2740 m, 28. Aug. 1921, *F. Kingdon-Ward 4869* (E). **Ningnan**: Xinjian, 2000 m, 19. Aug. 1978 *Ningnan Exped. 517* (SM). **Puge**: Jingyingsuo, thickets, 19. Aug. 1959, *Sichuan Econ. Plant Surv. (Pu) 5452* (KUN); Ju'andahuo Shan, 3400 m, 11. Aug. 1979, *Puge Exped 579* (SM). **Xiangcheng**: Shangyou Township, 3600 m, thickets, 22. Aug. 1973, *s. coll. (Xiang) 3230* (KUN); loc. eodem, 3800 m, thickets, grassy slopes, 23 Aug. 1972, *Z. G. Liu 1217* (CDBI). **Yanbian**: Yankou Township, 2800 m, 28. Jul. 2001, *C. M. Tan P315 B* (SZ). **Yanyuan**: Bailing, 3000 m, *Quercus* forests, 11. Aug. 1983, *Qinghai-Xizang Exped. 12902* (KUN, PE). **Zhaojue**: Poluo, 2420 m, 21. Aug. 1979, *s. coll. 893* (SM). **YUNNAN**, s. loc., *McLaren's Collectors 79A* (BM, E, K). Beyendjing, *S. Ten s. n.* (W). **Chuxiong**: To-Tsu, 1950 m, thickets, 20. Sep. 1939, *M. K. Li 51* (PE). **Dali**: Cangshan, 2500 m, Sept. 1914, *C. Schneider 3777* (G, PE); loc. eodem, 2200–2600 m, 27. Aug. 1994, *X. Cheng et al. s. n.* (KUN); Dali Range, 25°40′N, 2740–3350 m, Aug. 1906, *G. Forrest 3852* (E, P); loc. eodem, 25°40′N, 2440–2740 m, Jul. 1920, *G. Forrest 7079* (E, K); Dali Range, Sept. 1924, *G. Forrest 28101* (E, PE); Huadianba, 2900–3100 m, 27. Aug. 1984, *K. Iwatsuki et al. 984* (KUN); loc. eodem, 2900–3100 m, 28. Aug. 1984, *K. Iwatsuki et al. 1157* (KUN); Kunming to Xiaguan, 2210 m, 21. Sept. 1994, *ACE 1100* (E, PE); Moyi, Banshan, 16. Jul. 1944, *H. C. Wang 4456* (PE); Zhonghefeng, Banshan, Jul. 1941, *H. C. Wang 1096* (PE); "Dsolin-ho" (?), 2000–2300 m, 07. Sept. 1914, *Handel-Mazzetti 4917* (WU); s. loc., 1850–2150 m, 06. Sept. 1914, *Handel-Mazzetti 4871* (WU); s. loc., 2800 m, 10. Aug. 1913, *Limpricht 982* (WU). **Dêqên**: Yinmin Ganchong, grasses, 2700 m, *S. B. Lan 142* (PE); s. loc., 3660 m, Sept. 1911, *F. Kingdon-Ward 76* (E). **Dongchuan**: Fanze Forest Farm, 2880 m, bamboo thickets, 13. Aug. 1964, *NE Yunnan Exped. 532* (KUN, PE); Fazhe Township, 2600 m, thickets, 30. Jul. 2008, *H. Peng et al. 8886* (KUN); Fazhe Forest Farm, 2400 m, 31. Jul. 2008, *H. Peng et al. 9062* (KUN); Tong-tchouan, 2000–3000 m, Sept. 1912, *E. E. Maire* s. n. (G). **Heqing**: Houang-tsao-Pa, 1900–1920, *J. Cavalerie 4311* (K); Huangping, 2600 m, 16. Aug. 1963, *Jinshajiang Exped. 6565* (KUN, PE); loc. eodem, 2469 m, oak and pine forest, 17. Aug. 2010, *E. D. Liu et al. 2739* (KUN); Lianping, Nanxilu, 2800 m, over thickets, 21. Aug. 1929, *R. C. Ching 23859* (KUN, PE); Sanchang, mixed forests, 28. Jul. 1929, *R. C. Ching 23399* (KUN, PE). **Jianchuan**: Chünchuan-Mekong Divide, 26°40′N, 99°40′E, 3350–3660 m, Sept. 1922, *G. Forrest 22275* (E, K, W); Chunchuan-Mekong Divide, 26°30′N, 99°40′E, 3050–3350 m, Sept. 1922, *G. Forrest 22281* (E, K); s. loc. 1921-1922, *G. Forrest 22332* (E). **Jingdong**: Taizhong, 2000 m, 10. Oct. 1983, *H. Sun SH-140* (PE). **Kunming**: Dazi Shan, 03. Jul. 1938, *K. H. Liang 229* (KUN, PE); near Heijing, 26. Aug. 1948, *T. N. Liou 20738* (PE); West Hill, 2200 m, Aug. 1989, *T. J. Zhang 0019* (PE); loc. eodem, 26. Aug. 1991, *S. K. Wu et al. 945* (KUN); West Hill, Dahuasi, 2100 m, forests, 23. Aug. 1946, *T. N. Liou 16688* (KUN, PE); N of Kunming city, Balipo, 2300 m, pine-mixed forests, 14 Aug. 1963, *Z. H. Yang 101227* (KUN, PE); Longquan, Heping Township, 2100 m, sunny grasses, 15. Sep. 1958, *S. K. Wu 9* (PE); Panlong Distr., Shuangxiao Township, 2300 m, open pine woods/thickets by fields, 15. Sept. 2010, *D. Y. Hong et al. H10012* (PE); aihuasi, thickets, 20. Aug. 1953, *K. M. Feng 50003* (SZ); s. loc., 2200 m, 21. Aug. 1916, *O. Schoch 308* (G) & *309* (G). **Lanping**: 2800 m, sides of Mt Path, 17. Aug. 1933, *H. T. Tsai 53707* (KUN, PE, SZ). **Lijiang** (Yulong): Hutiaoxia, 1900 m, 08. Sept. 1987, *H. Ohba et al. 99* (KUN); behind Lashiban, 2800 m, thickets, 14. Jul. 1964, *S. W. Yu 64-019* (PE); Lijiang Snow Range, 3540 m, Sept. 1923, *J. F. Rock 10656* (E); between Lijiang, Yongning & Yongpei, on way to Muli, Aug. 1922, *J. F. Rock 5307* (PE); Ma-on-Shan, on tree trunk, 17. Aug. 1939, *R. C. Ching 21314* (KUN, PE); Shudi, 1500–1750 m, 07. Oct. 1987, *K. Iwatsuki et al. 1893* (KUN); Snow Range, mixed forests, 03. Sept. 1939, *R. C. Ching 30589* (PE); loc. eodem, 1922, *J. F. Rock 5899* (PE); loc. eodem, 3600 m, thickets, 15. Aug. 1942, *K. M. Feng 9039* (KUN, PE); Tamichung, edges of forests, 30. Aug. 1939, *R. C. Ching 21575* (KUN, PE); Wenfeng-cze, open thickets, 05. Sept. 1939, *R. C. Ching 21390* (KUN, PE); Xuesongcun, 3500 m, Aug. 1989, *T. J. Zhang 0028* (PE); Xuesongcun, 3500 m, Aug. 1989, *T. J. Zhang 0028* (PE) *& 0029* (PE); loc. eodem, 2900 m, edges of thickets, 01. Sept. 1958, *W. T. Wang* (Dali) *40* (PE); Yulong Snow Range,

Jul. 1922, *J. F. Rock 5476* (E, PE); loc. eodem, 1933, *H. D. McLaren's Native Collector 246B* (BM); Yulong Snow Range, 3800 m, edges of forests, 02. Oct. 1992, *J. R. Chen* 92529 (PE); s. loc., 3300 m, 29. Jun. 1914, *C. Schneider 2965* (G, K, PE); s. loc., 2800 m, 15. Aug. 1914, *C. Schneider 3389* (G, K, PE); s. loc., 3660 m, 1922, *J. F. Rock 6036* (K, PE); s. loc., 4000 m, 08. Sept. 1914, *C. Schneider 2456* (G); 27°25′N, 3350 m, Sept. 1913. *G. Forrest 11178* (E, K); 27°30′N, 3660 m, Sept. 1910, *G. Forrest 6706* (E, K). **Longling**: No 4 Distr., open woods, *J. S. Yang 1402* (KUN). **Luoci**: Jiaowoping, 2400 m, pine forests, 23. Oct. 1964, *B. Y. Qiu 596334* (KUN). **Luquan**: Wumeng Shan, 3500 m, open woods, 30. Oct. 1940, *Y. P. Chang 537* (KUN, PE). **Qiaojia**: Dacun to Ganghaizi, 2600 m, 16. Aug. 2003, *H. Wang et al. 03–0788* (KUN). **Shangri-la** (Zhongdian): Lamaso, 3100 m, bush woods, 05. Aug. 1937, *T. T. Yü 12644* (KUN, PE); Wucun, by Nabahai, 3300 m, thicket margins, 16. Sept. 1962, *S. W. Yu 62000* (PE); Wufeng Shan, 3300 m, thickets, 14. Aug. 1962, *Zhongdian Exped. 833* (KUN, PE); 3 km S of Zhongdian, 3240 m, alpine meadows with shrubs, 30. Sept. 1992, *J. R. Chen 92396* (PE); 4 km NW of Zhongdian town, 3400 m, grazed open thickets on dry slopes, limestones, 31. Aug. 2000, *E. von Raab-Straube, M. Smalla & H. Sun 417* (B); Zhongdian Valley, 3300 m, open dry hillsides, 09. Aug. 1939, *K. M. Feng 1950* (KUN, PE); ; Zhongdian Alpine Bot. Gard., 3350 m, tall thickets/waste fields, 17. Sept. 2010, *D. Y. Hong et al. H10015* (PE); 28°1′18″N, 99°46′48″E, 3180 m, 11. 09. 1995, *Roy. Bot. Gard. Edinburgh Exped. to Dêqên Pref. 127* (E). **Shuangbai**: Tuodian, 05. Oct. 1958, *Acad. Sin. Kunming Inst. Bot. Pl. Resources Integrative Surv. 173* (KUN, PE). **Songming**: Guodong, 2200, 10. Oct. 1952, *P. I. Mao 201* (KUN, PE); Longtan, 25°28′N, 102°46′E, Remnant, forests, 2100 m, 27–28. Jul. 1984, *Sino-Amer. Bot. Exped. 1324* (PE); s. loc., 2300 m, 10. Oct. 1955, *B. Y. Qiu 51191* (KUN). **Xundian**: Dapei Shan, 2800 m, open woods, 11. Nov. 1940, *Y. P. Chang 816* (KUN, PE); Yinghe-Township, 2300 m, forest margin, 28. Aug. 1958, *X. W. Li 22* (KUN, PE). **Yongping**: between Dali and Baoshan, 25°29′N, 99°36′E, 2200 m, 05. Sept. 1997, *Gaoligong Shan Exped. (1997) 8539* (E).

MYANMAR, **Maymyo Plateau**, 1070 m, 04. Nov. 1911, *J. H. Lace 5516* (K). **Taunggyi Reserve**, 1520 m, grasses of hill forests, 27. Sept. 1970, *U MG Gale -2* (PE).

2. **Pseudocodon vinciflorus** (Kom.) D. Y. Hong in Wang et al., J. Syst. Evol. 52(5): 548. 2014; Wang & Hong, Phytotaxa 204 (1): 61. 2015. ≡*Codonopsis vinciflora* Kom., Trudy Imp. S.-Peterburgsk. Bot. Sada 29(1): 103. 1908; Fletcher, Gard. Chron. 102: 102. 1937 & Bot. Mag. 166: pl. 59. 1949; Lammers & Klein, Bot. Stud. (Taipei) 51: 558. 2010. ≡*C. convolvulacea* var. *vinciflora* (Kom.) L. T. Shen in Hong, Fl. Reipubl. Popularis Sin. 73(2): 68. 1983. ≡*Codonopsis convolvulacea* subsp. *vinciflora* (Kom.) D. Y. Hong in Wu, Fl. Xizang. 4: 582. 1985; Hong & Ma in Wang & Wu, Vasc. Pl. Hengduan Mount. 2: 1961. 1994; Hong, Lammers & Klein in Wu, Raven & Hong, Fl. China 19: 525. 2011. Type: China, "Se-czuan (Sichuan) inter Va-sy-kou et Za-li in valle fl. Siao-tzin-che", 15. Jul. 1893, *G. N. Potanin s. n.* (holotype: LE).

Codonopsis tibetica Prains in sched. Type: China, Tibet, Lhasa, Aug. 1878, *Dungboo* s. n. (K).

Herbs twining, glabrous throughout except filaments and style. Roots globose. Stems from 20 cm to over 1 m long. Leaves all alternate; petiole 3–12 mm long; leaf blade membranous, ovate to lanceolate, rarely linear-lanceolate, 1.5–3.5 cm long, 0.4–2.2 cm broad, base rounded to broad-cuneate, margin usually dentate, rarely entire, apex acute; those on branches much smaller. Flowers solitary, terminal only on main stem or also on upper branches; pedicel 1.5–20 cm long. Calyx lobes subulate or subulate-lanceolate, 3–16 mm long, 1.5–6 mm broad. Corolla blue-purple, 2.5–5.0 cm wide; corolla lobes elliptic or oblong. Stamens 6 mm long, filaments shorter than anthers. Ovary mostly obconical, less often obovoid; style slightly longer than stamens. Fruit usually obconical or obconical-ovoid, 1–1.3 cm long, 0.8 cm in diameter.

Distribution: **endemic** to SW China: W Sichuan, C & SE Tibet, and NW Yunnan (Map IV 4:2).

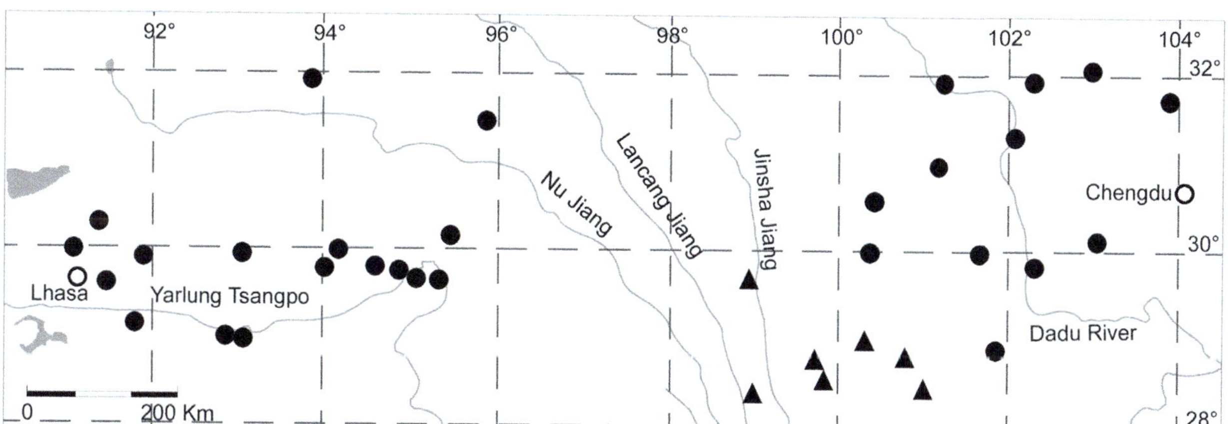

MAP IV 4:2 Distribution of *Pseudocodon vinciflorus* (Kom.) D. Y. Hong. Dots: subsp. *vinciflorus*; triangles: subsp. *dianchuanica* D. Y. Hong & Q. Wang.

1a. Calyx lobes 3–7 mm long, 1.5–3 mm broad; leaf blade 1.5–4 cm long, 0.3–2 cm broad ....2a. subsp. *vinciflorus*

1b. Calyx lobes 8–16 mm long, 3–6 mm broad; leaf blade 3–9.5 cm long, 0.7–3.3 cm broad..2b. subsp. *dianchuanicus*

2a. **Pseudocodon vinciflorus** subsp. **vinciflorus**

FIGURE IV 4:2A *Pseudocodon vinciflorus* (Kom.) D. Y. Hong subsp. *vinciflorus*.

Leaf blade 1.5–4 cm long, 0.3–2 cm broad; calyx lobes 3–7 mm long, 1.5–3 mm broad. Figures IV 4:2A & B.

Phenology: flowering from June to September; fruiting in September and October.

Habitats: thickets, forests, or grassy slopes; 2600–4600 m.

Distribution: confined to W and NW Sichuan and C and SE Tibet.

Additional specimens examined, CHINA, SICHUAN, s. loc., 1370 m, Jul. 1903, *E. H. Wilson 3986* (BM, K). **Barkam**: Dalangjiao, 2650 m, 13. Jul. 1957, *H. F. Zhou & T. Y. Chang 23003* (SZ). **Dawu**: Bonglong, 3500 m, 20. Jul.1979, *s. coll. 146* (SM). **Heishui**: Waboliangzi, 2500 m, thickets, 14. Jul. 1963, *X. Y. Tang & Z. L. Kuen 63011* (PE). **Jinchuan**: Anning, Muma Shan, 2800 m, grasses, 08. Jul. 1958, *X. Li 78087* (PE, SZ). **Jiulong**: Sanyanlong, 3500 m, 03. Sept. 1980, *Jiulong Exped. 299* (SM); loc. eodem, 3000 m, 22. Jul. 1979, *Jiulong Exped. 290* (SM); s. loc., 15. Aug. 1930, *s. coll. 2001* (KUN). **Kangding**: Sha Township, 3400 m, 08. Jul. 1974, *Sichuan Plant Exped. 4831* (CDBI, PE); Tongolo, 1893, *J. A. Soulie 2760* (P); Zierpo, 2800 m, 09. Aug. 1981, *Z. J. Zhao et al. 115211* (SZ); s. loc., 1893, *J. A. Soulié 202* (G, P). **Litang**: Lapo, 2500–3000 m, thickets, 10. Aug. 1964, *Y. T. Lei 009* (PE); loc. eodem, 3900 m, thickets, 04. Aug. 1973, *Sichuan Veget. Surv. 3692* (CDBI, PE). **Luding**: Erlang Shan, 2600 m, thickets, 14. Jun. 1959, *Sichuan Econ.Plant Surv. (Garzê) 0010* (PE); loc. eodem, 2600 m, 14. Jun. 1957, *Sichuan Agr. Colleage 1010* (CDBI); loc. eodem, Ganhaizi, 2780 m, grassy slopes, 18. Aug. 1982, *D. Y. Hu 46166* (CDBI); loc. eodem, Shixiazigou, 2400 m, 04. Sept. 1953, *W. K. Hu 35439* (SZ). **Maoxian**: Sanlong Township, 01. Jul. 1952, *Z. He & Z. L. Zhou 13063* (SZ). **Sêrtar**: 31°58′N, 100°56′E, 3110 m, 07. Aug. 2005, *D. E. Boufford et al. 34315* (PE). **Tianquan**: Erlang Shan, Ganhaizi, 2500 m, 01. Sept. 1953, *W. K. Hu 35419* (SZ). **Xinlong**: Longlaxi, 3400 m, thickets, 27. Jun. 1974, *Z. S. Qin 6347* (CDBI, PE). **Zamtang**: 31°54′52″N, 101°5′27″E, 2990–3050 m, heavily disturbed slopes, 01. Aug. 2007, *D. E. Boufford et al. 38874* (KUN). **TIBET, Bomi**: Yuren, 3300 m, waste fields, o2. Sept. 1973, *Qinghai-Xizang Exped. 73–1340* (KUN). **Dêngqên**: Damdoi, 2800 m, slopes, 15. Jul. 1961, *s. coll. 1665* (KUN). **Gongbo'gyamda**: W of Bahe Zeng, by highway 318, 3335 m, 14. Aug. 2000, *D. E. Boufford 30064* (A); Gongbo'gyamda to Nyingchi, 3300 m, thickets, 08. Aug. 2001, *H. N. Qin et al. 105* (PE); near the city, hillside, 3700 m, 15. Oct. 1989, *J. S. Yang et al. 89-074* (PE); loc. eodem, 3650 m, thickets, 27. Sept. 1990, *J. S. Yang 90–627* (KUN); near the city, S of Nyang Qu, 3465 m, thickets or pastures with scattered shrubs, 24. Aug. 2010, *D. Y. Hong & Q. Wang H10003* (PE); loc. eodem, N of Nyang Qu, 3440 m, thickets, 25. Aug. 2010, *D. Y. Hong & Q. Wang H10006* (PE); Pasum Lake, 3660 m, 15. Jul. 1947, *F. Ludlow et al. 14141* (BM, E); Pikang to Gongbo'gyamda, 4000 m, 16. Aug. 1974, *Qinghai-Xizang Exped. Vegetation Group 2738* (PE); Shoga Dzong, 3200 m, 16. Sept. 1947, *F. Ludlow et al. 15732* (BM); by 318 Road, forest edges, 3359 m, 12, Sept. 2012, *FLPH Tibet Exped. 12–0511* (PE). **Gyaca**: Lhasoi, 3400 m, thickets by road, 31. Aug. 2011, *D. Y. Hong et al. H11012* (PE); 10 km W of the city, 3500 m, thickets, 14. Sep. 1990, *J. S. Yang 90–352* (KUN). **Lhasa**: mountain near Kartosi, 4600 m, 19. Aug. 1965, *Y. T. Chang & K. Y. Lang 1995* (KUN, PE);

FIGURE IV 4:2B *Pseudocodon vinciflorus* (Kom.) D. Y. Hong subsp. *vinciflorus*. **a**, from Tibet: Mainling, 12. Aug. 2012, by Xin-Tang MA & De-Yuan HONG, **b** & **d**, from Tibet: Nyingchi, 24. Jul. 2010, by Hong-Ya GU; **c**, **e** & **f**, from Tibet: Gyaca, 31. Aug. 2011, by Xin-Tang MA & De-Yuan HONG; **g**, **h** & **i**, from Tibet: Gongbo'gyamda, 13. Aug. 2012, by De-Yuan HONG; **j**, from Tibet: Gongbo'gyamda, 21. Aug. 2012, by Xin-Tang MA & De-Yuan HONG; **k**, from the same locality as j, 25. Aug. 2010, by Qiang WANG & De-Yuan HONG. **i**, showing variation of leaves within population; **c–f**, showing colour patten at the base of corolla; **g**, showing lanate dilated base of filaments and half inferior ovary

hills N of Lhasa, 4270 m, 07. Aug. 1943, *F. Ludlow & G. Sherriff 9855* (BM); loc. eodem, 3960 m, 16. Aug. 1943, *F. Ludlow & Sherriff 9872* (BM, E); ca. 40 km E of Lhasa, 4520 m, 05. Aug. 1989, *B. Dickoré 3717* (K); Reting, 90 km N of Lhasa, 3960 m 24. Jul. 1942, *F. Ludlow & G. Sherriff 8872* (BM, E); loc. eodem, 06. Aug. 1944, *F. Ludlow & G. Sherriff 11107* (BM, E); near Reting N of Lhasa, 3660 m, 03. Aug. 1942, *F. Ludlow & G. Sherriff 8996* (BM); Tsangbo Valley near Lhasa, 3510 m, 13. Sept. 1944, *F. Ludlow & G. Sherriff 11155* (BM); s. loc., May – Jul. 1947, *J. Guthrie* s. n. (LD); s. loc., 3810 m, Jul. 1939, *H. E. Richardson 256* (BM) & *305* (BM); s. loc., 3660 m, 25. Aug. 1904, *L. A. Waddel* s. n. (E). **Lhünzhub:** Songpan Township to Poindo, 4175 m, 11. Sept. 2012, *FLPH Tibet Exped. 12–0422* (PE). **Mainling:** Gyala, 29°40′N, 95°E, 2840 m, 09. Jul. 1938, *F. Ludlow 7545* (BM); near Huilong Township, 3100 m, thickets by road, 31. Aug. 2011, *D. Y. Hong et al. H11013* (PE); near Jaggo, 3500 m, 26. Jul. 1975, *Qinghai-Xizang Suppl. Exped. 75843* (KUN, PE); loc. eodem, 3000 m, thickets, 06. Aug. 1982, *C. C. Ni et al. 2771* (PE); Lilung, Tsangbo Valley, 3050 m, 29°07′N, 93°54′E, 01. Oct. 1938, *F. Ludlow 5333A* (BM); Lilung Township, herbs at edges of forests, 3000–3300 m, *FLPH Tibet Exped. 12–2050* (PE) *& 12–2088* (PE); Pome, Paka, 2900 m, 06. Jul. 1947, *F. Ludlow et al. 15333* (BM); Shoka, 29°15′N, 94°15′E, 12. Jul. 1938, *F. Ludlow* et al. *5817* (BM, E); between Tripe and Gyala, 29°44′N, 94°57′E, 2890 m, 21. Jul. 1938, *F. Ludlow et al. 5333* (BM, E); Tsangbo Gorge, 2740 m, 19. Jul. 1924, *F. Kingdon-Ward 5958* (BM, E); Tsangbo Valley, 3050–3350 m, 19. Jul. 1935, *F. Kingdon-Ward 12018* (BM); loc. eodem, 4570 m, 26. Sept. 1904, *L. A. Waddel s. n.* (K); loc. eodem, Jul. 1904, *H. J. Walton* s. n. (K). **Maizhokunggar**: between the city and Mila Pass, 4150 m, thickets, 24. Aug. 2010. *D. Y. Hong & Q. Wang H10001* (PE). **Nangxian**: Jyêmdong to Laiyi Village, thickets, *FLPH Tibet Exped. 12–1123* (PE); Kyim Dzong, 3510 m, 29°N, 93°50′E, 22. Jul. 1936, *F. Ludlow & G. Sherriff 1938* (BM); Kyimdong Chu, 3440 m, 05. Aug. 1997, *S. Bowes Lyon 11156* (E). **Nêdong**: Gyêrba, 3805–3864 m, *Canagara* thickets, 09. Sept. 2012, *FLPH Tibet Exped. 12–0756* (PE). **Nyingchi**: ca.20 km W of Bayi Town, 22. Jul. 1990, *J. S. Yang 90–321* (KUN); Naixi, 3040 m, thickets, 28. Jul. 1965, *Y. T. Chang & K. Y. Lang 1076* (PE); near Nyingchi town, 3100 m, thickets 20. Sept. 1980, *C. C. Ni et al. 1744* (PE); Serji La, W slope, 3074 m, 06. Aug. 2008, *Liu JQ–08XZ–225* (KUN). **Sogxian**: Rongbo to Rabo, 3800–3900 m, forests, 20. Aug. 1976, *D. D. Tao 10995* (KUN). **Sangang**: s. loc., 4150 m, 11. Aug. 1936, *F. S. Chapman 1164* (PE).

2b. Pseudocodon vinciflorus subsp. **dianchuanicus** D. Y. Hong & Q. Wang, Phytotaxa 2014 (1): 62. 2015. Type: China, Yunnan, Dêqên, Tswalungla, 2600 m, thickets, Sept. 1937, *T. T. Yü 10286* (holotype: PE; isotype: KUN).

Codonopsis forrestii Diels var. *heterophylla* C. Y. Wu in Wu & Li, Yunnan Trop.
Subtrop. Fl. Res. Rep. 1: 80. pl. 30. fig. 1. 1965. Type: China, Yunnan, Dêqên, Hongpo, 3240 m, 03. Sept. 1937, *T. T. Yü 10173* (holotype: KUN; isotypes: PE).
Leaf blade 3–9.5 cm long, 0.7–3.3 cm broad. calyx lobes 8–16 mm long, 3–6 mm broad.

Phenology: flowering from late July to September; fruiting from September to October.

Habitats: sunny grassy slopes or forests, but mostly climbing up thickets; 2700–3900 m.

Distribution: endemic to SW China: SW Sichuan, SE Tibet, and NW Yunnan.

Notes: This subspecies is isolated from the typical one in geography. It is also rather distinct from the typical one in morphology (Wang & Hong, 2015, fig. 2); it may be proved as a distinct species.

Additional specimens examined, CHINA, SICHUAN, Daocheng: Gongga-ling, 3100 m, upon thickets, 23. Aug. 1937, *T. T. Yü 12881* (KUN, PE); Riwa, 3300 m, rocky slopes, 25. Aug. 1981, *Qinghai-Xizang Exped. 4291* (CDBI, PE); loc. eodem, 3200 m, 02. Sept. 1971, *Z. G. Liu 510* (CDBI); loc. eodem, Lamuge, edges of thickets, 3900 m, 30. Jul. 1973, *Sichuan Veget. Surv. 2417* (CDBI, PE); s. loc., 3000 m, dry grassland, 17. Jul. 1971, *Z. G. Liu 253* (CDBI). **Muli**: Bo'ao, Dingdonghai, 2800 m, thickets, 07. Sept. 1983, *Qinghai-Xizang Exped. 13657* (KUN); Lea-Lang-Keng-din, 2800 m, wood margins, 01. Aug. 1937. *T. T. Yü 7524* (KUN, PE); Mabam Pass to Donglang, *Picea* forests, 3100–3200 m, 08. Sept. 1983, *Qinghai-Xizang Exped. 14525* (KUN, PE). **Xiangcheng**: Qianjin, 3500 m, 16. Aug. 1979, *s. coll. 733* (SM); Rawu, 3500–3800 m, forests,16. Aug. 1981, *Qinghai-Xizang Exped. 4127* (CDBI, KUN, PE); loc. eodem, 2700 m, thickets, 31. Jul. 1981, *Qinghai-Xizang Exped. 3071* (CDBI); Shangyou Township, 3600 m, thickets, 22. Aug. 1973, *Sichuan Veget. Surv. 3231* (PE); No. 4 Distr., 3800 m, sunny grassy slopes, 23. Aug. 1972, *Sichuan Plant Exped. (Xiangcheng) 1217* (PE). **TIBET, Markam**: Chubarong, 29°44′N, 98°51′E, 3000 m, 19. Aug. 2010, *Kham Exped. 10–2401* (PE). **YUNNAN, Dêqên**: Baima Shan, 3000 m, forests, Sept. 1935, *C. W. Wang 69784* (PE); Huangpu, 3240 m, open slopes, 03. Sept. 1937, *T. T. Yü 10172* (KUN) ; s. loc., 2700 m, *C. W. Wang 70018* (KUN, PE).

3. Pseudocodon grey-wilsonii (J. M. H. Shaw) D. Y. Hong in Wang et al., J. Syst. Evol. 52(5): 548. 2014. ≡ *Codonopsis nepalensis* Grey-Wilson, Plantsman, 12(2): 99. 1990 & New Plantsman, 2(4): 219. 1995, non Hara (1978). ≡*Codonopsis grey-wilsonii* J. M. H. Shaw, New Plantsman, 3(2): 93. 1996; Lammers & Klein, Bot. Stud. (Taipei) 51: 559. 2010. ≡*Codonopsis*

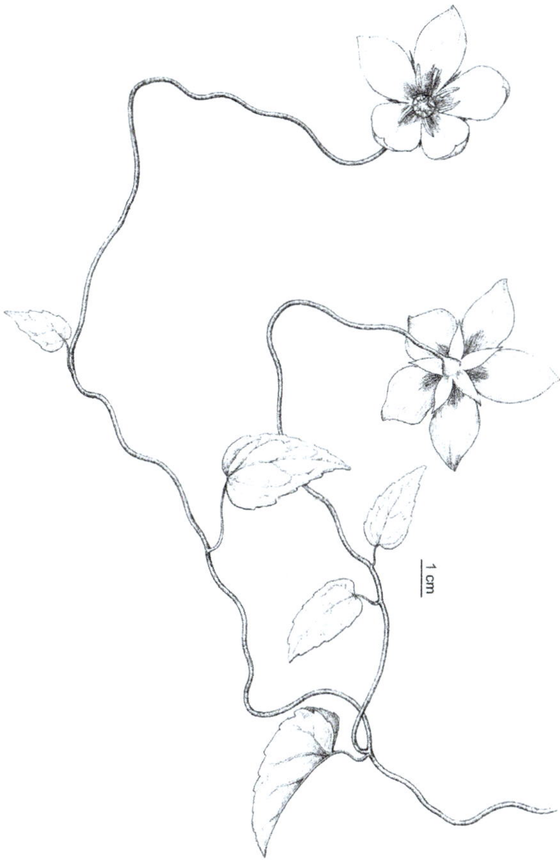

FIGURE IV 4:3A *Pseudocodon grey-wilsonii* (J. M. H. Shaw) D. Y. Hong: upper part of a twining stem.

convolvulacea subsp. *grey-wilsonii* (J. M. H. Shaw) D. Y. Hong, Novon 20(4): 421. 2010; Hong, Lammers & Klein in Wu, Raven & Hong, Fl. China 19: 525. 2011. Type: Nepal, Kali Gandaki Valley, W of Marpha, 3500 m, 06. Sept. 1973, *Grey-Wilson & Phillips 777* (holotype: K; isotype: K).

Herbs twining, glabrous throughout except filaments and style. Root tubers globose or ovoid. Stems over 1 m long. Leaves all alternate; petiole 1–3 cm long; leaf blade ovate to ovate-lanceolate, 2.5–4.5 cm long, 1.5–2.5 cm broad, base cordate to rounded, margin dentate, apex obtuse to acute; leaves at sterile branches much smaller. Flowers solitary and terminal on main stem and fertile branches; pedicel up to over 10 cm long. Calyx lobes lanceolate or ovate-lanceolate, 1–1.8 cm long, 0.4–0.7 cm broad. Corolla blue-purple, very rarely white, often with a wide dark red-purple ring at base, 3–6.5 cm across; corolla lobes ovate, apex acute-obtuse. Filaments ca. 1.3 cm long; anthers as long as filaments. Ovary obconical or obovoid; style nearly as long as stamens or slightly longer; stigma lobes ovate. Fruit obconical or obovoid. Figures IV 4:3A & B.

Phenology: flowering and fruiting from August to October.

Habitats: thickets or forests; 2100–4300 m.

Distribution: Bhutan, S Tibet of China, and Nepal. (Map IV 4:3).

Notes: This species was treated as a sister subspecies of *vinciflora* in *Codonopsis* (*Pseudocodon*) *convolvulacea* in "*Flora of China*" (19: 525. 2011). However, the species is distinctly different from *P.vinciflorus* in having calyx lobes much larger and lanceolate or ovate-lanceolate and petiole much longer. Furthermore, our molecular data indicate that three accessions from S Tibet form a clade sister to a clade of the accessions from N of the Himalaya. Therefore, *P. grey-wilsonii* is a distinct species, which is isolated from *P. vinciflorus* by the Himalaya.

Additional specimens examined, BHUTAN, Bumthang, 27°38′N, 90°53′E, 2900 m, 05. Sept. 1949, *F. Ludlow et al. 19647* (E) & *19674* (BM, E). **Ha** Dzong: Hatö, 27°25′N, 89°28′E, 2870 m, 17. Aug. 1949, *F. Ludlow et al. 19620* (BM, E); loc. eodem, 2740 m, 25. Aug. 1950, *F. Ludlow, & G. Sherriff 21486* (BM, E). **Hatey**: s. loc., 2900 m, 13. Aug. 1938, *B. J. Gould 1541* (K). **Paro**: Bela La, 2740 m, 14. Aug. 1937, *F.Ludlow & G. Sherriff 3568* (BM). **Thimphu**: Duke La 2740 m, 17. Aug. 1914, *R. E. Cooper & A. K. Bulley 2665* (E); East Himalaya, Galing, 3660 m, 12. Sept. 1912, *Rohmoo Lepcha 110* (E).

CHINA, TIBET, Cona: Lhakang, 28°10′N, 91°10′E, 3810–4110 m, 01. Sept. 1933, *F. Ludlow & G. Sherriff 506* (BM). **Gyirong**: near Gyirong Distr., 2150 m, thickets, 12. Sept. 1981, *C. C. Ni et al. 2180* (PE); Gyirong Township, 3100–3130 m, edges of forests, 06. Aug. 2010, *PE-Xizang Exped. 0185* (PE); Sale Township, Labi-Sale, on rocks, 3000 m, 01. Sept. 1990, *B. S. Li et al.* 13337 (PE); Zaxiang, *Pinus griffithii* thickets, 2900 m, 29. Aug. 1990, *B. S. Li et al.*13207 (PE). **Lhünzê** (Gyandro): Chayul-Charms Rv., 28°10′N, 92°50′E, 3960–4270 m, 26. Jul. 1936, *F. Ludlow & G. Sherriff 2431* (BM, E); loc. eodem, 28°25′N, 93°E, 3660 m, 01. Oct. 1936, *F. Ludlow & G. Sherriff 2714* (BM); Chayul Chu, 28°20′N, 93°E, 3350 m, *F. Ludlow & G. Sherriff 2423* (BM). **Nyalam**: Zham to Lixin, 2800 m, broad-leaved forests, 07. Oct. 1990, *B. S. Li & H. Li 14007* (PE); Zham, 2100 m, thickets by road, 27. Aug. 2011, *D. Y. Hong et al. H11005* (PE); Zham, Lixin, 2492 m, thickets at edges of forests, 19. Aug. 2010, *PE-Xizang Exped. 1500* (PE); loc. eodem, 2492 m, forests, 14. Aug. 2010, *PE-Xizang Exped. 0979* (PE).

NEPAL, Bagmati: Dhading Distr., near Nyilo, 3350 m, 21. Aug. 1973, *D. Lichter 120* (KATH); Rasuwa Distr., Kyangchin Gomba, 28°13′N, 85°34′E, 3600 m, 8. Sept. 1971, *J. F. Dobremez 1024* (G); Langtang: 28°12′N, 85°31′E, 3660 m, 16. Sept. 1970, *J. D. A. Stainton 6675* (BM, E); Langtang, 3660–3910 m, 01. Sept. 1991, *S. Noshiro et al. 9154547* (E); Langtang, below Kyangjin Ghyang, 3650 m,

FIGURE IV 4:3B *Pseudocodon grey-wilsonii* (J. M. H. Shaw) D. Y. Hong. **a**, **c**, **e** & **f**, from Tibet: Nyalam, 19. Aug. 2010, by Xin-Tang MA & De-Yuan HONG; **b**, from Tibet: Nyalam, 19. Aug. 2010, by De-Yuan HONG; **d**, from Tibet: Gyirong, 06. Aug. 2010, by Xin-Tang MA.

Aug. 1949, *O. Polunin 1708* (BM). **Dhawalagiri**: Baglung Distr., above Dogadi Khola , 4110 m, 01. Oct. 1954, *Stainton, Sykes* & *Williams 4662* (BM); Mustang Distr., Tukucha, Samargaon, 3960 m, 15. Aug. 1954, *Stainton, Sykes & Williams 7245* (BM); loc. eodem, 4270 m, 16. Aug. 1954, *Stainton, Sykes & Williams 7265* (BM); Mustang Distr., 3600 m, 20. Aug. 1999, *M. Mikag et al. 9967144* (KATH); Mustang Distr., N of Muktinath, 4110 m, 07. Sept. 1977,

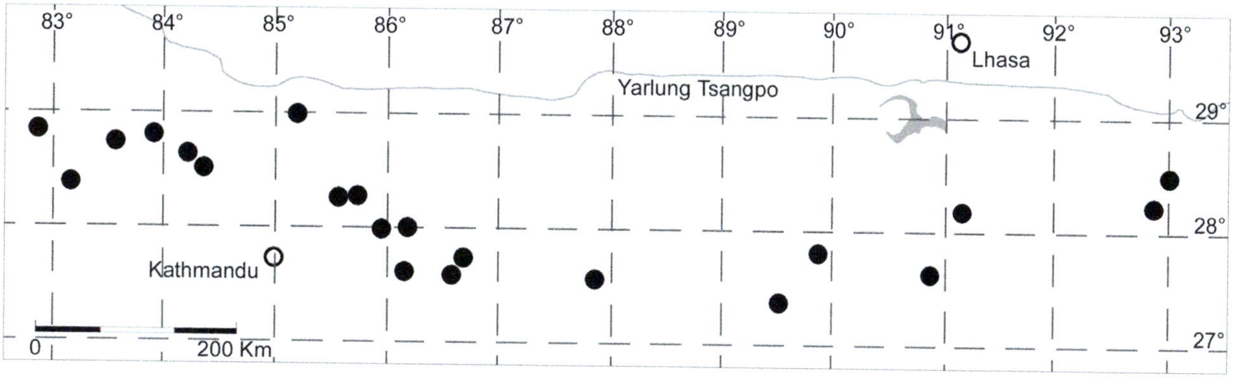

MAP IV 4:3 Distribution of *Pseudocodon grey-wilsonii* (J. M. H. Shaw) D. Y. Hong.

G. Miebe 617 (BM); Myagdi Distr., SW of Gurjakhani: 3200 m, 18. Oct. 1954, *Stainton, Sykes & Williams 9079* (BM); Tukucha, Kali Gandaki, 3200 m, 26. Aug. 1954, *Stainton, Sykes & Williams 7455* (BM, P); loc. eodem, 3510 m, 12. Sept. 1954, *Stainton, Sykes & Williams 7791* (BM); loc. eodem, 3500 m, 02. Sept. 1967, *H. J. Lange 61* (K). **Gandaki**: Manang Distr., Chame, 2700 m, 15. Aug. 1973, *D. P. Joshi & T. K. Bhattacherya 74–2483* (KATH); loc. eodem, 2590 m, 14. Aug. 1950, *D. G. Lowndes 1401* (BM, E, G); Manang Distr., Marsiandi Valley, 2130 m, 27. Aug. 1984, *J. D. A. Stainton 8889* (E). Bumra, Riyan Lekh, 3050 m, 28. Aug. 1952, *Polunin, Sykes & Williams 3057* (BM, E). Chantang (2250 m) – Redak (2400 m), 08. Aug. 1977, *H. Ohashi et al. 773585* (TI); loc. eodem, 08. Aug. 1977, *H. Ohashi et al. 772138* (TI). Redak (2400 m) – Chhurchathanga (3600 m), 09. Aug. 1977, H. *Ohashi et al. 773595* (TI). Chantang-Honkon (2400 m), 07. Aug. 1977, *H. Ohashi et al. 774925* (TI). Churchi Lekh, between Lumsa and Murma, 3050 m, 11. Aug. 1952, *Polunin, Sykes & Williams 5129* (BM, E). Rohagaon, Suli Gad, 2900 m, 14. Sept. 1952, *Polunin, Sykes & Williams 3374* (BM, E). **Janakpur**: Dolakha Distr., Prés de Thari, 27°49′N, 86°12′E, 2100 m, 13. Sept. 1954, *A. Zimmerman 1230* (G). **Karnali**: Dolpa Distr., Tarakot, 28°55′N, 82°50′E, 2700 m, 19. Aug. 1973, *Einarsson, Skärby & Wetterhall 3674* (BM, LD); Dolpa, Kaigaun, 2800 m, 25. Aug. 1973, *Einarsson, Skärby & Wetterhall 3756* (BM, LD). **Mechi**: Tamur Valley, Thapabu Khola, 2290 m, 02. Aug. 1956, *J. D. A. Stainton 1213* (BM). **Sagarmatha**: Solukhumbu Distr., 27°35′N, 86°35′E, 2660 m, 31. Aug. 1997, *M. Wakabayashi et al. 9715296* (E); loc. eodem, 27°32′N, 86°31′E, 2600–3500 m, 05. Sept. 1985, *H. Ohba et al. 8581374* (BM, E); loc. eodem, Jorsale, 27°46′47″N, 86°43′20″E, 2800 m, 12. Sept. 2005, *DNEP 3 AY19* (E). Naunigad-Chouta, 2800 m, bushes by fields, 15. Sept. 1995, *K. Shinozaki* s. n. (KATH).

4. **Pseudocodon hirsutus** (Hand.-Mazz.) D. Y. Hong in Wang et al., J. Syst. Evol. 52(5): 548. 2014; Wang & Hong, Phytotaxa 204 (1): 62. 2015. ≡*Codonopsis limprichtii* var. *hirsuta* Hand.- Mazz., Akad. Wiss. Wien Math.-Naturwiss. Kl. Anz. 61. 169. 1924. ≡*Codonopsis convolvulacea* var. *hirsuta* (Hand.-Mazz.) Nannf. in Handel- Mazzetti, Symb. Sin. 7(4): 1077. 1936. ≡*Codonopsis hirsuta* (Hand.-Mazz.) D. Y. Hong & L. M. Ma in Li & Zhu, Fl. Sichuan. 10: 546. 1992; Lammers & Klein, Bot. Stud. (Taipei) 51: 555. 2010; Hong, Lammers & Klein in Wu, Raven & Hong, Fl. China 19: 524. 2011. Type: China, "Prov. Setschwan austro-occ.: Ad terram abruptam calceam reg. temperatae supra vicum Lodjiahosan pr. oppidum Yenyüen"(Sichuan, Yanyuan), 3000 m, 30. Sept. 1914, *Handel-Mazzetti 5423* (lectotype designated by Hong in Wang & Hong, 2015: WU; isolectotype: W).

Roots globose or ovoid-globose, 1–2 cm in diameter. Stems up to 1 m long, decumbent below, twining only at upper part, hirsute at leafing section, very rarely glabrous. Cauline leaves nearly opposite or alternate, aggregated at lower part, sessile or with extremely short petiole up to 8 mm long; leaf blade ovate or narrowly oblong, 1.5–8 cm long, 0.3–3.2 cm broad, adaxially glabrous, rarely sparsely hirsute, abaxially densely hirsute, very rarely glabrous, base cuneate or rounded, margin entire or sinuous-crenulate, often revolute, apex acute or obtuse. Flowers solitary, terminal. Calyx tube adnate to hypanthium up to near top, obconical, 4–7 mm in wide, glabrous; calyx lobes subulate or linear-lanceolate, 6–18 mm long, 2–6 mm broad, apex acuminate or acute. Corolla pale blue or blue-purple; corolla lobes elliptic, 1–3.5 cm long, 0.5–1.6 cm broad. Capsule short-conical at superior part, obconical at inferior part. Seeds numerous, brown-yellow, smooth. Figures IV 4:4A & B.

Chromosomes: 2n=14 (Wang et al., 2014a).

Phenology: flowering in August and September; fruiting in October.

Habitats: open thickets, waste fields, sides of fields, or sunny grassy slopes in calcareous soils; 2400–3400 m.

Distribution: **endemic** to China and confined to SW Sichuan and Yunnan. Map IV 4:4.

Additional specimens examined, CHINA, SICHUAN, Jiulong: Tanggo, 3200 m, oak thickets, 08. Sept. 1979, *Q. Q.*

FIGURE IV 4:4A *Pseudocodon hirsutus* (Hand. - Mazz.) D. Y. Hong, showing hairy stem and leaves, of which lower ones are opposite.

Wang 21134 (CDBI). **Muli**: Liziping, Baicaoping Village, 2800 m, thickets by road, waste fields, 21. Sept.2011, *D. Y. Hong et al. H11034* (PE); No. 1 Distr., Ninglang to Eya, 2700–2900 m, pine forests, 27. Sept. 1983, *Qinghai-Xizang Exped. 14838* (PE); s. loc., 3000 m, 11. Oct. 1993, *Q. E. Yang 9306* (PE). **Yanbian**: Yankou Township, 2800 m, 28 Jul. 2001, *C. M. Tan P315 A* (SZ); loc. eodem, Dawan, 2657 m, pine-oak community, 23. Sept. 2011, *D. Y. Hong et al. H11043* (PE). **Yanyuan**: Yangjuan, 20. Aug. 2002, *2002 UW Biodiversity 2002–137* (SZ); Zhongxing Village, 2951 m, waste fields, waste clay walls, 20 Sept. 2011, *D. Y. Hong et al. H11023* (PE). **YUNNAN**, s. loc., 1920, *J. Cavalerie s. n.* (E); s. loc., 1920, *Siméon Tén 131* (E). **Lijiang** (Yulong): Ganhaizi, 2900 m, pine-oak forests, 18. Sept. 2010, *D. Y. Hong et al. H10027* (PE); loc. eodem, 2900 m, open pine woods, 18. Sept. 2010, *D. Y. Hong et al. H10029* (PE); Guluko [Xuesongcun Village], 3000 m, open thickets, 11. Sept. 1937, *T. T. Yü 15572* (KUN, PE); Wo-to-di, Sept. 1987, *B & L 12120* (E); Xuesongcun to Qiaomaidi 2800–3200 m, thickets or edges of thickets, limestones, 19. Sept. 2010, *D. Y. Hong et al. H10031* (PE); below Xuesongcun, 2550 m, edges of thickets & waste fields, 19. Sept. 2010, *D. Y. Hong et al. H10036* (PE). **Longling**: Huoshan, 2400 m, thickets by forests, 18. Aug. 1941, *C. W. Wang 90132* (KUN). **Shangri-la** (Zhongdian): Hong Shan, 28°1′18″N, 99°46′48″E, 3180 m, 11. Sept. 1995, *Roy. Bot. Gard. Edinburgh Exped. To Dêqên Pref. 131* (E). **Yanshan**: Ta-yai-kou, 1000 m, 26. Sept. 1940, *C. W. Wang 84011* (KUN, PE).

5. **Pseudocodon graminifolius** (H. Lév.) D. Y. Hong in Wang et al., J. Syst. Evol. 52(5): 548. 2014; Wang & Hong, Phytotaxa 204(1): 62. 2015. ≡*Codonopsis graminifolia* H. Lév., Cat. Pl. Yunnan 24. 1915; Hong & Ma in Wang & Wu, Vasc. Pl. Hengduan Mount. 2: 1961. 1994; Hong, Lammers & Klein in Wu, Raven & Hong, Fl. China 19: 526. 2011. Type : China, Yunnan, s. loc., 2600 m, Jul. 1911, *E. E. Maire s. n.* [lectotype designated by Hong in Wang & Hong, 2015: E (E00265634); isolectotype : E (E00265635), P].

Codonopsis limprichtii Lingelsh. & Borza var. *pinifolia* Hand.-Mazz., Akad. Wiss. Wien Math.- Naturwiss. Kl. Anz. 61: 170. 1924. ≡*Codonopsis convolvulacea* Kurz var. *pinifolia* (Hand.-Mazz.) Nannf. in Handel-Mazzetti, Symb. Sin. 7(4): 1077. 1936. ≡*Codonopsis pinifolia* (Hand.-Mazz.) Grey-Wilson, Plantsman 12(2): 89 & 76. fig. A. 1990, comb. invalid. Type: China, Yunnan, Yangtze-Mekong Divide, Weixi, 2800–3250 m, 22. Sept. 1916, *Handel-Mazzetti 10053* (lectotype designated by Hong in Wang & Hong, 2015: WU; isolectotypes: E, W); Yunnan, Lidjiang, Yulung-schan (Lijiang, Yulong Shan), 3350 m, 31. Jul. 1914, *Handel-Mazzetti 4370* (paratypes: W, WU); loc. eodem, *Handel- Mazzetti 3713* (paratypes: W ,WU).

Codonopsis limprichtii auct. non Lingelsh. & Borza: Lammers & Klein, Bot. Stud. (Taipei) 51: 555. 2010, p. p., quoad syn.

Plants glabrous throughout except filaments. Roots ovoid or oblong, 0.8–1.5 cm in diameter. Stems up to 1 m long, decumbent below, twining only at upper part, sometimes nearly ascending, simple or branched. Cauline leaves alternate, usually aggregated at lower part, sessile, pinoid, linear, rarely linear-lanceolate, 2–10.5 cm long, 0.1–0.5(–0.9) cm broad, base cuneate, margin entire, often revolute, apex acuminate. Flowers solitary, terminal mostly only on main stem, sometimes also on branches. Calyx tube adnate to hypanthium up to top, obconical; calyx lobes long-triangular, 4–10 mm long, 2–3 mm broad. Corolla blue or purple, often with a red ring at base; corolla lobes elliptic, 0.8–2.6 cm long, 0.4–1.2 cm broad. Anthers ca. 3 mm long. Capsule obconical, ca. 6 mm in diameter at inferior part, short-conical at superior part. Seeds brown-yellow. Figures IV 4:5A & B.

Phenology: flowering from July to September; fruiting in September and October.

FIGURE IV 4:4B *Pseudocodon hirsutus* (Hand. -Mazz.) D. Y. Hong. **a**, from Sichuan: Yanyuan, 20 Sept. 2011, by De-Yuan HONG; **b**, **c**, **e** & **f**, from Yunnan: Lijiang, 18. Sept. 2010, by De-Yuan HONG; **d**, from Sichuan: Muli, 21. Sept. 2011, by De-Yuan HONG.

Habitats: open grassy slopes, thickets, or forests; 1500–3300 m.

Distribution: **endemic** to China: W Guizhou, SW Sichuan, and N Yunnan (Map IV 4:5).

Notes: I have indicated under *Pseudocodon convolvulaceus* subsp. *forrestii* that *Codonopsis limprichtii* sensu Lammers & Klein is not Lingelsheim & Borza's *C. limprichtii*, but is *Codonopsis graminifolius* H. Lév.

Additional specimens examined, **CHINA**, **GUIZHOU**, **Weining**: Gali, Mab Dashan, 2400 m, 12.Jul. 1959, *Bijie Exped. 179* (KUN, PE), Shitou Distr., 1900 m, 11. Jul. 1959, *Bijie Exped. 262* (PE). **SICHUAN**, **Butou**: Jiaojihe Distr., 1500 m, 21. Aug. 1959, *Sichuan Econ. Plant Exped. (Liang) 5684* (PE); Jiaojihe Distr., Lianhe Township, 1600 m, s. d., *Sichuan Econ. Plant Exped. (Liang)* 5725 (PE); s. loc., thickets, 17. Aug. 1979, *s. coll. 750* (SM). **Dechang**: s. loc., 1800 m, 01. Sept. 1978, *Hu 356* (SM). **Huidong**: s. loc. 23. Jul. 1978, *Huidong Exped. 432* (SM); s. loc., 1778 m, 11. Sept. 1978, *X. C. Yi & Y. G. Li s. n.* (SM). **Huili**: s. loc., 2600 m, pine forests, 10. Sept. 1932, *T. T. Yü 1485* (PE). **Jinyang**: Tiantai, 1900 m, 10. Sept. 1979, *s. coll. 236* (SM). **Litang**: s. loc., 3600 m, grassy slopes, 30. Aug. 1979, *Litang General Surv. 004* (SM). **Mianning**: Mt Maoniushan, 3000 m, forests, 21. Sept. 1960, *J. S. Ying 4332* (PE); s. loc., 3500 m, 30. Aug. 1978, *Mianning Exped. 541* (SM). **Miyi**: s. loc., 2300 m, 04. Aug. 1978, *s. coll. 239* (SM). **Muli**: between Lijiang, Yongning, Yongpei to Muli, Aug. 1922, *J. F. Rock 5307* (E); Liziping,

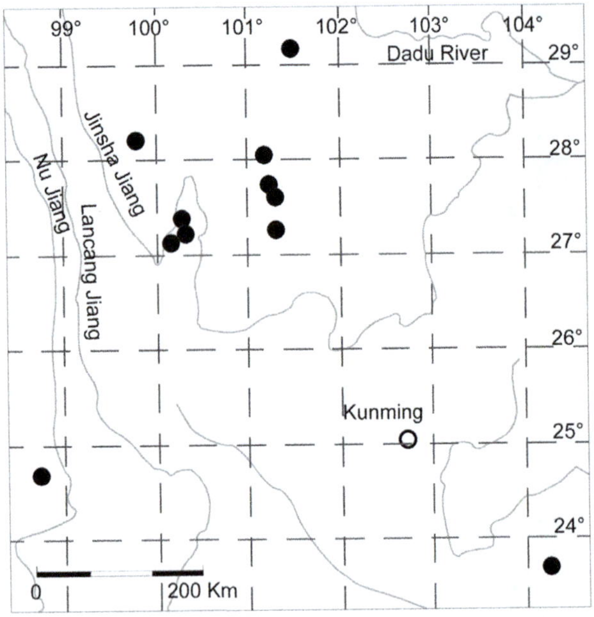

MAP IV 4:4 Distribution of *Pseudocodon hirsutus* (Hand.-Mazz.) D. Y. Hong.

FIGURE IV 4:5A *Pseudocodon graminifolius* (H. Lév.) D. Y. Hong, showing globose root and piniform leaves which are aggregated at lower part of stem.

Baicaoping, 2600 m, pine-oak community,21. Sept. 2011, *D. Y. Hong et al. H11024* (PE); 10 km S of Muli, 3050 m, 02. Aug. 1915, *F. Kingdon-Ward 4798* (E); 5 km, S of Muli, 3050 m, 21. Aug. 1927, *F. Kingdon-Ward 4864* (E); Ninglang Shan near Warzhong, 3000 m, *Quercus* forests, 22. Sept. 1983, *Qinghai-Xizang Exped. 14183* (KUN, PE); No.1 Distr., Meteorologic Station, 2700 m, thickets, 16. Aug. 1978, *C. S. Cao & G. Hu 8766* (CDBI). **Ningnan**: s. loc., 2000 m, 31. Aug. 1978, *Ningnan Exped. 639* (SM). **Xichang** (His-Chang): Lushan, 25. Jul. 1978, *Y. W. Zhu 362* (SM); s. loc., 2600 m, 09. Aug. 1932, *T. T. Yü 1263* (PE). **Xide**: Hongmo, 2000 m, 30. Aug. 1979, *s. coll. 827* (SM); s. loc., 2500 m, 17. Sept. 1979, *s. coll 910* (SM). **Yanbian**: Dapingzi, 2800 m, 24. Aug. 1978, *Yanbian Exped. 329* (SM); Yankou Township, Dawan, 2657 m, pine-oak community, 23. Sept. 2011, *D. Y. Hong et al. H11042* (PE). **Yanyuan**: Dalinsi 2600–2700 m, edges of pine forests, 25. Sept. 1965, *F. G. Zhu 006* (PE). **Zhaojue**: s. loc., 23. Aug. 1979, *s. coll. 927* (SM). **YUNNAN, Binchuan**: Jizu Shan, 23. Oct. 1946, *T. N. Liou 22183* (PE); loc. eodem, 3000 m, sparse woods, 16. Aug. 1982, *Y. L. Ma 218* (PE); Xiayang to Huxi, 14. Oct. 1946, *T. N. Liou 21512* (PE). **Dali**: Cangshan, 25°40′N, 2740–3050 m, Sept. 1905, *G. Forrest 886* (E); loc. eodem, 25°40′N, 2740–3050 m, Sept. 1913, *G. Forrest 11516* (BM, E, PE); Cangshan, Qingbi-si, 2400 m, Aug. 09. 1982, *H. Koyama et al. 883* (PE); Cangshan, Zhonghefeng, 2100 m, thickets, 11. Sept. 1983, *Sino-Amer. Exped. to Yunnan 0065* (KUN); Cangshan, between Sanyinfeng and Beifeng, 08. Sept. 1982, *H. Koyama et al. 1185* (KUN); Dali Range, 25°40′N, 2440–3050 m, Sept–Oct. 1906, *G. Forrest 3851* (E); Lanfeng Shan, half way to top, Jul, 1944, *H. C. Wang 4449* (PE); below Zhonghe-si, 2300 m, 30. Aug. 1989, *T. J. Zhang 027* (PE); on road to TV station, 2500 m, 29. Aug. 1984, *H. Hara & K. Y. Guan 392* (KUN); s. loc., 2800 m, Aug. 1914, *C. Schneider 2872* (G, PE); s. loc., *H. D. McLaren 246* (E, K). **Dongchuan**: Yingmin, 2800 m, meadows, 22. Jul. 1984, *S. B. Lan 019* (PE). **Heqing**: Baiyan, Shaxi, 2600 m, pine forests, 08. Sept. 1940, *R. C. Ching 24456* (PE); Huangping, 1700 m, thickets, 17. Sept. 1990, *X. F. Gao 518* (KUN); loe. eodem, 2469 m, oak and pine forests, 17. Aug. 2010, *E. D. Liu et al. 2740* (KUN); Yingyicun, ca. 2300 m, *Pinus yunnanensis* forests, Sept. 2010, *H. C. Wang H10037* (PE). **Jianchuan**: Shizhong Shan, forests, 27. Sept. 1975, *B. Y. Qiu 61092* (CDBI, PE); Yuhua Township, 2680 m, sparse pine woods, 16. Sept. 2010, *D. Y. Hong et al. H10013* (PE). **Kunming**: Heilongtan, Dongshanzuo, 1900 m, 19. Aug. 1989, *T. J. Zhang 07* (PE); Motianling, 2400 m, 12. Sept. 1957, *B. Y. Qiu 55067* (KUN, PE). **Lanping**: Biluo Shan, 3050 m, pine forests, 17. Sept. 1964, *S. K. Wu 8932* (KUN). **Lijiang** (Yulong): Ganhaizi, meadows, 3070 m, 02. Aug. 1960, *S. Jiang* 6338 (PE); loc. eodem, 2900 m, *Pinus yunnanensis* forests, limestones, 18. Sept. 2010, *D. Y. Hong et al. H10028* (PE); by Jinsha Jiang, thickets, 30. Sept. 1992, *J. R. Chen 92424* (PE); Licking Valley, 27°30′N, 2740–3050 m, Aug. 1910, *G. Forrest 6277* (E, BM, PE); near Shangfengkuo, *Pinus yunnanensis* forests 2400 m,

FIGURE IV 4:5B *Pseudocodon graminifolius* (H. Lév.) D. Y. Hong. **a–c**, from Yunnan: Jianchuan, 16. Sept. 2010, by De-Yuan HONG; **d** & **e**, from Sichuan: Muli, 21. Sept. 2011, by De-Yuan HONG.

03. Oct. 1983, *Qinghai-Xizang Exped. 14934* (KUN, PE); Tiejiashan, 2600 m, open *Pinus yunnanensis* forests, 18. Sept. 2010, *D. Y. Hong et al. H10025* (PE); Tong Shan, 27°20′N, Aug. 1914, *G. Forrest 13134* (BM, E, PE); 27°25′N, 2740 m, Sept. 1913, *G. Forrest 11179* (BM, E. K); s. loc., 3200 m, 24. Jul. 1914, *C. Schneider 1982* (G, K, PE); Wen Bi Shan, 03. Oct. 1990, *CLD-90 No. 648* (E, K); Xuesongcun to Qiaomaidi, 3100 m, edges of *Pinus yunnanensis* forests, limestones, 19. Sept. 2010, *D. Y. Hong et al. H10032* (PE); Yuhu Township, Xuesongcun, 2900 m, open woods, 01. Sept. 1955, *K. M. Feng 21304* (KUN, PE); loc. eodem, 2750–2900 m, 14. Aug. 1990, *J. Murata et al. 496* (A); Yulong Snow Range, 2900 m, meadows, 05. Aug. 1981, *Inst. Bot. CAS Hengduan Mts Exped. 02629* (PE); loc. eodem, 3100 m, thickets, 16. Aug. 1976, *NW Inst. Plateau Biol. & Chengdu Inst. Biol.0009* (CDBI); loc. eodem, near Wutuodi, in degraded *Pinus yunnanensis* forests, limestones, 2950 m, 23. Aug. 2000, *E. von Raab-Straube et al. 196* (B); loc. eodem, 3200 m, pine forests, 12. Aug. 1963, *C. Z. Bao 20039* (KUN); loc. eodem, Yunsaping, 3200 m, grassy slopes, 25. Jul. 1988, *Sino-Germ. Exped. 88-084* (KUN); loc. eodem, Yufeng Si, 2680 m, forests, 21. Aug. 1988, *J. Q. Li & Z. K. Zhou 179* (KUN); Yulong Snow Range, 2800 m, meadows, 10. Aug. 1981, *K. Y. Zhao 703* (PE); loc. eodem, 10. Aug. 1940, *R. C. Ching 30937* (PE); loc. eodem, 3300 m, among grasses, 22. Jul. 2001, *s. coll. s. n.* (PE); loc. eodem, below Mahuangba 3250 m, 30. Jul. 1962, *A. L. Zhang 100696* (PE); loc. eodem, Xianjiya, 3200 m, thickets, *Yunnan Univ. Vegetation Exped. 475* (PE); N of the city, 2750 m, stony places, 22. Aug. 1981, *X. W. Tian & al. 913* (PE); 27°12′N, 2740–3050 m, mixed or pine forests, Aug. 1906, *G. Forrest 2732*

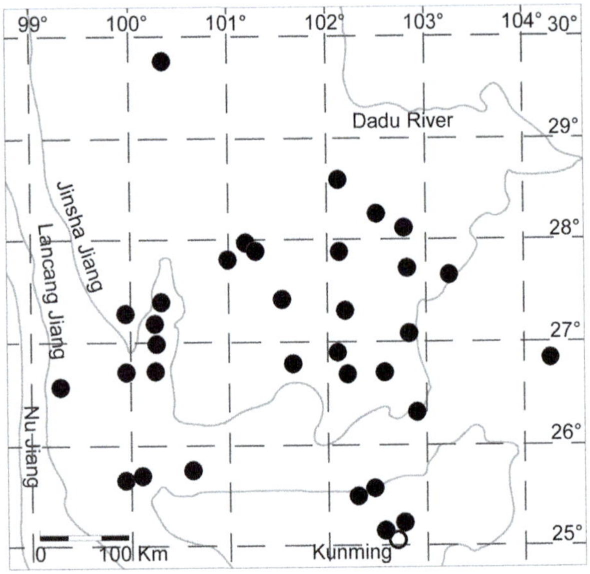

MAP IV 4:5 Distribution of *Pseudocodon graminifolius* (H. Lév.) D. Y. Hong.

(BM, E, PE); s. loc., 2400–2650 m, 06. Sept. 1987, *H. Ohba et al. 407* (KUN). **Luquan**: Emao Township, 2800 m, forests, 06. Nov. 1952, *P. I. Mao 1672* (KUN, PE). **Shangri-la** (Zhongdian): Annanchang to Zidi, *K. M. Feng 2050* (KUN); Haba, 2700 m, pine forests, 17. Sept. 1962, *Zhongdian Exped. 2114* (KUN). **Wuding**: Kuanpo, 2530 m, forests, 06. Nov. 2002, *H. Li et al. 0703* (KUN); loc. eodem, 2550 m, thickets, 27. Aug. 2002, *H. Li et al. 0635* (KUN). **Yangbi**: Zhongshan, 2400 m, grasslands, 25. Aug. 1984, *K. Iwatsuki et al. 1135* (KUN).

6. **Pseudocodon retroserratus** (Z. T. Wang & G. J. Xu) D. Y. Hong & Q. Wang, Phytotaxa 2014 (1): 62. 2015. ≡*Codonopsis retroserrata* Z. T. Wang & G. J. Xu, Acta Phytotax. Sin. 31(2): 186. fig. 2. 1993. Type: China, Sichuan, Yanyuan, 2400 m, 05. Sept. 1986, *Z. T. Wang 869382* (holotype: CPU; isotype: PE).

Plants glabrous throughout except ciliate filaments. Roots globose or short-conical, 2–3 cm long, 1.5–2.2 cm in diameter. Stems twining, mostly over 1 m long. Leaves alternate, nearly sessile or with petiole ca. 2 mm long; leaf blade papery, ovate or ovate-lanceolate, 1.5–4 cm long, 1–3 cm broad, base shallowly cordate or rounded, margin retroserrate and white-setose, apex acute. Flowers solitary, terminal only on main stem, or also on upper short branches; pedicel 5–30 cm long. Calyx tube entirely adnate to hypanthium, obconical, 5–7 mm long; calyx lobes lanceolate, 0.8–1.2 cm long, 3–5 mm broad. Corolla pale blue or violet; corolla lobes elliptic, 2.2–2.5 cm long, 1–1.5 cm broad, apex obtuse. Figure IV 4:6A.

Phenology: flowering in August and September.
Habitats: thickets; 2400–2600 m.
Distribution: **endemic** to China and confined to SW Sichuan (Map IV 4:6).

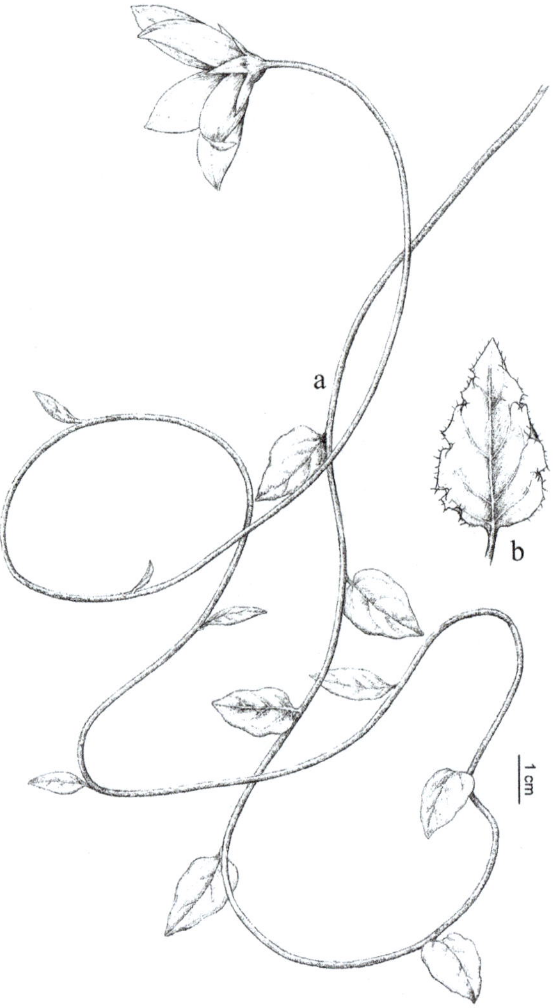

FIGURE IV 4:6A *Pseudocodon retroserratus* (Z. T. Wang & G. J. Xu) D. Y. Hong. **a**, middle and upper part of a twining stem; **b**, a leaf, showing retroserrate and spinulose margin.

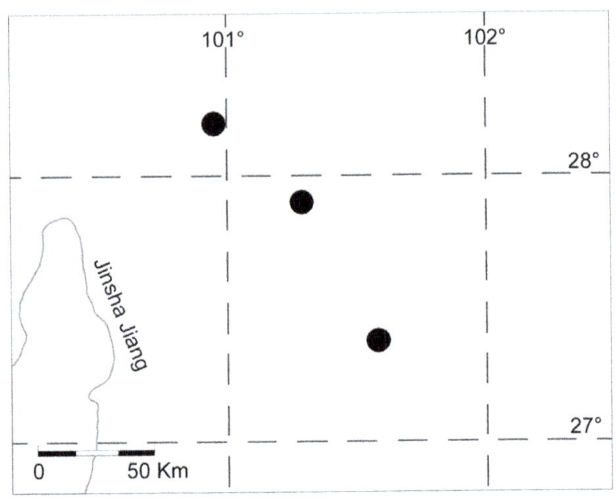

MAP IV 4:6 Distribution of *Pseudocodon retroserratus* (Z. T. Wang & G. J. Xu) D. Y. Hong.

Additional specimens examined: **CHINA, SICHUAN, Muli**: near the town, 2600 m, on thickets, 12. Aug. *T. T. Yü*

7669 (KUN, PE); Taoba Township, 2600 m, thickets, 22. Sept. 2011, *D. Y. Hong et al. H11038* (PE).

7. Pseudocodon petiolatus D. Y. Hong & Q. Wang, Phytotaxa 2014 (1): 62. 2015. Type: China, Sichuan, Muli, 913 Farm, Zamin, 2600 m, thickets, 23. Aug. 1983, *Qinghai-Xizang Exped. 13331* [holotype: PE (No. 1280549); isotype: PE (No. 1280611)].

Plants glabrous throughout except ciliate filament base. Roots elongated and attenuated with a constriction at middle part, over 15 cm long, 1.3 cm in diameter. Stems twining, slender, over 1 m long. Leaves alternate; petiole 1.5–2.5 cm long; leaf blade cordate to ovate, 2–3.5 cm long, 1.5–2.5 cm broad, base cordate to rounded, margin entire or shallowly crenulate, apex obtuse to acute. Flowers solitary, terminal on main stem. Calyx tube adnate entirely to hypanthium, obo-triangular; calyx lobes strongly reflexed even since anthesis, linear-lanceolate, 7–9 cm long, 3–3.5 cm broad. Corolla 5-lobed to near base, rotate, blue; corolla lobes elliptic, 2 cm long, 1 cm broad. Stamens free; filaments 4 mm long, base strongly dilated into horizontally rectangular, ciliate; anthers 4 mm long.Ovary inferior; style glabrous, 2 mm long; stigma 3- fid, stigma lobes elliptic- oblong, 4 mm long, 2.5mm wide. Fruit unknown. Figure IV 4:7A.

Phenology: flowering in August.
Habitats: thickets; ca. 2600 m.
Distribution: **endemic** to China; confined to Sichuan (Muli) (Map IV 4:7).

Notes: This species differs distinctly from all the other species of the genus in having roots elongated, petiole 1.5–2.5 cm long, nearly as long as leaf blade, calyx lobes strongly reflexed even since anthesis, and stigma lobes large.

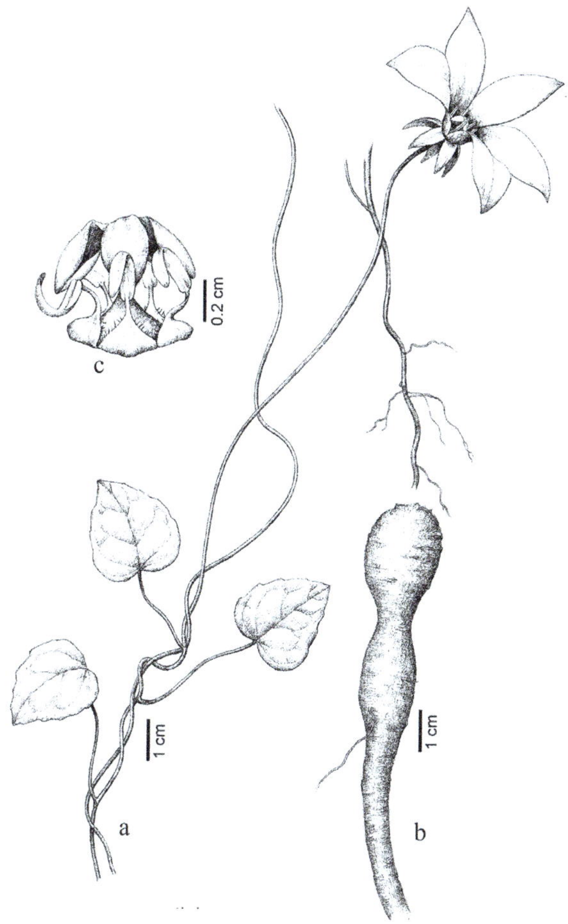

FIGURE IV 4:7A *Pseudocodon petiolatus* D. Y. Hong & Q. Wang. **a**, a section and terminal part of twining stem; **b**, root and basal part of stem; **c**, stamens, style and 3 oblong stigma segments.

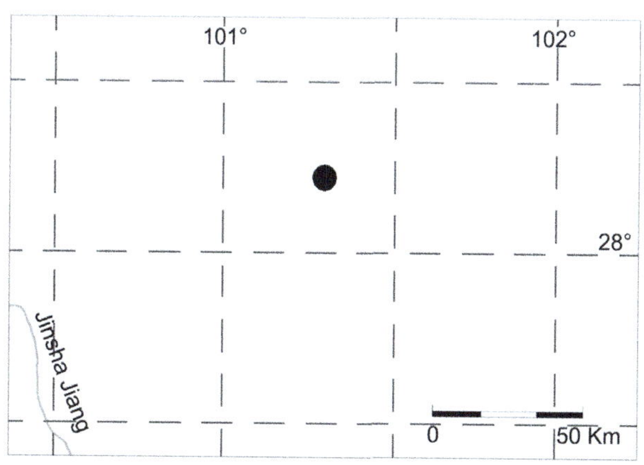

MAP IV 4:7 Distribution of *Pseudocodon petiolatus* D. Y. Hong & Q. Wang.

8. Pseudocodon rosulatus (W. W. Smith) D. Y. Hong in Wang et al., J. Syst. Evol. 52(5): 548. 2014. ≡*Codonopsis rosulata* W. W. Smith, Notes Roy. Bot. Gard. Edinburgh 13: 157. 1921; Nannfeldt in Handel-Mazzetti, Symb. Sin. 7: 1078. 1936; Hong, Lammers & Klein in Wu, Raven & Hong, Fl. China 19: 524. 2011. Type: China, Sichuan, Muli, 28°12′N, 3350 m, among heavy grasses, Aug. 1918, *G. Forrest 16856* (holotype: E)

Plants glabrous throughout except stamens. Roots globose or ovoid, 0.5–1.5 cm long, 0.5–1 cm in diameter. Caudex 2–4 cm long, less than 1 mm in diameter, with 2–4 alternate and very small scales. Stems single, or multiple and caespitose, erect or more or less decumbent, 10–50 cm tall, simple or with 2–3 branches. Basal leaves rosulate; petiole 0.1–3 cm long; leaf blade broad-ovate, ovate or elliptic, 0.5–7 cm long, 0.4–5 cm broad, base slightly cordate or rounded, margin crenate, apex obtuse; cauline leaves scale-like, lanceolate or linear-lanceolate, 1–5 mm long, 0.5–2 mm broad, base rounded or cuneate, margin sinuous-crenate, or nearly entire, apex acute. Flowers mostly 1–3, rarely up to 8, terminal on main stem and branches; pedicel 4–15 cm long. Calyx tube adnate to hypanthium up to near top, obconical, 2–5 mm long, 2–5 mm wide, 10-ribbed; calyx lobes ovate-triangular, 3–6 mm long, 1.5–3 mm broad, apex acute, 1 vein visible; sinus between calyx lobes narrow and pointed. Corolla pale blue, blue or blue-purple; corolla lobes ovate or broad-ovate, 7–14 mm long, 5–7 mm broad, apex acute or obtuse. Anthers oblong, ca. 2.5 mm long. Capsule obconical below, slightly truncate-obtuse above, ca. 1.2 cm long, 0.6 cm in diameter. Seeds compressed, ellipsoid, brown-yellow. Figures IV 4:8A & B.

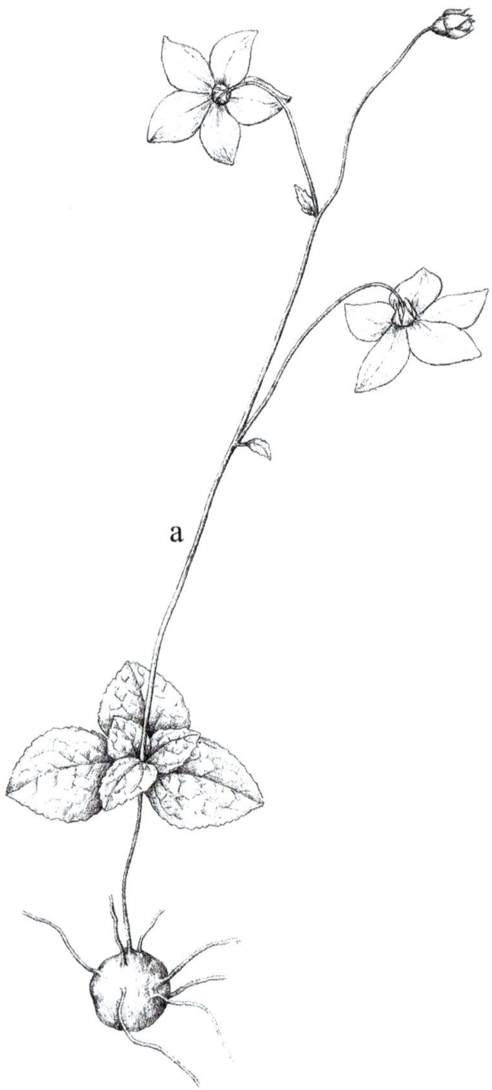

FIGURE IV 4:8A *Pseudocodon rosulatus* (W. W. Smith) D. Y. Hong. **a**, a plant, showing globose root, rosulate leaves, erect stem, and a terminal cyme of 3 flowers.

Phenology: flowering and fruiting in September and October.

Habitats: grassy slopes or edges of pine or pine-oak forests at an altitude of 2600–3400 m.

Distribution: **endemic** to China; confined to SW Sichuan (Map IV 4:8).

Additional specimens examined, **CHINA**, **SICHUAN**, **Muli**: Bowa, Liziping, 3180 m, thickets, 03. Sept. 1978, *C. S. Cao et al. 8390* (CDBI, SZ); Qing'o Forest Farm, 3400 m, pine forests, *S. K. Wu 2598* (CDBI, KUN). **Yanbian**: Dapingzi, 2700 m, 29. Aug. 1978, *Yanbian Exped. 372* (SM); Yankou Township, Dawan, 2700 m, thickets, 20. Sept. 2002, *X. F. Gao & Y. L. Peng 3669* (PE); loc. eodem, 2657 m, pine-oak community, 23. Sept. 2011, *D. Y. Hong et al. H11039* (PE). **Yanyuan**: s. loc., 3000 m, 30. Sept. 1914, *Handel-Mazzetti 5422* (W, WU); s. loc., 2600–2700 m, 20. Sept. 1965, *F. G. Zhu 005* (PE).

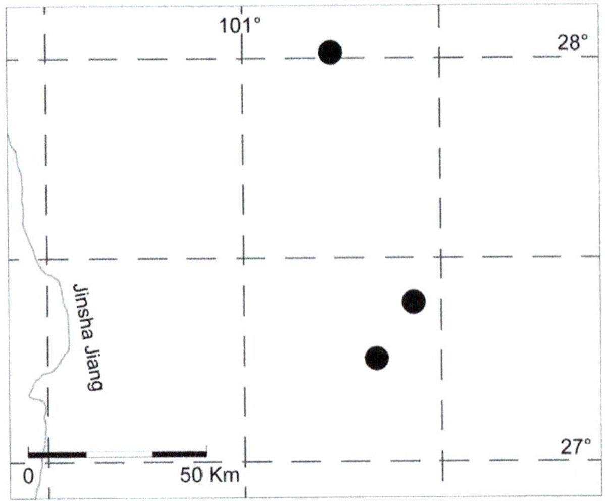

MAP IV 4:8 Distribution of *Pseudocodon rosulatus* (W. W. Smith) D. Y. Hong.

CYCLOCODON Griff. ex Hook. f. & Thomson

Hooker & Thomson, J. Proc. Linn. Soc. Bot. 2: 17. 1857; Griffith, Not. Pl. As. 4: 277. 1854, nom. nud.; Hong & Pan, Acta Phytotax. Sin. 36: 109. 1998; Lammers in T. C. Huang, Fl. Taiwan, ed. 2, 4: 784. 1998; Hong & Lammers in Wu, Raven & Hong, Fl. China 19: 527. 2011. ≡*Campanumoea* sect. *Cyclocodon* (Griff. ex Hook. f. & Thomson) C. B. Clarke in Hooker, Fl. Brit. Ind. 3:436. 1881; Hong, Fl. Reipubl. Popularis Sin. 73(2): 72, 1983. Type: *Cyclocodon parviflorus* (Wall. ex A. DC.) Hook. f. & Thomson (*Codonopsis parviflora* Wall. ex A. DC.).

Herbs perennial, without caudex, nor rhizome. Roots attenuate. Stems erect, sometimes woody at base, mostly branched. Leaves opposite, occasionally in whorls of 3. Flowers often in a dichasium or solitary, terminal or axillary; bracteoles often present, filiform or leaf-like. Calyx adnate to hypanthium, or completely free from it, and thus epigynous or hypogynous to ovary; calyx lobes 4–6, from nearly entire to ramiform. Corolla white, pale blue, pale lavender, or pale red, 4–6-merous, epigynous, infundibular, or rotate, lobed to 1/3 or 1/2. Stamens 4–6; filaments dilated below, dilated part glabrous or ciliate. Ovary 4–6-locular, inferior to corolla, from half-inferior to superior to calyx; stigma (4–)5–6 (–8)-fid, segments clavate; ovules extremely numerous. Fruit a berry. Seeds extremely numerous, ovoid or nearly globose, compressed in *C. parviflorus*; seed-coat irregularly reticulate, secondary ornamentation on radial walls spiral/rope-like. Pollen 3-colporate; sexine spinulose, spinules > 2μm long. Chromosomes 2n=18.

The genus *Cyclocodon* has changed greatly in its taxonomic status and systematic position. It was reduced by Clarke

FIGURE IV 4:8B *Pseudocodon rosulatus* (W. W. Smith) D. Y. Hong. All from Sichuan: Yanbian, 23. Sept. 2011, **a**, by Qiang WANG & De-Yuan HONG; **b–e**, by De-Yuan HONG.

(1881) to a section in *Campanumoea* Bl., *Campanumoea* sect. *Cyclocodon* (Griff. ex Hook. f. & Thomson) C. B. Clarke, and then the genus *Campanumoea* Bl. was merged into *Codonopsis* by Moeliono (1960). Erdtman (1952) revealed the sharp differences in pollen morphology between *Cyclocodon parviflorus* and three species of *Codonopsis*. Morris and Lammers (1997) recognized pollen grains of the two groups as two distinct pollen types: Type I for *Codonopsis*, Type II for the *Cyclocodon* group (then still treated as members of *Codonopsis*). Hong and Pan (1998) restored *Cyclocodon* at generic status based on the distinct differences in pollen, seed-coat and external morphology among *Cyclocodon*, *Codonopsis* and *Campanumoea*. The restoration is accepted in "*Fl. Taiwan*" (Lammers in Huang, 1998) and "*Flora of China*" (Hong et al., 2011). A recent molecular tree (Wang et al., 2013) shows that *Cyclocodon* forms a clade with the genus *Platycodon* (sharing colporate pollen), far from *Codonopsis*.

De Candolle (1839) recorded only two species under *Codonopsis*, while he excluded four species from *Codonopsis* (actually five, incl. *C. viridis* Wall., the type of *Codonopsis*). Actually the two species under *Codonopsis* belong to *Cyclocodon*, but four of the five excluded by him are real members of *Codonopsis*, *C. ovata* Benth., *C. rotundifolia* Benth., *C. thalictrifolia* Wall., and *C. viridis*, while *Codonopsis purpurea* Wall. has been treated as the only member of the new genus, *Pankycodon* D. Y. Hong & X. T. Ma (Wang et al. 2014a).

The genus *Cyclocodon* differs from *Codonopsis* in often having the flowers in a dichasium of 3, the stigma usually (4-)5- or 6-fid with clavate lobes, the calyx lobes ramiform to nearly entire and often inserted at a different level, distant from

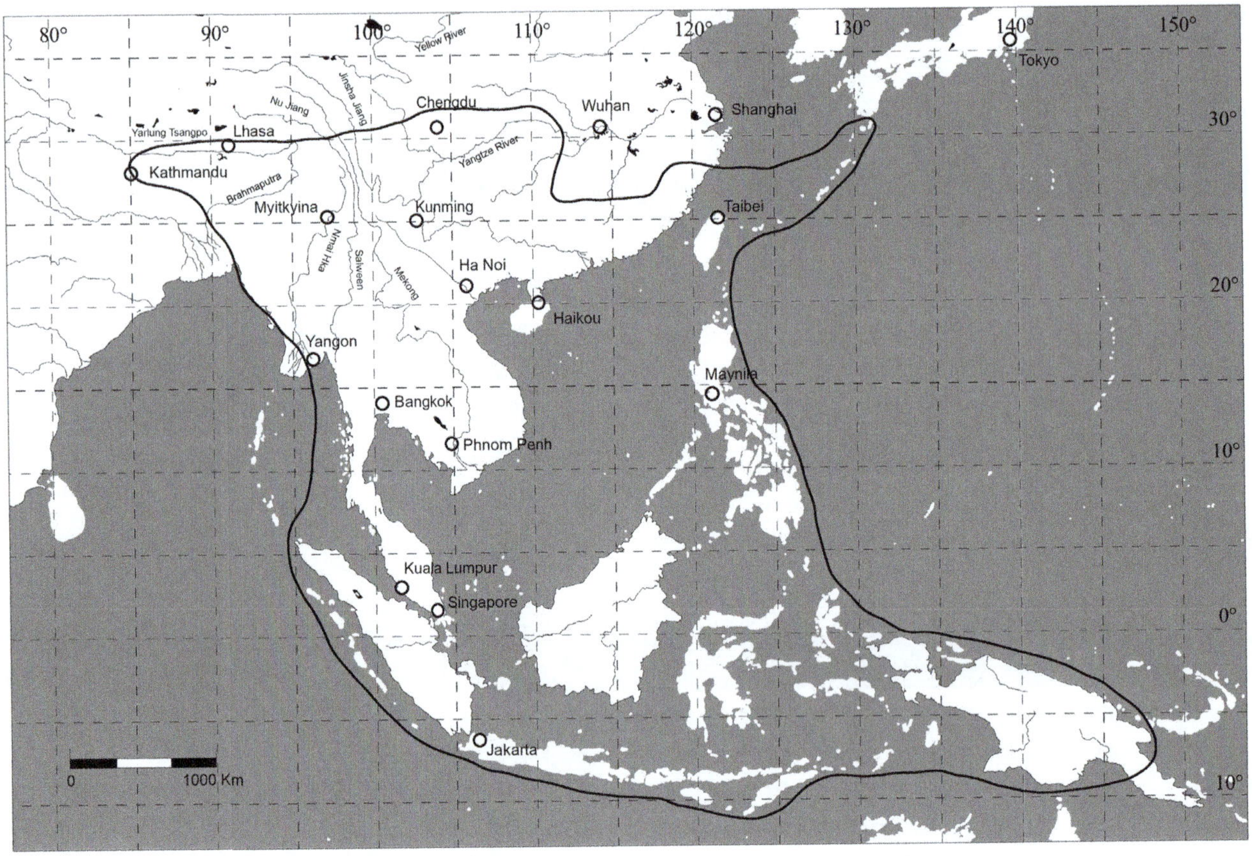

MAP IV 5:0 Distribution of the genus *Cyclocodon* Griff. ex Hook. f. & Thomson.

the other floral parts, the seed-coat irregularly reticulate with rope-like ornamentation on the radial walls, and the pollen 3-colporate, with long sexine spinules (>2 μm high).

Three species, distributed from the Himalaya eastwards to Japan (Kyushu and Ryukyu Islands), southeastwards to Philippines, Indonesia, and Papua New Guinea (Map IV 5:0).

1a. Flowers often 4-merous; calyx inserted on pedicel, 1–4 mm below hypanthium, resembling involucral bracts .... 1. *C. parviflorus*

1b. Flowers usually 5- or 6-merous; calyx at least partly adnate to hypanthium.

2a. Flowers all terminal; bracteoles often absent, rarely present and then stalked and more or less leaf-like; calyx lobes shortly denticulate to nearly entire; filaments glabrous; berry often white, less often yellow or pale red .... 2. *C. lancifolius*

2b. Flowers both axillary and terminal; bracteoles present and filiform; calyx lobes ramiform; filaments hairy below; berry dark purple... 3. *C. axillaris*

1. Cyclocodon parviflorus (Wall. ex A. DC.) Hook. f. & Thomson, J. Proc. Linn. Soc., Bot. 2: 18. 1857; Hong & Pan, Acta Phytotax. Sin. 36(2): 110. 1998; Hong & Lammers in Wu, Raven & Hong, Fl. China, 19: 528. 2011. ≡*Codonopsis parviflora* Wall. ex A. DC., Monogr. Camp. 123. 1830; Wallich, Cat. 1300, 1820, num. nud. ≡*Campanumoea parviflora* (Wall. ex A. DC.) Benth. & Hook. f. ex C. B. Clarke in Hooker, Fl. Brit. Ind. 3: 436. 1881; Hong, Fl. Reip. Popularis Sin. 73 (2): 74. 1983; Haridasan & Mukherjee, in Hajra & Sanjappa, Fasc. Fl. Ind. 22: 59. 1996. ≡*Campanula punduana* D. Dietr., Syn. Pl. 1: 758. 1839. Type: Bangladesh, Pundua, *Wallich 1300* [lectotype here designated: K (K000814184); isolectotypes: K (K000814182, K000814183, K000814185)].

Cyclocodon distans Griff., Not. Pl. As. 4: 277. pl. 481. fig. 1. 1854. Type: Griffith's pl. cited.

Cyclocodon adnatus Griff., Not. Pl. As. 4: 278. 1854. Type: not designated.

Plants 1–2 m tall. Stems terete, sometimes woody below, branched, glabrous. Leaves shortly petiolate; petiole up to

6 mm long; leaf blade lanceolate, 5–11cm long, 1.5–3 cm broad, those on branches smaller, abaxially somewhat glaucous, both surfaces glabrous, base rounded or broad-cuneate, margin serrate, teeth apiculate at apex, apex acuminate. Flowers terminal, often in a 3-flowered dichasium; pedicel 1–4 cm long, glabrous, with or without a pair of leaf-like or linear bracteoles. Calyx entirely free from hypanthium and ovary, and thus hypogynous, just below ovary or 1–4 mm distant from ovary, forming an anthophore; calyx lobes usually 4, free from each other, linear, 3.5–8 mm long, with 1–4 pairs of teeth. Corolla white, infundibular, 5–8 mm long; corolla lobes as long as tube. Stamens 4; filaments subulate, very slightly dilated below. Ovary globose or obovoid, 4-locular; stigma 4-fid. Berry globose, 4–6 mm in diameter. Seeds light brown, orbicular or circular, compressed. Figure IV 5:1A.

Chromosomes: 2n–18 (Zhang et al., 2010).
Phenology: flowering and fruiting from August to January.
Habitats: thickets or grassy slopes; 180–3000 m.
Distribution: S Bhutan (Deothang District), China (SE Tibet and Yunnan), India (Darjiling, Meghalaya, Nagaland, and Sikkim,), and Myanmar (Map IV 5:1).

Notes: Griffith indicated a collection from Khasia (Meghalaya) for his new species, *Cyclocodon distans*, but I failed in finding it. Therefore, I designate here his original plate as type.

Additional specimens examined, BHUTAN (S), **Deothang** Distr. (from Grierson & Long, *Fl. Bhutan*, 2(3): 1384. 2001).

FIGURE IV 5:1A *Cyclocodon parviflorus* (Wall. ex A. DC.) Hook. f. & Thomson. **a**, upper part of an erect stem, showing terminal dichasia; **b**, a flower; **c**, a berry.

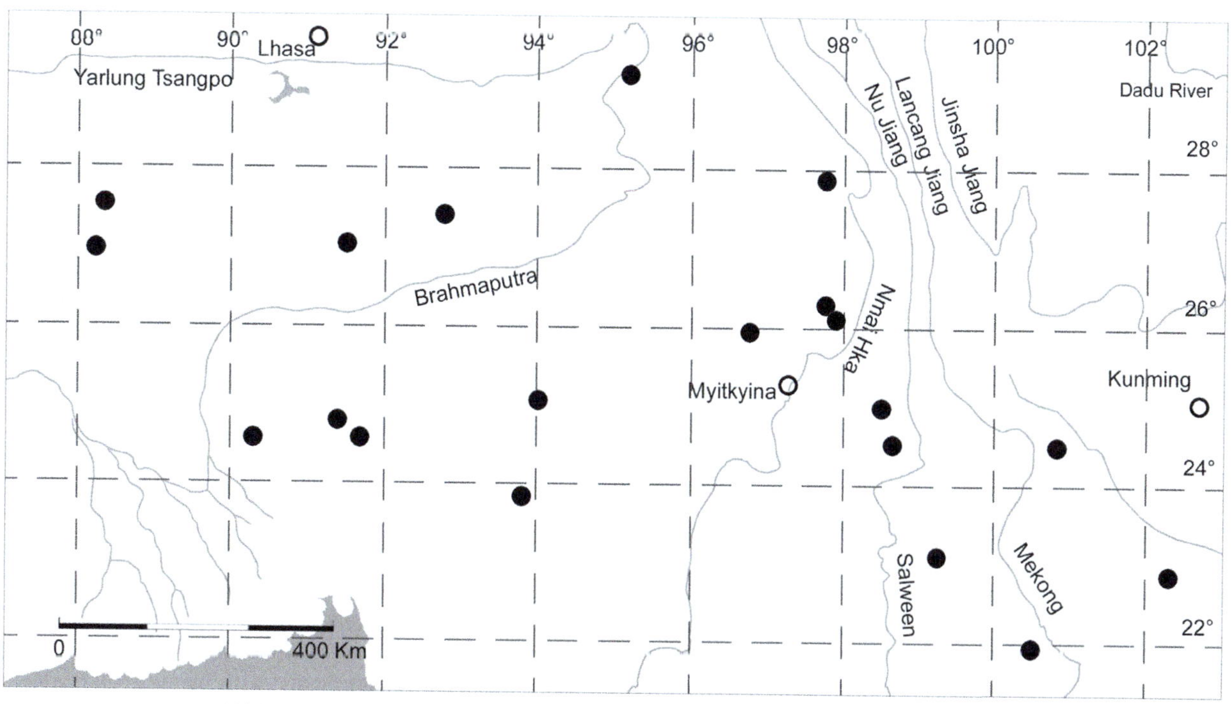

MAP IV 5:1 Distribution of *Cyclocodon parviflorus* (Wall. ex A. DC.) Hook. f. & Thomson.

CHINA, TIBET (SE), N Cachar Hills, Hafbry, 760 m, 19. Aug. 1908, *G. Craib 329* (K). **Mêdog**, Xiran, 600 m, grassland, fl. white, 24. Nov. 1992, *H. Sun et al. ETM. 1627* (KUN). **Kameng** Distr., Toppi, 250 m, 11. Oct. 1977, *S. S. R. Bennet & H. B. Naithani 2926* (PE). **YUNNAN, Cangyuan**: Banhong, 530 m, forests, 29. Oct. 1989, *D. D. Tao et al. 39967* (KUN). **Luchun**: near Erpu, 600 m, edges of forests, 19. Oct. 2000, *Y. M. Shui & W. H. Chen 13298* (KUN); Dayi Shan, 630–700 m, thickets by road, 29. Sept. 1973, *D. D. Tao 586* (KUN, PE). **Luxi**: Mangshi, 950 m, thickets by road, 19. Dec. 1958, *C. Chen 593* (KUN). **Mengla**: s. loc., 650 m, roadsides, 12. Oct. 1959, *Yunnan Trop. Biol. Resourses Surv. 59–13362* (KUN). **Tengchong**: s. loc., 1180–1350 m, ravines, 15. Oct. 1960, *W. Q. Yin 60–1473* (KUN).

INDIA, Darjiling: 1520 m, Sept. 1881. *J. S. Gamble 9872* (K); Rungbee, 1520 m, 19. Aug. 1869, *C. B. Clarke 8680* (K); loc. eodem, 1370 m, 04. Sept. 1869. *C. B. Clarke 9015C* (BM, K); loc. eodem, 3000 m, 24. Jan. 1923, *G. H. Cave* s. n. (PE). **Manipur**: s. loc. 1220–1520 m, Nov. 1907, *A. E. Meebold 6158* (K). **Meghalaya**: Anmoee, 910–1220 m, 26. Sept. 1850, *J. D. Hooker & Thomson* s. n. (K); Chitmang, Garo Hills, 1220 m, 18. Nov. 1930, *N. E. Parry 1000* (K); Jainta Hills, 610–910 m, Sept. 1872, *J. S. Tadul 207* (K); near Màirang, 610–1370 m, Oct. 1855, *Schlagintweit 208* (BM); Muhlow, 910–1220 m, 18. Oct. 1850, *J. D. Hooker & T. Thomson* s. n. (K); Tserra, 1220 m, 06. Oct. 1867, *C. B. Clarke 5599* (K); s. loc., *Griffith* s. n. (K); s. loc., *Th. Lobb* s. n. (K); s. loc., 1220 m, *J. D. Hooker & Thomson* s. n. (BM). **Nagaland:** Piekima, Naga Hills, 1370 m, 06. Aug. 1935, *N. L. Bor 6299* (K); Naga Hills, Neechrogard, 150 m, 17. Oct. 1885, *C. B. Clarke 40876E* (BM). **Sikkim**: Pachum, 1520–1830 m, 25. Sept. *J. D. Hooker* s. n. (K); 1220–1830 m, *J. D. Hooker* s. n. (BM, K, PE).

MYANMAR, Mahtum, 26°06′N, 97°58′E, 1370 m, 02. Sept. 1939, *R. Kaulback 378* (BM); loc. eodem, 26°06′N, 97°58′E, 460 m, 31. Aug. 1939, *R. Kaulback 368* (BM); Tara Hka, 26°09′N, 97°52′E, 460 m, 26. Aug. 1939, *R. Kaulback 363* (BM); **Nam Tamai Valley**, 27°42′N, 97°54′E, 1220 m, 14. Aug. 1938, *R. Kaulback* s. n. (BM); **Sumpra** Bum, 1370 m, 09. Jan. 1953, *F. Kingdon-Ward 20363* (BM).

2. **Cyclocodon lancifolius** (Roxb.) Kurz, Flora 55: 303. 1872. ≡*Campanula lancifolia* Roxb., Fl. Ind. 2: 96. 1824. ≡*Campanumoea lancifolia* (Roxb.) Merr., Enum. Philip. Pl. 3: 587. 1923. ≡*Codonopsis lancifolia* (Roxb.) Moeliono in van Steenis, Fl. Males. Ser. 1. 6: 120. 1960, p. p. Type: Bangladesh, Chittagong, Oct. 1810, *Roxburgh 217* (holotype: BM).

Cyclocodon celebicus (Bl.) D. Y. Hong, Acta Phytotax. Sin. 36: 109. 1998, syn. nov.; Hong & Lammers in Wu, Raven & Hong, Fl. China 19: 527. 2011; de Wilde & Duyfjes in Santisuk & Balslev, Fl. Thail. 11(4): 512. fig. 8, pl. 1: 4–8. 2014. syn. nov. *Campanumoea celebica* Bl., Bijdr. 727. 1826; De Candolle, Monogr. Camp. 120. 1830. *Campanula celebica* (Bl.) D. Dietr., Syn. Pl. 1: 758. 1839. ≡*Codonopsis celebica* (Bl.) Miq., Fl. Ind. Bat. 2: 565. 1835. ≡*Codonopsis lancifolia* subsp. *celebica* (Bl.) Moeliono in van Steenis, Fl. Males. Ser. 1, 6: 121. 1960. ≡*Cyclocodon lancifolius* subsp. *celebicus* (Bl.) K. E. Morris & Lammers, Novon 9: 387. 1999. Type: Indonesia, Sulawesi, *Reinwardt* s. n. (L, n. v.).

Codonopsis truncata Wall. ex A. DC., Monogr. Camp. 120. 1830; Wallich, Cat. 1301. 1820, nom. nud. ≡*Companumoea truncata* (Wall. ex A. DC.) Diels, Bot. Jahrb. Syst. 29: 606. 1901. ≡*Cyclocodon truncatus* (Wall. ex A. DC.) Hook. f. & Thomson, J. Proc. Linn. Soc. Bot. 2: 18. 1857. Type: India, Darjiling, *Wallich 1301* (lectotype here designated: K; isolectotype: BM).

Codonopsis leucocarpa Miq., Fl. Ind. Bat. 2: 565. 1857. Type: Indonesia, Sumatra, in de kloof van den Singalang. *Teijsmann* s. n. (holotype, n. v.).

Plants erect or scandent, 0.3–4 m long, sometimes woody at base. Stems terete, branched, sparsely hispidulous or nearly glabrous. Leaves shortly petiolate; petiole up to 5 mm long; leaf blade ovate-lanceolate to lanceolate, 5–10 cm long, 2–4 cm broad, those on branches smaller, glabrous or sparsely puberulent along veins on both surfaces, base usually rounded, margin serrate, teeth pointed towards apex, apex acuminate. Flowers terminal, usually in a dichasium of 3; pedicel 8–20 mm long, hispidulous, usually without bracteoles, rarely with stalked and more or less leaf-like bracteoles. Calyx adnate to basal ca. 1/3 of hypanthium; calyx tube glabrous; calyx lobes mostly 5 or 6, linear to linear-elliptic, 3–10 mm long, 1–3 mm broad, with 2–4 pairs of serrate teeth, rarely entire. Corolla infundibular, 7–10 mm long, 5- or 6-lobed to middle, pale blue, pale lavender, or white. Stamens 5 or 6, glabrous; filaments slightly dilated below. Ovary globose; stigma 5- or 6-fid. Berry white, oblate, 5- or 6-locular, up to 8 mm long, ca. 12 mm across. Seeds extremely numerous. Figures IV 5:2A & B.

Phenology: flowering and fruiting from July to February.

Habitats: forests or thickets, often by stream or river, mostly on granites, less often on marbles or serpentines; 600–2600 m.

Distribution: widely distributed in S and SE Asia: Bangladesh, China (SE Tibet and S Yunnan), N India, Indonesia, Laos, Malaysia, Myanmar, Philippines, Thailand, and Vietnam; also in Oceania: Papua New Guinea (Map IV 5:2).

Notes: 1) As the present author indicated in the key, in addition to *C. parviflorus* there are two other distinct entities in the genus. Moeliono (1960), Morris & Lammers (1999) and Lammers (2007) recognized these, but they treated them as two subspecies in *C. lancifolius*, subsp. *lancifolius* and subsp. *celebicus*. Hong & Pan (1998) and Hong & Lammers (2011) recognized them as two separate species, *C. lancifolius* and *C. celebicus*. In these two cases, "*lancifolius*" was used for the entity with many axillary flowers and ramiform calyx lobes, while "*celebicus*" for the other

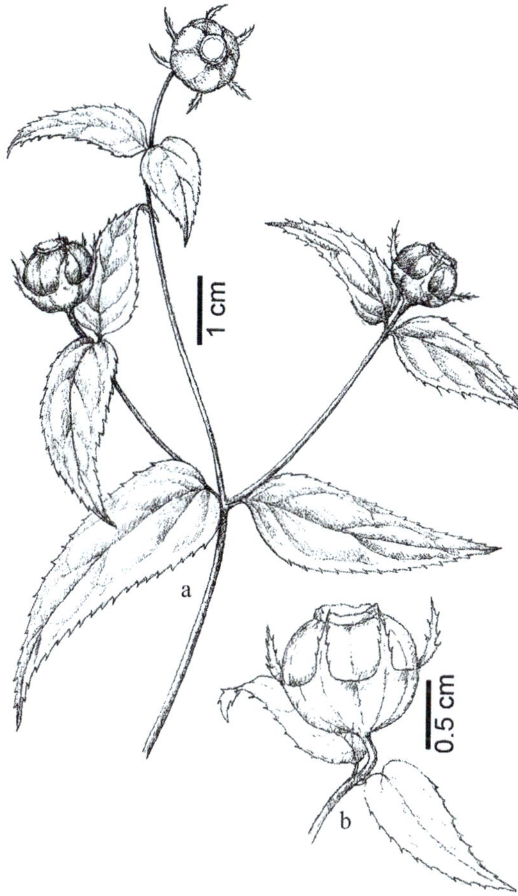

FIGURE IV 5:2A *Cyclocodon lancifolius* (Roxb.) Kurz. **a**, upper part of an erect stem; **b**, a berry, showing ovary half-inferior to calyx.

with strictly terminal flowers and acuminate, shallowly dentate or nearly entire calyx lobes. In December, 2012, I examined *Roxburgh 217* in BM, collected from Chittagong in October of 1810. In a letter to Mr. J. F. M. Cannon in 1961 (preserved in BM), Professor van Steenis stated that this specimen seems to be the type of *Campanula lancifolia* Roxb., because Roxburgh cited "Chittagong". According to the protologue of *Campanula lancifolia* Roxb., Lammers made a note on the specimen, "probable type material". However, this specimen, with only terminal flowers, represents the form hitherto named as '*celebicus*'. Furthermore, I examined three specimens of *Campanula lancifolia* Roxb. in K, which all represent this form. Therefore, "*lancifolius*" and "*celebicus*" are the same taxon and thus we have to find another name for the form with axillary flowers. *Codonopsis truncata* Wall. ex A. DC. and *Codonopsis leucocarpa* Miq. both represent the form with terminal flowers. The description of *Codonopsis albiflora* Griff. (1854) is ambiguous, and we could not find its type specimen. Oliver's *Campanumoea axillaris* is the oldest available name for the form with axillary flowers and ramiform calyx lobes. As it has an accurate figure and type material preserved in K and BM, I consider it reasonable to use the name *Cyclocodon axillaris* for this entity. The epithet is also very appropriate.

2) Moeliono's (1960) *Codonopsis lancifolia* (Roxb.) Moeliono includes two species, *Cyclocodon lancifolius* and *C. axillaris*, since he pulled *C. axillaris* Oliv. and *C. celebica* together as its synonyms.

Additional specimens examined, **BANGLADESH**, **Chittagong**: Rungamultea, 05. Feb. 1873, *C. B. Clarke 19516* (K); Bariadhala Forest Beat, 06. Oct. 1997, *M. A. Rahma et al. 1939* (K); Chittagong Hill tracts, 25. Jan. 1965, *M. S. Khan 1123* (K).

BHUTAN (C), **Punakha** Distr. (from Grierson & Long, *Fl. Bhutan* 2(3): 1384, 2001).

CHINA, TIBET, **Mêdog**: Dexing, 950 m, grasses by wild banana forests, 03. Aug. 2012, *D. Y. Hong et al. H12010* (PE); Dexing, 900 m, forests, 14. Jan. 1983, *B. S. Li & S. Z. Cheng 3467* (PE); Digdong, 800–1000 m, cliffs by forest, 26. Jul. 1973, *Biol. Inst. Tibet Exped. 1611* (HNWP); Jialasa to Suila, 1800 m, forests, 18. Nov. 1982, *S. Z. Cheng & B. S. Li 3033* (PE); Mêdog to Deergong, 1700 m, edges of broad-leaved forests, 20. Aug. 1974, *Qinghai-Xizang Exped. 74–4430* (KUN, PE). **Nyingchi**: Dongjiu to Layue, 21. Dec. 1953, *P. C. Tsoong 6259* (PE). **Zayü:** Samai, 1500 m, forests, 27. Jul. 1973, *Biol. Inst. Tibet Exped. 1184* (HNWP); near Samai, 1500 m, roadsides, 13. Jul. 1980, *C. C. Ni et al. 0650* (PE); Tha Chu Valley, Mishmi Hills, 1830 m, 20. Jul. 1950, *F. Kingdon-Ward 19663* (BM); Xiazayü, 1700 m, flood areas, 22. Jul. 1973, *Qinghai-Xizang Exped. 73–873* (KUN, PE). **YUNNAN, Fugong**: Jiakedi Township, 18. Aug. 2005, *Gaoligong Shan Biodiversity Survey 27764* (CAS); Lishadi Township, 1610 m, 14. Aug. 2005, *Gaoligong Shan Biodiversity Survey 27544* (CAS); Maji Township, 1700 m, 16. Aug. 2005, *Gaoligong Shan Biodiversity Survey 27682* (CAS); Shanpa, 1180–1280 m, 23. Aug. 2005, *Gaoligong Shan Biodiversity Survey 28923* (CAS); s. loc., 1800 m, woods, 17. Sept. 1933, *H. T. Tsai 54282* (KUN, PE, SZ); s. loc., 1500 m, 28. Oct. 1933, *H. T. Tsai 54966* (KUN, PE, SZ); s. loc., 2200 m, woods, 14. Oct. 1933, *H. T. Tsai 54425* (KUN, PE, SZ). **Gongshan**: Binzhongluo, 1740 m, 09. Nov. 2004, *Gaoligong Shan Biodiversity Survey 22204* (CAS); loc. eodem, 1630 m, 28. Sept. 2002, *Gaoligong Shan Biodiversity Survey 15581* (CAS); loc. eodem, 1780 m, 09. Nov. 2004, *Gaoligong Shan Biodiversity Survey 22280* (CAS); loc. eodem, 1530 m, 11. Aug. 2006, *Gaoligong Shan Biodiversity Survey 33025* (CAS); Bingzhongluo to Gouda, 1600–1700 m, grasses by road, 21. Jun. 1982, *Qinghai-Xizang Exped. 7318* (PE); Cikai, 1600–1800 m, thickets, 01. Oct. 1940, *K. M. Feng 8147* (KUN, PE); loc. eodem, forests, 26. Oct. 1940, *K. M. Feng 8658* (KUN, PE); loc. eodem, 1410 m, 14. Nov. 2004, *Gaoligong Shan Biodiversity Survey 23231* (CAS); loc. eodem, 1550 m, 25. Sept. 2002, *Gaoligong Shan Biodiversity Survey 15463* (CAS); loc. eodem, 1540 m, 12. Aug. 2006, *Gaoligong Shan Biodiversity Survey 33093*

FIGURE IV 5:2B *Cyclocodon lancifolius* (Roxb.) Kurz. All from Tibet: Mêdog, 03. Aug. 2012; **a**, **c** & **d**, by Xin-Tang MA & De-Yuan HONG; **b**, by De-Yuan HONG.

(CAS); loc. eodem, 1510–1570 m, 13. Jul. 2000, *H. Li et al. 12813* (A, CAS); Drungjiang, Dandang, 1400 m, evergreen forests, 17. Jan. 1991, *Drungjiang Exped. 3251* (KUN); Drungjiang, Gameilaihe, 800 m, thickets, 31. Dec. 1990, *Drungjiang Exped. 1417* (KUN); Drungjiang, Jimudang, 1400 m, forests by river bank, 06. Dec. 1990, *Drungjiang Exped. 890* (KUN); Drungjiang, Qênglangdang, 1300 m, 13. Aug. 1982, *Qinghai-Xizang Exped. 9300* (PE); Drungjiang, Shihuiyao, 1500 m, thickets, 20. Dec. 1990, *Drungjiang Exped, 1253* (KUN); Drungjiang Township, 1380 m, 27. Oct. 2004, *Gaoligong Shan Biodiversity Survey 21949* (CAS); loc. eodem, 1350 m, 20. Jul. 2002, *H. Li et al. 15134* (CAS); loc. eodem, 1410 m, 22 Aug. 2006, *Gaoligong Shan Biodiversity Survey 32725* (CAS); loc. eodem, 1400 m, 06. Dec. 1990, *Drungjiang Team 890* (CAS); loc. eodem, 1660 m, 31. Oct. 2004, *Gaoligong Shan Biodeversity Survey 21271* (CAS); loc. eodem, 1470 m, 19. Aug. 2006, *Gaoligong Shan Biodiversity Survey 32513* (CAS); loc. eodem, 1400 m, 15. Nov. 1990, *Drungjiang Team 331* (CAS); loc. eodem, 1500 m, 20. Dec. 1990, *Drungjiang Team 1253* (CAS); loc. eodem, 1350 m, 31. Oct. 2004, *Gaoligong Shan Biodiversity Survey 21200* (CAS); Drunjiang to Wenmaku, 1600 m, thickets, 05. Aug. 1982, *Qinghai-Xizang Exped. 8816* (PE); Gongshan to Drungjiang, E Flank of Gaoligong

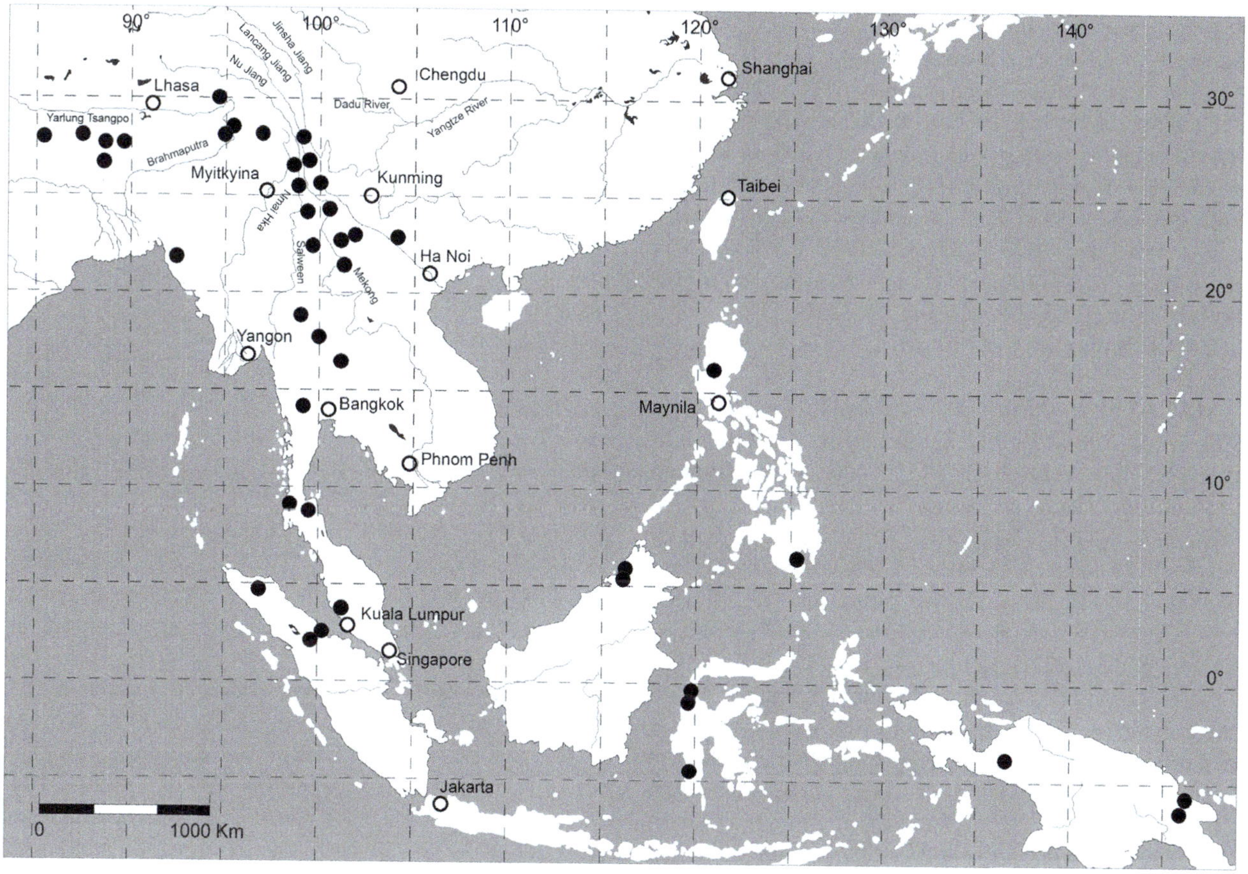

MAP IV 5:2 Distribution of *Cyclocodon lancifolius* (Roxb.) Kurz.

Shan, 1700–1800 m, evergreen broad-leaved forests, 17. Jul. 1982, *Qinghai-Xizang Exped. 8078* (PE); Longjiang, Dong'an, 1300 m, broad-leaved forests by river, 28. Nov. 1959, *K. M. Feng 24728* (KUN); by Lujiang, grasses by road, 10. Nov. 1959, *K. M. Feng 24514* (KUN, PE); Pengdang Township, 1420 m, 03. Oct. 2002, *Gaoligong Shan Biodiversity Survey 15719* (CAS); loc. eodem, 1640 m, 08. Oct. 2002, *Gaoligong Shan Biodiversity Survey 15841* (CAS); loc. eodem, 1520 m, 13. Aug. 2006, *Gaoligong Shan Biodiversity Survey 33228* (CAS); Salween Valley, Sekai, 1600 m, grassy slopes, 21. Sept. 1938, *T. T. Yü 23037* (KUN, PE). **Jingdong**: Hsiao-lou-shan, 2600 m, thickets, 28. Dec. 1940, *M. K. Li 1765* (KUN, PE); Lo-shui-tong, 2100 m, mixed forests, 07. Dec. 1939, *M. K. Li 2296* (PE); Wang-cha-ching, 2200 m, roadsides, 08. Jun. 1940, *M. K. Li 2869* (KUN). **Longling**: Zheng'an, 1530 m, 23. Aug. 2003, *Gaoligong Shan Biodiversity Survey 17685* (CAS). **Luchun**: Luchun to Huanglian Shan, edges of evergreen forests, 10. Nov. 2007, *T. Zhang et al. SCSB-B-000566* (KUN); Luchun to Laobian Township, 2000 m, thickets, 11. Oct. 1973, *D. D. Tao 723* (KUN) & *724* (KUN). **Mengla**: Longlin, 700 m, forests, 19. Oct. 1959, *H. T. Tsai 59 10979* (KUN); Nanxianhe, 800 m, river sides, Oct. 1936, *C. W. Wang 79489* (PE). **Pingbian**: Baiya Township, 1200 m, grasslands, 30. Jun. 1953, *P. I. Mao 2293* (KUN); Mawei Township, 880 m, grasslands, 26. Oct. 1954, *K. M. Feng 5112* (KUN, PE, SZ). **Simao**: Dakaihe, 1200 m, thickets, 14. Dec. 2001, *S. S. Zhou 410* (PE). **Tengchong**: Mingkwong Valley, 25°20′N, 1830 m, Jul. 1912, *G. Forrest 8806* (K). **Ximeng**: Mengbanhe River, 1000 m, forests, 23. Oct. 1989, *D. D. Tao 39808* (KUN). **Yangbi**: Huiminqiao to Dacun, by river, 09. Nov. 1946, *T. N. Liou 22728* (PE).

INDIA, Darjiling, Lebong, 1520 m, Sept. 1881, *J. S. Gamble 9784* (K); s. loc., 1830 m, 21. Aug. 1875, *C. B. Clarke 27084A* (K), *27084B* (K) & *27084C* (BM). **Sikkim**, Luycham, 1830 m, 21. Oct. 1875, *C. B. Clarke 25457* (K).

INDONESIA, Irian Java, Kamiri, 470 m, 27. Mar. 1973, *J. Raynal 16848* (K, P). **Java**: Besochi, ca. 1000 m, 06. Nov. 1895, *Thoorders 21109* (K). **Sulawesi** (Celebes): C Sulawesi, 24 km S of Pendolo, 500 m, 12. Dec. 1994, *K. Sidiyasa 1380* (K); Malino, baven Makassar, 1050 m, 19. Oct. 1931, *Rant 490* (K); Menado, Loewoek, between Pakoa and Pinapoeang, 19. Sept. 1938, *Eyma 3753* (K); Pass from Palu to central sulawesi, ca. 0°45′S, 120°E, ca. 800 m, 23. Feb. 1991, *G. Argent et al. 122* (E); Road to Lindau Lake, ca. 60 km SSE of Palu, 0°30′–1°30′S, 119°30′–120°30′E, 750 m, mixed rain forests, 30. May. 1979, *V. Balgooy 3561*

(K, PE); Tonsealama, near Tondano, 08. Jun. 1954, *A. H. G. Alston 15721* (BM, PE). **Sumatra**: East Coast, *H. S. Yates 1485* (B); East Coast, Asahan, vicinity of Tomoean Dolok, ca. 1000 m, 10–15. Jun. 1936, *R. S. Boeea 9056* (PE); East Coast, Asahan, vicinity of Aek Moente, ca. 500 m, 15. Jun–09. Jul. 1936, *R. S. Boeea 9104* (PE); Tapianoeli, Toetoepan Distr., 4–11. Nov. 1933, *R. S. Boeea 5996* (PE); N Sumatra, Gaju & Alas Lands, 800 m, 16. Feb. 1937, *C. G. G. J. van Steenis 8851* (K); NW Simelungun, N. V. Si-anakanak, Talwand, 16. Mar. 1925, *J. A. Lörzing 11488* (K, PE).

LAOS, **Nahay**: s. loc., 14–16. Dec. 1921. *B. Hayata* s. n. (TI).

MALAYSIA, Malay Peninsula, Selangor, Aug. 1904, *H. N. Ridley 12068* (K); Selangor, Track to Bukit Fraser, 22. Jan. 1921, *H. N. Ridley* s. n. (K). **N Borneo**: Kinabalu Mts, Kiau, 1070 m, 20. Aug. 1933, *J. & M. S. Clemens 40118* (BM, PE); Kinabalu Mts, Kibayo to Keung, 29. Oct. 1915, *M. S. Clemens 9834* (K, PE); Kinabalu Mts, B. N. B. Dallas, forest trail, above 1000 m, 24. Aug. 1931, *J. & M. S. Clemens 26043* (BM, K, PE); loc. eodem, 1220–1530 m, 15. Feb. 1933, *J. & M. S. Clemens 31653* (BM); Sabah, Tambunan Distr., Crocker Range, 5°46′N, 116°21′E, 1220 m, 09. Oct. 1983, *J. H. Beaman et al. 7176* (K); Sabah, Ranau Distr., Tenompok Ridge, 6°01′N, 116°30′E, 1400–1450 m, 03. Jan. 1984, *J. H. Beaman et al. 8206* (K); Sabah, Tambunan Distr., 21. Jul. 1984, *Amin et al. 60828* (K).

MYANMAR, N Mymnar, s. loc., 3050 m, Sept. 1924, *G. Forrest 24903* (K). **Amherst Distr.,** Spur to Mulegit, 610–1220m, 26. Jan. 1912, *J. H. Lace 5621* (K). **Tenasserim Division** (S Myanmar): Tavoy Distr., ca 14°N, 98°30′E, 670 m, Aug. 1961, *J. Keenan et al. 806* (K, PE).

NEPAL, **Bagmati**: Nuwakot, 27°58′N, 85°11′E, Sim Chotala, 900 m – 1800 m – Gram 1900 m, 20. Aug. 1972, *H. Kanai et al. 721668* (TI). **Kosi**: Arun Valley, N of Num, 87–88°E, 1370 m, edges of fields, 18. Aug. 1956, *J. D. A. Stainton 1366* (BM, PE); Arun Valley, Sabhaya Khola, 1530 m, 07. Sept. 1956, *J. D. A. Stainton 1558* (BM).

PAPUA NEW GUINEA, **BoRiDi**: ca. 1370 m, 27. Sept. 1935, *C. E. Carr 14235* (BM, K). **Central Division**: Mafulu, 1250 m, Sept–Nov. 1933, *L. J. Brass 5183* (K, PE). **Isuarava**: ca. 1070 m, secondary forests, 24. Feb. 1936, *C. E. Carr 15731* (BM, K, PE). **Morobe**: 6°55′S, 146°35′E, ca. 910 m, forests by streams, 24. Sept. 1970, *A. Kairo & H. Streimann 45242* (E, K, PE); S of Boana, Lae, 6°30′S, 146°50′E, 900 m, disturbed forests, 20. Feb. 1977, *Conn, Kairo & Masapuhafo 89* (K, PE); Boana, 760–1370 m, May to Nov. 1940, *M. S. Clemens 41634* (PE); loc. eodem, ca. 1000 m, fr. white, Jun. 14. 1938, *M. S. Clemens 8274A* (B); above Boana, open bushes, 1100–1500 m, fl. white or lavender tinge, fr. white, 23. Aug. 1938, *M. S. Clemens 8678* (B); Bulolo River gorge between Wau and Bulolo, 04. Apr. 1963, *Th. G. Hartley 11566* (K, PE).

PHILIPPINES, **Luzon**: Sierra Mts, NNW of Dingalan, 15°29′N, 121°23′E, 600–800 m, secondary forests, 13. Mar. 1968, *M. Jacobs 7727* (K, PE). **Mindanao**: SE Mindanao, Mt Apo, 1070 m, 28. Mar. 1905, *R. S. Williams 2583* (K); Davao, Mt Mayo, Apr.–May, 1927, *M Ramos Edano 2–107* (B). **Nueva Vizcaya** Prov., Jan. 1913, *R. C. MrGregor 20150* (K). **Panay**: Capiz Prov., Mt MaDiaas, Apr.–May. 1919, *M. Ramos & G. Edano 30654* (BM, K); Iloilo Prov., 1–11. Jan. 1913, *C. B. Robinson 18201* (K).

THAILAND, **Chiengmai,** Saket, 31. Oct. 1930, *A. P. G. Kerr 3277* (BM, K); Chê Sawn, Pa Huang, 1100 m, 19. Jun. 1914, *A. P. G. Kerr 3104* (BM, K); Kanchanaburi, E of Sangkhla, 280–880 m, 27. Mar. 1968, *C. F. van Beusekom & C. Phengkhlai 142* (K). **Kradung**: Mt., 16°53′N, 101°48′E, 02–03. Nov. 1984, *G. Murate et al. T-42890* (TI). **North Prov.**: Phrae, Mae Kray, 17°45′N, 100°15′E, 600 m, 10. Jan. 1972, *C. F. van Beusekom et al. 4786* (K). **Phangnga**: ca. 8°30′N, 99°E, 900 m, fr. white, 27. Feb. 1974, *R. Geesink et al. 7668* (B); Khao Paw, Ta Luang Keo, 9°20′N, 97°40′E, 1000–1300 m, evergreen forests, 24. Apr. 1974, *K. Larsen & S. S. Larsen 33469* (PE); Kao Katalawam, 1600 m, 07. Mar. 1930, *A. P. G. Kerr 18421* (BM, K). **Ranong**: Kho pawta luang keou, 8°50′N, 99°0′E, 700–900 m, 22. Jun. 1947, *R. Geesink et al. 7427* (K). **Soutep**: 670 m, 12. Jun. 1910, *A. P. G. Kerr 1217* (BM, K). **Wang Tao**: s. loc., 23. Apr. 1958, *Th. Sørensen et al. 2984* (PE). **Wiengchan**: Pu Tat, ca. 1000 m, 22. Apr. 1932, *A. P. G. Kerr 21194* (BM, K).

3. **Cyclocodon axillaris** (Oliv.) W. J. de Wilde & Duyfjes, Thai Forest Bull. Bot. 40: 22. 2012. ≡*Campanumoea axillaris* Oliv., Hook. Icon. Pl. 3: 8, t. 1775. 1888. Type: China, Sichuan, Omei (Emei), 1550 m, Dec. 1887, *R. E. Faber 253* [lectotype here designated: K (K000781227); isolectotype: K (K000781228)].

Codonopsis albiflora Griff., Not. Pl. As. 4: 279. 1854, ambiguous name. Type: India, Mergue, about Banlaw and Yengboo, Feb. 1835, *Griffith 1013* (n. v.; not preserved?).

Plants sometimes lignified at base of stem, usually glabrous throughout. Stems erect or ascending, up to 3 m tall, hollow; branches many, horizontal or pendulous. Leaves opposite, rarely in whorls of 3, shortly petiolate; petiole 3–10 mm long; leaf blade ovate to lanceolate, 6–16 cm long, 1–7 cm broad, base rounded, margin serrate, dentate, or crenate, apex acuminate. Flowers usually solitary, sometimes 3 in a dichasium, usually both terminal and axillary; pedicel or peduncle 1–10 cm long, with a pair of filiform bracteoles at middle to top. Calyx adnate to hypanthium up to lower part of hypanthium; calyx lobes (4–)5(–7), filiform or linear, margin with several ramiform teeth. Corolla white or pale red, rotate,

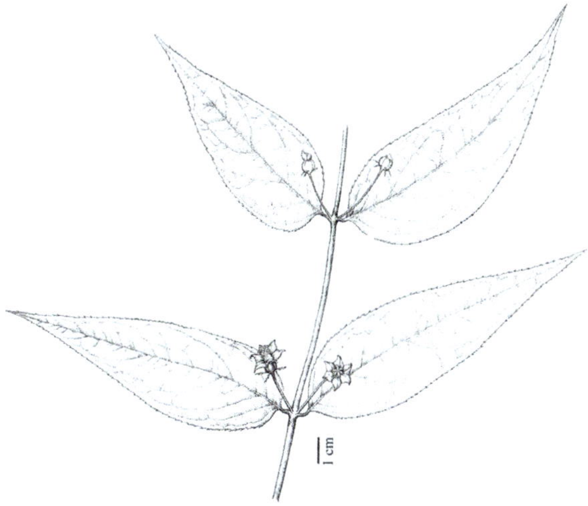

FIGURE IV 5:3A *Cyclocodon axillaris* (Oliv.) W. J. de Wilde & Duyfjes. **a**, a section of erect stem, showing solitary axillary flower and ovary which is nearly superior to calyx.

7–12 mm long, 5- or 6-cleft to middle; corolla lobes ovate to ovate-triangular. Stamens 5 or 6; filaments as long as anthers, dilated base with margin villous-ciliate. Style glabrous or hairy; stigma (4–)5- or 6-fid; ovary (4–)5- or 6-locular. Berry dark purple when mature, globose, (4–)5- or 6-locular, 5–10 mm in diameter. Seeds ovoid. Figures IV 5:3A & B.

Chromosomes: 2n=18 (Zhang et al., 2010).

Phenology: flowering and fruiting from July to November.

Habitats: forests, thickets, or grasslands, below 1670 m.

Distribution: widely distributed in E and SE Asia: South China, Indonesia, Japan (Ryukyu and Kyushu), Laos, Philippines, Thailand, and Vietnam (Map IV 5:3).

Notes: This taxon was for many years misidentified as *Campanumoea lancifolia* Roxb. (Merr., Enum. Philip. Pl. 3:587. 1923; Hong, Fl. Reip. Popularis Sin. 73(2): 72. 1983) or *Codonopsis lancifolia* (Roxb.) Moeliono subsp. *lancifolia* (Moeliono in van Steenis, Fl. Males. Ser. 1. 6: 120. 1960). The specific name of this entity has to be changed to *C. axillaris* (see the notes under *C. lancifolius* for details).

Additional specimens examined, **CHINA**, **CHONGQING**, **Beibei**: Jinyun Shan, 07. Nov. 1940, *C. Pei 7529* (PE). **Chengkou**: s. loc., *R. P. Farges 1563* (P, PE) ; Mingtongjing, 585 m, grasses, 10. Oct. 1958, *T. L. Dai 103607* (PE, SZ). **Fengjie**: Xinhe, 900 m, 14. Oct. 1964, *H. F. Zhou & H. Y. Su 111448* (PE, SZ); Xinzheng Township, Lianhe to Baota, 1029 m, 03. Sept. 2009, *Sanxia Exped. 3270* (PE); Zhuyuan, 750 m, 16. Oct. 1964, *H. F. Zhou & H. Y. Su 111629* (PE, SZ); loc. eodem, 950 m, edges of forests, 06. Oct. 1964, *H. F. Zhou & H. Y. Su 110727* (PE). **Nanchuan**: Jinfo Shan, Banzhulin, 950 m, grasslands, 05. Sept. 1957, *G. F. Li 64057* (KUN, SZ); Jinfo Shan, Leishuixi, 650 m, 19. Sept. 1957, *G. F. Li 64271* (KUN, PE, SZ). **Wushan**: Baolong, Daxihe, 272 m, 10. Sept. 2009, *Sanxia Exped. 3679* (PE); Duping Township, Xiaoxihe, 185–280 m, 08. Sept. 2009, *Sanxia Exped. 3523* (PE). **Xiushan**: Rongxi Township, 700 m, 26. Aug. 1988, *Z. C. Zhao 88–2151* (CDBI). **Youyang**: Xiaoxian, Daheiwan, 1650 m, 26. Aug. 1984, *S. X. Tan 368* (PE). **FUJIAN**, **Nanjing**: Hexi, Ludou Shan, 600 m, dense forests, 05. Dec. 1981, *G. D. Ye 2287* (PE); Hexi, 18. Nov. 1932, *Y. Ling 4593* (PE); loc. eodem, 08. Sept. 1974, *236–6 1322* (PE); Shuhaichashan, 400 m, 18. Nov. 1963, *Xiamen Univ. Exped. 743* (PE). **Shanghang**: Guanhuban, 700 m, forests, 06. Sept. 1987, *L. K. Ling 6977* (PE); Qiushan, 800 m, edges of forests, 09. Sept. 1987, *L. K. Ling 3579* (PE). **Shunchang**: Jihui Township, 262 m, roadsides, 17. Aug. 1986, *M. S. Li & Z. Y. Li et al. 6435* (PE); loc. eodem, 262 m, forests, 18. Aug. 1986, *M. S. Li & Z. Y. Li et al. 6446* (PE); Lanxia, 350 m, dense forests, 30. Dec. 1979, *G. S. He 0343* (PE); s. loc., 306 m, roadsides, 23. Sept. 1988, *M. S. Li & Z. Y. Li 6629* (PE). **Wuyi**: Daling, Tianping, 550 m, edges of mixed forests, 16. Aug. 1964, *Xiamen Univ. Exped. 1223* (PE); Yanghuangkeng, 700 m, dense forests, 08. Sept. 1979, *Wuyi Exped. 1398* (PE). **GUANGDONG**, **Conghua**: Lutian Township, 500–800 m, open woods, 15. Nov. 1958, *L. Deng 8524* (PE). **Dabu**: Pucheng Township, Gaoqian, dense forests, 16. Aug. 1958, *X. G. Li 202501* (PE). **Huaiji**: Douyue Shan, 16. Nov. 1933, *W. T. Tsang 23201* (P, PE); Beizhong, Sanyue Shan, 500 m, 03. Nov. 1958, *Y. G. Liu 2922* (PE). **Lecang**: Hongliankeng, by stream, 25. Oct. 1930, *N. Q. Chen 41996* (PE, SZ); Zhuangyuan Shan, 2–30. Nov. 1932, *W. T. Tsang 20765* (PE). **Longmen**: Nankun Shangping, 28, Oct. 1958, *C. F. Wei 122027* (KUN). **Meixian**: Jiaying, Yingna Shan, 4–31. Aug. 1932, *W. T. Tsang 21495* (P, PE); loc. eodem, grasses, roadsides, s. d. *W. T. Tsang 21457* (P, PE). **Renhua**: Changjiang Township, forests, 31. Aug. 1958, *L. Deng 7374* (KUN, PE, SZ). **Wengyuan**: Qingyun Shan, forests, 27. Sept. 1933, *S. K. Lau 2424* (PE); Wengyuan to Yingde, forests, Aug.–Oct. 1933, *S. K. Lau 989* (P, PE). **Xinfeng**: Huangbo Township, 700 m, forests, 09. Oct. 1958, *L. Deng 7896* (KUN, PE, SZ); s. loc., 11. Aug. 1931, *S. P. Ko 51774* (SZ). **Yangshan**: Chengjia Nature Reserve, 24. Oct. 1985, *Guangdong & Guangxi Exped. 0324* (PE). **Yingde** (Ying-Tak): Wentongshan, 28 Aug. 1931, *H. Y. Liang 60900* (PE, SZ). **GUANGXI**, Chen Pien Distr. (?), woodside, 09. Oct. 1935, *S. P. Ko 55874* (PE). **Bose**: vicinity of the town, thickets, 1937, *Y. W. Taam 40* (PE). **Damiao Shan**: grasses by road, 25. Aug. 1958, *S. H. Chun 16490* (KUN). **Guilin**: Qifeng Shan, 1–11, Oct. 1937, *W. T. Tsang 28418* (PE). **Hexian**: Heisongpian, thickets, 10. Oct. 1959, *Z. T. Li 604106* (PE). **Lingui**: Yuanfen Shan, forests, 18. Nov. 1953, *C. F. Liang 31062* (KUN). **Longsheng**: Pingshui Township, 14. Dec. 1955, *Guangfu Forests Farm Exped. 1185* (PE, SZ). **Luocheng**: Qiaoshan Township, 250 m, edges of forests, 05. Nov. 1999, *X. C. Zhang & X. H. Jin 1371* (PE). **Napo**: 20 km NW of Napo, 1200 m, 20. Oct. 2002, *H. Wang 6582* (PE). **Ningming**: s. loc., forests, 09. Oct. 1958, *C. C. Chang 12022* (KUN). **Pingxiang**: Fubo Forest Farm, thickets by forest, 550 m, 03. Sept. 1986, *Beijing Youth*

FIGURE IV 5:3B *Cyclocodon axillaris* (Oliv.) W. J. de Wilde & Duyfjes. **a**, **d** & **e**, from Sichuan: Emei, 19. Sept. 2011, by De-Yuan HONG; **b** & **c**, from Chongqing Three Gorges, 03–08. Sept. 2009, by Xin-Tang MA.

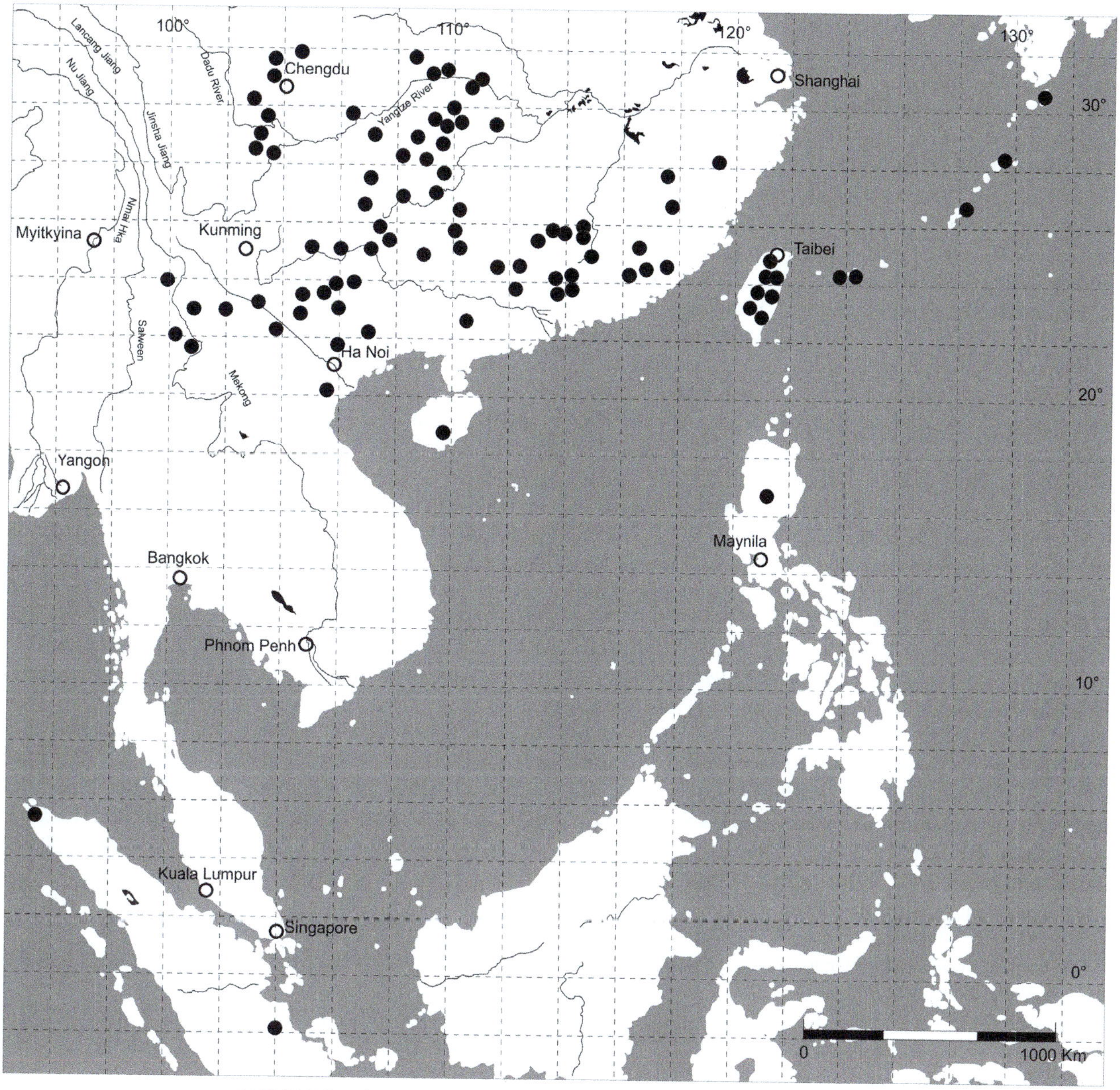

MAP IV 5:3 Distribution of *Cyclocodon axillaris* (Oliv.) W. J. de Wilde & Duyfjes.

Exped. 1237 (PE). **Rongxian**: (Yung Hsien): Ta Tseh Shan, 540 m, forests, 29. Sept. 1933, *A. N. Steward & H. C. Cheo 893* (P, PE). **Tian'e**: Dashan Forest Farm, 06. Sept. 1989, *Beijing Exped. 896272* (PE). **GUIZHOU**, **Ceheng**: Rongju, Lujiadong, 1500 m, 02. Sept. 1958, *Z. Y. Cao 452* (PE); Zhelou, 560 m, by stream, 07. Aug. 2004, *S. C. Wang 0347* (PE). **Dejiang**: Qinglong Township, Chazi Shan, 840 m, forests, 13. Sept. 2003, *M. X. An 3818* (PE); Wenping, 780 m, thickets in ravines, 06. Oct. 2003, *M. X. An 3920* (PE). **Dushan**: Pingtang, 1000 m, dense forests, 05. Sept. 1959, *Libo Exped. 1502* (PE). **Guiyang**: Shibanshao, banbianshan, 1200 m, 24. Oct. 1956, *Sichuan-Guizhou Veget. Exped. 2057* (SZ). **Libo**: Maolan Nature Reserve, 680 m, forests, 14. Aug. 2003, *Y. Xiang 2–335* (PE). **Shibing**: Yuntai Shan, 1000 m, edges of forests, 31. Aug. 1988, *Wulingshan Exped. 3581* (KUN,PE). **Wangmo**: Leyuan, 750 m, open woods, 20. Aug. 2003, *G. F. Wang 1–0729* (PE); Youmai Township, 750 m, open woods, 22. Aug. 2003, *G. F. Wang 1–0739* (PE). **Xingyi**: Jingnan, 10. Aug. 1959, *Anshun Exped. 768* (KUN, PE). **Zunyi**: Liangfenya, 1150 m, 12. Aug. 1931, *A. N. Steward et al. 291* (PE). **HAINAN, Wuzhi Shan**: 700 m, shrubs, 21. Dec. 1921, *F. A. McClure 2133* (PE). **HUBEI, Badong:** Yanhedu, 600 m, broad-leaved forests, 30. Jul. 2009, *C. M. Zhao et al. Ex 2321* (PE). **Hefeng**: s. loc., 250 m, 20. Oct. 1955, *H. J. Li 8124* (SZ); s. loc., *H. J. Li 6510* (SZ). **Laifeng**: s. loc., 700 m, forests, s. d. *H. J. Li 7051* (KUN, SZ). **Zigui**: Maoping, Sixi, 370 m, broad-leaved forests, 19. Sept.

2007, *C. M. Zhao EX1270* (PE). **HUNAN, Baojing**: Quanpong, 500 m, forests, 10. Sept. 1958, *L. H. Liu 9732* (KUN). **Fenghuang**: Liangdouyang Township, 700 m, ravines, 20. Sept. 1988, *Wulingshan Exped.* 1375 (PE). **Jianghua**: Hexikou, 500 m, 20. Oct. 1975, *B. G. Li & S. B. Wan 5102* (KUN). **Sangzhi**: Longtanping, 500 m, ravines, s. d., *Sangzhi Forestry Inst. 1615* (KUN); Tianping Shan, forests, 27. Aug. 1988, *Beijing Exped. to Hunan 4142* (PE). **Shimen**: Jiangping, 450–800 m, 05. Oct. 1987, *Hupingshan Exped. A16* (PE). **Suining**: Huangsang Nature Reserve, 530 m, edges of forests, 14. Aug. 2002, *Q. Z. Lin 2002222* (PE). **Yizhang**: Mangshan, 520 m, 10. Oct. 2005, *B. Z. Xiao 4824* (PE). **Yongshun**: Xiaoxi, 300–500 m, evergreen forests, 22. Oct. 1982, *H. Li & H. Y. Yu 4460* (KUN); loc. eodem, 300 m, forests, 02. Aug. 1983, *Hunan-Guizhou Exped. 2832* (KUN); loc. eodem, 300 m, 24. Oct. 1982, *H. Li et al. 2074* (KUN) & *2075* (KUN); loc. eodem, 200 m, broad-leaved forests, 24. Oct. 1982, *H. Li et al. 1834* (KUN); Xiaoxi Nature Reserve, 250–350 m, evergreen broad-leaved forests, 29. May 1988, *Beijing Exped. to Hunan 352* (PE). **Zhijiang**: Dashuao Township, 500 m, 15. Oct. 1988, *Wulingshan Exped. 2082* (PE). **JIANGXI, Chongyi**: Ridu, Zhongdong, 400 m, 14. Sept. 1984, *D. C. Wu 840336* (PE); Niedu, 17. Sept. 2011, *X. X. Huang et al. 10901* (PE). **Dayou**: Lannijing Forest Farm, 19. Sept. 1984, *D. C. Wu 840400* (PE). **Jiang'an**: Zhongyuan Township, 490 m, evergreen broad-leaved forests, 08. Nov. 2004, *Lushan Bot. Gard. Exped. WN 200400489* (PE). **Shangyou**: Shuangxi, forests, 24. Sept. 1971, *Jiangxi Exped. 798* (PE); Wuzhifeng, Luojia Shan, 01. Sept. 1971, *Jiangxi Exped. 622* (PE). **SICHUAN, Beichuan**: Bodi, 1000 m, 02. Sept. 1984, *C. L. Tang et al. 676* (CDBI). **Chongzhou**: Jiulonggou, 700 m, 30. Aug. 2007, *X. J. He et al. 131330* (SZ); Wanjia, 1000 m, 26. Jul. 2008, *X. J. He & C. S. Cao 163402* (SZ). **Dujiangyan**: Qingcheng Shan, 1000 m, 07. Nov. 1938, *T. P. Wang 9999* (PE); loc. eodem, 650 m, 02. Nov. 1956, *D. P. He 47196* (SZ); loc. eodem, 700 m, 19. Jul. 2006, *C. S. Cao 293* (SZ); s. loc., 18. Aug. 1953, *L. Z. Hu 21303* (SZ). **Ebian**: Daping to Shanping, 28. Oct. 1938, *T. N. Liou 12643* (PE). **Emei:** Baozhangfong, 1400 m, 22. Oct. 1941, *W. P. Fang 19075* (SZ); Baozhang Shan, 1400 m, 25. Nov. 1956, *G. F. Li 53299* (SZ); below Baozhang Shan, 950 m, 23. Aug. 1957, *G. H. Yang 56852* (KUN, PE, SZ); Chudian, 27. Sept. 1942, *C. L. Chow 6796B* (SZ); Chunyangdian, 800 m, 03. Sept. 1956, *Y. H. Tao 51921* (SZ); Da'e Si, 02. Sept. 1938, *Z. W. Yao 4962* (PE); Emei Shan, 1100 m, 22. Aug. 1938, *C. Y. Chiao & C. S. Fan 360* (PE); Fuhu-si, 550 m, 18. Dec. 1956, *G. F. Li 53400* (SZ); Heilongjiang, 750 m, 30. Aug. 1956, *Y. H. Tao 51836* (SZ); Hongchunping, 1000 m, 22. Aug. 1957, *G. H. Yang 56825* (KUN); loc. eodem, 14. Sept. 1952, *J. H. Xiong et al. 32540* (SZ); loc. eodem, 13. Sept. 1952, *J. H. Xiong et al. 32457* (SZ); loc. eodem, 1150 m, thickets, 27. Aug. 1964, *K. C. Kuan et al. 1831* (CDBI, PE); loc. eodem, 22. Aug. 1940, *L. Y. Tai 380* (PE); loc. eodem, 900–1000 m, 22. Aug. 1938, *H. C. Chow 8182* (PE); loc. eodem, 05. Aug. 1957. *S. Y. Chen et al. 4109* (SZ); loc. eodem, 14. Sept. 1957, *S. Y. Chen et al. 4299* (SZ); Niuxin-si, 950 m, 22. Aug. 1957, *G. H. Yang 56834* (PE, SZ); loc. eodem, 21. Sept. 1952, *J. H. Xiong et al. 32840* (SZ) ; Qingyinge to Hongchunping, 900 m, 12. Oct. 1956, *G. F. Li 52648* (SZ); loc. eodem, 650 m, 17. Sept. 1956, *Y. H. Tao 52115* (SZ); below Qingyinge, 29 Jul. 1956, *L. Z. Hu 51088* (SZ); Wannian-si, 19. Jul. 1973, *Y. B. Yang 013* (CDBI) ; Wannian-si, 1040 m, 31. Aug. 1941, *W. P. Fang 17628* (SZ); loc. eodem, 01. Nov. 1952, *H. L. Tsiang & S. S. Chang 33421* (SZ); loc. eodem, 1000 m, 19. Sept. 2011, *D. Y. Hong et al. H11021* (PE); s. loc. 1700 m, 10. Oct. 1940, *T. C. Lee 3778* (KUN, SZ); s. loc., Oct. 1940, *C. L. Zhou 4640* (SZ); s. loc., 05. Sept. 1957, *S. Y. Chen et al. 4271* (SZ); s. loc., 04. Oct. 1958, *S. Y. Chen et al. 61918* (SZ) ; s. loc., 950 m, 08. Sept. 1940, *L. Y. Tai 459* (PE); s. loc., Sept. 1985, *Y. L. Chen 8530* (CDBI). **Hongya**: Hongya Forest Farm, 770 m, 24. Sept. 1993, *W. K. Bao 1692* (CDBI). **Leibo**: Xiaogou, 1200 m, by rock, 22. Aug. 1983, *C. S. Cao & M. Y. He121663* (CDBI, SZ); loc. eodem, 1200 m, 29. Oct. 1983, *C. S. Cao & M. Y. He 122987* (SZ); s. loc., 1180 m, edges of forests, 11 Sept. 1983, *C. S. Cao & C. M. Tan 118015* (CDBI, SZ). **Lushan**: Siyan Township, Shizipo, 06. Aug. 1983, *D. Y. Hu 47989* (CDBI). **Pengxian**: Dabaoshan, 1315 m, 31. Jul. 2007, *X. J. He et al.137632* (SZ). **Wenchuan**: Sanjiang, 1050 m, 25. Jul. 2007, *X. J. He et al. 129841* (SZ). **TAIWAN,** s. loc., *A. Henry 1458* (K). **Botel Tobago** (isl.)**:** 04–17. Aug. 1968, *G. Ikeda 2310* (TI). **Hualian**: Tushatuo, 23°22′N, 121°09′E, 1600 m, 20. Aug. 2002, *T. T. Chen 12100* (PE); Daluge-Tianxiang, 20. Aug. 1968, *T. Namba et al. 2220* (TI). **Jiayi**: Dabanshe, 09. Oct. 1939, *s. coll. 1780* (TI). **Miaoli**: Tai'an Township, 24°25′N, 120°58′E, 1250–1600 m, mixed coniferous and broad-leaved forests, 17. Oct. 2002. *P. J. Lin 90* (PE). **Nanhudashan**: Nanshan, 1200 m, 19. Aug. 1969, *T. Yamazaki et al. 917* (TI); Nanshan, 1140 m, 19. Aug. 1969, *C. C. Hsu 5808* (TI). **Nantou**: Chitou to God Tree, 1000 m, *C. C. Hsu 6141* (TI); Puli-Wushe, 31. Oct. 1929, *S. Saito 5860* (TI); Chitou, 1300 m, 02. Sept. 1969, *T. Yamazaki & F. Yamazaki 199* (TI). **Taibei**: Urai, thickets, 11. Dec. 1961, *M. T. Kao & T. I. Chuang 4678* (PE); Urai, Taihoku-shu, 17. Oct. 1932, *T. Tanaka & Y. Shimada 13426* (B, P, PE). **Taidong**: s. loc., 02. Nov. 1939, *s. coll. 2162* (TI). Yangming Shan, 300–700 m, 17. Dec. 1997, *H. L. Chiang 486* (PE). Mt. Taiheizan, 19. Aug. 1925, *S. Suzuki 295* (PE). **Xinzhu**: Wufeng, 24°39′N, 121°08′E, 565 m, 13. Jan. 2003, *C. M. Wang 06460* (PE). **YUNNAN, Funing**: s. loc., 1410 m, edges of forests, 15. Oct. 1965, *Wenshan (65) Group 187* (KUN). **Lincang**: Mayidui, 1350 m, open woods, 24. Aug. 1957, *J. S. Xin 335* (KUN, PE). **Luchun**: Dahei Shan, 510 m, thickets, 04. Oct. 1973, *D. D. Tao 507* (KUN, PE); loc. eodem, 530 m, thickets, 05. Oct. 1973, *D. D. Tao 539* (KUN, PE); loc. eodem, 700 m, thickets, 30. Sept. 1973, *D. D. Tao 612* (KUN). **Malipo**: Kwan-Kao, *C. W. Wang 86839* (KUN, PE); Hwang-jin-yin, 1100 m, near woods, 21. Jan. 1940, *C. W. Wang 86335* (KUN, PE); Hwang-jin-in, 1300–1600 m, mixed forests, 07. Nov. 1947, *K. M. Feng 12984* (KUN, PE). **Mengla**: Mengxing, monsoon forests, 09. Jan. 1982, *C. Y. Wu et al. 82020* (KUN); Mengnen, 800 m, forest edges, Oct. 1936, *C. W. Wang 78853* (KUN, PE); Longhu, 880 m, ravines, Oct. 1936, *C. W. Wang 80183* (KUN,PE).

Pingbian: s. loc., 1600 m, border of dense woods, 15. Oct. 1939, *C. W. Wang 82500* (PE); Shiban Township, 400 m, ravines, 29. Sept. 1954, *K. M. Feng 4714* (KUN, PE, SZ). **Simao**: s. loc., 1520 m, s. d. *A. Henry 12590* (PE); Maliping, 1000 m, open woods, 03. Oct. 1955. *P. I. Mao 6398* (KUN). **Xishuangbanna**: Puwen, 760–930 m, ravines, 16. Oct. 1955, *P. I. Mao 6745* (KUN, PE, SZ). **Xichou**: Fadou, 1300 m, open forests, 08. Dec. 1939, *C. W. Wang 85488* (KUN, PE); loc. eodem, 1350–1450 m, open thickets, 11 Sept. 1947, *K. M. Feng 11641* (KUN, PE); Xichou to Houzidong, 29. Nov. 1943, *C. P. Chien* s. n. (PE). **ZHEJIANG, Longquan**: Dasai to Guanuyang, Shizikou, thickets, 01. Oct. 1961, *s. coll. 8927* (PE).

INDONESIA, Boeroe: Nal'Besi, 800 m, 09. Jun. 1921, *Toxopeus 185* (K). **Ceram:** between Piliana (400 m) and Hatumete (sea level), 14. Feb. 1985, *M. Kate et al. C-6659* (TI). **Sumatra**: East Coast, Soelakli, Karolank, 21. Jun. 1928, *C. Hamel & R. S. Toroes 763* (PE); Berastaqi, Baltak Valley, Feb. 1921, *H. N. Ridley* s. n. (K); N Sumatra, Atjeh, 120–180 m, 15. Jul. 1972, *Wilde & Wilde-Duyfjes 13611* (K).

JAPAN, Kagoshima: Naze, 04. Dec. 1900, *T. Uchiyama* s. n. (TI); Tanegashima (from Iwatsuki et al., *Fl. Japan* IIIa, 1993). **Okinawa**: Yabe, Katsudake, 400 m, evergreen forests, 13. Oct. 1990, *T. Yamazaki* s. n. (TI). **Yaeyamaretto**: s. loc., Aug. 1887, *Tashiro* s. n. (TI); Iriomote Island, 31 Aug. 1951, *H. Walker et al. 6890* (TI); Ishigaki, Yonehara, 29. Aug. 1972, *M. Furuse 949* (K); loc. eodem, 17. Sept. 1972, *M. Furuse 1197* (K); Ishigaki, Mt Banna-dake, 200–300 m, 26. Dec. 1983, *J. Murata 1702* (K); Yaeyama Islands, 04. Feb. 1973, *M. Furuse 2298* (K).

LAOS, 240 km from Vinh to Trauinh, 1400 m, 30. Aug. 1929, *M. E. Poilane 16787* (P).

PHILIPPINES, Luzon: Bontoc, 01. Sept. 1914, *F. M. Vanoverbergh 965* (P).

THAILAND, Ching Dam, 01. Dec. 1957, *T. Tuyama T57312* (TI).

VIETNAM, Lao Cai: Lao to Shaba, 22–25 km from Lao Cai, 750–800 m, granites, secondary forests, 25. Dec. 1964, *Sino-Vietnamese Exped. 882* (KUN, PE). **Ninh Binh**: 20°23′N, 105°34′E, 250 m, 15. Oct. 2001, *N. M. Cuong NMC 1495* (P); s. loc., 460–590 m, forests, 05. Feb. 1965, *Sino-Vietnamese Exped. 2308* (KUN). S of **Nnê**, 1200 m, 14. Sept. 1938, *M. E. Poilane 27813* (P). **Thai Nguyen**: Vo Nhai Distr., 21°49′N, 105°58′E, limestone forests, 22. Oct. 2004, *L. Q. Li et al. 0205* (PE). **Thua Thien-Hue**: Bach Ma Natl. Park, 200 m, dense forests, 19. Oct. 1998, *H. L. Qing 196* (PE). **Tonkin**: Lankor, Mt Ozavi, Oct. 1887, *B. Balansa 3616* (K, P); Chapa. Aug. 1927, *M. Petelot 5095* (P).

ECHINOCODON D. Y. Hong

Hong, Acta Phytotax. Sin. 22(3): 183. 1984; Lammers in Kubitzki, The Families and Genera of Vascular Plants 8: 38. 2007; Hong & Lammers in Wu, Raven & Hong, Fl. China 19: 528. 2011. Type: *Echinocodon draco* (Pamp.) D. Y. Hong (*Echinocodon lobophyllus* D. Y. Hong).

Herbs perennial, diffuse. Roots slightly thickened, carrot-shaped. Leaves alternate, petiolate, pinnately lobed. Flowers solitary, terminal, or in a cyme of 2 or 3. Calyx (2–) 4 (–5)-lobed; calyx lobes with 2–4 spinose lobules. Corolla tubular, 3–5 (often 4)-lobed to middle. Stamens 3–5, mostly 4; filaments free, dilated at base, dilated part ciliate; anthers connivent. Ovary almost entirely inferior, 3–5-locular; stigma lobes equal in number to ovary locules, clavate, recurved; ovules numerous. Fruit a capsule, globose, superior part conical, loculicidal, valves alternate with calyx lobes. Seeds numerous, ellipsoid, 3-ribbed, ribs obtuse. Pollen short-colpate, colpi short with colpus length/polar axis length ≤ 0.4; spinules on sexine divided at base. Chromosome number 2n=16.

Monospecific genus endemic to China.

The author who first found this plant is an Italian botanist, Pampanini, but he placed it in the genus *Codonopsis*. However, this plant differs clearly from all members of *Codonopsis* in gross morphology: flowers mostly 4-merous (vs. 5-merous) and calyx lobes with 2–4 spinose lobules (vs. entire or obscurely dentate). Hong and Pan's (2012) palynological observation shows that pollen in *Echinocodon* is short-colpate with colpus length / polar axis length ≤ 0.4 (vs. ≥ 0.7) and spine base divided (vs. entire). Molecular data (Wang et al. 2013; Wang et al. 2014a) show that *Echinocodon* and *Codonopsis* are in different major clades; the former forms a clade with *Platycodon*, *Canarina* and *Cyclocodon*. All the data indicate that *Echinocodon* is not closely related to *Codonopsis*.

1. **Echinocodon draco** (Pamp.) D. Y. Hong, Plant Divers. Resourc. 36(3): 301. 2014. ≡*Codonopsis draco* Pamp., Nuov. Giorn. Bot. Ital. 17: 733. fig. 19. 1910. Type: Pampanini's figure cited (lectotype designated by Hong, 2014).

Echinocodon lobophyllus D. Y. Hong, Acta Phytotax. Sin. 22(3) : 183. pl. 1. 1984 ; Hong & Lammers in Wu, Raven & Hong, Fl. China 11 : 528. 2011. Type : China, Hubei Prov., Yunxi CO, near the town, 300 m, 01. Jul. 1983, *Y. J. Ma & D. Y. Hong PB83006* (holotype: PE; isotype: EDC).

Plants glabrous throughout. Roots up to 5 mm in diameter. Stems up to 40 cm long, multi-branched. Petiole 5–10 mm long; leaf blade elliptic, 5–20 mm long, 3–15 mm broad, adaxially green-gray, abaxially green, base narrowly cuneate, apex obtuse, margin pinnatipartite nearly to middle vein or to middle. Pedicel 1–5 cm long. Calyx lobes ovate-lanceolate, 2.6 mm long, 1–3 mm broad (lobules excluded). Corolla purple-blue, 3–4.5 mm long. Stamens ca. 1.5 mm long; anthers narrowly oblong, ca. 1 mm long. Style ca. 1 mm long. Capsule globose, 3–5 mm in diameter, superior part conical, up to 2 mm long. Seeds minute, ca. 0.3 mm long. Figure IV 6:1.

FIGURE IV 6:1 *Echinocodon draco* (Pamp.) D. Y. Hong. **a**, habit; **b**, a flower with corolla opened, showing calyx, stamens, style, and stigma; **c**, a 4-loculicidal capsule. **a** & **d**, from Shaanxi: Ankang, 04. May 2013, by Yi REN; **b** & **c**, from Shaanxi: Shangyang, 22. May 2011, by Bin LI.

Chromosomes: 2n=16 (Hong, 1984).

Phenology: flowering and fruiting in June and July.

Habitats: among rocks and grasses by fields; ca. 300 m.

Distribution: **endemic** to China; confined to NW Hubei and SE Shaanxi (Map IV 6:1).

Additional specimens examined, **CHINA**, **HENAN**, **Yunxi**, near the town, 11. Jul. 1978, *Y. J. Ma 3921* (paratypes of *E. lobophylla*: EDC, PE). **SHAANXI**. **Ankang**, Xiangxi Dong, at bottom of cliffs in ravine, 372 m, fl. blue, 03 May 2013, Y. Ren 2013581 (PE).

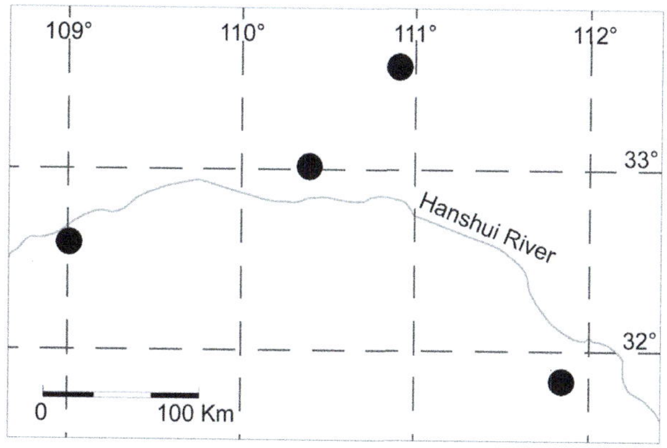

MAP IV 6:1 Distribution of *Echinocodon draco* (Pamp.) D. Y. Hong.

Bibliography

Ali, S. S., Yu, Y., Pfosser, M. & Wetschnig, W. (2012). Inferences of biogeographical histories within subfamily Hyacinthoideae using S-DIVA and Bayesian binary MCMC analysis implemented in RASP (Reconstruct Ancestral State in Phylogenies). *Annals of Botany*, *109*, 95–107.

Anthony, J. (1926). A key to the genus *Codonopsis* Wall. with an account of two undescribed species. *Notes from the Royal Botanic Garden, Edinburgh*, *15*, 173–190.

Antonelli, A., Verola, C. F., Parisod, C. & Gustafsson, A. L. S. (2010). Climate cooling promoted the expansion and radiation of a threatened group of South American orchids (Epidendroideae: Laeliinae). *Botanical Journal of the Linnean Society*, *100*, 597–607.

Avetisjan, E. M. (1967). *Morfologiya Pyl'cy sem. Campanulaceae i blizkikh k nemu semeystv (Sphenocleaceae, Lobeliaceae, Cyphiaceae) v svyazi s voprosami ikh sistematiki i filogenii –Flora i rastitel "nost" armyanskoy SSR* (Vol. 16). Erevan: Trudy Botaniczeskogo Instituta.

Avetisjan, E. M. (1986). Palynomorphology of the families Campanulaceae, Sphenocleaceae and Pentaphragmataceae. *Botanicheskii Zhurnal*, *71*, 1003–1010.

Axelrod, D. I., Al, S. I. & Raven, P. H. (1996). History of the modern flora of China. In A. L. Zhang & S. G. Wu (eds.), *Floristic characteristics and diversity of East Asian plants* (pp. 43–55). Hongkong: Springer.

Ballard, F. (1939). *Codonopsis convolvulacea* var. *forrestii*. *Botanical Magazine*, *162* tab. 9581.

Belyayev, A. A. (1986). The features of anatomy and surface ultrastructure of the seed coat in some representatives of critical genera of the Campanulaceae family. *Botanicheskii Zhurnal*, *71*(10), 1371–1375.

Bentham, G. (1834). Codonopsis rotundifolia. In *Royle, Ill. Bot. Himal. Mts.* (p. 254). London: WM. H. Allen & Co. t. 62.

Bentham, G. & Hooker, J. D. (1876). *Genera plantarum* (Vol. 4). London: A. Black.

Black, M., Bewley, D. & Halmer, P. (2006). *The encyclopedia of seeds: Science, technology and uses* (pp. 682–688).

Borsch, T., Korotkova, N., Raus, T., Lobin, W. & Löhne, C. (2009). The *petD* group II intron as a species level marker: utility for tree inference and species identification in the diverse genus *Campanula (Campanulaceae)*. *Willdenowia*, *39*, 7–33.

Böcher, T. W. (1936). Cytological studies on *Campanula rotnndilolia*. *Hereditas*, *22*, 269–277.

Böcher, T. (1964). Chromosome connections and aberrations in the *Campanula persicifolia* group. *Svensk Botanisk Tidskrift*, *58*, 1–18.

Brummitt, R. K. (2007). Campanulaceae. In V. H. Heywood et al. (ed.), *Flowering plant families of the world* (pp. 82–84). London: Royal Botanic Gardens, Kew.

Candolle, A. de (1830). *Monographie des Campanulées*. (Viii+384pp. + pl. 1–20). Paris: Veuve Desray. Fr. La.

Candolle, A. de (1839). Campanulaceae. In A. P. de Candolle (Ed.), *Prodromus syst. nat.* (Vol. 7) (pp. 414–496). Paris: Treuttel & Würtz.

Cellinese, N., Smith, S. A., Edwards, E. J., Kim, S. T., Haberle, R. C., Avramakis, M. & Donoghue, M. J. (2009). Historical biogeography of the endemic Campanulaceae of Crete. *Journal of Biogeography*, *36*, 1253–1269.

Chamberlain, D. F. (1977). Catalogue of the names published by Hector Léveille: X, Campanulaceae. *Notes from the Royal Botanic Garden, Edinburgh*, *35*, 248–253.

Chapman, J. L. (1967). Comparative palynology in Campanulaceae. *Transactions of the Kansas Academy of Science*, *69*, 197–200.

Chen, S. P., Wang, L. & Song, X. Q. (1991). Study on the polyploid breeding of *Codonopsis pilosula* (Franch.) Nannf. *Chinese Traditional and Herbal Drugs*, *22*(5), 224–227.

Chen, S. T., Xing, Y. W., Su, T., Zhou, Z. Q., Dilcher, E. D. L. & Soltis, D. E. (2012). Phylogeographic analysis reveals significant spatial genetic structure of *Incarvillea sinensis* as a product of mountain building. *BMC Plant Biology*. http://dx.doi.org/10.1186/1471-2229-12-58.

Cheng, Y. C., Liu, B. K., Jiang, Z. S. & Duan, Y. C. (1991). Observations of chromosome numbers of several medical plants. *Journal of Hunan Agricultural College*, *11*(2), 166–170.

Chipp, T. F. (1908). A revision of the genus *Codonopsis* Wall. *Journal of the Linnean Society of London, Botany*, *38*, 374–391.

Chung, S. L., Lo, C. H., Lee, T. Y., Zhang, Y. Q., Xie, Y. W., Li, X. H. et al. (1998). Diachronous uplift of the Tibetan plateau starting 40 Myr ago. *Nature*, *394*, 769–773.

Clarke, C. B. (1881). Campanulaceae. In J. D. Hooker (ed.), *Fl. Brit. Ind.* (Vol. 3) (pp. 421–442).

Contandriopoulos, J. (1984). Differentiation and evolution of the genus *Campanula* in the Mediterranean region. In W. F. Grant (ed.), *Plant biosystematics* (pp. 141–158). Canada, Ontario: Academic Press.

Cosner, M. E., Jansen, R. K. & Lammers, T. G. (1994). Phylogenetic relationships in the Campanulales based on *rbcL* sequences. *Plant Systematics and Evolution*, *190*, 79–95.

Cronquist, A. (1981). *An integrated system of classification of flowering plants*. New York: Columbia University Press.

Cun, Y. Z. & Wang, X. Q. (2010). Plant recolonization in the Himalaya from the southeastern Qinghai-Tibetan Plateau: geographical isolation contributed to high population differentiation. *Molecular Phylogenetics and Evolution*, *56*, 972–982.

Dahlgren, R. (1983). General aspects of angiosperm evolution and macrosystematics. *Nordic Journal of Botany*, *3*, 119–149.

Davis, M. E. & Thompson, L. G. (2006). An Andean ice-core record of a Middle Holocene mega-drought in North Africa and Asia. *Annals of Glaciology*, *43*(1), 34–41.

Diels, L. (1901). Die Flora von Central-China. *Botanische Jahrbücher für Systematik*, *29*, 603–607.

Diels, L. (1912). Plantae Chinenses Forrestianae: new and imperfectly known species, Campanulaceae. *Notes from the Royal Botanic Garden, Edinburgh*, *5*, 170–177.

Dietrich, D. (1839). Syn. Pl. 1, 758.

Don, D. (1825). *Prodromus Florae Nepalensis*. Londini: Veneunt Apud J. Gale.

Dunbar, A. (1975a). On pollen of Campanulaceae and related families with special reference to the surface ultrastructure I. Campanulaceae subfam. Campanuloideae. *Botaniska Notiser*, *128*(1), 71–101.

Dunbar, A. (1975b). On pollen of Campanulaceae and related families with special reference to the surface ultrastructure II. Campanulaceae subfam. Cyphioideae and subfam. Lobelioideae; Goodeniaceae; Sphenocleaceae. *Botaniska Notiser*, *128*, 102–118.

Dunbar, A. (1984). Pollen morphology in Campanulaceae IV. *Nordic Journal of Botany*, *4*, 1–19.

Eddie, W. M. M., Shulkina, T., Gaskin, J., Haberle, R. G. & Jansen, R. K. (2003). Phylogeny of Campanulaceae s. str. inferred from ITS sequences of nuclear ribosomal DNA. *Annals of the Missouri Botanical Garden*, *90*, 554–575.

Erdtman, G. (1952). Pollen morphology and plant taxonomy. *Geologiska Föreningen i Stockholm Förhandlingar*, *77*(4), 526–527.

Ezcurra, M. D. & Agnolin, F. (2012). A new global palaeobiogeographical model for the late Mesozoic and early tertiary. *Systematic Biology*, *61*, 553–566.

Fedorov, A. A. (1957). Campanulaceae. In V. L. Komarov (ed.), *Flora URSS* (Vol. 24) (pp. 126–450). Moscow: Akademia Nauk. 459–475.

Fletcher, H. R. (1937). The genus *Codonopsis* in cultivation. *Gardeners Chronicle (Ser. 3)*, *102*(80–83), 120–121.

Fletcher, H. R. (1950a). Codonopsis macrocalyx. *Botanical Magazine*, *167* tab. 94 & fig. 1.

Fletcher, H. R. (1950b). *Codonopsis rotundifolia* var. *angustifolia* (Campanulaceae). *Botanical Magazine*, *167* tab. 131.

Frajman, B. & Schneeweiss, G. M. (2009). A Campanulaceous fate: the Albanian stenoendemic *Asyneuma comosiforme* in fact belongs to isophyllous *Campanula*. *Systematic Botany*, *34*, 595–601.

Gadella, T. W. J. (1966). Some notes on the delimitation of genera in the Campanulaceae I. *Proceedings of the Series C. Biology Medical Science*, *69*(4), 502–521.

Ge, C. J. & Li, Y. K. (1989). Observation on the chromosome numbers of medicinal plants of Shangdong province (II). *Chinese Traditional Herbal Drugs*, *20*(6), 34–35.

Ge, C. J., Li, Y. K., Zhou, Y. & Hsu, P. S. (1987). Observations on the chromosome numbers of medicinal plants of Shangdong Province (III). *Acta Botanica Yunnanica*, *9*, 333–338.

Ge, C. J. & Wang, A. W. (1990). Cytological study on *Codonopsis lanceolata* Benth. & Hook. *Acta Phytobiologica Sinica*, *1*, 41–44.

Givnish, T. J., Barfuss, M. H. J., Ee, B. V., Riina, R., Schulte, K., Horres, R., Gonsiska, P. A., Jabaily, R. S., Crayn, D. M., Smith, J. A. C., Winter, K., Brown, G. K., Evans, T. M., Holst, B. K., Luther, H., Till, W., Zizka, G., Berry, P. E. & Sytsma, K. J. (2011). Phylogeny, adaptive radiation, and historical biogeography in Bromeliaceae: Insights from an eight-locus plastid phylogeny. *American Journal of Botany*, *98*, 872–895.

Grey-Wilson, C. (1990). A survey of *Codonopsis* in cultivation. *The Plantsman*, *12*(2), 64–99.

Grey-Wilson, C. (1995). *Codonopsis convolvulacea* and its allies. *The New Plantsman*, *2*(4), 213–225.

Griffith, W. (1854). *Campanulaceae. Notulae Plantas Asiaticas*. 227–281. Calcutta: Charles A. Serrao.

Gustafsson, M. H. G., Backlund, A. & Bremer, B. (1996). Phylogeny of the Asterales sensu lato based on *rbcL* sequences with particular reference to the Goodeniaceae. *Plant Systematics and Evolution*, *199*, 217–242.

Gustafsson, M. H. G. & Bremer, K. (1995). Morphology and phylogenetic interrelationships of the Asteraceae, Calyceraceae, Campanulaceae, Goodeniaceae and related families (Asterales). *American Journal of Botany*, *82*, 250–265.

Haberle, R. C., Dang, A., Lee, T., Peñaflor, C., Cortes-Burns, H., Oestreich, A. et al. (2009). Taxonomic and biogeographic implications of a phylogenetic analysis of the Campanulaceae based on three chloroplast genes. *Taxon*, *58*(3), 715–734.

Handel-Mazzetti, H. F. von (1924). Plantae novae Sinenses. *Anzeiger der Akademie der Wissenschaften in Wien. Mathematische-naturwissenchaftliche Klasse. Wien*, *61*(20), 162–170.

Hara, H. (1978). New or noteworthy flowering plants from eastern Himalaya (21). *Journal of Japanese Botany*, *53*(5), 134–140.

Haridasan, V. K. & Mukherjee, P. K. (1988). Seed surface features of some members of the Indian Campanulaceae. *Phytomorphology*, *37*(4), 277–285.

Hasskarl, J. K. (1856). *Natuurkundig tijdschrift voor Nederlandsch-Indië*, *10*, 9 Batavia: Lange & Co.

Hedberg, O. (1961). Monograph of the genus *Canarina* L. (Campanulaceae). *Svensk Botanisk Tidskrift*, *55*, 17–62.

Hemsley, W. B. (1889). Enumeration of all the plants known from China Proper, Formosa, Hainan, the Corea, the Luchu Archipelago and the Island of Hong Kong: Campanulaceae. *Journal of the Linnean Society Botany*, *26*, 2–14.

Hong, D. Y. (1980). New taxa of Campanulaceae from Xizang. *Acta Phytotaxonomica Sinica*, *18*(2), 245–247.

Hong, D. Y. (1983a). On pollen shape in some groups of dicotyledons. *Grana*, *22*, 73–78.

Hong, D. Y. (1983b). Campanulaceae. *Flora Reipublicae Popularis Sinicae*, *73*(2), 1–177 Beijing: Science Press.

Hong, D. Y. (1984). *Echinocodon* Hong, a new genus of Campanulaceae and its systematic position. *Acta Phytotaxonomica Sinica*, *22*(3), 181–184 pl. 1&2.

Hong, D. Y. (1985). Campanulaceae: *Codonopsis*. In C. Y. Wu (ed.), *Flora Xizang*. (Vol. 4) (pp. 568–583). Beijing: Science Press.

Hong, D. Y. (1990). *Plant cytotaxonomy*. Beijing: Science Press. (pp. 1–439). (in Chinese).

Hong, D. Y. (2010). Taxonomic notes on Chinese Campanulaceae. *Novon*, *20*, 420–425.

Hong, D. Y. (2014a). Six new species and one new subspecies of *Codonopsis* (Campanulaceae) from the Pan-Himalaya. *Plant Diversity and Resources*, *36*(3), 285–295.

Hong, D. Y. (2014b). The identity of *Codonopsis draco* Pamp. *Plant Diversity and Resources*, *36*(3), 301–302.

Hong, D. Y., Lammers, T. G. & Klein, L. L. (2011). Codonopsis. In Wu, Raven & Hong (eds.), *Flora of China* (Vol. 19) (pp. 513–526). Beijing & St. Louis: Science Press and Missouri Bot. Gard.

Hong, D. Y. & Ma, L. M. (1991). Systematics of the genus *Cyananthus* Wall. ex Royle. *Acta Phytotaxonomica Sinica*, *29*(1), 25–51.

Hong, D. Y. & Ma, L. M. (1992). Campanulaceae. In Li, & Zhu (Eds.), *Fl. Sichuan* (Vol. 10) (pp. 515–586). Chengdu: Sichuan Nationality Press.

Hong, D. Y. & Ma, L. M. (1994). Campanulaceae; 3. *Codonopsis*. In Wang, & Wu (eds.), *Vasc. Pl. Henduan Mountains* (Vol. 2) (pp. 1954–1961). Beijing: Science Press.

Hong, D. Y. & Pan, K. Y. (1998). The restoration of the genus *Cyclocodon* (Campanulaceae) and its evidence from pollen and seed-coat. *Acta Phytotaxonomica Sinica*, *36*(2), 106–110 pl. I & II.

Hong, D. Y. & Pan, K. Y. (2012). Pollen morphology of the platycodonoid group (Campanulaceae s. str.) and its systematic implication. *Journal of Integrative Plant Biology*, *54*(9), 773–789.

Hong, D. Y. & Wang, Q. (2015). A new taxonomic system of the Campanulceae s. s. *Journal of Systematics and Evolution*, *53*(3), 203–209.

Hooker, J. D. (1856). Codonopsis rotundifolia. *Botanical Magazine*, *12* tab. 4942.

Hooker, J. D. (1857). *Codonopsis rotundifolia* var. *grandiflora*. *Botanical Magazine*, *13* tab. 5018.

Hooker, J. D. & Thomson, T. (1855). *Ill. Himal. Pl.* t. 16A & B.

Hooker, J. D. & Thomson, T. (1857). Ad Fl. Ind.- Campanulaceae, *Codonopsis* Wall. *Journal of the Proceedings of the Linnean Society London*, *2*, 11–17.

Hou, Z. E., Sketb, B., Fišerb, C. & Li, S. Q. (2011). Eocene habitat shift from saline to freshwater promoted Tethyan amphipod diversification. *Proceedings of the National Academy of Sciences of the United States of America*, *108*, 14533–14538.

Huang, T. C. (1972). *Pollen flora of Taiwan*. Taipei: National Taiwan University.

Huang, S. H. (1984). New taxa of the genus *Codonopsis* from Yunnan. *Acta Botanica Yunnanica*, *6*(4), 393–394.

Hutchinson, J. (1973). *The families of flowering plants* (3rd ed.). Oxford: Clarendon Press.

Jee, V., Dhar, U. & Kachroo, P. (1989). Cytogeography of some endemic taxa of Kashmir Himalaya. *Proceedings of the Indian National Science Academy Part B, Biology Science*, *55*, 177–184.

Jia, D. R., Abbott, R. J., Liu, T. L., Mao, K. S., Bartish, I. V. & Liu, J. Q. (2012). Out of the Qinghai–Tibet plateau: evidence for the origin and dispersal of Eurasian temperate plants from a phylogeographic study of *Hippophaë rhamnoides* (Elaeagnaceae). *New Phytologist*, *194*, 1123–1133.

Kadereit, J. W., Licht, W. & Uhink, C. H. (2008). Asian relationships of the flora of the European Alps. *Plant Ecology & Diversity*, *2*, 171–179.

Kolakovsky, A. A. (1987). System of the Campanulaceae family from the old World. *Botanicheskii Zhurnal*, *72*(2), 1572–1579.

Kolakovsky, A. A. (1994). The conspectus of the system of the old world Campanulaceae. *Botanicheskii Zhurnal*, *79*(1), 109–124.

Komarov, V. L. (1908). Prolegomena ad Floras Chinae nec non Mongoliae–*Codonopsis* Wall. *Trudy Imperatorskago S. -Peterburgskago Botanicheskago Sada*, *29*(1), 102–124.

Kovanda, M. (1978). Campanulaceae. In V. H. Heywood (ed.), *Flowering plants of the world* (pp. 254–256). New York: Mayflower Books.

Krasnoborov, L. M., Rostovtseva, T. S. & Ligus, S. A. (1980). Chromosome numbers of some plant species of south Siberia and the far east. *Botanicheskii Zhurnal*, *65*(5), 659–668.

Kumar, V. & Chauhan, K. P. S. (1975). Cytology of *Cyananthus linifolius* Wall. *Proceedings of the Indian Science Congress Association*, *62*(3), 130 (Group B).

Kuntze, O. (1891). *Revis. Gen. Pl.* (Vol. 2). Leipzig: Arthur Felix. Ge. 378–379.

Lammers, T. G. (1992a). Circumscription and phylogeny of the Campanulales. *Annals of the Missouri Botanical Garden*, *79*, 388–413.

Lammers, T. G. (1992b). New combinations for Asian campanulaceae. *Botanical Bulletin of Academia Sinica*, *33*, 285–287 (Taipei).

Lammers, T. G. (1992c). Systematics and biogeography of the campanulaceae of Taiwan. In C. I. Peng (ed.), *Phytogeography and botanical inventory of Taiwan Inst. bot. Acad. Sin. Monograph Series No. 12:* (pp. 43–61). (Taipei).

Lammers, T. G. (1998). Campanulaceae. In (2nd ed.) T. C. Huang (ed.), *Flora Taiwan* (Vol. 4) (pp. 775–802). Taipei: Editorial Committee of the Flora of Taiwan, Depart. Bot., Nat. Taiwan Univ.

Lammers, T. G. (1999). A new *Lobelia* from Mexico, with additional new combination in World Campanulaceae. *Novon*, *9*, 381–389.

Lammers, T. G. (2001). Further nomenclatural innovation in world Campanulaceae. *Novon*, *11*, 67–68.

Lammers, T. G. (2007a). Campanulaceae. In K. Kubitzki (Ed.), *The families and genera of vascular plants Flowering plants. Eudicots. Asterales: Vol. 8*. (pp. 26–56). Berlin: Springer.

Lammers, T. G. (2007b). *World Checkl. Bibliogr. Campanulaceae*. (pp. 262–273). London: Kew Publisher.

Lammers, T. G. & Klein, L. L. (2010). A synopsis of *Codonopsis* subg. *Pseudocodonopsis* (Campanulaceae: Campanuloideae), with description of a new species of uncertain provenance. *Botanical Studies*, *51*, 553–561 (Taipei).

Lancucka-Srodoniowa, M. (1979). Macroscopic plant remains from the freshwater Miocene of the Nowy-Sacz Basin (West Carpathians, Poland). *Acta Palaeobotanica*, *20*, 3–117.

Lee, S., Chung, Y. J. & Lee, J. (1988). A palynotaxonomic study of the Korean Campanulaceae. *Korean Journal of Plant Taxonomy*, *18*, 115–131.

Léveillé, A. A. H. (1914). *Codonopsis Bodinieri*. Fl. Kouy-Tchéou, 57. LE MANS.

Léveillé, A. A. H. (1915). Campanulacées. Cat. Pl. Yunnan, 24. LE MANS.

Léveillé, A. A. H. (1916). Cucurbitacées. Cat. Pl. Yunnan, 65. LE MANS.

Li, L. C. (1986a). Chromosome observations of some plants in China. *Guihaia*, *6*, 99–105.

Li, L. C. (1986b). A karyotype study of *Platycodon grandiflora*. *Journal of Wuhan Botanical Research*, *4*, 343–346.

Li, Q. Q., Zhou, S. D., He, X. J., Yu, Y., Zhang, Y. C. & Wei, X. Q. (2010). Phylogeny and biogeography of *Allium* (Amaryllidaceae: Allieae) based on nuclear ribosomal internal transcribed spacer and chloroplast rps16 sequences, focusing on the inclusion of species endemic to China. *Annals of Botany*, *106*, 709–733.

Lindley, J. (1839). *Codonopsis lurida*. Bot. Miss. 82.

Lingelsheim, A. von & Borza, A. (1914). Plantae novae Limprichtianae in Yunnan collectae. *Repertorium novarum specierum regni vegetabilis*, *13*, 385–392.

Ludlow, F. (1972). Codonopsis bhutanica. *Journal of Royal Horticultural Society*, *97*, 127.

Lundberg, J. & Bremer, K. (2003). A phylogenetic study of the order Asterales using one morphological and three molecular data sets. *International Journal of Plant Sciences*, *164*, 553–578.

Ma, X. H., Qin, R. L. & Xing, W. B. (1984). Chromosome observation of twenty species of drug plants in Xinjiang. *Acta Bot Boreali-Occidential Sinica*, *5*(2), 149–154.

Mabberley, D. J. (2008). *Mabberley's plant-book* (3rd ed.). Cambridge: Cambridge University Press.

Mao, K. S., Hao, G., Liu, J. Q., Adams, R. P. & Milne, R. I. (2010). Diversification and biogeography of *Juniperus* (Cupressaceae): variable diversification rates and multiple intercontinental dispersals. *New Phytologist*, *188*, 254–272.

Maximowicz, C. J. (1881). *Bulletin de l'Academie Imperiale des Sciences de Saint-Pétersbourg*, *27*, 496.

Moeliono, B. & Tuyn, P. (1960). Campanulaceae. In C. G. G. J. van Steenis (ed.), *Fl. Malesiana, Ser. 1* (pp. 107–141). Djakarta: Noordhoff-Kolff. 6(1).

Morris, K. E. & Lammers, T. G. (1997). Circumscription of *Codonopsis* and the allied genera *Campanumoea* and *Leptocodon* (Campanulaceae: Campanuloideae). I. Palynological data. *Botanical Bulletin of Academia Sinica, 38*, 277–284.

Murthy, G. V. S. (1982). Pollen morphology of Indian *Campanumoea* Bl. – a revision of the genus. *Journal of Palynology, 18*(1&2), 55–59.

Nannfeldt, J. A. (1930). Plantae Sinenses. XXI. Campanulaceae. *Acta Horti Gothoburgensis, 5*, 13–32, pl. III–XIV.

Nannfeldt, J. A. (1931). Some notes on the genus *Codonopsis* Wall. *Notes from the Royal Botanic Garden, Edinburgh, 16*, 149–160. pl. 227, 228.

Nannfeldt, J. A. (1936). Campanulaceae. In Handel-Mazzetti, *Symb. Sin.* (Vol. 7) (pp. 1067–1081).

Nannfeldt, J. A. (1940). Bemerkungen über einige morphologische eigen'tümlichkeiten in der gattung *Codonopsis* Wall., nebst beschreibung drei neuer arten. *Svensk Botanisk Tidskrift, 34*(4), 381–388 pl. IV & V.

Nauheimer, L., Boyce, P. C. & Renner, S. S. (2012). Giant taro and its relatives: a phylogeny of the large genus *Alocasia* (Araceae) sheds light on Miocene floristic exchange in the Malesian region. *Molecular Phylogenetics and Evolution, 63*, 43–51.

Nishikawa, T. (1985). Chromosome counts of flowering plants of Hokkaido (8). *Journal of Hokkaido University of Education. Section 2B, 35*, 97–111.

Nishikawa, T. (1986). Chromosome counts of flowering plants of Hokkaido (10). *Journal of Hokkaido University of Education. Section 2B, 37*, 5–17.

Nowicke, J. W., Shetler, S. G. & Morin, N. (1992). Exine structure of pantoporate *Campanula* (Campanulaceae) species. *Annals of the Missouri Botanical Garden, 79*, 65–80.

Nylander, J. A. A., Wilgenbusch, J. C., Warren, D. L. & Swofford, D. L. (2008). AWTY (are we there yet?): a system for graphical exploration of MCMC convergence in Bayesian phylogenetics. *Bioinformatics, 24*, 581–583.

Oliver, D. (1888). *Campanumoea axillaris. Hook. Icon. Pl.* Ser. 3. 8: pl. 1775.

Oliver, D. (1891). *Codonopsis tangshen & C. henryi. Hook. Icon. Pl.* Ser. 3. 20; pl. 1966 & pl. 1967.

Pampanini, R. (1910). Campanulaceae. *Nuovo Giornale Botanico Italiano*, nuovo ser. 17, 732–735.

Pandey, R. M. & Pal, M. (1980a). Chromosome number in *Platycodon grandiflorum. Science and Culture, 46*, 301–302.

Pandey, R. M. & Pal, M. (1980b). In chromosome number report LXVIII. *Taxon, 29*, 544–545.

Pax, F. A. & Hoffman, K. (1922). "II Aufzählung der von Dr. Limpricht in Ostasien gesammelten Pflanzen."–Campanulaceae. *Repertorium Specierum Novarum Regni Vegetabilis Beihefte, 12*, 498–502.

Probatova, N. S., Kazanovsky, S. G., Rudyka, E. G., Barkarov, V. Y., Seledes, V. P. & Nechaev, V. A. (2011). IAPT/IOPB chromosome data 12. *Taxon, 60*(6), 1284–1296.

Probatova, N. S., Seledes, V. P. & Rudyka, E. G. (2008). IAPT/IOPB chromosome data 5. *Taxon, 57*(2), 553–562.

Qiao, C. Y., Ran, J. H., Li, Y. & Wang, X. Q. (2007). Phylogeny and biogeography of *Cedrus* (Pinaceae) inferred from sequences of seven paternal chloroplast and maternal mitochondrial DNA regions. *Annals of Botany, 100*, 573–580.

Qin, A. L., Wang, M. M., Cun, Y. Z., Yang, F. S., Wang, S. S., Ran, J. H. & Wang, X. Q. (2013). Phylogeographic evidence for a link of species divergence of *Ephedra* in the Qinghai-Tibetan plateau and adjacent regions to the Miocene Asian aridification. *PLoS ONE, 8*, e56243.

Ran, J. H., Wei, X. X. & Wang, X. Q. (2006). Molecular phylogeny and biogeography of *Picea* (Pinaceae): Implications for phylogeographical studies using cytoplasmic haplotypes. *Molecular Phylogenetics and Evolution, 41*, 405–419.

Raven, P. H. (1975). The bases of angiosperm phylogeny: cytology. *Annals of Missouri Botanical Garden, 62*, 724–764.

Reichinger, K. H. & Schiman-Czeika, H. (1965). Codonopsis. In K. H. Reichinger (ed.), *Flora Iranica* (Vol. 13) (pp. 3–5).

Roquet, C., Sanmartin, I., Garcia-Jacas, N., Saez, L., Susanna, A., Wikström, N. et al. (2009). Reconstructing the history of Campanulaceae with a Bayesian approach to molecular dating and dispersal-vacariance analyses. *Molecular Phylogenetics and Evolution, 52*, 575–587.

Rosatti, T. J. (1986). The genera of Sphenocleaceae and Campanulaceae in the southeastern United States. *Journal of the Arnold Arboretum, 67*, 1–64.

Roxburgh, W. (1824). *Flora Indica.* Serampore: Mission Press.

Royden, L. H., Burchfie, B. C. & van der Hilst, R. D. (2008). The geological evolution of the Tibetan Plateau. *Science, 321*, 1054–1058.

Royle, J. F. (1834). *Illustrations of the botany and other branches of the natural history of the Himalayan Mountains and of the flora of Cashmere,*. 450. London: Wm. H. Alland and Co.

Ruprecht, F. J. & Maximowicz, C. J. (1857). *Bulletin de la Class Physico-Mathematique de l'Academie Imperiale des Sciences de Saint-Pétersbourg, 15*, 209.

Schönland, S. (1889). Campanulaceae. In Engler, & Prantl (eds.), *Die Natürlichen Pflanzenfamilien* (Vol. 4–5) (pp. 40–70). Leipzig: Engelmann.

Schrenk, A. G. von (1841). In Fischer & Meyer. *Enum. Pl. Nov. Songar, 1*, 38.

Shen, L. T. & Hong, D. Y. (1983) *Codonopsis*. In D. Y. Hong (ed.), *Fl. Reipubl. Popularis Sin. Vol. 73*(2) (pp. 32–89). Beijing: Science Press.

Shen, L. T., Tang, S. Y., Yuen, S. G. & Hsieh, C. K. (1975). The genus *Codonopsis* Wall. in Szechuan. *Acta Phytotaxonomica Sinica, 13*(3), 51–68.

Shetler, S. G. & Morin, N. R. (1986). Seed morphology in North American campanulaceae. *Annals of Missouri Botanical Garden, 73*, 653–688.

Shrestha, K. K. (1997). Taxonomic revision of the Sino-Himalayan genus *Cyananthus (Campanulaceae). Acta Phytotaxonomica Sinica, 35*(5), 396–433.

Shrestha, K. K. & Kravtsova, T. I. (1992). Seed coat anatomy and ultrastructure in the genus *Cyananthus* (Campanulaceae) in relation to its systematics. *Botanicheskii Zhurnal, 77*(6), 18–29.

Shrestha, K. K. & Tarasevich, V. F. (1992). Comparative pollen morphology of genus *Cyananthus* in relation to its systematics and its pollen within the family Campanulaceae. *Botanicheskii Zhurnal, 77*, 1–13.

Siebold, P. F. de & Zuccarinii, J. G. (1835). *Flora Japonica.* Lugduni Batavorum.

Skan, S. A. (1906). Codonopsis tangshen. ser. 4 *Botanical Magazine*, 2 tab. 8090.

Smith, W. W. (1913a). *Records of the Botanical Survey of India, 4*, 388.

Smith, W. W. (1913b). Diagnoses specierum novarum Chinensium-Convolvulaceae. *Notes from the Royal Botanic Garden, Edinburgh, 8*, 107–109.

Song, B. H., Wang, X. Q., Wang, X. R., Ding, K. Y. & Hong, D. Y. (2003). Cytoplasmic composition in *Pinus densata* and population establishment of the diploid hybrid pine. *Molecular Ecology, 12*, 2995–3001.

Sprengel, K. (1825). *Cura Post, 2*, 77.

Stapf, O. (1928). Codonopsis meleagris. *Botanical Magazine, 154* tab. 9237.

Sui, W. B., Cui, Q. H. & Jia, Y. F. (1985). Karyotype analyses of *Carthamus tinctorius* L. and *Codonopsis pilosula* Nannf. *Journal of Wuhan Botanical Research, 3*(4), 459–460.

Sun, H. (2002). Tethys retreat and Himalayas–Hengduanshan Mountains uplift and their significance on the origin and development of the Sino-Himalayan elements and alpine flora. *Acta Botanica Yunnanica, 24*, 273–288.

Sun, J. M. & Liu, T. S. (2000). Stratigraphic evidence for the uplift of the Tibetan Plateau between ~1.1 and ~0.9 myr ago. *Quaternary Research*, *54*, 309–320.

Takhtajan, A. (1997). Campanulales Reichenbach 1828. In *Diversity and classification of flowering plants* (pp. 405–412). New York: Columbia University Press.

Tarasevich, V. F. & Shrestha, K. K. (1992). Palynological data on the position of the genus *Ostrowskia* within the Campanulaceae family. *Botanicheskii Zhurnal*, *77*(9), 27–30.

Thulin, M. (1975). The genus *Wahlenbergia* s. lat. (Campanulaceae) in tropical Africa and Madagascar. *Acta Universitatis Upsaliensis Symbolae Botanicae Upsalienses*, *21*(1), 1–223.

Trautvetter, E. R. (1879). Catalogus Campanulaceanum Rossicarum (*Codonopsis*). *Trudy Imperatorskago S.-Peterburgskago Botanicheskago Sada*, *6*, 46–48.

Tsoong, P. C. (1935). Preliminary study on Chinese Campanulaceae. *Contributions from the Institute of Botany, National Academy of Peiping*, *3*(3), 61–118, pl. IX–XIV.

Tu, T. Y., Volis, S., Dillon, M. O., Sun, H. & Wen, J. (2010). Dispersals of Hyoscyameae and Mandragoreae (Solanaceae) from the New World to Eurasia in the early Miocene and their biogeographic diversification within Eurasia. *Molecular Phylogenetics and Evolution*, *57*, 1226–1237.

Vetaas, O. R. & Grytnes, J. (2002). Distribution of vascular plant species richness and endemic richness along the Himalayan elevation gradient in Nepal. *Global Ecology and Biogeography*, *11*, 291–301.

Wallich, N. (1824). Codonopsis. In W. Roxburgh (ed.), *Fl. India* (Vol. 2) (pp. 103–107).

Wang, C. S., Zhao, X. X., Liu, Z. F., Lippert, P. C., Graham, S. A., Coe, R. S., et al. (2008). Constraints on the early uplift history of the Tibetan Plateau. *Proceedings of the National Academy of Sciences of the United States of America*, *105*, 4987–4992.

Wang, L. Y., Abbott, R. J., Zheng, W., Chen, P., Wang, Y. J. & Liu, J. Q. (2009). History and evolution of alpine plants endemic to the Qinghai-Tibetan plateau: *Aconitum gymnandrum* (Ranunculaceae). *Molecular Ecology*, *18*, 709–721.

Wang, Q. & Hong, D. Y. (2014). *Codonopsis gongshanica* (Campanulaceae), a new species from NW Yunnan based on morphology and molecular phylogenetic analysis. *Phytotaxa*, *188*(3), 145–152.

Wang, Q. & Hong, D. Y. (2015). Taxonomic revision of the genus *Pseudocodon* (Campanulaceae) based on character analysis and molecular phylogeny. *Phytotaxa*, *204*(1), 49–64.

Wang, Q., Zhou, S. L. & Hong, D. Y. (2013). Molecular phylogeny of the platycodonoid group (Campanulaceae s. str.) with special reference to the circumscription of *Codonopsis*. *Taxon*, *62*(3), 498–504.

Wang, Q., Ma, X. T. & Hong, D. Y. (2014a). Phylogenetic analyses reveal three new genera of the Campanulaceae. *Journal of Systematics and Evolution*, *52*(5), 541–550.

Wang, Q., Wang, X. O., Sun, H., Yu, Y., He, X. J. & Hong, D. Y. (2014b). Evolution of the platycodonoid group with particular references to biogeography and character evolution. *Journal of Integrative Plant Biology*, *56*(10), 995–1008.

Wang, Z. T. & Xu, G. J. (1993). Two new species of *Codonopsis* from China. *Acta Phytotaxonomica Sinica*, *31*(2), 184–187.

Wei, Z. X. (2001). Pollen morphology of some *Codonopsis* and related species of Campanulaceae. *Acta Botanica Yunnanica*, *23*(3), 335–338.

Wettstein, R. von (1924). *Hansdbuch der systematischen Botanik* (3. Aufl). Wien.

Wikström, N., Savolainen, V. & Chase, M. (2001). Evolution of the angiosperms: calibrating the family tree. *Proceedings of the Royal Society B: Biological Sciences*, *268*, 2211–2220.

Wu, C. Y. (1965). Campanulaceae. In C. Y. Wu, & X. W. Li (eds.), *Yunnan Trop. Subtrop. Fl. Res. Rep.* (Vol. 1) (pp. 57–97). Beijing: Science Press.

Wu, Z. Y. (1988). Hengduan mountains flora and its significance. *Journal of Japanese Botany*, *63*, 297–311.

Yang, F. S., Li, Y. F., Ding, X. & Wang, X. Q. (2008). Extensive population expansion of *Pedicularis longiflora* (Orobanchaceae) on the Qinghai-Tibetan plateau and its correlation with the quaternary climate change. *Molecular Ecology*, *17*, 5135–5145.

Yeo, P. F. (1993). Platycodoneae, a new tribe in Campanulaceae. *Taxon*, *42*, 109.

Yoo, K. O. & Lee, W. T. (1989). A taxonomic study of the genus *Codonopsis* in Korea. *Korean Journal of Taxonomy*, *19*(2), 81–102.

Yoo, K. O. & Lee, W. T. (1995). Seed morphology of Korean campanulaceae. *Korean Journal of Taxonomy*, *25*, 61–70.

Yu, Y., Harris, A. J. & He, X. J. (2010). S-DIVA (statistical dispersal-vicariance analysis): a tool for inferring biogeographic histories. *Molecular Phylogenetics and Evolution*, *56*, 848–850.

Zakharyeva, O. I. & Astanova, S. B. (1968). Chromosome numbers of some wild species of flowering plants of Middle Asia. *Reports Academy of Sciences Tadzhik SSR*, *11*, 72–75.

Zhang, P., Chen, Y. Q., Zhou, H., Liu, Y. F., Wang, X. L., Papenfuss, T. J., Wake, D. B. & Qu, L. H. (2008). Phylogeny, evolution, and biogeography of Asiatic Salamanders (Hynobiidae). *Proceedings of the National Academy of Sciences of the United States of America*, *103*, 7360–7365.

Zhang, L. B., Comes, H. P. & Kadereit, J. W. (2001). Phylogeny and quaternary history of the European montane/alpine endemic *Soldanella* (Primulaceae) based on ITS and AFLP variation. *American Journal of Botany*, *88*, 2331–2345.

Zhang, L. B., Comes, H. P. & Kadereit, J. W. (2004). The temporal course of quaternary diversification in the European high mountain endemic *Primula* sect. *Auricula (Primulaceae)*. *International Journal of Plant Sciences*, *165*, 191–207.

Zhang, M. L., Kang, Y., Zhou, L. H. & Podlech, D. (2009). Phylogenetic origin of Phyllolobium with a further implication for diversification of *Astragalus* in China. *Journal of Integrative Plant Biology*, *51*, 889–899.

Zhang, M. L., Uhink, C. H. & Kadereit, J. W. (2007). Phylogeny and biogeography of Evpimedium/Vancouveria (Berberidaceae): Western North American–East Asian disjunctions, the origin of European mountain plant taxa, and East Asian species diversity. *Systematic Botany*, *32*, 81–92.

Zhang, Y. F., Zhang, C. Y., Zhang, T., Guan, H. L. & Yan, S. Q. (2010). A cyto-evolutional study of *Campanumoea* Blume (Campanulaceae) and a possible pathway for secondary karyotype formation. *Plant Systematics and Evolution*, *285*, 245–257.

Zhao, C., MA, X. G., Liang, Q. L., Wang, C. B. & He, X. J. (2013). Phylogeography of an alpine plant (*Bupleurum smithii*, Apiaceae) endemic to the Qinghai–Tibetan plateau and adjacent regions inferred from chloroplast DNA sequence variation. *Journal of Systematics and Evolution*, *51*, 382–395.

Zheng, W., Wang, L. Y., Meng, L. H. & Liu, J. Q. (2008). Genetic variation in the endangered *Anisodus tanguticus* (Solanaceae), an alpine perennial endemic to the Qinghai-Tibetan plateau. *Genetica*, *132*, 123–129.

Zhou, Z., Hong, D. Y., Niu, Y., Li, G. D., Nie, Z. L., Wen, J. & Sun, H. (2013). Phylogenetic and biogeographic analyses of the Sino-Himalayan endemic genus *Cyananthus* (Campanulaceae) and implications for the evolution of its sexual system. *Molecular Phylogenetics and Evolution*, *68*, 482–497.

Index to Specimens Cited

The figures in square parentheses represent the species number given in this monograph, e.g. [C.1] = Codonopsis lixianica; [Him.1] = Himalacodon dicentrifolius; [Pan.1] = Pankycodon purpureus; [Pse.1] = Pseudocodon conolvulaceus; [Cyc.1] = Cyclocodon parviflorus; [E.1] = Echirocodon draco

A. Type Specimens

Aitchison, J. E. T. 748 (lectotype here designated: BM; isolectotypes: CAL, G, K, PE) [C.10]

Bao, W. K. et al. 3165 [holotype: CDBI (CDBI0141276); isotype: CDBI (CDBI0141275)] [C.7b]

Beichuan Exped. 0383 (holotype: SM; isotype: PE) [C.2]

Blume s. n. (holotype: P) [C.45]

Bodinier 2437 (lectotype of *C. bodinieri* here designated: E); 1180 (syntype of *C. bodinieri*: E) [C.41]

Chang, Y. T. & Lang, K. Y. 889 (paratypes: PE) [C.44]

Clarke, C. B. 16304 (syntype of *C. viridis var. hirsuta*: BM) [C.32]; 25722A (holotype: K) [Him.1]; 44896A (syntypes of *C. viridis var. hirsuta*: BM, CAL) [C.32]

David, A. 2333 (lectotype designated by Hong 2010: P) [C.23]

Ducloux, P. 513 (holotype: K) [C.40]

Faber, R. E. 253 [lectotype here designated: K (K000781227); isolectotype: K (K000781228)] [Cyc.3]

Farrer 1144 (syntype: E) [C.33]

Feng, K. M. 8295 (holotype of *C. gombalana*: KUN; isotypes of *C. gombalana*: PE) [C.33]

Forrest, G. 48 (lectotype designated by Hong in Wang & Hong, 2015: E; isolectotype: PE) [Pse.1b]; *2674* (lectotype here designated: E; isolectotypes: BM, CAL) [C.4]; *2731* (holotype: E) [C.19]; *3008* (lectotype of *C. macrocalyx* designated by Hong, 2010: E isolectotype *C. macrocalyx:* K, PE) [C.34]; *3854* (lectotype of *C. macrocalyx var. parviloba* designated by Hong 2010: K; isolectotype of *C. macrocalyx var. parviloba*: E) [C.34]; *6258* (holotype: E) [Pse.1b]; *6684* (lectotype here designated: E; isolectotypes: BM, K) [C.37]; *14659* (holotype: E; isotypes: K, W) [C.13]; *16856* (holotype: E) [Pse.8]; *18949* (holotype: E; isotypes: K, W) [C.18b]; *24723* (syntypes of C. *macrocalyx var. parviloba:* E, K, W) [C.34]; *24885* (lectotype here designated: E; isolectotypes: K, W) [C.33]; *24887* (lectotype here designated: E; isolectotypes: K, W) [C.3], *26973* (syntypes: E, K, W) [C.33]; *26998* (syntypes: E, K) [C.3]; *27223* (syntypes: E, K, W) [C.3]

Fu, K. X. 605 (paratype of *C. xizangensis*: PE) [C.15]

Grey-Wilson & Phillips 777 (holotype: K; isolectotype: K) [Pse.3]

Grey-Wilson, C. et al. s. n. July 1989, (holotype: K) [C.38]

Grey-Wilson, C. s. n. Jul. 1989, (holotype: K) [C.12]

Griffith 1013 (holotype; n. v.; not preserved?) [Cyc.3]; *3435* (syntypes: K, PE) [C.10]; *3439* (holotype of *affinis var. burmanica*: K) [C.31]

Hancock, W. 379 (holotype: K) [C.35]

Handel-Mazzetti 3713 (syntypes of *C. pinifolia*: W, WU) [Pse.5]; *4370* (syntypes of *C. pinifolia*: W, WU) [Pse.5]; *4777* (lectotype here designated: WU; isolectotype: W) [C.23c]; *5423* (lectotype designated by Hong in Wang & Hong, 2015: WU; isolectotype: W) [Pse.4]; *9618* (lectotype of *C. macrocalyx var. coerulescens* here designated: WU; isolectotype of *C. macrocalyx var. coerulescens*: E) [C.34]; *9638* (lectotype of *C. foetens var. major* here designated: WU; isolectotypes: K, W) [C.13]; *10053* (lectotype of *C. pinifolia* designated by Hong in Wang & Hong, 2015: WU; *isolectotypes* of *C. pinifolia*: E, W) [Pse.5]

Henry, A. 6468 (lectotype here designated: K (K000229965); isolectotypes: K (K000229964; K000229966; K000229969), E, G, LE, P, PE, TI) [C.23b]; *6468A* (syntypes: K, PE) [C.23b]; *6468B* (syntypes: K, P, PE) [C.23b]; *6651* [lectotype here designated: K (K000575336; BM, CAL, K (K000575337), LE. P, PE) [C.30]; *9425* (neotype designated by Hong, 2010: PE; isoneotypes: E, K, LE) [Pse.1]; *10167* (lectotype here designated: K; isolectotypes: A, E, PE) [C.35]

Hooke, J. D. & Thomson s. n. (fr.) (syntypes of *C. viridis var. hirsute:* BM, G, K, W) [C.32]; s. n. 11. Aug. 1850, (fl.) (lectotype of *C. viridis var. hirsute* here designated: K; isolectotypes of *C. viridis var. hirsute*: BM, G, W) [C.32]

Hooker J. D. s. n. 31. Jul. 1849, [lectotype designated by Hong 2010: K (K000814213)] [C.34]; s. n. 02. Oct. 1849, (syntypes: K, PE, W) [C.31]; s. n. 03. Aug. 1849, 2740m. (lectotype here designated: K) [C.31]; s. n. 03. Aug. 1849, 2130-2740m. (syntype: K) [C.31]; s. n. 03. Aug. 1849, 2750-3050m. [syntype: K (K000814212)] [C.34]; s. n. 03. Oct. 1849, [syntype: K (K000781243)] [C.43]; s. n. 15. Aug. 1849, [lectotype here designated: K (K000781242)] [C.43]; s. n. 15. Jul. 1849, (lectotype here designated: K) [C.11]; s. n. 24. Jul. 1849, (lectotype here designated: K; isolectotype: BM) [C.18]; s. n. 30. Jul. 1849, (syntypes: CAL, G, K, W) [C.11]; s. n. May. 1849, [syntype: K (K000781244)] [C.43]; s. n. Sikkim 1830m, (syntype: K [K000814188]) [C.46]; s. n. Sikkim, 1520–2130m (lectotype here designated: K [K000814187]; isolectotypes: CAL, P, PE) [C.46]; s. n. Sikkim, 2750–3350m, s. d. [syntype: K(K000814210)] [C.34]; s. n. 11. Sept. 1849 (syntype: K) [C.18]; s. n. 15. Jul. 1849, (syntype: K) [C.18]; s. n. Sikkim Lachung (syntype: K [K000814190]) [C.46]

Hu, W. K. & Deng, D. Z. 1988 (holotype: SZ) [C.1]

Hu, W. K. 37900 (paratype: PE) [C.30]

India Orient., Lobb, s. n. (syntype: W) [C.32]

Jin, X. H. et al. ST1992 (holotype: PE; isotype: PE) [C.5]

Kingdon-Ward, F. 1100 (paratype of *C. macrantha*: E) [C.18b]; *1655* (syntype of *C. macrocalyx var. parviloba*: E) [C.34]

Komarov s. n. 05. Aug.1896, (holotype of *C. silvestris*: LE; isotype of *C. silvestris*: LE) [C.23]

Non-Type Specimens

Index to Botanical Names and Synonyms in *Codonopsis* and Allied Genera

Codonopsis Wall. (1824) (*Continued*)
efilamentosa W. W. Smith (1913) = *Pseudocodon convolvulaceus* subsp. *forrestii* (Diels) D. Y. Hong
elliptica D. Y. Hong (2014)
farreri J. Anthony (1926)
var. *grandiflora* S. H. Huang (1984) = *C. farreri* J. Anthony
foetens Hook. f. & Thomson (1857)
subsp. ***foetens***
subsp. ***nervosa*** (Chipp) D. Y. Hong (2010) = *Codonopsis nervosa* (Chipp) Nannf. (1930) = *Codonopsis ovata* Benth. var. *nervosa* Chipp (1908)
var. *major* Hand.-Mazz. (1924) = *C. alpina* Nannf.
forrestii Diels (1912) = *Pseudocodon convolvulaceus* subsp. *forrestii* (Diels) D. Y. Hong
var. *heterophylla* C. Y. Wu (1965) = *Pseudocodon vinciflorus* (Kom.) D. Y. Hong
var. *hirsuta* P. C. Tsoong & L. T. Shen ex L. T. Shen (1975) = *Pseudocodon convolvulaceus* subsp. *forrestii* (Diels) D. Y. Hong
glaberrima Nannf. (1940) = *C. pilosula* (Franch.) Nannf.
gombalana C. Y. Wu (1965) = *C. farreri* J. Anthony
gongshanica Q. Wang & D. Y. Hong
gracilis Hook. f. & Thomson (1855)
graminifolia H. Lév. (1915) = *Pseudocodon graminifolius* (H. Lév.) D. Y. Hong
grey-wilsonii J. M. H. Shaw (1996) = *Pseudocodon grey-wilsonii* (J. M. H. Shaw) D. Y. Hong
griffithii C. B. Clarke (1881) = *C. viridis* Wall.
handeliana Nannf. (1936) = *C. pilosula* subsp. *handeliana* (Nannf.) D. Y. Hong & L. M. Ma
hemisphaerica P. C. Tsoong ex D. Y. Hong (2014)
henryi Oliv. (1891)
hirsuta (Hand.-Mazz.) D. Y. Hong & L. M. Ma (1992) = *Codonopsis limbrichtii* var. *hirsuta* Hand.-Mazz. (1924) = *Pseudocodon hirsutus* (Hand.- Mazz.) D. Y. Hong
hirsutus (D. Y. Hong) K. E. Morris & Lammers (1997), non D. Y. Hong & L. M. Ma (1992) = *Codonopsis hongii* Lammers
hongii Lammers (2001) = *Leptocodon hirsutus* D. Y. Hong (1980)
inflata Hook. f. & Thomson (1855)
javanica (Bl.) Hook. f. (1855) = *Campanumoea javanica* Bl. (1826)
subsp. ***japonica*** (Makino) Lammers (1992) = *Campanumoea javanica* var. *japonica* Makino (1908)
kawakamii Hayata (1911)
lanceolata (Sieb. & Zucc.) Trautv. (1879) = *Campanumoea lanceolata* Sieb. & Zucc. (1835)
var. *ohmurae* T. Koyama (1957) = *Codonopsis lanceolata* (Sieb. & Zucc.) Trautv.
lancifolia (Roxb.) Moeliono (1960) = *Cyclocodon axillaris* + *Cyclocodon lancifolius*
subsp. *lancifolia* = *Cyclocodon axillaris* (Oliv.) W. J. de Wilde & Duyfjes
subsp. *celebica* (Bl.) Moeliono (1960) = *Campanumoea celebica* Bl. (1826) = *Cyclocodon lancifolius* (Roxb.) Kurz
leucocarpa Miq. (1857) = *Cyclocodon lancifolius* (Roxb.) Kurz
levicalyx L. T. Shen (1975) = *C. henryi* Oliv.
var. *hirsuticalyx* L. T. Shen (1983) = *C. henryi* Oliv.
limprichtii Lingelsh. & Borza (1914) = *Pseudocodon convolvulaceus* subsp. *forrestii* (Diels) D. Y. Hong
var. *hirsuta* Hand.-Mazz. (1924) = *Pseudocodon hirsutus* (Hand.-Mazz.) D. Y. Hong
var. *pinifolia* Hand.-Mazz. (1924) = *Pseudocodon graminifolius* (H. Lév.) D. Y. Hong
lixianica D. Y. Hong (2014)
longifolia D. Y. Hong (1980) = *C. rotundifolia* Benth.
lurida Lindl. (1839) = *C. rotundifolia* Benth.
macrantha Nannf. (1931) = *C. foetens* subsp. *nervosa* (Chipp) D. Y. Hong
macrocalyx Diels (1912) = *C. benthamii* Hook. f. & Thomson
var. *coerulescens* Hand.-Mazz. (1924) = *C. benthamii* Hook. f. & Thomson
var. *parviloba* J. Anthony (1926) = *C. benthamii* Hook. f. & Thomson
macrophylla Lemmers & L. L. Klein (2010) = *Pseudocodon convolvulaceus* subsp. *forrestii* (Diels) D. Y. Hong
mairei H. Lév. (1915) = *Pseudocodon convolvulaceus* subsp. *forrestii* (Diels) D. Y. Hong
meleagris Diels (1912)
micrantha Chipp (1908)
microtubulosa Z. T. Wang & G. J. Xu (1993)
minima Nakai (1915) = *C. ussuriensis* (Rupr. & Maxim.) Hemsl.
modesta Nannf. (1930) = *C. pilosula* (Franch.) Nannf.
mollis Chipp (1908) = *C. thalictrifolia* subsp. *mollis* (Chipp) D. Y. Hong
nepalensis Grey-Wilson (1990) = *Pseudocodon grey-wilsonii* (J. M. H. Shaw) D. Y. Hong
nepalensis Hara (1978) = C. *rotundifolia* Benth.
nervosa (Chipp) Nannf. (1930) = *Codonopsis ovata* var. *nervosa* Chipp (1908) = *C. foetens* subsp. *nervosa* (Chipp) D. Y. Hong
obtusa (Chipp) Nannf. (1930) = *C. ovata* var. *obtusa* Chipp (1908)
ovata Benth. (1834)
var. *nervosa* Chipp (1908) = *C. foetens* subsp. *nervosa* (Chipp) D. Y. Hong
var. *cuspidata* Chipp (1908) = *C. clematidea* (Schrenk) C. B. Clarke (India: Kashmir) + *C. pilosula* subsp. *handeliana* (China: Sichuan)
var. *obtusa* Chipp (1908) = *C. obtusa* (Chipp) Nannf.
parviflora Wall. ex A. DC. (1830) = *Cyclocodon parviflorus* (Wall. ex A. DC.) Hook. f. & Thomson
pianmaensis S. H. Huang (1984) = *C. benthamii* Hook. f. & Thomson
pilosa Chipp (1908) = *C. tubulosa* Kom.
pilosula (Franch.) Nannf. (1930) = *Campanumoea pilosula* Franch. (1884)
subsp ***handeliana*** (Nannf.) D. Y. Hong & L. M. Ma (1992) = *C. handeliana* Nannf. (1936)
subsp. ***tangshen*** (Oliv.) D. Y. Hong (2010) = *C. tangshen* Oliv. (1891)
pinifolia (Hand.-Mazz.) Grey-Wilson (1990) = *Codonopsis limprichtii* var. *pinifolia* Hand.-Mazz. (1924) = *Pseudocodon graminifolius* (H. Lév.) D. Y. Hong
purpurea Wall. (1824) = *Pankycodon purpureus* (Wall.) D. Y. Hong & X. T. Ma
reflexa D. Y. Hong (2014)
restroserrata Z. T. Wang & G. J. Xu (1993) = *Pseudocodon retroserratus* (Z. T. Wang & G. J. Xu) D. Y. Hong
rosulata W. W. Smith (1921) = *Pseudocodon rosulatus* (W. W. Smith) D. Y. Hong
rotundifolia Benth. (1834)
var. *angustifolia* Nannf. (1950) = *C. bomiensis* D. Y. Hong
var. *grandiflora* Hook. f. (1857) = *C. rotundifolia* Benth.
silvestris Kom. (1901) = *C. pilosula* (Franch.) Nannf.
subglobosa W. W. Smith (1913)
subscaposa Kom. (1908)
subsimplex Hook. f. & Thomson (1857)
tangshen Oliv. (1891) = *C. pilosula* subsp. *tangshen* (Oliv.) D. Y. Hong
thalictrifolia Wall. (1824)
subsp. ***thalictrifolia***
subsp. ***mollis*** (Chipp) D. Y. Hong (2010) = *Codonopsis mollis* Chipp (1908)
var. *mollis* (Chipp) L. T. Shen (1983) = *Codonopsis mollis* Chipp (1908) = *Codonopsis thalictrifolia* subsp. *mollis* (Chipp) D. Y. Hong
tibetica Prains (in sched.) = *Pseudocodon vinciflorus* (Kom.) D. Y. Hong
truncata Wall. ex A. DC. (1830) = *Cyclocodon lancifolius* (Roxb.) Kurz
tsinlinensis Pax & K. Hoffm. (1922)
tubulosa Kom. (1908)
subsp. ***vadsea*** (S. S. Dash & A. A. Mao) D. Y. Hong, stat. nov. = *Codonopsis vadsea* S. S. Dash & A. A. Mao
ussuriensis (Rupr. & Maxim.) Hemsl. (1889) = *Glossocomia ussuriensis* Rupr. & Maxim. (1857)
f. *viridiflora* J. Ohara (1985) = *C. ussuriensis* (Rupr. & Maxim.) Hemsl.
vadsea S. S. Dash & A. A. Mao (2011) = *Codonopsis tubulosa* subsp. *vadsea* (S. S. Dash & A. A. Mao) D. Y. Hong

vinciflora Kom. (1908) = *Pseudocodon vinciflorus* (Kom.) D. Y. Hong
viridiflora Maxim. (1881).
var. *chlorocodon* (C. Y. Wu) S. H. Huang (1991) = *Codonopsis chlorocodon* C. Y. Wu
viridis Wall, (1824)
var. *hirsuta* Chipp (1908) = *C. viridis* Wall.
volubilis Nannf. (1940) = *C. pilosula* (Franch.) Nannf.
xizangensis D. Y. Hong (1980) = *C. bhutanica* Ludlow
yesoensis Nakai (1930) = *C. lanceolata* (Sieb. & Zucc.) Trautv
Cyclocodon Griff. ex Hook. f. & Thomson (1854)
adnatus Griff. (1854) = *Cyclocodon lancifolius* (Roxb.) Kurz
axillaris (Oliv.) W. J. de Wilde & Duyfjes (2012) = *Campanumoea axillaris* Oliv. (1888)
celebicus (Bl.) D. Y. Hong (1998) = *Campanumoea celebica* Bl. (1826) = *Cyclocodon lancifolius* (Roxb.) Kurz
distans Griff. (1854) = *Cyclocodon parviflorus* (Wall. ex A. DC.) Hook. f. & Thomson
lancifolius (Roxb.) Kurz (1872) = *Campanula lancifolia* Roxb. (1824)
subsp. *celebicus* (Bl.) K. E. Morris & Lammers (1999) = *Campanumoea celebica* Bl. (1826) = *Cyclocodon lancifolius* (Roxb.) Kurz
parviflorus (Wall. ex A. DC.) Hook. f. & Thomson (1857) = *Codonopsis parviflora* Wall. ex A. DC. (1830)
truncatus (Wall. ex A. DC.) Hook. f. & Thomson (1857) = *Codonopsis truncata* Wall. ex A. DC. (1830) = *Cyclocodon lancifolius* (Roxb.) Kurz
Echinocodon D. Y. Hong (1984)
draco (Pamp.) D. Y. Hong (2014) = *Codonopsis draco* Pamp. (1910)
lobophyllus D. Y. Hong (1984) = *Echinocodon draco* (Pamp.) D. Y. Hong (2014)
Glossocomia D. Don (1825)
clematidea (Schrenk) Fisch. ex Regel (1856) = *Wahlenbergia clematidea* Schrenk (1841) = *Codonopsis clematidea* (Schrenk) C. B. Clarke
hortensis Rupr. (1857) = *Codonopsis lanceolata* (Sieb. & Zucc.) Trautv.
lanceolata (Sieb. & Zucc.) Maxim. (1886) = *Campanumoea lanceolata* Sieb. & Zucc. (1835) = *Codonopsis lanceolata* (Sieb. & Zucc.) Trautv.
var. *obtusa* Regel (1857) = *Codonopsis ussuriensis* (Rupr. & Maxim.) Hemsl.
var. *ussuriensis* (Rupr. & Maxim.) Regel (1886) = Glossocomia ussuriensis Rupr. & Maxim. (1857) = *Codonopsis ussuriensis* (Rupr. & Maxim.) Hemsl.
lurida (Lindl.) Lindl. (1842) = *Codonopsis lurida* Lindl. (1839) = *Codonopsis rotundifolia* Benth.
purpurea (Wall.) Rupr. (1857) = *Codonopsis purpurea* Wall. (1824) = *Pankycodon purpureus* (Wall.) D. Y. Hong & X. T. Ma
rotundifolia (Benth.) Rupr. (1857) = *Codonopsis rotundifolia* Benth
tenera D. Don (1825) = *Codonopsis thalictrifolia* Wall.
thalictrifolia (Wall.) Wall. (1828) = *Codonopsis thalictrifolia* Wall.
ussuriensis Rupr. & Maxim. (1857) = *Codonopsis ussuriensis* (Rupr. & Maxim.) Hemsl.
viridis (Wall.) Rupr. (1857) = *Codonopsis viridis* Wall.
Himalacodon D. Y. Hong & Q. Wang (2014)
dicentrifolius (C. B. Clarke) D. Y. Hong & Q. Wang (2014) = *Wahlenbergia dicentrifolia* C. B. Clarke (1881)
Leptocodon (Hook. f. & Thomson) Lem. (1856)
gracilis (Hook. f. & Thomson) Lem. (1856) = *Codonopsis gracilis* Hook. f. & Thomson
hirsutus D. Y. Hong (1980) = *Codonopsis hongii* Lammers.
Namaeucampa Gagnep. (1948) = *Codonopsis* (see Takhtajan, 1997; Mabberley, 2008); = ?(see Lammers, 2007b); = *Lycianthes* (Dunal) Hassl. (a genus of Solanaceae, see Wang &; Hong, unpublished)
kerrii Gagnep. (1948) = ?(see Lammers, 2007b); = *Lycianthes lysimachioides* (Wall.) Bitter (see Wang & Hong, unpublished)
Pankycodon D. Y. Hong & X. T. Ma (2014)
purpureus (Wall.) D. Y. Hong & X. T. Ma (2014) = *Codonopsis purpurea* Wall. (1824)
Pseudocodon D. Y. Hong & H. Sun (2014)
convolvulaceus (Kurz) D. Y. Hong & H. Sun (2014) = *Codonopsis convolvulacea* Kurz (1873)
subsp. ***convolvulaceus***
subsp. ***forrestii*** (Diels) D. Y. Hong (2014) = *Codonopsis forrestii* Diels (1912)
graminifolius (H. Lév.) D. Y. Hong (2014) = *Codonopsis graminifolia* H. Lév. (1915)
grey-wilsonii (J. M. H. Shaw) D. Y. Hong (2014) = *Codonopsis grey-wilsonii* J. M. H. Shaw (1996)
hirsutus (Hand.-Mazz.) D. Y. Hong (2014) = *Codonopsis limprichtii* var. *hirsuta* Hand.-Mazz. (1924)
petiolatus D. Y. Hong & Q. Wang (2015)
retroserratus (Z. T. Wang & G. J. Xu) D. Y. Hong & Q. Wang (2015) = *Codonopsis retroserrata* Z. T. Wang & G. J. Xu (1993)
rosulatus (W. W. Smith) D. Y. Hong (2014) = *Codonopsis rosulata* W. W. Smith (1921)
vinciflorus (Kom.) D. Y. Hong (2014) = *Codonopsis vinciflora* Kom. (1908)
subsp. ***vinciflorus***
subsp. ***dianchuanicus*** D. Y. Hong & Q. Wang (2015)
Wahlenbergia Shrad, ex Roth (1821)
clematidea Schrenk (1841) = *Codonopsis clematidea* (Schrenk) C. B. Clarke
dicentrifolia C. B. Clarke (1881) = *Himalacodon dicentrifolius* (C. B. Clarke) D. Y. Hong & Q. Wang
lurida (Lindl.) G. Manetti (1842) = *Codonopsis lurida* Lindl. (1839) = *Codonopsis rotundifolia* Benth. (1834)
purpurea (Wall.) A. DC. (1839) = *Codonopsis purpurea* Wall. (1824) = *Pankycodon purpureus* (Wall.) D. Y. Hong & X. T. Ma
rotundifolia (Benth.) A. DC. (1839) = *Codonopsis rotundifolia* Benth. (1834)
roylei A. DC. (1839) = *Codonopsis ovata* Benth.
thalictrifolia (Wall.) A. DC. (1839) = *Codonopsis thalictrifolia* Wall. (1824)
viridis (Wall.) A. DC. (1839) = *Codonopsis viridis* Wall.

Index

Note: Page numbers followed by "f" and "t" indicate figures and tables, respectively.

Printed and bound by CPI Group (UK) Ltd, Croydon, CR0 4YY
22/07/2026
02174536-0003